BIRMINGHAM CITY
UNIVERSITY
DISCARDED

Risk Assessment in
Environmental Management

Risk Assessment in Environmental Management

A Guide for Managing Chemical Contamination Problems

Kofi Asante-Duah

Environmental Consultant, California, USA
and
Research Assistant Professor, Center for Environmental Engineering, Science and Technology, University of Massachusetts–Lowell, USA

JOHN WILEY & SONS
Chichester • New York • Weinheim • Brisbane • Singapore • Toronto

Copyright © 1998 by D. Kofi Asante-Duah

Published 1998 by John Wiley & Sons Ltd,
Baffins Lane, Chichester,
West Sussex PO19 1UD, England

National   01243 779777
International (+44) 1243 779777
e-mail (for orders and customer service enquiries): cs-books@wiley.co.uk
Visit our Home Page on http://www.wiley.co.uk
or http://www.wiley.com

All Rights Reserved. No part of this publication may be reproduced, stored in a retrieval system, or transmitted, in any form or by any means, electronic, mechanical, photocopying, recording, scanning or otherwise, except under the terms of the Copyright, Designs and Patents Act 1988 or under the terms of a licence issued by the Copyright Licensing Agency, 90 Tottenham Court Road, London, UK W1P 9HE, without the permission in writing of publisher and the copyright owner.

Other Wiley Editorial Offices

John Wiley & Sons, Inc., 605 Third Avenue,
New York, NY 10158-0012, USA

WILEY-VCH Verlag GmbH, Pappelallee 3,
D-69469 Weinheim, Germany

Jacaranda Wiley Ltd, 33 Park Road, Milton,
Queensland 4064, Australia

John Wiley & Sons (Asia) Pte Ltd, 2 Clementi Loop #02-01,
Jin Xing Distripark, Singapore 129809

John Wiley & Sons (Canada) Ltd, 22 Worcester Road,
Rexdale, Ontario M9W 1L1, Canada

Library of Congress Cataloging-in-Publication Data

Asante-Duah, D. Kofi
  Risk assessment in environmental management : a guide for managing chemical contamination problems / Kofi Asante-Duah.
    p. cm.
  Includes bibliographical references and index.
  ISBN 0-471-98147-8
  1. Environmental management. 2. Environmental risk assessment.
3. Environmental monitoring. 4. Pollution—Environmental aspects.
5. Environmental health. I. Title.
GE300.A76 1998
363.738—dc21   98-16174
                                                                    CIP

British Library Cataloguing in Publication Data

A catalogue record for this book is available from the British Library

ISBN 0-471-98147-8

Typeset in 10/12pt Times from authors' disks by Mayhew Typesetting, Rhayader, Powys.
Printed and bound in Great Britain by Biddles Ltd, Guildford and King's Lynn.
This book is printed on acid-free paper responsibly manufactured from sustainable forestry, in which at least two trees are planted for each one used for paper production.

To my families at Abaam, Kade and Nk4
To mom – Alice Adwoa Twumwaa
To dad – George Kwabena Duah
To the Duah brothers and sisters
To the truly good friends

Contents

Preface		xvii

PART I: A GENERAL OVERVIEW

1 Introduction   3
- 1.1 An Agenda for Global Environmental Protection   3
  - 1.1.1 Environmental Management in a Socio-economic Context   4
  - 1.1.2 Harmonizing Global Environmental Management Policies   5
- 1.2 Minimizing Environmental Contamination Problems   6
  - 1.2.1 Components of a Good Waste Management Program   7
  - 1.2.2 Pollution Abatement Strategies and Policies   9
- 1.3 Facing Up to Environmental Chemicals in Society   10
- 1.4 Risk Assessment as a Facilitator of Cost-Effective Environmental Management Decisions   11
- Suggested Further Reading   13
- References   13

2 The nature of environmental contamination problems   15
- 2.1 Sources of Environmental Pollutants   15
  - 2.1.1 Industry as a Major Generator of Environmental Pollutants   17
  - 2.1.2 Hazardous Materials Release and Behavior in the Environment   18
- 2.2 Health, Environmental, and Socio-Economic Implications of Environmental Contamination Problems   18
  - 2.2.1 Human Health and Ecological Effects of Environmental Contaminants   27
- 2.3 A Hazard Categorization System for Environmental Contamination Problems   30
- Suggested Further Reading   32
- References   32

3 Characterization of environmental contamination problems   35
- 3.1 The Environmental Characterization Process   35
  - 3.1.1 The Sampling and Analysis Plan   35
  - 3.1.2 The Health and Safety Plan   41
  - 3.1.3 The Quality Assurance/Quality Control Plan   43
- 3.2 Designing an Environmental Characterization Program   44

3.3	Implementation of an Environmental Characterization Program	45
	Suggested Further Reading	47
	References	47

4 Legislative–regulatory control needs for the management of environmental contamination problems — 51
4.1 The Nature of Environmental Regulations — 51
    4.1.1 Environmental Laws in North America — 52
    4.1.2 Environmental Laws within the European Community Bloc — 55
    4.1.3 Environmental Policies and Regulations in Different Parts of the World — 59
4.2 A Paradigm of Contemporary International Environmental Law — 60
Suggested Further Reading — 62
References — 62

PART II: PRINCIPLES OF RISK ASSESSMENT

5 Fundamentals of hazard, exposure, and risk assessment — 67
5.1 The Nature of Hazard and Risk — 67
    5.1.1 Basis for Measuring Risks — 70
5.2 Risk Assessment Defined — 72
5.3 Purpose and Attributes of Risk Assessment — 73
    5.3.1 The Purpose — 74
    5.3.2 The Attributes — 75
Suggested Further Reading — 76
References — 77

6 The risk assessment process — 79
6.1 Elements of the Risk Assessment Process — 79
    6.1.1 Hazard Identification and Accounting — 80
    6.1.2 Exposure–Response Evaluation — 81
    6.1.3 Exposure Assessment and Analysis — 81
    6.1.4 Risk Characterization and Consequence Determination — 82
6.2 Risk Assessment as a Diagnostic Tool — 82
    6.2.1 Baseline Risk Assessments — 83
6.3 Risk Assessment in Practice — 84
6.4 Risk Assessment as an Holistic Tool for Environmental Management — 87
Suggested Further Reading — 88
References — 88

7 Basic concepts in risk assessment practice — 91
7.1 Individual versus Group Risks — 91
7.2 What Constitutes an 'Acceptable' Risk? — 92
    7.2.1 The Risk Acceptability Criteria: *de Minimis* versus *de Manifestis* Risks — 92

Contents   ix

	7.2.2 Consideration of Risk Perception Issues	94
7.3	Conservatisms in Risk Assessments	95
7.4	Recognition of Uncertainty as an Integral Component of Risk Assessment	96
7.5	Risk Assessment versus Risk Management	97
	Suggested Further Reading	97
	References	98

PART III: PRINCIPAL ELEMENTS OF A RISK ASSESSMENT

8 Determination of contaminant fate and behavior in the environment — 103
- 8.1 The Cross-Media Transfer of Contaminants between Environmental Compartments — 103
  - 8.1.1 Phase Distribution of Environmental Contaminants — 104
- 8.2 Important Fate and Transport Properties, Processes, and Parameters of Environmental Contaminants — 107
  - 8.2.1 Physical State — 107
  - 8.2.2 Water Solubility — 107
  - 8.2.3 Diffusion — 108
  - 8.2.4 Dispersion — 108
  - 8.2.5 Volatilization — 108
  - 8.2.6 Partitioning and the Partition Coefficients — 109
  - 8.2.7 Sorption and the Retardation Factor — 112
  - 8.2.8 Degradation — 114
- 8.3 The Modeling of Contaminant Migration in a Contaminant Fate and Transport Assessment — 115
  - 8.3.1 Utility and Application of Environmental Models — 116
  - 8.3.2 Model Selection — 117
  - 8.3.3 Selected Environmental Models Potentially Applicable to Risk Assessment Studies and Environmental Management Programs — 117
- 8.4 Selected Environmental Fate Algorithms for Estimating Contaminant Concentrations — 137
  - 8.4.1 Estimation of Contaminant Concentrations in Air — 137
  - 8.4.2 Estimation of Contaminant Concentrations in Soils — 145
  - 8.4.3 Estimation of Contaminant Concentrations in Water — 145
  - 8.4.4 Estimation of Contaminant Concentrations in Vegetation — 146
  - 8.4.5 Estimation of Contaminant Concentrations in Animal Products — 147
  - 8.4.6 Estimation of Contaminant Concentrations in Fish Products — 148
- 8.5 Factors Affecting Contaminant Fate and Transport in the Environment — 149
  - Suggested Further Reading — 150
  - References — 150

9 Hazard identification, data collection, and data evaluation   155
9.1 Sources of Environmental Hazards   155
9.2 Hazard Accounting and Environmental Investigation   156
    9.2.1 Development of Data Quality Objectives   158
    9.2.2 Design of an Environmental Investigation Program   159
9.3 Data Collection and Analysis   161
    9.3.1 Data Collection and Analysis Strategies   162
    9.3.2 Background Sampling Considerations   163
    9.3.3 Evaluation of Quality Control Samples   165
9.4 Evaluation of Environmental Sampling Data   166
    9.4.1 General Approach to the Statistical Analysis of Environmental Sampling Data   167
    9.4.2 Treatment of Censored Data in Environmental Samples   172
9.5 Selecting the Chemicals of Potential Concern   174
Suggested Further Reading   177
References   177

10 Design of conceptual models and exposure analysis   181
10.1 Design of Conceptual Models for Environmental Contamination Problems   181
    10.1.1 Elements of a Conceptual Site Model   182
10.2 Development of Exposure Scenarios   185
    10.2.1 The Nature and Spectrum of Exposure Scenarios   185
10.3 The Exposure Assessment Process   187
    10.3.1 Multimedia and Multipathway Exposure Modeling   190
    10.3.2 Chemical Intake versus Dose   193
    10.3.3 Chronic versus Subchronic Exposures   193
10.4 A Generic Exposure Estimation Model   194
    10.4.1 Receptor Age Adjustments to Human Exposure Factors   195
    10.4.2 Incorporating Contaminant Degradation into Exposure Calculations   195
    10.4.3 Averaging Exposure Estimates   197
10.5 Utility of the Exposure Characterization   198
Suggested Further Reading   199
References   199

11 The toxicology of environmental contaminants and hazard effects determination   203
11.1 Identification of Toxic Substances   203
    11.1.1 Manifestations of Toxicity   203
    11.1.2 Dose–Response Relationships   204
11.2 Categorization of Human Toxic Effects   207
    11.2.1 Basis for the 'Threshold' versus 'Nonthreshold' Concepts   208

		11.2.2 Determination of Chemical Carcinogenicity	208

	11.2.2	Determination of Chemical Carcinogenicity	208
	11.2.3	Carcinogen Classification Systems	209
11.3	Evaluation of Chemical Toxicity	213
	11.3.1	Hazard Effects Assessment	213
	11.3.2	Dose–Response Assessment	216
11.4	Utility of the Hazard Effects Assessment	218
	Suggested Further Reading	219
	References	219

12 Risk characterization and uncertainty analysis — 223

12.1	Risk Presentation and Summarization	223
	12.1.1 Graphical Presentation of the Risk Summary Information	224
	12.1.2 Presenting and Managing Uncertain Risks	224
12.2	Uncertainty and Variability Issues in Risk Assessment	229
	12.2.1 The Need for Uncertainty and Variability Analyses	229
12.3	Types and Nature of Uncertainty	230
	12.3.1 Common Sources of Uncertainty in Endangerment Assessments	231
12.4	Analysis of Uncertainties and Variability	233
	12.4.1 Qualitative Analysis of Uncertainties	234
	12.4.2 Quantitative Analysis of Uncertainties	234
12.5	Utility of a Risk Analysis	237
	Suggested Further Reading	238
	References	238

PART IV: RISK ASSESSMENT TECHNIQUES AND METHODS OF APPROACH

13 Human health risk assessments — 243

13.1	The Health Risk Assessment Methodology	243
13.2	Potential Receptor Exposures to Environmental Contaminants	245
	13.2.1 Potential Receptor Inhalation Exposures	247
	13.2.2 Potential Receptor Ingestion Exposures	250
	13.2.3 Potential Receptor Dermal Exposures	254
	13.2.4 Computation of 'Intake Factors' for Exposure Assessment: Illustration of the Calculation of Chemical Intakes and Doses	256
13.3	Determination of Toxicological Parameters Used in Human Health Risk Assessments	267
	13.3.1 Toxicity Parameters for Noncarcinogenic Effects	267
	13.3.2 Toxicity Parameters for Carcinogenic Effects	273
	13.3.3 The Use of Surrogate Toxicity Parameters	276
13.4	Risk Characterization	279
	13.4.1 Absorption Adjustments	280
	13.4.2 Aggregate Effects of Chemical Mixtures	282

13.4.3 Estimation of Carcinogenic Risks to Human Health 283
13.4.4 Estimation of Noncarcinogenic Hazards to Human Health 286
13.4.5 Risk Computations: Illustration of the Processes for Calculating Carcinogenic Risks and Noncarcinogenic Hazards 288
13.5 Human Health Risk Assessment in Practice 295
Suggested Further Reading 296
References 296

14 Ecological risk assessments 299
14.1 The Ecological Risk Assessment Methodology 299
14.1.1 Ecological Hazard Evaluation 302
14.1.2 Exposure Assessment: The Characterization of Ecological Exposures 307
14.1.3 Ecotoxicity Assessment: The Characterization of Ecological Effects 310
14.1.4 Ecological Risk Characterization 311
14.2 A General Framework for Ecological Risk Assessments 313
14.2.1 The Problem Formulation Phase 314
14.2.2 The Problem Analysis Phase 314
14.2.3 The Risk Characterization Phase 315
14.3 Ecological Risk Assessment in Practice 315
14.3.1 The General Purpose 318
14.3.2 General Considerations in Ecological Investigations 319
14.4 A Comparative Look at the Parallel Nature of Ecological and Human Health Endangerment Assessments 320
Suggested Further Reading 322
References 322

15 Probabilistic risk assessments 325
15.1 The Probabilistic Risk Assessment Methodology 325
15.1.1 Event-Tree Modeling and Analysis 326
15.1.2 Fault-Tree Modeling and Analysis 330
15.1.3 Other PRA Techniques and Tools 335
15.2 Probabilistic Risk Assessment in Practice 336
15.2.1 A PRA Demonstration Problem 338
15.3 Utility of Probabilistic Risk Assessments 340
Suggested Further Reading 341
References 341

PART V: DETERMINATION OF ACCEPTABLE RISK-BASED LIMITS FOR ENVIRONMENTAL CHEMICALS

16 General protocols for establishing acceptable chemical concentrations and environmental quality criteria 345

16.1	Requirements and Criteria for Establishing Risk-Based Target Levels for Environmental Chemicals	345
16.2	Derivation of Health-Based Action Levels for Environmental Chemicals	347
	16.2.1 Action Levels for Carcinogenic Chemicals	348
	16.2.2 Action Levels for Noncarcinogenic Chemicals/Systemic Toxicants	349
16.3	Health-Protective Risk-Based Chemical Concentrations	350
	16.3.1 *RBC*s for Carcinogenic Constituents	351
	16.3.2 *RBC*s for Noncarcinogenic Effects of Chemical Constituents	352
16.4	A 'Preferable' Health-Protective Chemical Level	353
	16.4.1 The Modified *RBC* for Carcinogenic Chemicals	354
	16.4.2 The Modified *RBC* for Noncarcinogenic Constituents	355
16.5	Health-Protective Chemical Concentrations in Consumer Products	355
	16.5.1 Formulation of Potential Consumer Exposures	356
	16.5.2 Assessing the Chemical Safety of Consumer Products	358
16.6	Miscellaneous Methods for Establishing Environmental Quality Goals	359
16.7	Incorporating Degradation Rates into the Estimation of Environmental Quality Criteria	362
16.8	Utility of Environmental Benchmarks	363
	Suggested Further Reading	364
	References	364

17 Development of risk-based remediation goals — 367

17.1	Factors Affecting the Development of Risk-Based Site Restoration Goals	367
17.2	Risk-Based Soil Cleanup Levels	369
	17.2.1 Soil Cleanup Level for Carcinogenic Contaminants	370
	17.2.2 Soil Cleanup Criteria for the Noncarcinogenic Effects of Site Contaminants	371
17.3	Risk-Based Water Cleanup Levels	373
	17.3.1 Water Cleanup Level for Carcinogenic Contaminants	374
	17.3.2 Water Cleanup Level for the Noncarcinogenic Effects of Site Contaminants	375
17.4	The Cleanup Decision in Site Restoration Programs	377
	Suggested Further Reading	378
	References	379

PART VI: THE ROLE OF RISK ASSESSMENT IN ENVIRONMENTAL MANAGEMENT DECISIONS

18 Illustrative examples of risk assessment practice — 383

18.1 Evaluation of Human Health Risks Associated with Airborne
    Exposures to Asbestos   383
    18.1.1 Study Objective   384
    18.1.2 Summary Results of the Environmental Sampling and
           Analysis   384
    18.1.3 The Risk Estimation   384
    18.1.4 A Risk Management Decision   386
18.2 A Diagnostic Human Health Risk Assessment for a
    Contaminated Site Problem   386
    18.2.1 Introduction and Background   386
    18.2.2 Objective and Scope   387
    18.2.3 Technical Elements of the Diagnostic Risk Assessment
           Process   387
    18.2.4 Identification of Site Contaminants   389
    18.2.5 Screening for Chemicals of Potential Concern   389
    18.2.6 Risk Characterization for Site-Specific Exposure
           Scenarios   390
    18.2.7 A Risk Management Decision   400
18.3 A Risk-Based Strategy for Developing a Corrective Action
    Response Plan for Petroleum-Contaminated Sites   400
    18.3.1 Evaluation of Petroleum Product Constituents   402
    18.3.2 The Fate and Behavior of Petroleum Constituent
           Releases   403
    18.3.3 Contaminant Release Analysis   404
    18.3.4 The Corrective Action and Risk Management Decision
           Process   405
18.4 General Scope of Risk Assessment Practice   407
    Suggested Further Reading   407
    References   408

19 Design of risk management programs   409
19.1 The General Nature of Risk Management Programs   409
    19.1.1 A System for Establishing Risk Management Needs   412
    19.1.2 Interim Corrective Action Programs   414
    19.1.3 Conditions for 'No-Further-Action' Decisions   415
19.2 A Framework for Risk Management Programs   416
19.3 Application of Decision Analysis Methods to Environmental
    Management Programs   418
    19.3.1 Risk–Cost–Benefit Optimization and Tradeoffs
           Analysis   420
    19.3.2 Multi-attribute Decision Analysis and Utility Theory
           Applications   421
19.4 Utilization of GIS in Risk Assessment and Environmental
    Management Programs   426
    19.4.1 Integration of Environmental and Risk Models with
           GIS   426

		19.4.2	The Role of GIS Applications in Environmental Management	427
	19.5		Risk Communication as a Facilitator of Risk Management	428
		19.5.1	Designing an Effective Risk Communication Program	429
	19.6		Managing Risks	431
			Suggested Further Reading	431
			References	432
20	Risk assessment applications to environmental management problems			435
	20.1		General Scope of the Practical Application of Risk Assessment to Environmental Problems	435
	20.2		Risk Assessment as a Cost-Effective Tool in the Formulation of Environmental Management Decisions	445
	20.3		A Concluding Note	447
			Suggested Further Reading	449
			References	449

PART VII: APPENDICES

Appendix A	Selected abbreviations and acronyms	453
Appendix B	Glossary of selected terms and definitions	457
Appendix C	Some basic definitions and concepts in probability theory and statistics	475
Appendix D	Toxicological information for selected environmental chemicals	481
Appendix E	Selected environmental tools and databases of potential relevance to risk assessment and environmental management programs	485
	E.1 Selected Decision Support Tools and Logistical Computer Software	485
	E.2 Selected Databases and Information Libraries with Important Risk Information for Risk Assessment and Environmental Management	492
Appendix F	Selected units of measurements and noteworthy expressions	499
Appendix G	Suggested reference journals	501
Index		505

Preface

Environmental contamination issues are a complex problem with worldwide implications. Risks to both human and ecological health as a result of toxic materials present or introduced into the environment are a matter of grave concern to modern society. The effective management of environmental contamination problems has therefore become an important environmental priority that will remain a growing social challenge for years to come. But it has also become evident that the proper management of environmental contamination problems poses several challenges. Risk assessment seems to be one of the fastest evolving tools that promises a way for developing appropriate strategies to aid environmental management decisions in this arena.

Risk assessment is a tool used to organize, structure, and compile scientific information in order to help identify existing hazardous situations or problems, anticipate potential problems, establish priorities, and provide a basis for regulatory controls and/or corrective actions. It may also be used to help gage the effectiveness of corrective measures or remedial actions. A key underlying principle of risk assessment is that some risks are tolerable – a reasonable and even sensible view, considering the fact that nothing is wholly safe *per se*. In fact, whereas large amounts of a toxic substance may be of major concern, simply detecting a hazardous chemical in the environment should not necessarily be a cause for alarm. In order to be able to make a credible decision on the cut-off between what really constitutes a 'poisonous dose' and a 'no-effect dose' or 'tolerable dose', systematic scientific tools – such as those afforded by risk assessment – become important in the design of environmental management programs. In this regard, therefore, risk assessment seems to represent an important foundation in the development of effective environmental management strategies.

This book is an attempt to provide a concise, yet comprehensive overview of the many facets/aspects relating to chemical risk assessments. It presents some very important tools and methodologies that can be used to help resolve environmental contamination problems in a consistent, efficient, and cost-effective way. Overall, the book represents a collection and synthesis of the principal elements of the risk assessment process that pertain to chemical contamination problems. The review includes an elaboration of pertinent risk assessment concepts and techniques/methodologies for performing human health and ecological risk assessments, as well as methods for the safety evaluation of chemical containment systems. The main emphasis of this book, however, is on the use of human health risk assessment principles to support risk management and corrective action decisions relevant to addressing environmental contamination problems. A number of illustrative example problems – consisting of a variety of typical environmental management problems – are interspersed throughout the book, in order to help present the book in an easy-to-follow, pragmatic manner.

The principal goals for this book are: to present concepts and techniques in risk assessment that may be applied to environmental contamination problems; to provide a guidance framework for the formulation of risk assessments in relation to environmental management issues; to describe methods for conducting human health and environmental risk assessments; and to elaborate strategies for developing environmental quality goals and corrective action plans – with the focus of this volume being on environmental management problems associated with chemical contaminants. Even though the main focus of this title is on risk assessment of the potential human health effects associated with chemical exposures, it is noteworthy that the same principles may be extrapolated to deal with other forms of environmental contamination (such as exposures to radionuclides and pathogens). Thus, the chemical risk assessment framework may be applied to exposures to other agents – albeit many unique issues may have to be addressed for exposures to the new agent under consideration.

Overall, this book will introduce risk assessment to anyone who needs to have knowledge of how to conduct, evaluate, or review risk assessment for environmental toxicants. It offers an understanding of the scientific basis of risk assessment and its applications in the environmental industry – including a portrayal of how the risk assessment process fits into regulatory policies. The book is organized into seven parts – consisting of 20 chapters, together with a set of seven appendices.

The subject matter of this book can indeed be used to aid the resolution of a variety of environmental management problems. The book should serve as a useful reference for many a professional encountering risk assessment in environmental management programs. The specific intended audience would include environmental consulting professionals, environmental managers for industrial facilities, environmental attorneys serving various chemical industries, environmental regulatory officials, public health and environmental health professionals, policy analysts, and a miscellany of environmental interest groups. The book is also expected to serve as a useful *educational and training resource for both students and professional consultants* dealing with risk assessment issues and/or environmental management problems. This book, written for both the novice and the experienced, is an attempt at offering a simplified and systematic presentation of risk assessment methods and application tools, that will carefully navigate the user through the major processes involved.

I am indebted to a number of people for both the direct and indirect support afforded me during the period that I worked on this book project. Sincere thanks are due the Duah family (of Abaam, Kade, and Nkwantanang), and several friends and colleagues who provided much-needed moral and enthusiastic support throughout preparation of the manuscript for this book. I thank the Publishing, Editorial and Production staff at John Wiley & Sons (Chichester, UK) who helped bring this book project to a successful conclusion. I also wish to thank every author whose work is cited in this volume, for having provided some pioneering work to build on. Finally, it should be acknowledged that this book also benefited greatly from review comments of several anonymous individuals, as well as from discussions with a number of professional colleagues. Any shortcomings that remain are, however, the sole responsibility of the author.

D. Kofi Asante-Duah
California, USA
November, 1997

PART I
A GENERAL OVERVIEW

This part of the book comprises general introductory materials and background discussions pertaining to the subject matter offered by this title. Specifically, it consists of the following chapters:

- Chapter 1, *Introduction*, presents a general discussion on the global dimensions of environmental contamination problems – including global environmental management issues; practices that may help minimize environmental contamination problems; and the trend towards a use of risk assessment methodologies in environmental management programs.
- Chapter 2, *The Nature of Environmental Contamination Problems*, describes the sources of chemical pollutants entering the environment. It also offers a discussion of the health, environmental and socio-economic implications of environmental contamination problems. A conceptual system for hazard and risk classification is also presented here.
- Chapter 3, *Characterization of Environmental Contamination Problems*, deals with the investigatory and assessment processes involved in mapping out the types and extents of contamination at potentially impacted locales. It elaborates on the types of activities that are usually undertaken in the conduct of environmental assessments; this includes a discussion of the general tasks in a typical field investigation schedule or workplan.
- Chapter 4, *Legislative–Regulatory Control Needs for the Management of Environmental Contamination Problems*, furnishes an annotated overview of selected environmental legislations that are fair representations of the general regulatory considerations relevant to environmental management programs.

Chapter One

Introduction

Risks to both human and ecological health as a result of toxic materials present or introduced into the environment are a matter of grave concern to modern society. In fact, environmental contamination problems may pose significant risks to the general public because of the potential health and environmental effects, and to other potentially responsible parties because of possible financial liabilities that could result from their effects. The effective management of environmental contamination problems has therefore become an important environmental priority that will remain a growing social challenge for years to come. However, it has also become evident that the proper management of environmental contamination problems poses great challenges. Risk assessment seems to be one of the fastest evolving tools that promises a way for developing appropriate strategies to aid environmental management decisions.

Risk assessment is a tool used to organize, structure, and compile scientific information in order to help identify existing hazardous situations or problems, anticipate potential problems, establish priorities, and provide a basis for regulatory controls and/or corrective actions. It may also be used to help gage the effectiveness of corrective measures or remedial actions. A key underlying principle of risk assessment is that some risks are tolerable – a reasonable and even sensible view, considering the fact that nothing is wholly safe *per se*. In fact, whereas large amounts of a toxic substance may be of major concern, simply detecting a hazardous chemical in the environment should not necessarily be a cause for alarm. In order to be able to make a credible decision on the cut-off between what really constitutes a 'poisonous dose' and a 'no-effect dose' or 'tolerable dose', however, systematic scientific tools – such as those offered by risk assessment – become important in the design of environmental management programs. In this regard, therefore, risk assessment seems to represent an important foundation in the development of effective environmental management strategies.

Indeed, to ensure public health and environmental sustainability, decisions relating to environmental management should be based on systematic and scientifically valid processes, such as via the use of risk assessment. This book will focus on the application of risk assessment concepts and principles in the development of effective environmental management programs, in relation to environmental contamination problems.

1.1 AN AGENDA FOR GLOBAL ENVIRONMENTAL PROTECTION

Many present-day environmental problems can only be solved on the basis of international understanding, which is far from straightforward, given that adequate frameworks for cooperation in the processes involved are yet to be developed and/ or implemented (Asante-Duah and Nagy, 1998). In fact, the very nature of the environment, its complexity, its multiple interconnections, and its in-built delay mechanisms render environmental problems difficult to understand – and thus easy to ignore. This scenario has resulted in uncertainties and disagreements about the fields of greatest importance, and conflicts over boundaries of interest. Views vary between individuals and organizations, depending upon their particular concerns and interests, and also the time-scale of their outlook. As a consequence, there is a paucity of sound environmental information. Many of the relevant data are still missing and those which do exist are not always known or accessible. In general, insufficient use is made of existing data, but also those available are often inadequate – being spatially and temporally patchy, incomplete, and inconsistent. The underlying cause of many of these problems is the absence of an appropriate institutional framework to address environmental concerns directly and fully. Furthermore, the fact that environmental problems tend to vary significantly from one region to another makes it more challenging to develop consistent strategies that are applicable to the global community.

Even so, the problem of the environment should be an object of both national and international cooperation; only this combination of efforts can contribute to an effective solution to the global environmental problems the world faces. What Ananichev (1976) noted over two decades ago – that international cooperation in the field of environmental protection is an objective necessity for the peoples of the world not so much as a form of rational division of labor, but as an inevitable consequence of the progressive deterioration of the biosphere – is still very much true today.

Solutions to environmental problems do indeed require innovative methods of approach, as well as the cooperation and participation of all sectors of society for their implementation. This should include: regulation (through legal and economic instruments), control and management, international cooperation and agreements, and monitoring and assessment. Such problems may therefore be addressed by the application of macro and micro procedures to the various aspects of the issues involved. Among other issues, the development and international harmonization of risk assessment procedures and protocols presents a particularly important challenge to the global community.

In any case, to ensure that resources are properly husbanded and the world's wealth more equitably distributed, there needs to be a radical reappraisal of the wasteful lifestyle that modern societies have grown accustomed to, especially in the industrialized countries.

1.1.1 Environmental Management in a Socio-economic Context

After the 1992 United Nations Conference on Environment and Development (UNCED) – held in Rio de Janeiro, Brazil – whose explicit purpose was to reconcile economic development with environmental protection, the immense complexity of the task of achieving sustainable world development became clear to the international community. Of particular interest, Agenda 21 (which is a comprehensive statement of concerns, findings, and recommended actions on specific environmental areas) fundamentally committed the UN member nations to a 'cradle-to-grave' approach to environmental management.

Of special interest in environmental management issues are the problems associated with environmental contamination problems. In fact, a myriad of environmental health issues – such as may be caused by pollutants in air, chemical contaminants in drinking water, pesticides in food, chemical additives in consumer products, and toxic chemicals at hazardous waste sites – have increased public concern about the development of chemical products and byproducts of modern industrial society. Overall, releases of toxic chemicals into the environment seem to have become one of the major environmental issues of today. It is no coincidence, therefore, that several aspects of Agenda 21 of the UNCED allude to the importance and relevance of using specialized tools – that include hazard and risk assessment – in the management of chemical risks.

Whereas many factors may indeed contribute to various social conflicts in society, it is now apparent that environmental stress and security problems are a very important contributor to such situations. Environmental stress is a result of a complex set of processes that have to be carefully identified and sorted out before a full understanding of the intricate inter-connections can be gained. This must be considered in economic, socio-political, and developmental terms because it has the potential to cause multiple and complex consequences that engender social disruptions, political chaos, and potential regional/inter-state conflicts. In fact, as environmental stress grows in different parts of the world, it may cause global insecurity and developmental crises. In this regard, therefore, it may be concluded that inappropriate environmental management policies can accelerate environmental degradation such that, in the long term, they will engender social disruption and political instability.

To help avert possible disastrous situations and/or a state of paranoia, the UNCED process has come to affirm that national environmental legislation and related institutions are among the critical elements in capacity building for sustainable development. In this regard, there is increasing focus on long-term environmental planning – with the objective to integrate it into overall national development programs. The need to incorporate environmental planning into national socio-economic programs is therefore widely recognized at this time – with the environmental impact assessment (EIA) process having become (since the 1970s) the predominant *modus operandi* for such integration (see, e.g., Gilpin, 1995; Glasson et al., 1994; Morris and Therivel, 1995).

1.1.2 Harmonizing Global Environmental Management Policies

Waste management issues appear to be at center-stage in many countries' environmental concerns (OECD, 1994). Traditionally, society's responses have been mainly directed towards the collection, treatment, and disposal of such wastes. Increasingly, however, efforts are now being aimed at waste minimization – which has become a fundamental goal of emerging environmental policies; this can be achieved through waste prevention, recycling and recovery, and more broadly through a better integration of environmental concerns in consumption and production patterns (OECD, 1994).

Many existing and anticipated environmental problems can indeed be traced to hazardous material processing or handling facilities. Notwithstanding the fact that potential environmental problems tend to vary in dimension and logistics according to the degree of development and the level of industrialization of a nation, there is one fundamental need for all countries – better programs and policies are required globally for these problems. Without adequate environmental management programs, extensive environmental degradation affecting several areas globally could threaten the security of all countries in the world.

In any case, differences exist, and probably will always exist, in environmental management policies adopted by various countries and/or regions, consistent with each nation's overall development and sustainability goals. For example, from the beginning of the 1970s, a strong divergence of policies appeared between Eastern and Western Europe, in which environmental concerns did not receive significant attention in the Eastern European countries (Kara, 1992). Since environmental issues were given less priority, it is not surprising that the importation of hazardous wastes from Western Europe, for instance, was subsequently embraced by the East. However, by the end of the 1980s, the growing global environmental ecstasy advanced the desire to pursue a pan-European approach to environmental restoration and improvement. Among other issues, the safe handling or processing of hazardous materials has become one of the particularly critical environmental themes globally. In fact, recent years have seen the development of another aspect of the environmental movement, characterized by concerns that encompass both the national and international issues affecting the widespread environmental problems.

1.2 MINIMIZING ENVIRONMENTAL CONTAMINATION PROBLEMS

In the past, hazardous waste management practices were tantamount and synonymous to the simplistic rule of 'out of sight, out of mind'. This resulted in the creation of several environmental contamination problems that need to be addressed today. With the improved knowledge linking environmental contaminants to several potential human health problems and also to some ecological disasters, the 'new society' has come to realize the urgent need to clean up for the 'past sins'. Proactive actions are being taken to minimize the continued creation of more environmental contamination problems. Indeed, in recent years, and in several countries, social awareness of environmental problems and the need to reduce sources of contamination, as well as the urge to restore contaminated locales, has been increasing. Nonetheless, some industries continue to generate and release large quantities of contaminants into the environment – either via the production processes or as a result of their waste management plans.

In fact, large quantities of wastes have always been generated by many industries, to the point that there is a near-crisis situation with managing such wastes.

Consequently, waste/materials recycling and re-use, waste exchange, and waste minimization are becoming more prominent in the general waste management practices of several countries. For instance, the recovery of waste oils, solvents, and waste heat from incinerators has become a common practice in several countries. Also, operations exist in a number of countries to recycle heavy metals from various sources (such as silver from photo-finishing operations; lead from lead-acid batteries; mercury from batteries and broken thermometers; heavy metals from metal finishing wastes; etc.). Indeed, waste/materials recycling has become an integral part of many modern industrial processes. Furthermore, waste exchange schemes exist in some countries, to promote the use of one company's byproduct or waste as another's raw material (Asante-Duah and Nagy, 1998).

In conclusion, it is apparent that the possibility of creating environmental contamination problems will remain an inevitable characteristic of industrial activities in modern societies. However, their effects and extent can be minimized by taking several proactive measures – such as those identified below.

1.2.1 Components of a Good Waste Management Program

Several types of regulatory decisions can shape or affect the implementation of hazardous waste management programs; such decisions are often affected by political factors. As Davis (1993) notes, in the US and probably several other places, 'political criteria have clearly assumed a more prominent role in the hazardous waste policymaking process than economic or technical factors'. Indeed, hazardous waste policy problems can become entangled with and complicated by political as well as economic and technical factors; but the key concern is the degree of commitment of the policy-makers responsible for such decision-making.

In fact, regulatory agencies usually are left in rather delicate positions because of the need to balance technical and political factors in reaching a hazardous waste policy decision. A case in point concerns the siting of hazardous waste management facilities (i.e., TSDFs); this presents one of the most controversial aspects of hazardous waste policy-making. This is because the so-called NIMBY (not-in-my-backyard) and related syndromes continue to pose a significant political hurdle for policy-makers seeking TSDFs for certain regions. Whereas these forms of opposition are difficult to overcome under any set of circumstances, it is recognized that there are both political factors and institutional forms that may contribute to a less adversarial forum for site-selection decisions and the development of a policy consensus (Davis, 1993).

In any case, Figure 1.1 illustrates the basic components of a typical waste management program for an industrial facility. The general trend of choice is for on-site waste management – with more emphasis for the future directed at waste minimization. On-site waste minimization strategies that are applied to waste management programs will usually comprise:

- Waste recyling (i.e., the recovery of materials used or produced by a process for separate use or direct re-use in-house).

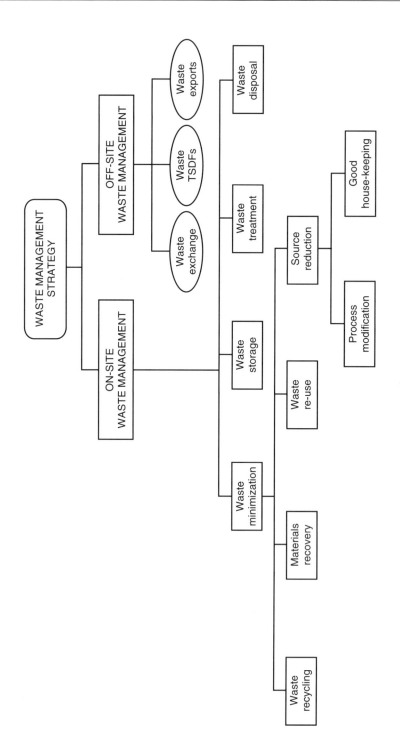

Figure 1.1 Essential elements of a typical good waste management program

Introduction 9

> **Box 1.1 Typical elements of the process modification aspect of waste reduction at source**
>
> - Modification of the process pathway
> - Improved control measures
> - Equipment modification
> - Changes in operational setting/environment
> - Increased automation
> - Product substitution

> **Box 1.2 Typical tasks for a good and efficient house-keeping**
>
> - Material handling improvements
> - Spill and leak monitoring, collection, and prevention
> - Preventative maintenance
> - Inventory control
> - Waste stream segregation

- Materials recovery (i.e., the processing of waste streams to recover materials which can be used as feedstock for conversion to another product).
- Waste re-use (i.e., the direct re-use of the wastes as is, without further processing).
- Reduction at source (that may comprise process modifications and/or the introduction of good house-keeping methods).

All these aspects of the waste management program will likely dominate the scene in industrial waste management planning of the future. In particular, waste reduction at source will gain more prominence in the long term. Reduction at source may be accomplished through process modifications (Box 1.1) and through efficient or good house-keeping (Box 1.2).

Ultimately, even when extensive efforts have been made at minimizing wastes, there still may be some 'residual' wastes remaining when all is said and done. Such wastes can be permitted for long-term storage on site; or for permanent disposal at on-site landfills, surface impoundments, etc.; or for an end-of-pipe treatment on site in order to reduce waste quantity or contaminant concentrations prior to off-site disposal.

1.2.2 Pollution Abatement Strategies and Policies

In the past few decades, environmental technology has gone through a rapid evolutionary process. But an innovative approach for product development and design that allows environmental problems to be completely tackled at their sources in the production process still remains to be born.

At present, the development of an end-of-pipe directed environmental technology has catalyzed much process and product innovation. In fact, many new technologies aimed at waste prevention are being researched into or developed by industry. Other than the apparent financial burdens suffered in attempts to meet environmental regulations, these cleaner technologies promise significant tradeoffs in terms of improved product quality, enhanced production capacity, better process controllability, and increased product reliability. Unfortunately, some industries still do not appreciate the fact that, due to the product- and process-specific nature of the cleaner technology approach, substantial environmental and economic benefits could be realized in the long term. In fact, a crude comparison of the economic parameters for operating an existing industrial plant versus the installation of a new plant will tend to show several indicators of an apparent economic barrier – thus hindering the introduction of new and improved facilities in a number of situations (Neubacher, 1988). As a result, several existing plants with obsolete technology and high potential for pollution and hazardous waste generation have remained in operation. Nevertheless, with the serious commitments by the global community towards harmonizing environmental policies, there are great opportunities for even the skeptics to come to appreciate the innovative strategies being pursued.

Overall, most of the developed economies of the world have made significant progress in reducing emissions and waste stream discharges into the environment. Such levels of environmental protection have been reached by imposing strict emission and disposal standards on environmental pollutants from production processes. However, by the very nature of the currently available technologies and the companion regulations, complete abatement of pollution problems is not possible in production processes.

1.3 FACING UP TO ENVIRONMENTAL CHEMICALS IN SOCIETY

An annotation of selected chemicals in social contexts will show that hazards from several commonly encountered environmental chemicals are major concerns with respect to human and ecological health impacts. In fact, environmental and health problems are generally a logical derivation of human activities in concert with prevailing natural conditions – as illustrated by the limited number of examples for select chemicals given below.

- Lead – believed to be neurotoxic – has been used in water supply systems, gasolines, automobile batteries, and paints for a long time in modern human history; this has resulted in extensive releases into the environment.
- Mercury – also a nervous system toxin, though far less common than lead – is a significant environmental pollutant in several geographical regions/areas because of its use in measuring instruments (e.g., thermometers), in medicines (as antiseptics), in dental practice, in lamps, and in fungicides.
- Chlorinated aromatic hydrocarbons – such as PCBs (which have been widely used in electrical transformers) and DDT (which has been widely used as a

potent pesticide/insecticide) – have proven to be notoriously persistent in the environment. These have indeed become environmental disaster stories, especially in view of their potential to cause severe health problems.

This list – illuminating the 'two-edged sword' nature of a variety of environmental chemicals – could be continued for several different families of both naturally occurring and synthetic groups of chemicals or their derivatives. All these type of situations represent very important public health management problems that call for appropriate decisions on what levels of exposure may indeed pose significant danger – i.e., 'what dose makes the poison?'.

In fact, the 16th century Swiss physician-alchemist, Paracelsus, indicated once upon a time that only the dose of a substance determines its toxicity. This notion makes it even more difficult to ascertain what constitutes hazardous environmental exposure to chemicals. Careful application of risk assessment and risk management principles and tools should generally help remove some of the fuzziness in defining the cut-off line between what may be considered a 'safe level' and what apparently is a 'dangerous level' for most chemicals. Indeed, some chemicals (e.g. arsenic and chromium) are even believed to be essential nutrients with therapeutic properties in small amounts but are extremely toxic in slightly larger amounts.

1.4 RISK ASSESSMENT AS A FACILITATOR OF COST-EFFECTIVE ENVIRONMENTAL MANAGEMENT DECISIONS

There are several health, environmental, political and socio-economic implications associated with environmental contamination problems. It is therefore generally important to use systematic and technically sound methods of approach in the relevant environmental management programs. Risk assessment provides one of the best mechanisms for completing the tasks involved. In fact, a systematic and accurate assessment of current and future risks associated with a given environmental contamination problem is crucial to the development and implementation of a cost-effective corrective action plan. Consequently, risk assessment should generally be considered as an integral part of most environmental management programs.

The primary focus of a risk appraisal is the assessment of whether existing or potential receptors are presently, or in the future may be, at risk of adverse effects as a result of exposure to conditions from potentially hazardous situations. This evaluation then serves as a basis for developing mitigation measures in risk management and risk prevention programs. Typically, the risk assessment will help define the level of risk as well as set performance goals for various response alternatives. The application of risk assessment can indeed provide for prudent and technically feasible and scientifically justifiable decisions about corrective actions that will help protect public health and the environment in a most cost-effective manner.

In order to arrive at cost-effective environmental management decisions, answers will typically have to be generated for several pertinent questions when one is

> Box 1.3   Major issues important to making cost-effective environmental management decisions for an environmental contamination problem
>
> - What is the nature of contamination?
> - What are the sources of, and the 'sinks' for, the contamination?
> - What is the current extent of contamination?
> - What population groups are potentially at risk?
> - What are the likely and significant exposure pathways and scenarios that connect contaminant source(s) to potential receptors?
> - What is the likelihood of health and environmental effects resulting from the contamination?
> - What interim measures, if any, are required as part of a risk management and/or risk prevention program?
> - What corrective action(s) may be appropriate to remedy the prevailing situation?
> - What level of residual contamination will be tolerable or acceptable for the situation?

confronted with a potential environmental contamination problem (Box 1.3). In a number of situations, it becomes necessary to implement interim corrective measures prior to the development and full implementation of a comprehensive remedial action and/or risk management program. Ultimately, a thorough investigation – incorporating a risk assessment – that establishes the nature and extent of contamination may become necessary, in order to arrive at an appropriate and realistic corrective action and/or risk management decision.

To date, risk assessment has been used in Europe for a relatively constrained set of purposes – chiefly to assess new and existing chemical substances (including pesticides), pharmaceutical products, cosmetics, and food additives. There also are proposals and established plans for its application in the occupational health and safety field, as well as for possible use in site remediation decisions in some countries (see, e.g., Cairney, 1995; Ellis and Rees, 1995; HSE, 1989a, 1989b; Smith, 1996).

By contrast, risk assessment principles and methodologies have found extensive and various applications in the US for several years. They have typically been used to evaluate many forms of new products (e.g., foods, drugs, cosmetics, pesticides, consumer products); to set environmental standards (e.g., for air and water); to predict the health threat from contaminants in air, water, and soils; to determine when a material is hazardous (i.e., to identify hazardous wastes and toxic industrial chemicals); to set occupational health and safety standards; and to evaluate soil and groundwater remediation efforts (see, e.g., Asante-Duah, 1996; ASTM, 1995; McTernan and Kaplan, 1990; Millner et al., 1992; NRC, 1993, 1995; Shere, 1995; Sittig, 1994; Smith, 1996; Smith et al., 1996; Tsuji and Serl, 1996).

For now, risk assessment applications in most of the other parts of the world appear to be limited and sporadic. But that is expected to change before too long, as the world continues to search for cost-effective and credible environmental management tools. In fact, in the wake of the June 1992 UN Conference on Environment and Development in Rio de Janeiro, the global/international community's reliance on risk assessment as an effective environmental management tool is likely to grow well into the future – despite skepticism expressed by some (see, e.g., Shere, 1995) who consider the art and science of risk assessment more as a mythical subject rather than real – albeit the process may be fraught with several sources of uncertainty.

In any event, risk assessment should be recognized as a multidisciplinary process that draws on data, information, principles, and expertise from many scientific disciplines – including biology, chemistry, earth sciences, engineering, epidemiology, medicine and health sciences, physics, statistics, and toxicology, among others. Indeed, risk assessment may be viewed as bringing a wide range of subjects – from 'archaeology to zoology' – together.

SUGGESTED FURTHER READING

Allen, D. 1996. *Pollution Prevention for Chemical Processes*. J. Wiley, New York.
Fischer, F. and M. Black (eds). 1995. *Greening Environmental Policy: The Politics of a Sustainable Future*. St Martin's Press, New York.
Kolluru, R. (ed.). 1994. *Environmental Strategies Handbook: A Guide to Effective Policies and Practices*. McGraw-Hill, New York.
Masters, G.M. 1998. *Introduction to Environmental Engineering and Science*, 2nd edition. Prentice-Hall, Upper Saddle River, NJ.
Mathews, J.T. 1991. *Preserving the Global Environment: The Challenge of Shared Leadership*. W.W. Norton & Co., New York.
Moeller, D.W. 1997. *Environmental Health*, Revised edition. Harvard University Press, Cambridge, MA.
Schleicher, K. (ed.). 1992. *Pollution Knows No Frontiers: A Reader*. Paragon House Publishers, New York.
Woodside, G. 1993. *Hazardous Materials and Hazardous Waste Management: A Technical Guide*. J. Wiley, New York.
Worster, D. 1993. *The Wealth of Nature: Environmental History and the Ecological Imagination*. Oxford University Press, New York.

REFERENCES

Ananichev, K. 1976. *Environment: International Aspects*. Progress Publishers, Moscow, Russia.
Asante-Duah, D.K. 1996. *Managing Contaminated Sites: Problem Diagnosis and Development of Site Restoration*. J. Wiley, Chichester, UK.
Asante-Duah, D.K. and I.V. Nagy. 1998. *International Trade in Hazardous Waste*. E. & F.N. Spon/Chapman & Hall/Routledge, London, UK.
ASTM (American Society for Testing and Materials). 1995. *Standard Guide for Risk-Based Corrective Action Applied at Petroleum-Release Sites*. ASTM (E1739-95), Philadelphia, PA.
Cairney, T. 1995. *The Re-use of Contaminated Land: A Handbook of Risk Assessment*. J. Wiley, Chichester, UK.
Davis, C.E. 1993. *The Politics of Hazardous Waste*. Prentice-Hall, Englewood Cliffs, NJ.
Ellis, B. and J.F. Rees. 1995. Contaminated land remediation in the UK with reference to risk assessment: two case studies. *Journal of the Institute of Water and Environmental Management*, 9(1): 27–36.

Gilpin, A. 1995. *Environmental Impact Assessment (EIA): Cutting Edge for the Twenty-First Century*. Cambridge University Press, Cambridge, UK.

Glasson, J., R. Therivel, and A. Chadwick. 1994. *Introduction to Environmental Impact Assessment*. UCL Press, London, UK.

HSE (Health and Safety Executive). 1989a. *Risk Criteria for Land-Use Planning in the Vicinity of Major Industrial Hazards*. HMSO, London, UK.

HSE. 1989b. *Quantified Risk Assessment – Its Input to Decision Making*. HMSO, London, UK.

Kara, J. 1992. Geopolitics and the environment: the case of Central Europe. *Environmental Politics*, 1(2): 18–36.

McTernan, W.F. and E. Kaplan (eds). 1990. *Risk Assessment for Groundwater Pollution Control*. American Society of Civil Engineers, New York.

Millner, G.C., R.C. James, and A.C. Nye. 1992. Human health-based soil cleanup guidelines for diesel fuel No.2. *Journal of Soil Contamination*, 1(2): 103–157.

Morris, P. and R. Therivel (eds). 1995. *Methods of Environmental Impact Assessment*. UCL Press, London, UK.

Neubacher, F.P. 1988. *Policy Recommendations for the Prevention of Hazardous Waste*. Elsevier Science Publishers, Amsterdam, The Netherlands.

NRC (National Research Council). 1993. *Issues in Risk Assessment*. National Academy Press, Washington, DC.

NRC. 1995. *Science and Judgment in Risk Assessment*. National Academy Press, Washington, DC.

OECD (Organization for Economic Cooperation and Development). 1994. *Environmental Indicators: OECD Core Set*. OECD, Paris, France.

Shere, M.E. 1995. The myth of meaningful environmental risk assessment. *Harvard Environmental Law Review*, 19(2): 409–492.

Sittig, M. 1994. *World-Wide Limits for Toxic and Hazardous Chemicals in Air, Water and Soil*. Noyes Publications, Park Ridge, NJ.

Smith, A.H., S. Sciortino, H. Goeden, and C.C. Wright. 1996. Consideration of background exposures in the management of hazardous waste sites: a new approach to risk assessment. *Risk Analysis*, 16(5): 619–625.

Smith, T.T., Jr. 1996. Regulatory reform in the USA and Europe. *Journal of Environmental Law*, 8(2): 257–282.

Tsuji, J.S. and K.M. Serl. 1996. Current uses of the EPA lead model to assess health risk and action levels for soil. *Environmental Geochemistry and Health*, 18: 25–33.

Chapter Two

The Nature of Environmental Contamination Problems

Environmental contamination by a variety of toxic chemicals has become a major environmental issue in a number locations around the world. Industrial activities and related waste management facilities apparently are responsible for most environmental contamination problems. The contributing waste management activities may relate to industrial wastewater impoundments, land disposal sites for solid wastes, land spreading of sludges, accidental chemical spills, leaks from chemical storage tanks and piping systems, septic tanks and cesspools, disposal of mine wastes, or indeed a variety of waste treatment, storage, and disposal (TSD) facilities. Ultimately, once a chemical constituent is released into the environment, it may be transported and/or transformed in a variety of complex ways. These types of situations have, in several ways, contributed to the widespread occurrence of environmental contamination problems globally.

Environmental contamination will generally be due to the presence or release of hazardous materials or wastes at any given locale. In a broad sense, a hazardous material is one which is capable of producing some adverse effects and/or reactions in potential biological receptors; toxic substances generally present 'unreasonable' risk of harm to human health and/or the environment, and need to be regulated. Specifically, hazardous wastes include those byproducts with the potential to cause detrimental effects on human health and/or the natural environment; such wastes may be toxic, bioaccumulative, persistent (i.e., nondegradable), radioactive, carcinogenic (i.e., cancer-causing), mutagenic (i.e., causing gene alterations), and/ or teratogenic (i.e., capable of damaging a developing fetus). The nature of the hazardous constituents will determine the potential for detrimental impacts to human health and the environment. The identification of the types of potentially hazardous contaminants present in the environment is therefore important in the investigation of potential risks attributable to environmental contaminants.

2.1 SOURCES OF ENVIRONMENTAL POLLUTANTS

The primary sources of pollutants entering the environment may be divided into two general categories (Hemond and Fechner, 1994; Petts et al., 1997):
- Point sources, which refer to discrete, localized, and often readily measurable discharges of pollutants into the environment. Examples of this category include industrial outfall pipes, sewage outfalls, stack emissions from industrial

chimneys, accidental spillages or leakages of chemicals at a manufacturing or storage site, and land disposal of wastes.
- Nonpoint (or diffuse) sources, which refer to pollutants covering large areas or that are a composite of numerous and diffused point sources, and often are more difficult to measure. Examples of this category include pesticide and fertilizer runoff from agricultural fields, emissions from automobile traffic, and deposition of sulfur dioxide and other acidic chemicals emitted into the atmosphere in industrial areas.

In general, point sources are easier to control by regulatory and similar actions than are diffuse (or nonpoint) sources – albeit both types can result in extensive environmental contamination problems under any set of circumstances.

Environmental contamination occurs when chemicals are detected where such constituents are not expected and/or not desired. Environmental contamination problems may arise in a number of ways, many of which are the result of manufacturing and other industrial activities or operations. In fact, many of the environmental contamination problems encountered in a number of places are the result of waste generation associated with various forms of industrial activities. Wastes are generated from several operations associated with industrial (e.g., manufacturing and mining), agricultural, military, commercial (e.g., automotive repair shops, utility companies, fueling stations, dry-cleaning facilities, transportation centers and food processing industries), and domestic activities. In particular, the chemicals and allied products manufacturers are generally seen as the major sources of industrial hazardous waste generation.

Wastes are indeed generated at various stages of human activities, and their composition and amounts depend largely on consumption patterns and on industrial and economic structures (OECD, 1994). In fact, waste production is an inevitable characteristic of just about any modern society. Several views of the inevitable concomitance of waste generation from industrialization have been presented by various authors (e.g., Asante-Duah and Nagy, 1998; Bhatt et al., 1986; Kempa, 1991); in one case summary, hazardous waste management facilities have been portrayed as the kidneys of industrial societies (Bhatt et al., 1986). A responsible system for dealing with hazardous wastes is therefore essential to sustain the modern way of life, in much the same way as a well-functioning kidney is necessary to rid the human body of certain toxins.

In fact, the effective management of hazardous wastes, and the associated treatment, storage, and disposal facilities (TSDFs) are of major concern not only to the industry producing such material, but also to governments and individual citizens alike due to the nature and potential impacts of such wastes on the environment and public health. The actual impacts do indeed depend on the waste handling and disposal practices, which in turn are influenced by the waste composition and origin (OECD, 1994). Consequently, the improvement of hazardous waste and materials management practices is a very important environmental issue for the global community. But it is also evident that the proper management of hazardous wastes and materials poses great challenges, which has in a way contributed to the extensive interest in global environmental contamination problems.

> Box 2.1 Major industries and manufacturers potentially contributing to environmental contamination problems. (Source: Asante-Duah, 1996)
>
> - Aerospace
> - Ammunitions
> - Automobile
> - Batteries
> - Beverages
> - Chemical production
> - Computer manufacture
> - Electronics and electrical
> - Electroplating and metal finishing
> - Explosives manufacture
> - Food and dairy products
> - Herbicides, insecticides, and pesticides
> - Ink formulation
> - Inorganic pigments
> - Iron and steel
> - Leather tanning and finishing
> - Medical care facilities
> - Metal smelting and refining
> - Mineral exploration and mining
> - Paint products
> - Perfumes and cosmetics
> - Petroleum products
> - Pharmaceutical products
> - Photographic equipment and supplies
> - Printing and publishing
> - Pulp and paper mills
> - Rubber products, plastic materials, and synthetics
> - Shipbuilding
> - Soap and detergent manufacture
> - Textile products
> - Wood processing and preservation

2.1.1 Industry as a Major Generator of Environmental Pollutants

Wastes from industrial processes include a wide range of materials which may have varied chemical compositions. They may contain varying proportions of organic and inorganic compounds. The heterogeneity makes their disposal all the more difficult. Major categories of industrial wastes which are generally considered as hazardous include: solvents, waste paint, waste containing heavy metals, acids and oily wastes; wastes from mining activities which may occur as mine tailings that are contaminated by metals and chemicals; and large amounts of ash that are often the product of energy generation processes.

Box 2.1 lists the major industrial sectors that are potential sources of waste generation. These industries generate several waste types, such as organic waste sludges and still bottoms (containing chlorinated solvents, metals, oils, etc.); oil and grease (contaminated with polychlorinated biphenyls [PCBs], polyaromatic hydrocarbons (PAHs), metals, etc.); heavy metal solutions (of arsenic, cadmium, chromium, lead, mercury, etc.); pesticide and herbicide wastes; anion complexes (containing cadmium, copper, nickel, zinc, etc.); paint and organic residuals; and several other chemicals and byproducts that need special handling/management.

Table 2.1 summarizes the typical hazardous waste streams potentially generated by selected major industrial operations. Among the most persistent of the organic-based contaminants are the organo-halogens; organo-halogens are rarely found in nature and therefore relatively few biological systems can break them down, in contrast to compounds such as petroleum hydrocarbons which are comparatively more easily degraded (Fredrickson et al., 1993). All of these waste streams have the potential to enter the environment by several mechanisms.

2.1.2 Hazardous Materials Release and Behavior in the Environment

Wide variations on environmental contamination scenarios can generally be anticipated from the release of environmental pollutants at any given locale. A classic example of an environmental contamination situation involves a contaminated site problem (Figure 2.1). Contaminated site problems typically are the result of soil contamination due to placement of wastes on or in the ground; as a result of accidental spills, lagoon failures or contaminated runoff; and/or from leachate generation and migration. Contaminants released to the environment are indeed controlled by a complex set of processes that include various forms of transport and cross-media transfers, transformation, and biological uptake. For instance, atmospheric contamination may result from emissions of contaminated fugitive dusts and volatilization of chemicals present in soils; surface water contamination may result from contaminated runoff and overland flow of chemicals (from leaks, spills, etc.) and chemicals adsorbed onto mobile sediments; groundwater contamination may result from the leaching of toxic chemicals from contaminated soils, or the downward migration of chemicals from lagoons and ponds. Indeed, several different physical and chemical processes can affect contaminant migration from a contaminated site, as well as the cross-media transfer of contaminants at any given site. Consequently, contaminated soils can potentially impact several other environmental matrices.

In general, environmental pollutants will tend to move from the source matrix into other receiving media, which could in turn become secondary release sources for such pollutants or their transformation products.

2.2 HEALTH, ENVIRONMENTAL, AND SOCIO-ECONOMIC IMPLICATIONS OF ENVIRONMENTAL CONTAMINATION PROBLEMS

Human populations and ecological receptors are continuously in contact with varying amounts of environmental contaminants present in air, water, soil, and food. Thus, methods for linking contaminant sources in the multiple environmental

Table 2.1 Typical potentially hazardous waste-streams from various industrial sectors

Sector/source	Typical hazardous waste-stream
Agricultural and food production	Acids and alkalis; fertilizers (e.g., nitrates); herbicides (e.g., dioxins); insecticides; unused pesticides (e.g., aldicarb, aldrin, DDT, dieldrin, parathion, toxaphene)
Airports	Hydraulic fluids; oils
Auto/vehicle servicing	Acids and alkalis; heavy metals; lead-acid batteries (e.g., cadmium, lead, nickel); solvents; waste oils
Chemical/pharmaceuticals	Acids and alkalis; biocide wastes; cyanide wastes; heavy metals (e.g., arsenic, mercury); infectious and laboratory wastes; organic residues; PCBs; solvents
Domestic	Acids and alkalis; dry-cell batteries (e.g., cadmium, mercury, zinc); heavy metals; insecticides; solvents (e.g., ethanol, kerosene)
Dry-cleaning/laundries	Detergents (e.g., boron, phosphates); dry-cleaning filtration residues; halogenated solvents
Educational/research institutions	Acids and alkalis; ignitable wastes; reactives (e.g., chromic acid, cyanides; hypochlorites, organic peroxides, perchlorates, sulfides); solvents
Electrical transformers	PCBs
Equipment repair	Acids and alkalis; ignitable wastes; solvents
Leather tanning	Inorganics (e.g., chromium, lead); solvents
Machinery manufacturing	Acids and alkalis; cyanide wastes; heavy metals (e.g., cadmium, lead); oils; solvents
Medical/health services	Laboratory wastes; pathogenic/infectious wastes; radionuclides; solvents
Metal treating/manufacture	Acids and alkalis; cyanide wastes; heavy metals (e.g., antimony, arsenic, cadmium, cobalt); ignitable wastes; reactives; solvents (e.g., toluene, xylenes)
Military training grounds	Heavy metals
Mineral processing/extraction	High-volume/low-hazard wastes (e.g., mine tailings); red muds
Motor freight/railroad terminals	Acids and alkalis; heavy metals; ignitable wastes (e.g., acetone; benzene; methanol); lead-acid batteries; solvents
Paint manufacture	Heavy metals (e.g., antimony, cadmium, chromium); PCBs; solvents; toxic pigments (e.g., chromium oxide)
Paper manufacture/printing	Acids and alkalis; dyes; heavy metals (e.g., chromium, lead); inks; paints and resins; solvents
Petrochemical industry/petrol stations	Benzo-a-pyrene (BaP); hydrocarbons; oily wastes; lead; phenols; spent catalysts
Photofinishing/photographic industry	Acids; silver; solvents
Plastic materials and synthetics	Heavy metals (e.g., antimony, cadmium, copper, mercury); organic solvents
Shipyards and repair shops	Heavy metals (e.g., arsenic, mercury, tin); solvents
Textile processing	Dyestuff; heavy metals and compounds (e.g., antimony, arsenic, cadmium, chromium, mercury, lead, nickel); halogenated solvents; mineral acids; PCBs
Timber/wood preserving industry	Heavy metals (e.g., arsenic); nonhalogenated solvents; oily wastes; preserving agents (e.g., creosote, chromated copper arsenate, pentachlorophenol)

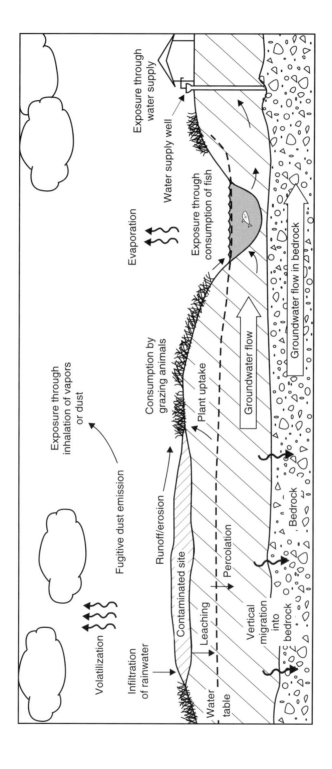

Figure 2.1 A diagrammatic representation of a contaminated site problem. (Source: Asante-Duah, 1996)

media to human and ecological receptor exposures are often necessary to facilitate the development of sound environmental programs.

In fact, the mere existence of environmental contamination can lead to contaminant releases and possible receptor exposures, resulting in both short- and long-term effects on a variety of populations potentially at risk. For instance, lessons from the past, as recorded in several US locations, in Europe, in Japan, and in other places in Asia, clearly demonstrate the dangers that may result from the presence of environmental contamination problems within or near residential communities (Table 2.2) (Alloway and Ayres, 1993; BMA, 1991; Brooks et al., 1995; Canter et al., 1988; Gibbs, 1982; Grisham, 1986; Kletz, 1994; Levine, 1982; Long and Schweitzer, 1982; Meyer et al., 1995; Petts et al., 1997).

Exposure to chemical constituents present in the environment can indeed produce several adverse effects in both human and ecological receptors. For example, human exposures to certain environmental contaminants may result in such diseases as allergic reactions, anemia, anxiety, asthma, blindness, bronchitis, various cancers, contact dermatitis, convulsions, embryotoxicity, emphysema, heart disease, hepatitis, obstructive lung disease, memory impairment, nephritis, neuropathy, and pneumonoconiosis; and ecological exposures to certain chemicals may result in such toxic manifestations as bioaccumulation and/or biomagnification in aquatic organisms. In general, any chemical present in the environment can cause severe health impairment or even death if taken by organisms (including humans) in sufficiently large amounts. On the other hand, there are those chemicals of primary concern which can cause adverse impacts, even from limited exposures.

The existence of unregulated environmental contamination problems in society can indeed be perceived as a potential source of several health, environmental, and additionally possible socio-economic problems. In fact, this situation can actually reduce the future development potentials of a community. Conversely, it has been noted that sensible policies and actions that protect the environment can at the same time contribute to long-term economic progress (Schramm and Warford, 1989). Several analysts have indeed concluded that environmental problems generally are inseparable from socio-economic development problems, and that long-term economic growth depends on protecting the environment (World Bank, 1989). Economic development and environmental protection should therefore complement each other, since improving one has the potential to enhance the other.

In any case, a cultural set-up in which issues relating to public safety are completely ignored is apparently a thing of the past. All peoples are becoming increasingly aware and sensitive to the dangers of inadequacies in environmental management practices. The notion of 'what you don't know doesn't hurt' is no longer tolerated, since even the least educated of populations have some conscious awareness of the potential dangers associated with hazardous materials. Consequently, in addition to managing the actual risks associated with environmental pollution, professionals may also have to deal with the psycho-social effects that the mere presence of toxic materials may have on communities (Peck, 1989).

In general, potential health, ecological, and socio-economic problems are averted by carefully implementing substantive corrective action and risk management programs for environmental contamination problems.

Table 2.2 Selected examples of potential human exposures to hazardous materials resulting from environmental contamination

Site location	Source/nature of problem	Contaminants of concern	Nature of exposure scenario	General comments
Love Canal, Niagara Falls, New York, USA	Section of an abandoned excavation for a canal was used as industrial waste landfill. Site received over 20 000 tonnes of chemical wastes containing more than 80 different chemicals	Various carcinogenic and volatile organic chemicals – including hydrocarbon residues from pesticide manufacture	Potential human exposure routes are by direct contact and also various water pathways	Section of an abandoned excavation for a canal that lies within suburban residential setting had been used as industrial waste landfill Industrial waste dumping occurred from the 1940s through the 1950s; this subsequently caused entire blocks of houses to be rendered unhabitable Problem first uncovered in 1976 Several apparent health impairments – including birth defects and chromosomal abnormalities to residents living in vicinity of the contaminated site
Bloomington, Indiana, USA	Industrial waste entering municipal sewage system. Sewage material was used for garden manure/fertilizer	PCBs	Direct human contacts and also exposures via the food chain (as a result of human ingestion of contaminated food)	PCB-contaminated sewage sludge used as fertilizer – resulting in plant uptakes Also discharges and runoff into rivers resulted in potential fish contamination
Triana, Alabama, USA	Industrial wastes dumped in local stream by a pesticide plant	DDT and other compounds	Potential for human exposure via food chain – i.e., resulting from consumption of fish	High DDT metabolite residues detected in fish consumed by community residents

Site	Source	Pathways	Details

Woburn, Massachusetts, USA	Abandoned waste lagoon with several dumps	Arsenic compounds, various heavy metals, and organic compounds	Potential for generated leachate to contaminate groundwater resources and also for surface runoff to carry contamination to surface water bodies
			
			Potential human receptors and ecosystem exposure via direct contacts and water pathways
				Problem came to light in 1979 when construction workers discovered more than 180 large barrels of waste materials in an abandoned lot alongside a local river
				  High levels of carcinogens found in several local wells – which were then ordered closed
				  Inordinately high degree of childhood leukemia. This apparent excess of childhood leukemia was linked to contaminated well water in the area
				  In general, leukemia and kidney cancer in the area were found to be higher than normal

Times Beach, Missouri, USA	Dioxins (tetrachlorinated dibenzo(p)dioxin – TCDD) in waste oils sprayed on public access areas for dust control	Dioxins (TCDD)	Direct contacts, inhalation, and probable ingestion of contaminated dust and soils	Waste oils contaminated with dioxins (TCDD) were sprayed in several public areas (residential, recreational, and work areas) for dust control of dirt roads, etc. in the late 1960s and early 1970s Problem deemed to present extreme danger in 1982

Chemical Control, Elizabeth, New Jersey, USA	Fire damage to drums of chemicals – resulting in leakages	Various hazardous wastes from local industries	Potential exposures mostly via inhalation of airborne contaminants in the plume of smoke from the fire that blew over surrounding communities	The Chemical Control site was adjacent to an urban receptor community
				  The site was located at the confluence of two rivers. Leaked chemicals contaminated water (used for fire fighting) that subsequently entered adjacent rivers
				  Plume of smoke from fire deposited ash on homes, cars, and playgrounds

continues overleaf

Table 2.2 (continued)

Site location	Source/nature of problem	Contaminants of concern	Nature of exposure scenario	General comments
Santa Clarita, California, USA	Runoff from an electronics manufacturing industry resulted in the contamination of drinking water	Trichloroethylene (TCE) and other volatile organic compounds (VOCs)	TCE and other VOCs contaminated drinking water in this community (due to runoff from industrial facility)	An excess of adverse reproductive outcomes and excess of major cardiac anomalies among infants suspected
Three Mile Island, Pennsylvania, USA	Overheating of nuclear power station in March 1979	Radioactive materials	Emission of radioactive gases Potential for exposure to radioactivity	A small amount of radioactive materials escaped to the atmosphere Unlikely that anyone was harmed by radioactivity from incident. Apparently, the discharge of radioactive materials was too small to cause any measurable harm
Chernobyl, Ukraine (then part of the USSR)	Overheating of a water-cooled nuclear reactor in April 1986	Radioactive materials	Exposure to radioactivity	Nuclear reactor blew out and burned, spewing radioactive debris over much of Europe About 30 people reported killed immediately or died within a few months that may be linked to the accident. It has further been estimated that several thousands more may/could die from cancer during the next 40 years or so as a result of incident
Flixborough, England, UK	Explosion in nylon manufacturing factory in June 1974	Mostly hydrocarbons	Explosive situation – i.e., vapor cloud explosion	Hydrocarbons processed in reaction vessels/ reactors (consisting of oxidation units, etc.) Destruction of plant in explosion, causing death of 28 men on site and extensive damage and injuries in surrounding villages

Location	Description	Chemicals	Exposure pathway	Consequences
Lekkerkirk (near Rotterdam), The Netherlands	Residential setting/development on land raised by layer of household demolition waste and covered with relatively thin layer of sand. Housing project spanned 1972–1975	Various chemicals – comprised mainly of paint solvents and resins (containing toluene, lower boiling point solvents, antimony, cadmium, lead, mercury, and zinc)	Rising groundwater carried pollutants upwards from underlying wastes into the foundations of the houses. This caused deterioration of plastic drinking water pipes, contamination of the water, noxious odors inside the houses, and toxicity symptoms in garden crops	Problem of severe soil contamination was discovered in 1978. Several houses had to be abandoned, while the waste materials were removed and transported by barges to Rotterdam for destruction by incineration. Polluted water was treated in a physico-chemical purification plant. Evacuation of residents commenced in the summer of 1980
Seveso (near Milan), Italy	Discharge containing highly toxic dioxin	Dioxin and caustic soda	Mostly dermal contact exposures (resulting from vapor-phase/gas-phase deposition on the skin) – especially from smoke particles containing dioxin falling on skins, etc.	Discharge containing highly toxic dioxin contaminated a neighboring village over a period of approximately 20 minutes in July 1976. About 250 people developed the skin disease, chloracne, and about 450 were burned by caustic soda A large area of land was contaminated – with part of it being declared uninhabitable
Union Carbide Plant, Bhopal, India	Leak of methyl isocyanate from storage tank in December 1984	Methyl isocyanate (MIC)	MIC vapor discharged into the atmosphere	Exposure to high concentrations of MIC can cause blindness, damage to lungs, emphysema, and ultimately death Leak of over 25 tonnes of MIC from storage tank occurred at Bhopal Spread beyond plant boundary, killing well over 2000 people and injuring several tens of thousands more

continues overleaf

Table 2.2 (continued)

Site location	Source/nature of problem	Contaminants of concern	Nature of exposure scenario	General comments
Kamioka Zinc Mine, Japan	Contaminated surface waters	Cadmium	Ingestion of contaminated water and consumption of rice contaminated by crop uptake of contaminated irrigation water	Water containing large amounts of cadmium discharged from the Kamioka Zinc Mine into river used for drinking water, and also for irrigating paddy rice. Long-term exposures resulted in kidney problems for population
Minamata Bay and Agano River at Niigata, Japan	Effluents from wastewater treatment plants entering coastal waters by a plastics factory	Mercury, giving rise to the presence of highly toxic methylmercury	Human consumption of contaminated seafood	Accumulation of methylmercury in fish and shellfish. Human consumption of the contaminated seafood resulted in health impairments, particularly severe neurological symptoms

2.2.1 Human Health and Ecological Effects of Environmental Contaminants

Several health effects may arise when human and ecological receptors are exposed to some agent or stressor present in the environment. The following represent the major categories of human health and ecological effects that could result from exposure to environmental contaminants (Andelman and Underhill, 1988; Asante-Duah, 1996; Bertollini et al., 1996; Brooks et al., 1995; Grisham, 1986; Lippmann, 1992):

- Human health

    - Carcinogenicity (i.e., capable of causing cancer in humans and/or laboratory animals).
    - Heritable genetic and chromosomal mutation (i.e., capable of causing mutations in genes and chromosomes that will be passed on to the next generation).
    - Developmental toxicity and teratogenesis (i.e., capable of causing birth defects or miscarriages, or damage to developing fetus).
    - Reproductive toxicity (i.e., capable of damaging the ability to reproduce).
    - Acute toxicity (i.e., capable of causing death from even short-term exposures).
    - Chronic toxicity (i.e., capable of causing long-term damage other than cancer).
    - Neurotoxicity (i.e., capable of causing harm to the nervous system).
    - Alterations of immunobiological homeostasis.
    - Congenital abnormalities.

- Ecological

    - Environmental toxicity (i.e., capable of causing harm to wildlife and vegetation).
    - Persistence (i.e., does not breakdown easily, thus persisting and accumulating in portions of the environment).
    - Bioaccumulation (i.e., can enter the bodies of plants and animals but is not easily expelled, thus accumulating over time through repeated exposures).

In fact, several different symptoms, health effects, and other biological responses may be produced from exposure to various specific toxic chemicals commonly encountered in the environment. A number of environmental chemicals are known or suspected to cause cancer; several others may not have carcinogenic properties, but are nonetheless of significant concern due to their systemic toxicity effects. Table 2.3 lists typical symptoms, health effects and other biological responses that could be produced from some representative toxic chemicals commonly encountered in the environment.

The potential for adverse health effects on populations contacting hazardous chemicals present in the environment can indeed involve any organ system. The target and/or affected organ(s) will depend on several factors (Box 2.2) – especially

Table 2.3 Some typical health and ecologic effects due to the presence of selected toxic chemicals in the environment. (Sources: various including: Blumenthal, 1985; Grisham, 1986; Lave and Upton, 1987; Rowland and Cooper, 1983).

Chemical	Typical health effects/symptoms and toxic manifestions/responses
Arsenic and compounds	Acute hepatocellular injury, anemia, angiosarcoma, cirrhosis, developmental disabilities, embryotoxicity, heart disease, hyperpigmentation, peripheral neuropathies
Antimony	Heart disease
Asbestos	Asbestosis (scarring of lung tissue)/fibrosis (lung and respiratory tract), emphysema, irritations, pneumonia/pneumoconioses
Benzene	Aplastic anemia, CNS depression, embryotoxicity, leukemia and lymphoma, skin irritant
Beryllium	Granuloma (lungs and respiratory tract)
Cadmium	Developmental disabilities, kidney damage, neoplasia (lung and respiratory tract), neonatal death/fetal death, pulmonary edema; bioaccumulates in aquatic organisms
Carbon tetrachloride	Narcosis, hepatitis, renal damage, liver tumors
Chromium and compounds	Asthma, cholestasis (of liver), neoplasia (lung and respiratory tract), skin irritant
Copper	Gastrointestinal irritant, liver damage; toxic to fish
Cyanide	Asthma, asphyxiation, hypersensitivity, pneumonitis, skin irritant; toxicity to fish
Dichlorodiphenyl trichloroethane (DDT)	Ataxic gait, convulsions, human infertility/reproductive effects, kidney damage, neurotoxin, peripheral neuropathies, tremors; bioaccumulation in aquatic organisms
Dieldrin	Convulsions, kidney damage, tremors; bioaccumulates in aquatic organisms
Dioxins and furans (PCDDs/PCDFs)	Hepatitis, neoplasia, spontaneous abortion/fetal death; bioaccumulative
Formaldehyde	Allergic reactions; gastrointestinal upsets; tissue irritation
Lead and compounds	Anemia, bone marrow depression, CNS symptoms, convulsions, embryotoxicity, neoplasia, neuropathies, kidney damage, seizures; biomagnifies in food chain
Lindane	Convulsions, coma and death, disorientation, headache, nausea and vomiting, neurotoxin, paresthesias
Lithium	Gastroenteritis, hyperpyrexia, nephrogenic diabetes, Parkinson's disease
Manganese	Bronchitis, cirrhosis (liver), influenza (metal-fume fever), pneumonia
Mercury and compounds	Ataxic gait, contact allergen, CNS symptoms; developmental disabilities, neurasthenia, kidney and liver damage, Minamata disease; biomagnification of methyl mercury
Methylene chloride	Anesthesia, respiratory distress, death
Naphthalene	Anemia

Table 2.3 (continued)

Chemical	Typical health effects/symptoms and toxic manifestions/responses
Nickel and compounds	Asthma, CNS effects, gastrointestinal effects, headache, neoplasia (lung and respiratory tract)
Nitrate	Methemoglobinemia (in infants)
Organo-chlorine pesticides	Hepatic necrosis, hypertrophy of endoplasmic reticulum, mild fatty metamorphosis
Pentachlorophenol (PCP)	Malignant hyperthermia
Phenol	Asthma, skin irritant
Polychlorinated biphenyls (PCBs)	Embryotoxicity/infertility/fetal death, dermatoses, hepatic necrosis, hepatitis, immune suppression; toxicity to aquatic organisms
Silver	Blindness, skin lesions, pneumonoconiosis
Toluene	Acute renal failure, ataxic gait, CNS depression, memory impairment
Trichloroethylene	CNS depression, deafness, liver damage, paralysis, respiratory and cardiac arrest, visual effects
Vinyl chloride	Leukemia and lymphoma, neoplasia, spontaneous abortion/fetal death, tumors, death
Xylene	CNS depression, memory impairment
Zinc	Corneal ulceration, esophagus damage, pulmonary edema

Box 2.2 Pertinent factors that influence human response to toxic chemicals present in the environment

- Dosage (because large dose may mean more immediate effects)
- Age (since the elderly and children are more susceptible to toxins)
- Sex (since each sex has hormonally controlled hypersensitivities)
- Body weight (which is inversely proportional to toxic responses/effects)
- Psychological status (because stress increases vulnerability)
- Genetics (because different metabolic rates affect receptor responses)
- Immunological status and presence of other diseases (because health status influences general metabolism)
- Weather conditions (since temperature, humidity, barometric pressure, season, etc., potentially affect absorption rates)

---

the specific chemicals contacted; the extent of exposure (i.e., dose or intake); the characteristics of the exposed individual (e.g., age, gender, body weight, psychological status, genetic make-up, immunological status, susceptibility to toxins, hypersensitivities); the metabolism of the chemicals involved; weather conditions (e.g., temperature, humidity, barometric pressure, season); and the presence or absence of confounding variables such as other diseases (Brooks et al., 1995; Grisham, 1986).

Invariably, exposures to chemicals escaping into the environment can result in a reduction of life expectancy and possibly a period of reduced quality of life (e.g., caused by anxiety from exposures, diseases, etc.). The presence of environmental contaminants can therefore create potentially hazardous situations and pose significant risks of concern to society at large.

2.3 A HAZARD CATEGORIZATION SYSTEM FOR ENVIRONMENTAL CONTAMINATION PROBLEMS

It is important to recognize the fact that there may be varying degrees of hazards associated with different environmental contamination problem situations, and that there are good technical and economic advantages for ranking such problems according to the level of hazard they potentially pose. A typical categorization scheme for potential environmental contamination problems may comprise defining the 'candidate' problems on the basis of potential risks associated with such situations – (e.g., high-, intermediate- and low-risk problems, conceptually depicted by Figure 2.2). Such a classification will facilitate the development and implementation of efficient environmental management, risk management, or corrective action programs.

In general, the high-risk problems will prompt the most concern, requiring immediate and urgent attention or corrective measures. A problem may be designated as 'high-risk' when contamination represents a real or imminent threat to human health and/or to the environment. In this case, an immediate action will generally be required to reduce the threat. Thus, in order to ensure the development of adequate and effective risk management, environmental management, or corrective action strategies, potential environmental contamination problems should preferably be categorized in a similar or other appropriate manner.

For example, in the investigation of potentially contaminated site problems, a broad categorization scheme may be used for a variety of contaminated area designations – and this will facilitate the planning, development, and implementation of appropriate environmental restoration programs. Sites may indeed be clustered into different groups in accordance with the level of effort required to implement the appropriate or necessary response actions. Typical designations, which are by no means exhaustive, are enumerated below.

- *Low-risk areas*, represented by locations with no confirmed contamination, may consist of areas (e.g., suspected sources) where the results of records search and environmental investigations show that no hazardous substances were stored for any substantial period of time, released into the environment or location structures, or disposed of on the property; or areas where the occurrence of such storage, release, or disposal is not considered to have been probable. The determination of a low-risk area can be made at any of several decision stages during the environmental assessment.
- *Intermediate-risk areas*, represented by locations with limited contamination, may consist of impacted areas where no response or remedial action is necessarily required to ensure protection of human health and the environment. Such areas typically will include locations where an environmental assessment may have demonstrated that hazardous materials have been released, stored, or

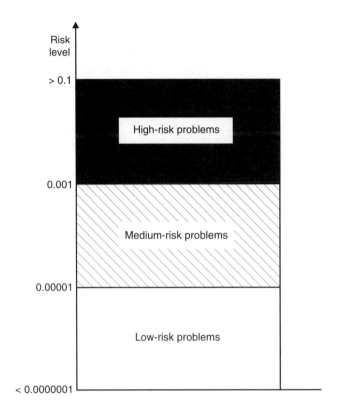

Figure 2.2 A conceptual representation of typical risk categories for environmental contamination problems. (Source: Asante-Duah, 1996)

disposed of, but are present in quantities that probably require only limited to no response or remedial action to protect human health and the environment. Such designation means that levels of hazardous substances detected in a given area *do not* exceed media-specific action levels; *do not* result in significant risks; nor otherwise exceed requisite regulatory standards.
- *High-risk areas*, represented by locations with extensive contamination, may consist of areas where the location records indicate that hazardous materials are known to have been released or disposed of. Typically, a significant level of corrective action response will be required for such areas.

Environmental contamination problems may indeed pose different levels of risk, depending on the type and extent of contamination present at any given location. The degree of hazard posed by the contaminants involved will generally be dependent on several factors such as: physical form and composition of contaminants; quantities of contaminants; reactivity; biological and ecological toxicity effects; mobility in various environmental media; persistence or attenuation in the environment; and local conditions and environmental setting (e.g., temperature, soil type, groundwater flow conditions, humidity, and light). Consequently, it is very important to adequately characterize environmental contamination problems – that is if cost-effective solutions are to be found for dealing with the prevailing hazards.

Ultimately, depending on the problem category, different corrective action response strategies may be utilized in the requisite environmental management program.

SUGGESTED FURTHER READING

Colten, C.E. and P.N. Skinner. 1996. *The Road to Love Canal: Managing Industrial Waste before EPA*. University of Texas Press, Austin, TX.

Francis, B.M. 1994. *Toxic Substances in the Environment*. J. Wiley, New York.

Mitchell, P. and D. Barr. 1995. The nature and significance of public exposure to arsenic: a review of its relevance to South West England. *Environmental Geochemistry and Health*, 17: 57–82.

Patnaik, P. 1992. *A Comprehensive Guide to the Hazardous Properties of Chemical Substances*. Van Nostrand Reinhold, New York.

Rail, C.C. 1989. *Groundwater Contamination: Sources, Control and Preventive Measures*. Technomic Publishing Co., Inc., Lancaster, PA.

Zoller, U. (ed.). 1994. *Groundwater Contamination and Control*. Marcel Dekker, New York.

REFERENCES

Alloway, B.J. and D.C. Ayres. 1993. *Chemical Principles of Environmental Pollution*. Blackie Academic & Professional/Chapman & Hall, London, UK.

Andelman, J.B. and D.W. Underhill. 1988. *Health Effects from Hazardous Waste Sites*. Lewis Publishers, Chelsea, MI.

Asante-Duah, D.K. 1996. *Managing Contaminated Sites: Problem Diagnosis and Development of Site Restoration*. John Wiley, Chichester, UK.

Asante-Duah, D.K. and I.V. Nagy. 1998. *International Trade in Hazardous Waste*. E. & F.N. Spon/Chapman & Hall, London, UK.

Bertollini, R., M.D. Lebowitz, R. Saracci, and D.A. Savitz (eds). 1996. *Environmental Epidemiology (Exposure and Disease)*. Lewis Publishers/CRC Press, Boca Raton, FL.

Bhatt, H.G., R.M. Sykes, and T.L. Sweeney (eds). 1986. *Management of Toxic and Hazardous Wastes*. Lewis Publishers, Inc., Chelsea, MI.

Blumenthal, D.S. (ed.). 1985. *Introduction to Environmental Health*. Springer Publishing Co., New York.

BMA (British Medical Association). 1991. *Hazardous Waste and Human Healh*. Oxford University Press, Oxford, UK.

Brooks, S.M. et al. 1995. *Environmental Medicine*. Mosby, Mosby-Year Book, Inc., St Louis, MO.

Canter, L.W., R.C. Knox, and D.M. Fairchild. 1988. *Ground Water Quality Protection*. Lewis Publishers, Inc., Chelsea, MI.

Fredrickson, J.K., H. Bolton, Jr, and F.J. Brockman. 1993. In situ and on-site bioreclamation. *Environmental Science and Technology (ES&T)*, 27(9): 1711–1716.

Gibbs, L.M. 1982. *Love Canal: My Story*. State University of New York Press, Albany, New York.

Grisham, J.W. (ed.). 1986. *Health Aspects of the Disposal of Waste Chemicals*. Pergamon Press, Oxford, UK.

Hemond, H.F. and E.J. Fechner. 1994. *Chemical Fate and Transport in the Environment*. Academic Press, San Diego, CA.

Kempa, E.S. (ed.). 1991. *Environmental Impact of Hazardous Wastes*. PZITS Publishing Dept, Warsaw, Poland.

Kletz, T. 1994. *Learning from Accidents*, 2nd edition. Butterworth–Heinemann, Oxford, UK.

Lave, L.B. and A.C. Upton (eds). 1987. *Toxic Chemicals, Health, and the Environment*. Johns Hopkins University Press, Baltimore, MD.

Levine, A.G. 1982. *Love Canal: Science, Politics, and People*. Lexington Books, Lexington, MA.

Lippmann, M. (ed.). 1992. *Environmental Toxicants: Human Exposures and their Health Effects*. Van Nostrand Reinhold, New York.

Long, F.A. and G.E. Schweitzer (eds). 1982. *Risk Assessment at Hazardous Waste Sites*. American Chemical Society, Washington, DC.

Meyer, P.B., R.H. Williams, and K.R. Yount. 1995. *Contaminated Land: Reclamation, Redevelopment and Reuse in the United States and the European Union*. Edward Elgar Publishing Ltd, Aldershot, UK.

OECD (Organization for Economic Co-operation and Development). 1994. *Environmental Indicators: OECD Core Set*. OECD, Paris, France.

Peck, D.L. (ed.). 1989. *Psychosocial Effects of Hazardous Toxic Waste Disposal on Communities*. Charles C. Thomas Publishers, Springfield, IL.

Petts, J., T. Cairney, and M. Smith. 1997. *Risk-Based Contaminated Land Investigation and Assessment*. J. Wiley, Chichester, UK.

Rowland, A.J. and P. Cooper. 1983. *Environment and Health*. Edward Arnold, London, UK.

Schramm, G. and J.J. Warford (eds). 1989. *Environmental Management and Economic Development*. A World Bank Publication. Johns Hopkins University Press, Baltimore, MD.

World Bank. 1989. *Striking a Balance – The Environmental Challenge of Development*. IBRD/World Bank, Washington, DC.

Chapter Three

Characterization of Environmental Contamination Problems

The characterization of potential environmental contamination is a process used to establish the presence or absence of contamination or hazards, to delineate the nature and extent of the contamination or hazards, and to determine possible threats posed by the contamination or hazard situation to human health and/or the environment.

When there is a source release, contaminants may reach potential receptors via several environmental media (e.g., air, soils, groundwater and surface water). A complexity of processes may indeed affect contaminant migration in the environment, resulting in possible threats to human and ecological receptors outside the source area. Thus, it is imperative to adequately characterize a source area and vicinity through a well-designed environmental assessment program – in order to arrive at appropriate and cost-effective environmental management decisions.

3.1 THE ENVIRONMENTAL CHARACTERIZATION PROCESS

The process involved in the characterization of environmental contamination problems typically consists of the collection and analysis of a variety of environmental data. Box 3.1 enumerates several important elements of environmental characterization activities that are used to support environmental management decisions. In general, credible environmental characterization programs tend to involve a complexity of activities that require careful planning. To facilitate this process, workplans are usually needed to specify the administrative and logistic requirements of environmental characterization activities. A typical workplan used to guide the investigation of environmental contamination problems will generally include: a sampling and analysis plan; a health and safety plan; and a quality assurance/quality control plan. The major components and tasks required of most environmental characterization workplans are briefly discussed in the following sections – with greater details offered elsewhere in the literature (e.g., Asante-Duah, 1996; ASTM, 1997b; Boulding, 1994; CCME, 1993; CDHS, 1990; Keith, 1988, 1991; USEPA, 1985a, 1985b, 1987a, 1987b, 1988a, 1988b, 1989a, 1989b, 1989c, 1989d).

3.1.1 The Sampling and Analysis Plan

The sampling and analysis plan (SAP) is an essential component of any environmental investigation – generally being required to specify sample types, numbers,

> Box 3.1 Elements of an environmental characterization activity
>
> - Identify and describe the physical setting and/or geographical location of problem situation
> - Determine sources of contamination, differentiating between naturally-occurring and human-made sources
> - Design a sampling plan, and conduct field data collection and analysis
> - Determine the nature (i.e., contaminant types and their characteristics) and extent of contamination
> - Assess the contaminant fate and transport characteristics in the various environmental matrices
> - Delineate potential migration pathways
> - Prepare a data summary that contains pertinent sampling information for the problem situation
> - Identify and compile regulatory limits for the target contaminants being investigated
> - Develop a conceptual model for the case problem – to include patterns of potential exposures
> - Conduct a risk assessment for the case problem
> - Identify risk management and corrective action needs for the problem situation

locations, and procedures. Its purpose is to ensure that sampling and data collection activities will be comparable to, and compatible with previous (and possible future) data collection activities. SAPs provide a mechanism for planning and approving field activities (USEPA, 1988a, 1988b, 1989b). The required level of detail and the scope of the planned investigation generally determines the data quality objectives (DQOs). It is important that the sampling and analysis strategies are planned in such a manner as to minimize the costs associated with achieving the DQOs. Data necessary to meet the project objectives should be specified, including the selection of sampling methods and analytical protocols for the particular situation; this will also include an evaluation of multiple-option approaches that will ensure timely and cost-effective data collection and evaluation. Box 3.2 enumerates a checklist of items that should be ascertained in the development of a SAP (CCME, 1993; Keith, 1988, 1991).

Important issues to consider when one is making a decision on how to obtain reliable samples relate to the sampling objective and approach; sample collection methods; chain-of-custody documentation; sample preservation techniques; sample shipment methods; and sample holding times (Box 3.3). A detailed discussion of sampling considerations and strategies for various environmental matrices can be found in the literature elsewhere (e.g., ASTM, 1997a; BSI, 1981a, 1981b, 1987, 1991a, 1991b, 1993, 1996a, 1996b; BSI/ISO, 1995a, 1995b, 1995c, 1995d; CDHS, 1990; DOE, 1994; Keith, 1988, 1991; Lave and Upton, 1987; Petts et al., 1997; USEPA, 1988b, 1989b). Overall, the methods by which data of adequate quality and quantity are to be obtained to meet the overall project goals should be specified and fully documented in the SAP developed as part of a detailed environmental characterization workplan. It is noteworthy that the use of appropriate sample collection methods can be as important as the use of appropriate analytical methods for sample analyses. Consequently, effective protocols in both the

Characterization of environmental contamination problems 37

Box 3.2 Checklist for developing sampling and analysis protocols

- What observations at sampling locations are to be recorded?
- Has information concerning data quality objectives, analytical methods, analytical detection limits, etc., been included?
- Have instructions for modifying protocols in case of unanticipated problems been specified?
- Has a list of all likely sampling equipment and materials been prepared?
- Are there instructions for cleaning equipment before and after sampling?
- Have instructions for each type of sample collection been prepared?
- Have instructions for completing sample labels been included?
- Have instructions for preserving each type of sample (such as preservatives to use, and also maximum holding times of samples) been included?
- Have instructions for packaging, transporting, and storing samples been included?
- Have instructions for chain-of-custody procedures been included?
- Is there a waste management plan to deal with wastes generated during the environmental investigation activities?

Box 3.3 Elements of a sampling plan

- Background information about locale (that includes a description of the problem location and surrounding areas, and a discussion of known and suspected contamination sources, probable migration pathways, and other general information about the physical and environmental setting)
- Sampling objectives (describing the intended uses of the data)
- Sampling location and frequency (that also identifies each sample matrix to be collected and the constituents to be analyzed)
- Sample designation (that establishes a sample numbering system for the specific project, and should include the sample number, the sampling round, the sample matrix, and the name of the case property)
- Sampling equipment and procedures (including equipment to be used and material composition of equipment, along with decontamination procedures)
- Sample handling and analysis (including identification of sample preservation methods, types of sampling jars, shipping requirements, and holding times)

---

sampling and laboratory procedures should be adequately specified by the SAP, in order to help minimize uncertainties associated with the data collection and analysis activities.

Sampling and analysis of environmental pollutants is indeed a very important part of the decison-making process involved in the management of environmental contamination problems. Yet, sampling and analysis could become one of the most expensive and time-consuming aspects of an environmental management program. Even of greater concern is the fact that errors in sample collection, sample handling, or laboratory analysis can invalidate environmental characterization projects or add to the overall project costs. All environmental samples that are intended for use in environmental characterization programs must therefore be

> **Box 3.4 Sampling plan checklist**
>
> - What are the DQOs, and what corrective measures are planned if DQOs are not met (e.g., re-sampling or revision of DQOs)?
> - Do program objectives need exploratory, monitoring, or both sampling types?
> - Have arrangements been made for access to the case property?
> - Is specialized sampling equipment needed and/or available?
> - Are field crew who are experienced in the required types of sampling available?
> - Have all analytes and analytical methods been listed?
> - Have required good laboratory practice and/or method QA/QC protocols been listed?
> - What type of sampling approach will be used (i.e., random, systematic, judgmental, or combinations thereof)?
> - What type of data analysis methods will be used (e.g., geostatistical, control charts, hypothesis testing, etc.)?
> - Is the sampling approach compatible with data analysis methods?
> - How many samples are needed?
> - What types of QC samples are needed, and how many of each type of QC samples are needed (e.g., trip blanks, field blanks, equipment blanks, etc.)?

collected, handled, and analyzed properly, in accordance with all applicable/relevant methods and protocols. Box 3.4 provides a convenient checklist of the issues that should be verified when planning a sampling activity for an environmental contamination problem (CCME, 1993).

In general, an initial evaluation of an environmental contamination problem should provide insight into the types of contaminants, the populations potentially at risk, and possibly the magnitude of the risk. These factors can be combined to design a sampling plan and to specify the size of sampling unit or support addressed by each sample or set of samples.

3.1.1.1 Sampling protocols

Sampling protocols are written descriptions of the detailed procedures to be followed in collecting, packaging, labeling, preserving, transporting, storing, and documenting samples. The selection of analytical methods is also an integral part of the processes involved in the development of sampling plans, since this can strongly affect the acceptability of a sampling protocol. For example, the sensitivity of an analytical method could directly influence the amount of a sample needed in order to be able to measure analytes at pre-specified minimum detection (or quantitation) limits. The analytical method may also affect the selection of storage containers and preservation techniques (Holmes et al., 1993; Keith, 1988).

Regardless of the medium sampled, data variability problems may arise from temporal and spatial variations in field data. That is, sample composition may vary depending on the time of the year and weather conditions when the sample is collected. Ideally, samples from various media should be collected in a manner that accounts for temporal factors and weather conditions. If temporal fluctuations cannot be characterized in the investigation, details of meteorological, seasonal,

> Box 3.5 Minimum requirements for documenting environmental sampling
>
> - Sampling date
> - Sampling time
> - Sample identification number
> - Sampler's name
> - Sampling location
> - Sampling conditions or sample type
> - Sampling equipment
> - Preservation used
> - Time of preservation
> - Auxiliary data (i.e., relevant observations at sample location)

and climatic conditions during sampling must be well documented. Choosing an appropriate sampling interval that spans a sufficient amount of time to allow one to obtain, for example, an independent sample will generally help reduce the effects of autocorrelation. Also, sampling both background and 'compliance' locations at the same point in time should reduce temporal effects. Consequently, the ideal sampling scheme will incorporate a full annual sampling cycle. If this strategy cannot be accommodated in an investigation, at least two sampling events should be considered that take place during opposite seasonal extremes (such as high-water versus low-water, high-recharge versus low-recharge, etc.).

The overall sampling protocol must identify sampling locations and include all of the equipment and information needed for sampling, such as: the types, number, and sizes of containers; labels; field logs; types of sampling devices; numbers and types of blanks, sample splits, and spikes; the sample volume; any composite samples; specific preservation instructions for each sample type; chain-of-custody procedures; transportation plans; field preparations (such as filter or pH adjustments); field measurements (such as pH, dissolved oxygen, etc.); and the reporting requirements (Keith, 1988). The sampling protocol should also identify those physical, meteorological, and related variables to be recorded or measured at the time of sampling (Keith, 1988). In addition, information concerning the analytical methods to be used, minimum sample volumes, desired minimum levels of quantitation, and analytical bias and precision limits may help sampling personnel make better decisions when unforeseen circumstances require changes to the sampling protocol.

In general, the devices used to collect, store, preserve, and transport samples must *not* alter the sample in any manner. In this regard, it is noteworthy that special procedures may be needed to preserve samples during the period between collection and analysis. In any case, the more specific a sampling protocol is, the less chance there will be for errors or erroneous assumptions. Box 3.5 lists the minimum documentation needed for environmental sampling activities (CCME, 1993).

3.1.1.2 Sampling strategies

There are three basic sampling approaches: random, systematic, and judgmental. There are also three primary combinations of each of these, i.e., stratified–

(judgmental)–random, systematic–random, and systematic–judgmental (CCME, 1993; Keith, 1991). Additionally, there are further variations that can be found among the three primary approaches and the three combinations thereof. For example, the systematic grid may be square or triangular; samples may be taken at the nodes of the grid, at the center of the spaces defined by a grid, or randomly within the spaces defined by a grid (CCME, 1993). A combination of judgmental, systematic, or random sampling is often the most feasible approach to employ in the investigation of potential enviromental contamination problems. However, the sampling scheme should be flexible enough to allow relevant adjustments/ modifications during field activities.

In general, several different methods are available for acquiring data to support environmental characterization programs. The methodology used for sampling can indeed affect the accuracy of subsequent evaluations. It is therefore imperative to select the most appropriate methodology possible in order to obtain the most reliable results attainable; Holmes et al. (1993) enumerate several factors that should be considered when selecting a sampling method. Standard sampling practices often employed in the investigation of commonly impacted environmental media include the following (Holmes et al., 1993):

- Three main types of water sampling schemes – i.e., grab samples, composite samples, and continuous flowing samples. For the different sampling approaches, samples are collected with different types of water sampling equipment or tools.
- Two main types of soil sampling schemes – i.e., surface sampling (0–15 cm) and soil samples from depth (>15 cm) (i.e., shallow subsurface sampling, and deep soil samples). Samples can be collected with some form of core sampling or auger device, or they may be collected by use of excavations or trenches.
- Two main types of air sampling schemes – i.e., instantaneous or grab samples (usually taken in a period of less than 5 min.) and integrated air sampling. The choice of procedure for the air sampling is dependent on the contaminant to be measured.

Additional discussion on sampling strategies and requirements is given in Chapter 9 of this volume. Further elaboration of sampling schemes, together with several techniques and equipment that can be used in the sampling of contaminated air, soils, sediments, and waters, are enumerated in the literature elsewhere (e.g., CCME, 1993; Holmes et al., 1993; Petts et al., 1997).

3.1.1.3 Laboratory and analytical program requirements

Analytical protocol and constituent parameter selection are usually carried out in a way that balances costs of analysis with adequacy of coverage. Oftentimes, the initial analyses of environmental samples may be performed with a variety of field methods used for screening purposes. The purpose of using initial field screening methods is to decide whether the level of pollution at a given locale is high enough to warrant more expensive (and more specific and accurate) laboratory analyses. Methods that screen for a wide range of compounds, even if determined as groups

> **Box 3.6 Minimum requirements for documenting laboratory work**
>
> - Method of analysis
> - Date of analysis
> - Laboratory and/or facility carrying out analysis
> - Analyst's name
> - Calibration charts and other measurement charts (e.g., spectral)
> - Method detection limits
> - Confidence limits
> - Records of calculations
> - Actual analytical results

or homologs, are useful because they allow more samples to be measured faster and less expensively than with conventional laboratory analyses. In the more detailed assessment, environmental sample analysis is generally performed by laboratory programs that consist of routine and nonroutine standardized analytical procedures and associated quality control requirements managed under a broad quality assurance program; these services are provided through routine analytical services and special analytical services.

In general, effective analytical programs and laboratory procedures are necessary to help minimize uncertainties in environmental investigation activities. Guidelines for the selection of analytical methods are offered elsewhere in the literature (e.g., CCME, 1993). Usually there are several methods available for most environmental analytes of interest. Some analytes may have up to a dozen methods to select from. On the other hand, some analytes may have no proven methods available. In the latter case, it usually means that some of the specific isomers that were selected as representative compounds for environmental pollution have not been verified to perform acceptably with any of the commonly used methods (CCME, 1993).

Box 3.6 lists the minimum requirements for documenting laboratory work performed to support environmental characterization activities (CCME, 1993; USEPA, 1989c). The applicable analytical procedures, the details of which are outside the scope of this book, should be strictly adhered to.

3.1.2 The Health and Safety Plan

To minimize risks to environmental investigation personnel (and possible nearby populations) as a result of exposure to environmental contamination, health and safety issues must always be addressed as part of any environmental characterization activity. The purpose of a health and safety plan (HSP) is to identify, evaluate, and control health and safety hazards, and to provide for emergency response during environmental characterization activities. Proper planning and execution of safety protocols will help protect an investigation team from accidents and needless exposure to hazardous or potentially hazardous chemicals. Protecting the health and safety of an investigation team, as well as the general public, is indeed a major concern during environmental investigation activities.

> **Box 3.7 Relevant elements of a health and safety plan**
>
> - Description of known hazards and risks associated with the planned environmental characterization activities (i.e., a health and safety risk analysis for existing conditions, and for each task and operation)
> - Listing of key personnel and alternates responsible for safety, emergency response operations, and public protection in the project location
> - Description of the levels of protection to be worn by investigative personnel and visitors to the locale
> - Delineation of work and rest areas
> - Establishment of procedures to control access to the case property
> - Description of decontamination procedures for personnel and equipment
> - Establishment of emergency procedures, including emergency medical care for injuries and toxicological problems
> - Development of medical monitoring program for personnel (i.e., medical surveillance requirements)
> - Establishment of procedures for protecting workers from weather-related problems
> - Specification of any routine and special training required for personnel responding to environmental or health and safety emergencies
> - Definition of entry procedures for 'confined' or enclosed spaces
> - Description of requirements for environmental surveillance program
> - Description of the frequency and types of air monitoring, personnel monitoring, and environmental sampling techniques and instrumentation to be used

The principal objective of the HSP is to specify safety precautions needed to protect the populations potentially at risk during environmental characterization activities. Consequently, a case-specific HSP should be prepared and implemented prior to the commencement of any environmental characterization activity. The HSP should be developed to conform with all the requirements for occupational saftey and health, and also with applicable national, state, and local laws, rules, regulations, statutes, and orders as necessary to protect all populations potentially at risk. Furthermore, all personnel involved with environmental characterization activities should have received adequate training, and there should be a contingency plan in place that meets all safety requirements. For instance, in the US, the HSP developed and implemented in the investigation of a potentially contaminated site should be in full compliance with all the requirements of the Occupational Safety and Health Administration (OSHA) (i.e., OSHA: 29 CFR 1910.120); the requirements of USEPA (i.e., EPA: Orders 1420.2 and 1440.3); as well as any other relevant state or local laws, rules, regulations, statutes, and orders necessary to protect the populations potentially at risk. In addition, all personnel involved with on-site activities should have received a 40-hour OSHA Hazardous Waste Operations and Emergency Response Activities (HAZWOPER) training, including the 8-hour refresher course, where necessary.

Box 3.7 contains the relevant elements of a HSP that will satisfy the general requirements of a safe work activity (CDHS, 1990; USEPA, 1987a). It is important that these elements are fully evaluated, to ensure compliance with local health and safety regulations and/or to avert potential health and safety problems during the environmental characterization activities. Health and safety data are generally required to establish the level of protection needed for an environmental investigation crew. Such data are also used to determine whether there should be immediate concern for any population living in proximity to the problem locale. Details of specific items of required health and safety issues and equipment are discussed elsewhere (e.g., Cheremisinoff and Graffia, 1995; Martin et al., 1992; OBG, 1988).

In general, a health and safety officer establishes the level of protection required during the environmental characterization program, and then determines whether the level should be advanced or reduced at any point in time during the investigation. Typically, every environmental characterization project should start with a health and safety review (at which all personnel sign a review form); a tailgate safety meeting (to be attended by all environmental investigation personnel); and a safety compliance agreement (that should be signed by all persons entering the case property).

3.1.3 The Quality Assurance/Quality Control Plan

Quality assurance (QA) refers to a system for ensuring that all information, data, and resulting decisions compiled from an investigation (e.g., monitoring and sampling tasks) are technically sound, statistically valid, and properly documented. The QA program consists of a system of documented checks used to validate the reliability of a data set.

Quality control (QC) is the mechanism through which quality assurance achieves its goals. Quality control programs define the frequency and methods of checks, audits, and reviews necessary to identify problems and corrective actions, thus verifying product quality. All QC measures should be performed for at least the most sensitive chemical constituents from each sampling event/date.

A detailed quality assurance/quality control (QA/QC) plan, describing specific requirements for QA and QC of both laboratory analysis and field sampling/analysis, should be part of the environmental characterization project workplan. The plan requirements will typically relate to, but not be limited to the following: the use of blanks, spikes, and duplicates; sample scheduling and sampling procedures; cleaning of sampling equipment; storage; transportation; DQOs; chain-of-custody; reporting and documentation; audits; and methods of analysis. The practices to be followed by the environmental investigation team and the oversight review – which will ensure that DQOs are met – must be clearly described in the QA/QC plan.

Several aspects of the environmental characterization program can, and should indeed be subjected to a quality assessment survey. In part, this is accomplished by submitting sample blanks (alongside the environmental samples) for analysis on a regular basis. The various blanks and checks that are recommended as part of the quality assurance plan include the following (CCME, 1994):

- *Trip blank*, required to identify contamination of bottles and samples during travel and storage. To prepare the trip blank, the laboratory fills containers

with contaminant-free water and delivers them to the sampling crew; the field sampling crew subsequently ship and store these containers with the actual samples obtained from the environmental investigation activities. It is recommended to include one trip blank per shipment, especially where volatile contaminants are involved.
- *Field blank*, required to identify contamination of samples during collection. This is prepared in the same manner as the trip blank (i.e., the laboratory fills containers with contaminant-free water and delivers to the sampling crew); subsequently, however, the field sampling crew expose this water to air in the locale (just like the actual samples obtained from the environmental investigation activities). It is recommended to include one field blank per locale or sampling event/day.
- *Equipment blanks*, required to identify contamination from sampling equipment. To obtain an equipment blank, sampling devices are flushed with contaminant-free water, which is then analysed. Typically, equipment blanks become important only if a problem is suspected (such as using a bailer to sample from multiple wells).
- *Blind replicates*, required to identify laboratory variability. To prepare the blind replicate, a field sample is split into three containers and labeled as different samples before shipment to the laboratory for analyses. It is recommended to include one blind replicate per day, or an average of one per 10 to 25 samples where large numbers of samples are involved.
- *Spiked samples*, required to identify errors due to sample storage and analysis. To obtain the spiked sample, known concentration(s) are added to the sample bottle and then analysed. It is recommended to include one spiked sample per locale, or an average of one per 25 samples where large numbers of samples are involved.

The development and implementation of a good QA/QC program during a sampling and analysis activity is indeed critical to obtaining reliable analytical results for the environmental characterization program. The soundness of the QA/QC program has a particularly direct bearing on the integrity of the environmental sampling and also the laboratory work. Thus, the general design process for an adequate QA/QC program, as discussed elsewhere in the literature (e.g., CCME, 1994; USEPA, 1987a, 1987b), should be followed religiously.

3.2 DESIGNING AN ENVIRONMENTAL CHARACTERIZATION PROGRAM

The scope and detail of an environmental characterization program should generally be adequate to determine the the following:

- primary and secondary sources of contamination, or hazards;
- amount and extent of contamination, or degree of hazard;
- fate and behavior characteristics of the contaminants, or hazards;

- pathways of contaminant migration, or hazard propagation;
- types of exposure scenarios associated with the hazard situation, or contamination problem;
- risks to human health and the environment; and
- feasible solutions to mitigate potential receptor exposures to the hazards or contaminants.

Ideally, a preliminary identification of the types of contaminants, the chemical release potentials, and also the potential exposure pathways should be made very early in the environmental characterization, because these are crucial to decisions on the number, type, and location of environmental samples to be collected.

In general, it should recognized that the investigation of huge complex problems may present increased logistical problems, making the design of environmental characterization programs for such situations more complicated. To allow for a manageable situation in the investigation of extensive or large areas, it usually is prudent to divide the case problem up into 'risk management units' (RMUs) or 'environmental management units' (EMUs). The RMUs (or EMUs) would define the areas of concern, which are used to guide the identification of the general sampling locations. The areas of concern, defined by the individual RMUs (or EMUs), will typically include sections or portions of a facility/property or locale that are believed to have different chemical constituents; have different anticipated concentrations or 'hot spots'; are a major contaminant release source; differ from each other in terms of the anticipated spatial or temporal variability of contamination; or must be sampled using different field procedures and/or equipment and tools (USEPA, 1989d). All of the areas of concern (designated as RMUs or EMUs) together should account for, or be representative of, the entire case facility/property or locale. In any case, sufficient information should be obtained that will reliably show the identity, areal and vertical extent, and the magnitude of contamination associated with the problem facility/property or locale.

3.3 IMPLEMENTATION OF AN ENVIRONMENTAL CHARACTERIZATION PROGRAM

The initial step in the implementation of environmental characterization programs usually involves a data collection activity to compile an accurate property description, history, and chronology of significant events. This may consist of such background information as the following: location of facility or property; property boundaries; regulatory situation and related issues; physical characteristics of facility; environmental setting; zoning profile; current and historical land uses; facility operations; hazardous substances used or generated – together with waste management practices – at facility; and land uses in vicinity and/or proximity of facility that might influence conditions at the locale. Subsequently, the most important functions are sampling and laboratory analyses. Field data are collected to help define the nature and extent of contamination associated with a locality.

Consequently, it is important to use proven methods of approach for the sampling and analysis programs designed for the locale.

The collection of representative samples generally involves different procedures for different situations. Several methods of choice that can be used to effectively complete environmental characterization programs in various types of situations are described elsewhere in the literature (e.g., ASTM, 1997a, 1997b; Boulding, 1994; BSI, 1988; Byrnes, 1994; CCME, 1993, 1994; CDHS, 1990; Csuros, 1994; DOE, 1994; Driscoll, 1986; Holmes et al., 1993; Jolley and Wang, 1993; Keith, 1988, 1991, 1992; Millette and Hays, 1994; Nielsen, 1991; OBG, 1988; O'Shay and Hoddinott, 1994; Perket, 1986; Petts et al., 1997; Singhroy et al., 1996; USEPA, 1985a, 1985b, 1987a, 1987b, 1988a, 1988b, 1989a, 1989b, 1989c, 1991a, 1991b, 1991c; USEPA–NWWA, 1989; WPCF, 1988). In every situation, all sampling equipment should be cleaned using a nonphosphate detergent, a tap water rinse, and a final rinse with distilled water prior to a sampling activity. Decontamination of equipment is necessary so that sample results do not show false positives. Decontamination water generated from the environmental investigation activities is transferred into containers and sampled for analysis.

In general, all sampling should be conducted in a manner that maintains sample integrity and encompasses adequate quality assurance and control. Specific sample locations should be chosen such that representative samples can be collected. Also, samples should be collected from locations with visual observations of contamination, so that possible worst-case conditions may be identified. The use of field blanks and standards, and also spiked samples can account for changes in samples which occur after sample collection. Sampling equipment should be constructed of inert materials; solid materials/samples collected are typically placed in resealable plastic bags, and fluid samples are placed in air-tight glass or plastic containers. When samples are to be analyzed for organic constituents, glass containers are required. The samples are then labeled with an indelible marker. Each sample bag or container is labeled with a sample identification number, sample depth (where applicable), sample location, date and time of sample collection, preservation, and possibly a project number and the sampler's initials. A chain-of-custody form listing the sample number, date and time of sample collection, analyses requested, a project number, and persons responsible for handling the samples is then completed. Samples are generally kept on ice prior to and during transport/shipment to a certified laboratory for analysis; completed chain-of-custody records should accompany the samples to the laboratory. Further details of the appropriate technical standards for sampling and sample handling procedures can be found in the literature elsewhere (e.g., CCME, 1994; CDHS, 1990; Holmes et al., 1993).

Ultimately, the characterization of contaminant or hazard sources, contaminant migration or hazard propagation pathways, and potential receptors (that are most likely to be at risk) probably form the most important basis for determining the nature of response or environmental management action required for any one problem situation. The completion of an adequate environmental characterization is therefore considered a very important component of any environmental management program that is designed to effectively remedy an environmental contamination problem.

SUGGESTED FURTHER READING

Acar, Y.B. and D.E. Daniel (eds). 1995. *Geoenvironment 2000: Characterization, Containment, Remediation, and Performance in Environmental Geotechnics*, Vols 1 and 2. ASCE Geotechnical Special Publication No. 46, American Society of Civil Engineers, New York.
Bartram, J. and R. Ballance (eds). 1996. *Water Quality Monitoring*. E. & F.N. Spon/ Chapman & Hall, London, UK.
Fetter, C.W. 1993. *Contaminant Hydrogeology*. Macmillan Publishing Co., New York.
Jain, R.K., L.V. Urban, G.S. Stacey, and H.E. Balbach. 1993. *Environmental Assessment*. McGraw-Hill, New York.
Kreith, F. (ed.). 1994. *Handbook of Solid Waste Management*. McGraw-Hill, New York.
NRC (National Research Council). 1993. *Ground Water Vulnerability Assessment: Predicting Relative Contamination Potential Under Conditions of Uncertainty*. National Academy Press, Washington.

REFERENCES

Asante-Duah, D.K. 1996. *Managing Contaminated Sites: Problem Diagnosis and Development of Site Restoration*. John Wiley, Chichester, UK.
ASTM. 1997a. *ASTM Standards on Environmental Sampling*, 2nd edition. American Society for Testing and Materials, Philadelphia, PA.
ASTM. 1997b. *ASTM Standards Related to Environmental Site Characterization*. American Society for Testing and Materials, Philadelphia, PA.
Boulding, J.R. 1994. *Description and Sampling of Contaminated Soils (A Field Manual)*, 2nd edition, Lewis Publishers/CRC Press, Boca Raton, FL.
BSI (British Standards Institution). 1981a. *Code of Practice for Site Investigations*. BS5930. BSI, London, UK.
BSI. 1981b. *Water Quality: Part 6, Sampling; Section 6.1, Guidance on the Design of Sampling Programmes*. BS6068, Part 6, Section 6.1:1981 [ISO 5667-1:1980]. BSI, London, UK.
BSI. 1987. *Water Quality: Part 6, Sampling; Section 6.4, Guidance on Sampling from Lakes, Natural and Man Made*. BS6068, Part 6, Section 6.4 1987 [ISO 5667-4:1987]. BSI, London, UK.
BSI. 1988. *Draft for Development, DD175: 1988 Code of Practice for the Identification of Potentially Contaminated Land and its Investigation*. BSI, London, UK.
BSI. 1991a. *Water Quality: Part 6, Sampling; Section 6.2, Guidance on Sampling Techniques*. BS6068, Part 6, Section 6.2:1991 [ISO 5667-2:1991]. BSI, London, UK.
BSI. 1991b. *Water Quality: Part 6, Sampling; Section 6.6, Guidance on Sampling of Rivers and Streams*. BS6068, Part 6, Section 6.6:1991 [ISO 5667-6:1991]. BSI, London, UK.
BSI. 1993. *Water Quality: Part 6, Sampling; Section 6. 11, Guidance on Sampling of Groundwaters*. BS6068, Part 6, Section 6. 11: 1993 [ISO 5667-11: 1993]. BSI, London, UK.
BSI. 1996a. *Water Quality: Part 6, Sampling; Section 6.3, Guidance on the Preservation and Handling of Samples*. BS6068, Part 6, Section 6.3:1996 [ISO 5667-3:1985]. BSI, London, UK.
BSI. 1996b. *Water Quality: Part 6, Sampling; Section 6. 12, Guidance on Sampling of Bottom Sediments*. BS6068, Part 6, Section 6.12:1996 [ISO 5667-12:1995]. BSI, London, UK.
BSI/ISO (British Standards Institution/International Organisation for Standardisation). 1995a. *Draft BS7755, Part 2, Section X: Sampling – Sampling Technique*. [ISO DIS 10381-2]. BSI, London, UK.
BSI/ISO. 1995b. *Draft BS7755, Part 2, Section X: Sampling – Guidance on Safety*. [ISO DIS 10381-3]. BSI, London, UK.

BSI/ISO. 1995c. *Draft BS7755. Part 2, Section X: Soil Quality – Sampling – Design of Sampling Programmes.* [ISO DIS 10381-1]. BSI, London, UK.

BSI/ISO. 1995d. *Draft BS7755, Part 2, Section X: Soil Quality – Sampling – Guidance on the Procedures for Investigation of Natural, Near Natural and Cultivated Sites.* [ISO DIS 10381-4]. BSI, London, UK.

Byrnes, M.E. 1994. *Field Sampling Methods for Remedial Investigations.* Lewis Publishers/ CRC Press, Boca Raton, FL.

CCME (Canadian Council of Ministers of the Environment). 1993. *Guidance Manual on Sampling, Analysis, and Data Management for Contaminated Sites.* Volume I: Main Report (Report CCME EPC-NCS62E), and Volume II: Analytical Method Summaries (Report CCME EPC-NCS66E). CCME, The National Contaminated Sites Remediation Program, Winnipeg, Manitoba, Canada.

CCME. 1994. *Subsurface Assessment Handbook for Contaminated Sites.* CCME, The National Contaminated Sites Remediation Program (NCSRP), Report No. CCME-EPC-NCSRP-48E (March, 1994), Ottawa, Ontario, Canada.

CDHS (California Department of Health Services). 1990. *Scientific and Technical Standards for Hazardous Waste Sites.* Prepared by CDHS, Toxic Substances Control Program, Technical Services Branch, Sacramento, CA.

Cheremisinoff, N.P. and M.L. Graffia. 1995. *Environmental and Health and Safety Management: A Guide to Compliance.* Noyes Publications, Park Ridge, NJ.

Csuros, M. 1994. *Environmental Sampling and Analysis for Technicians.* Lewis Publishers/ CRC Press, Boca Raton, FL.

DoE (Department of the Environment). 1994. *Sampling Strategies for Contaminated Land.* CLR Report No. 4, Department of the Environment, London, UK.

Driscoll, F.G. 1986. *Groundwater and Wells.* Johnson Division, St Paul, MN.

Holmes, G., B.R. Singh, and L. Theodore. 1993. *Handbook of Environmental Management and Technology.* J. Wiley, New York.

Jolley, R.L. and R.G.M. Wang (eds). 1993. *Effective and Safe Waste Management: Interfacing Sciences and Engineering with Monitoring and Risk Analysis.* Lewis Publishers, Boca Raton, FL.

Keith, L.H. (ed.). 1988. *Principles of Environmental Sampling.* American Chemical Society (ACS), Washington, DC.

Keith, L.H. 1991. *Environmental Sampling and Analysis – A Practical Guide.* Lewis Publishers, Boca Raton, FL.

Keith, L.H. (ed.). 1992. *Compilation of E.P.A.'s Sampling and Analysis Methods.* Lewis Publishers/CRC Press, Boca Raton, FL.

Lave, L.B. and A.C. Upton (eds). 1987. *Toxic Chemicals, Health, and the Environment.* Johns Hopkins University Press, Baltimore, MD.

Martin, W.F., J.M. Lippitt, and T.G. Prothero. 1992. *Hazardous Waste Handbook for Health and Safety*, 2nd edition. Butterworth–Heinemann, London, UK.

Millette, J.R. and S.M. Hays. 1994. *Settled Asbestos Dust Sampling and Analysis.* Lewis Publishers/CRC Press, Boca Raton, FL.

Nielsen, D.M. (ed.). 1991. *Practical Handbook of Groundwater Monitoring.* Lewis Publishers, Chelsea, MI.

OBG (O'Brien & Gere Engineers, Inc.). 1988. *Hazardous Waste Site Remediation: The Engineer's Perspective.* Van Nostrand Reinhold, New York.

O'Shay, T.A. and K.B. Hoddinott (eds). 1994. *Analysis of Soils Contaminated with Petroleum Constituents.* ASTM Publication, STP 1221, ASTM, Philadelphia, PA.

Perket, C.L. (ed.). 1986. *Quality Control in Remedial Site Investigation: Hazardous and Industrial Solid Waste Testing.* Fifth Volume, ASTM Publication, STP 925, ASTM, Philadelphia, PA.

Petts, J., T. Cairney, and M.R. Smith. 1997. *Risk-Based Contaminated Land Investigation and Assessment.* J. Wiley, Chichester, UK.

Singhroy, V.H., D.D. Nebert, and A.I. Johnson (eds). 1996. *Remote Sensing and GIS for Site*

Characterization: Applications and Standards. ASTM Publication STP 1279, ASTM, Philadelphia, PA.

USEPA (US Environmental Protection Agency). 1985a. *Characterization of Hazardous Waste Sites: A Methods Manual*, Volume 1, Site Investigations. USEPA, Environmental Monitoring Systems Laboratory, Las Vegas, EPA/600/4-84/075.

USEPA. 1985b. *Practical Guide to Ground-water Sampling.* Robert S. Kerr Environmental Research Lab., Office of Research and Development, USEPA, Ada, OK. EPA/600/2-85/104.

USEPA. 1987a. *RCRA Facility Investigation (RFI) Guidance.* Washington, DC. EPA/530/SW-87/001.

USEPA. 1987b. *Quality Assurance Program Plan.* Quality Assurance Management Staff, USEPA, Las Vegas, NV. EPA/600/X-87/241.

USEPA. 1988a. *Guidance for Conducting Remedial Investigations and Feasibility Studies Under CERCLA.* OSWER Directive 9355.3-01, Office of Emergency and Remedial Response, Washington, DC. EPA/540/G-89/004.

USEPA. 1988b. *Interim Report on Sampling Design Methodology.* Environmental Monitoring Support Laboratory, Las Vegas, NV. EPA/600/X-88/408.

USEPA. 1989a. *Ground-water Sampling for Metals Analyses.* Office of Solid Waste and Emergency Response, Washington, DC. EPA/540/4-89/001.

USEPA. 1989b. *Soil Sampling Quality Assurance User's Guide*, 2nd edition, Environmental Monitoring Systems Laboratory, Las Vegas, NV. EPA/600/8-89/046.

USEPA. 1989c. *User's Guide to the Contract Laboratory Program.* Office of Emergency and Remedial Response, Washington, DC. OSWER Directive 9240.0-1.

USEPA. 1989d. *Risk Assessment Guidance for Superfund. Volume I – Human Health Evaluation Manual (Part A).* Office of Emergency and Remedial Response, Washington, DC. EPA/540/1-89/002.

USEPA. 1991a. *Conducting Remedial Investigations/Feasibility Studies for CERCLA Municipal Landfill Sites.* Office of Emergency and Remedial Response, Washington, DC. EPA/540/P-91/001 (OSWER Directive 9355.3-11).

USEPA. 1991b. *Guidance for Performing Site Inspections Under CERCLA – Interim Version.* Draft Publication, Office of Emergency and Remedial Response, Washington, DC. OSWER Directive 9345.1-06.

USEPA. 1991c. *Management of Investigation-Derived Wastes During Site Inspections.* Office of Emergency and Remedial Response, Washington DC. EPA/540/G-91/009.

USEPA–NWWA. 1989. *Handbook of Suggested Practices for the Design and Installation of Ground Water Monitoring Wells.* National Water Well Association, Dublin, OH.

WPCF (Water Pollution Control Federation). 1988. *Hazardous Waste Site Remediation: Assessment and Characterization.* A Special Publication of the WPCF, Technical Practice Committee, Alexandria, VA.

Chapter Four

Legislative–Regulatory Control Needs for the Management of Environmental Contamination Problems

A disproportionately large part of environmental contamination problems may be attributed to waste management – or rather, waste mismanagement – cases. In one classic example of a waste mismanagement situation, when a committee of the British Parliament reported on waste disposal in 1981, reference was made to the fact that, seven years earlier, an operator at a landfill site near London had died as a result of breathing toxic fumes resulting from the accidental disposal of two incompatible wastes at the same location. Another classic example a world away involved waste disposal at Love Canal in New York, USA; Love Canal was a disposal site for chemical wastes for about three decades. Subsequent use of the site apparently resulted in the residents in a township in the area suffering severe health impairments; it is believed that several children in the neighborhood apparently were born with serious birth defects. As alluded to earlier on in Chapter 2, similar and/or comparable problems have occurred in several other locations globally.

To help abate potential problems to public health and the environment, several items of legislation have been formulated and implemented in most industrialized countries to deal with the regulation of toxic substances present in our modern societies (Forester and Skinner, 1987). Some of the well-established environmental laws found in the different regions of the world are introduced in this chapter, with further examination of related legislations and regulations alluded to in the appropriate sections of subsequent chapters. Depending on the type of program under evaluation, one or more of the existing regulations may dominate the decision-making process in an environmental management program. It is noteworthy that several of the regulations are subject to changes/amendments. Also, depending on the specific situation or application, a combination of several regulatory specifications may be employed. For further and up-to-date discussions on region-specific environmental regulations, the reader is encouraged to consult with their local regulatory agencies, or applicable literature pertaining to the case region(s).

4.1 THE NATURE OF ENVIRONMENTAL REGULATIONS

Generally, most industrialized countries incorporate a system for 'cradle-to-grave' control in their hazardous materials management programs. The 'cradle-to-grave' type of system monitors and regulates the movement of hazardous materials from manufacture through usage to the ultimate disposal of any associated hazardous wastes. The use of some kind of manifest system helps minimize abuses and violations of established national or regional control systems associated with hazardous materials movements and management. The manifest system serves as an identification form that accompanies each shipment of hazardous materials; the manifest is signed at each stage of transfer of responsibility, with each responsible person in the chain of custody keeping a record that is open to scrutiny and inspection by regulatory officials or other appropriate authorities, and also any interested parties. It is expected that the use of such effective control systems will help minimize the creation of extensive environmental contamination problems.

Several of the newly industrializing and other developing economies are also moving towards the establishment of workable environmental regulations, to help protect their often fragile environments. Invariably, there tend to be improved environmental management practices in countries or regions where regulatory programs are well established – in comparison to those without appropriate regulatory and enforcement programs. The process used in the design of effectual environmental management programs must therefore incorporate several elements of all relevant environmental regulations. In fact, the establishment of national and/or regional regulatory control programs within the appropriate legislative–regulatory and enforcement frameworks is a major step in developing effective environmental management programs. In any case, it is noteworthy that variations in national legislations and controls do affect options available for the management of environmental contamination problems in different regions of the world. Such variances also affect the restoration goals and costs necessary to achieve the appropriate environmental management program objectives. The need to consult with local environmental regulations when faced with an environmental contamination problem cannot, therefore, be over-emphasized.

4.1.1 Environmental Laws in North America

North America, and the United States (US) in particular, has over the past few years come up with mountains of environmental legislation. Starting with a National Environmental Protection Act (NEPA) in 1969 that empowered the US Environmental Protection Agency (EPA) and a National Environmental Quality Control Council (NEQCC) to lay the basis for regulations, a stream of environmental legislation followed in the next two decades and into the present time. Some of the relevant fundamental statutes and regulations which, directly or indirectly, may affect environmental management decisions, policies, and programs in the US are briefly enumerated below, with further discussion offered elsewhere in the literature (e.g., Forester and Skinner, 1987; Holmes et al., 1993; USEPA, 1974, 1985, 1987, 1988, 1989).

- *Clean Air Act.* The Clean Air Act (CAA) was originally conceived and adopted in 1963; however, a sweeping amendment that was effected in 1970 is widely recognized as the more powerful and important piece of this environmental legislation. The objective of the CAA of 1970 is to protect and enhance air quality resources, in order to promote and maintain public health and welfare and the productive capacity of the population. This covers all pollutants that may cause significant risks. The CAA of 1970 has indeed undergone several amendments since its inception. In particular, the CAA Amendments of 1990 introduced further sweeping changes – including, for the first time, specific provisions addressing global air pollution problems.

- *Safe Drinking Water Act.* The Safe Drinking Water Act (SDWA) was enacted in 1974 in order to assure that all people served by public water systems would be provided with a supply of high quality water. The SDWA Amendments of 1986 established new procedures and deadlines for setting national primary drinking water standards, and established a national monitoring program for unregulated contaminants, among others. The statute covers public water systems, drinking water regulations, and the protection of underground sources of drinking water.
- *Clean Water Act.* The Clean Water Act (CWA) was enacted in 1977 and an amendment to this introduced the Water Quality Act (WQA) of 1987. The objective of the CWA is to restore and maintain the chemical, physical, and biological integrity of the nation's waters. This objective is achieved through the control of discharges of pollutants into navigable waters – implemented through the application of federal, state and local discharge standards. The statute covers the limits on waste discharge to navigable waters; standards for discharge of toxic pollutants; and the prohibition on discharge of oil or hazardous substances into navigable waters. Water quality criteria developed by the EPA under CWA authority are not by themselves enforceable standards, but can be used in the development of enforceable standards. Another closely associated legislation, the Federal Water Pollution Control Act (FWPCA), deals solely with the regulation of effluent and water quality standards.
- *Resource Conservation and Recovery Act.* The Resource Conservation and Recovery Act (RCRA) was enacted in 1976 (as an amendment to the Solid Waste Disposal Act of 1965, later amended in 1970 by the Resource Recovery Act) to regulate the management of hazardous waste, to ensure the safe disposal of wastes, and to provide for resource recovery from the environment by controlling hazardous wastes 'from cradle to grave.' This is a federal law that establishes a regulatory system to track hazardous substances from the time of generation to disposal. The law requires safe and secure procedures to be used in treating, transporting, sorting, and disposing of hazardous substances. Basically, RCRA regulates hazardous waste generation, storage, transportation, treatment, and disposal. Of special interest, the 1984 Hazardous and Solid Waste Amendments (HSWA) to RCRA, Subtitle C, covers a management system that regulates hazardous wastes from the time they are generated until the ultimate disposal – the so-called 'cradle-to-grave' system. Thus, under RCRA, a hazardous waste management program is based on a 'cradle-to-grave' concept, that allows all hazardous wastes to be traced and equitably accounted for.
- *Comprehensive Environmental Response, Compensation, and Liability Act.* The Comprehensive Environmental Response, Compensation, and Liability Act of

1980 (CERCLA or 'Superfund') establishes a broad authority to deal with releases or threats of releases of hazardous substances, pollutants, or contaminants from vessels, containments, or facilities. This legislation deals with the remediation of hazardous waste sites, by providing for the cleanup of inactive and abandoned hazardous waste sites. The objective is to provide a mechanism for the federal government to respond to uncontrolled releases of hazardous substances to the environment. The statute covers reporting requirements for past and present owners/operators of hazardous waste facilities; and the liability issues for owners/operators for cost of removal or remedial action and damages, in case of release or threat of release of hazardous wastes. The Superfund Amendments and Reauthorization Act of 1986 (SARA) strengthens and expands the cleanup program under CERCLA; focuses on the need for emergency preparedness and community right-to-know; and changes the tax structure for financing the Hazardous Substance Response Trust Fund established under CERCLA to pay for the cleanup of abandoned and uncontrolled hazardous waste sites.

- *Toxic Substances Control Act.* The Toxic Substances Control Act (TSCA) of 1976 provides for a wide range of risk management actions to accommodate the variety of risk–benefit situations confronting the USEPA. The risk management decisions under TSCA would consider not only the risk factors (such as probability and severity of effects), but also nonrisk factors (such as potential and actual benefits derived from use of the material and availability of alternative substances). Broadly, TSCA regulates the manufacture, use and disposal of chemical substances. It authorizes the USEPA to establish regulations pertaining to the testing of chemical substances and mixtures; pre-manufacture notification for new chemicals or significant new uses of existing substances; control of chemical substances or mixtures that pose an imminent hazard; and record-keeping and reporting requirements.
- *Federal Insecticide, Fungicide, and Rodenticide Act.* The Federal Insecticide, Fungicide, and Rodenticide Act (FIFRA) of 1978 deals with published procedures for the disposal and storage of excess pesticides and pesticide containers. The USEPA has also promulgated tolerance levels for pesticides and pesticide residues in or on raw agricultural commodities under authority of the Federal Food, Drug, and Cosmetic Act (40 CFR Part 180). FIFRA does indeed provide the USEPA with broad authorities to regulate all pesticides.
- *Endangered Species Act.* The Endangered Species Act (ESA) of 1973 (re-authorized in 1988) provides a means for conserving various species of fish, wildlife, and plants that are threatened with extinction. The ESA considers an endangered species as that which is in danger of extinction in all or significant portions of its range; a threatened species is that which is likely to become an endangered species in the near future. Also, the ESA provides for the designation of critical habitats (i.e., specific areas within the geographical area occupied by the endangered or threatened species) on which are found those physical or biological features essential to the conservation of the species in question.
- *Pollution Prevention Act.* The Pollution Prevention Act of 1990 calls pollution prevention a 'national objective' and establishes a hierarchy of environmental

protection priorities as national policy. Under this Act, it is the US national policy that pollution should be prevented or reduced at the source whenever feasible; where pollution cannot be prevented, it should be recycled in an environmentally safe manner. In the absence of feasible prevention and recycling options or opportunities, pollution should be treated, and disposal should be used only as a last resort.

Comparable laws also exist in Canada. However, in Canada, the implementation and enforcement of environmental regulations – with a few exceptions – have been principally the responsibilities of the provinces. In many ways, the Canadian system is less formal than in the US, but not any less stringent.

Other laws and regulations exist in the US and Canada that also work towards preventing or limiting the potential impacts of environmental contamination problems. For example, in the US, the Hazardous Materials Transport Act (HMTA) of 1975 provides a legislation that deals with the regulation of transport of hazardous materials; the Fish and Wildlife Conservation Act of 1980 requires States to identify significant habitats and develop conservation plans for these areas; the Marine Mammal Protection Act of 1972 protects all marine mammals – some of which are endangered species; the Migratory Bird Treaty Act of 1972 implements many treaties involving migratory birds – in order to protect most species of native birds in the US; the Wild and Scenic Rivers Act of 1972 preserves select rivers declared as possessing outstanding remarkable scenic, recreational, geologic, fish and wildlife, historic, cultural, or other similar values; etc. Also, laws similar or comparable to those annotated above, together with similar legal provisions and regulations can be found in the legislative requirements for several other advanced economies, as reflected in the brief discussions below.

4.1.2 Environmental Laws within the European Community Bloc

The European Community (EC) was created in 1957 by the Treaty of Rome. The European Union (EU) was more recently created by the Treaty of Maastricht, and came into existence as an umbrella organization in November 1993. The EU presently consists of the following 15 member nations: Austria, Belgium, Denmark, Finland, France, Germany, Greece, Italy, Luxembourg, the Netherlands, Portugal, Republic of Ireland, Spain, Sweden, and the United Kingdom. Fundamentally, the EC is a structural member of the EU.

The two most common forms of EU legislation are regulations and directives; regulations become law throughout the EU as of their effective date, generally enforceable in each Member State, whereas, in general, directives are not necessarily enforceable in Member States but are meant to set out goals for the Member States to achieve through national legislation. In fact, of the five/major legislative acts enumerated in Article 189 of the EC Economic Treaty (i.e., Treaty Establishing the European Economic Community, Rome, 25 March, 1957) – viz.: regulations, directives, decisions, recommendations, and opinions – the directive seems to play a unique role that affords the European Parliament the ability to respect indigenous national legislative traditions and philosophies while guiding the EU as a whole toward a single common goal. The directive sets forth an objective and provides that each national government achieve this objective by the national means considered best able to accomplish the goal embodied in the directive.

The EU directive was designed to respect indigenous legislative traditions while providing a means for unification of goals. Specifically, the EU directives play an important role in environmental regulations within the member nations. For instance, the 1975 framework (Council Directive 75/442/EEC on Waste) established the general obligations of Member States for waste management, to ensure that wastes are disposed of without endangering human health or harming the environment. Toxic and dangerous wastes were covered under a 1978 directive (Council Directive 78/319/EEC on Toxic and Dangerous Waste) that required these materials to be properly stored, treated, and disposed of in authorized facilities governed by proper regulatory authorities; this directive has more recently been replaced with the Directive on Hazardous Wastes (Council Directive 91/689/EEC), adopted under the provisions of the framework Directive on Waste, and this new directive took effect at the start of 1994. Also, enacted in 1985, the environmental assessment directive (Council Directive 85/337/EEC of 27 June, 1985, on the Assessment of the Effects of Certain Public and Private Projects on the Environment) provides the basis for environmental impact statements to be produced for those development projects that are likely to have significant effects on the environment by virtue of their nature, size, or location. The environmental directives are designed to establish uniform results with respect to environmental integrity throughout the Member States. While striving toward a common goal of environmental consistency, the environmental directive respects national policy for the management, control, and regulation of environmental pollutants.

A key element of the relevant EU environmental directives is the recent concept of using the 'best available techniques not entailing excessive cost' (BATNEEC) to prevent or reduce pollution and its effects. BATNEEC is interpreted as the technology for which operating experience had adequately demonstrated it to be the best commercially available to minimize releases, providing it had been proven to be economically viable.

It is also interesting to note the EU issuance of its Council Regulation No.1836/93 on the voluntary participation of commercial enterprises in a common environmental management and audit scheme (EMAS). The regulation entered into force on 13 July, 1993, with its ultimate application having been dependent on the issuing of national regulations in Member States. In these regulations, the prevention, reduction, and elimination of environmental burdens – possibly at their source – results in a sound management of resources on the basis of the 'polluter pays principle', and the promotion of market-oriented environmental management is required. All of the aforementioned are meant to stimulate and foster responsibility among commercial and industrial enterprises. The core elements of the regulation are the introduction of an environmental management scheme at specific industrial sites and the execution of an environmental audit by independent auditors. In contrast to the environmental impact assessment which is carried out before a facility is installed, environmental audits take place during plant operations. After a successful participation in EMAS, the company obtains a participation certificate.

It is noteworthy that the EMAS requires a public statement – i.e., transparency in environmental terms.

For the most part, the individual European countries have consistently approached the issues of environmental management legislation from very different angles. Notwithstanding the inter-state differences in environmental management legislation in Europe, however, there is a general tendency for the development of hazardous waste minimization policies in most of the countries in the region. In fact, several European countries have already adopted, or are developing regulatory instruments that will directly affect the amount of hazardous waste produced – and therefore the concomitant environmental contamination problems.

Overall, the countries within the European Union have aggressively worked on producing consistent environmental regulatory programs that also recognize the sovereign rights of the individual countries. As part of this process, the European Environment Agency was created in 1990 by a Community Regulation (Council Regulation 1210/90/EEC on the Establishment of the European Environment Agency and the European Environment Information and Observation Network), with the main purposes of collecting and disseminating information – rather than to have any enforcement authority. In fact, it is apparent that a vast and comprehensive legislative arsenal has been created within the EU and therefore in each of the Member States, albeit the implementation of these legislations is yet to achieve the complete anticipated results.

Some of the apparently unique legislative measures and relevant statutes and regulations which, directly or indirectly, may affect environmental management decisions, policies, and programs within specific/selected EU nations are briefly annotated below – recognizing that similar or comparable frameworks exist in the other member nations. Further discussion of several aspects of environmental regulatory programs in the region appear elsewhere in the legislative control literature (e.g., Garbutt, 1995a, 1995b; Lister, 1996).

- *United Kingdom (UK)*. Waste management has always been an issue visited by governments and policy-makers, past and present alike. Before the 19th century, however, pollution problems arising from the disposal of waste or refuse was frequently resolved through private litigation rather than by public prosecution under the Sanitary Acts. During the current era, the Public Health Act of 1936 enacted that any accumulation or deposition of refuse prejudicial to health or a nuisance is a Statutory Nuisance. The regulatory authorities have the power to serve Abatement Notices and to seek public prosecution; this power will remain extant, even after the full implementation of the Deposit of Poisonous Waste Act (DPWA) of 1972 and the Control of Pollution Act (CoPA) in 1974.

    Early in 1972, the Deposit of Poisonous Waste Act (DPWA) was issued, penalizing the deposit on land of poisonous, noxious, or polluting wastes that could give rise to an environmental hazard, and making offenders criminally and civilly liable for any resultant damage. The UK Control of Pollution Act of 1974 allows the government to restrict the production, importation, sale, or use of chemical substances in the UK; the Act controls the disposal of wastes on land by means of site licensing. Subsequently, the 1980 Control of Pollution (Special Waste) Regulations, which also repealed the DPWA, were established to ensure pre-notification for a limited range of the most hazardous types of wastes; to keep a 'cradle-to-grave' record of each disposal of special wastes; and to keep long-term records of locations of special waste disposal landfill sites.

The 1990 Environmental Protection Act (EPA 90) marked an important milestone in the development of the legislative philosophy and framework in the UK. The Act is seen as being of major importance, since it largely established Britain's strategy for pollution control and waste management for the foreseeable future (D. Slater, in Hester and Harrison, 1995).

Most recently, the Environment Act of 1995 sought to develop a new breed of environmental regulations in the UK. The significance of this Act includes – among other issues/things – a re-organization and concentration/integration of environmental regulatory functions, whose adoption is complemented by pivotal reliance on cost–benefit and risk assessment techniques as mechanisms for its implementation (Jewell and Steele, 1996). The Act brings together in a single body, the existing environmental protection and pollution control functions of a number of regulators.

- *The Netherlands*. The Netherlands has adopted a general ban on land disposal of hazardous waste. When exemptions from this ban are granted, the government may require the recipient of the exemption to conduct research on alternative technologies or management methods aimed at preventing potential needs for future land disposal. The ban itself acts, to some extent, as an incentive for waste minimization, although the relatively easy access to waste export to other countries for treatment or disposal has so far limited the impact of such a ban.

- *France*. In France, the main goals of environmental policy in relation to the management of special industrial wastes are to develop an effective national treatment and disposal infrastructure, reduce the amount of wastes generated at source, and clean up sites contaminated by hazardous wastes. The French Ministry of the Environment's plan to attain these goals has been via providing clear definitions of the different classes of wastes, and imposing regulatory and financial constraints on waste generators.

Measures provided for in recent legislation include restrictions on the types of waste that can be landfilled, and a tax on the off-site treatment and disposal of special industrial waste – that will be used to fund the cleanup of abandoned contaminated sites. The special industrial wastes will include medical wastes and hazardous domestic wastes. The relevant Decree lists 51 groups of compounds, as well as 14 chemical and biological properties that are likely to render a waste hazardous.

The French government encourages waste avoidance and reduction by including waste production and management practices among the factors that are evaluated when the government has to make a decision on whether or not to authorize the operation of new or modified industrial facilities. An impact study is required prior to any authorization. However, the potential impact of these requirements on the reduction of hazardous waste is probably not being realized for the time being, because of an insufficient number of enforcement personnel and inadequate technical expertise among regulators and industry concerning alternative technologies and management methods.

From the year 2002, disposal by landfill in France is only to be permitted for wastes or treatment residues for which there is no commercially available recycling or treatment technology, and that require disposal by landfill. The present legislation requires that, by 1998, all special industrial wastes must be stabilized prior to disposal.

- *Federal Republic of Germany.* In Germany, industrial waste is managed under the Federal Waste Disposal Act, which defines a category of hazardous waste as 'special' – requiring particular/specific handling procedures. If a facility does not generate or handle 'special' wastes by definition, the agency may authorize the facility to handle the waste by itself, with only minimal regulatory requirements.

A regulatory instrument adopted by the German Parliament in 1986 provides an accelerated waste reduction policy, and authorizes the government to regulate the use, collection, and disposal of products likely to cause environmental problems when discarded as waste. The regulation is coordinated with other efforts such as public financial support for developing and demonstrating new waste avoidance technologies, increasing the technical competence of regulators, and discouraging reliance on treatment and disposal as methods of choice for managing hazardous waste.

Overall, the federal system for the management of industrial waste consists of a basic legislative framework that delegates significant authority to the state for the management and implementation of industrial waste regulations.

In general, the EU directly affects Member States through various legal instruments and several guiding principles. The key principles that shape environmental policies of the EU's Member States include the 'polluter pays' principle; the need to avoid environmental damage at the source; and the need to take the environmental effects of energy and other EU policies into consideration.

4.1.3 Environmental Policies and Regulations in Different Parts of the World

The legislative measures and relevant statutes and regulations which, directly or indirectly, affect environmental management decisions, policies, and programs generally tend to be similar in their degree of sophistication according to the level of technological advancement. This means that developed economies the world over usually will have regulatory systems comparable to those found in Western Europe or North America (as exemplified by the brief reference that follows for Japan) – although this may not always be true, as for instance is the case with the former Soviet bloc countries of Central and Eastern Europe.

To illustrate the case for Japan, the Waste Management Law of 1970 requires anyone undertaking the collection, transport, or disposal of industrial wastes to have obtained a permit; waste disposal sites are also required to be designed and operated in a manner that prevents trespassing, and the disposal sites must be isolated from both surface and groundwater resources, and measures must be taken to prevent leakage. In addition, the 1973 Chemical Substances Control Act requires a manufacturer or importer of a new chemical (i.e., one that is not enumerated in a 1973 pre-listed compilation) to provide the government with all available information concerning such a chemical; the substance can then be evaluated and classified as being dangerous or safe. As part of the nation's overall waste management program, Japan has put in place laws and regulations, as well as enforcement mechanisms, that are comparable to those found in most other industrialized countries.

On the other hand, many developing economies tend to lack laws and/or enforcement tools that regulate environmental problems in an effective manner – albeit several nations in the developing regions of Latin America, the Caribbean, and South Africa often have sophisticated environmental legislations in place, even if they are with inadequate regulatory systems. In fact, information and environmental management regulations have been almost nonexistent and/or nonenforceable in most developing economy or newly industrializing countries. This is because most of these countries with economies in transition lack the laws and governmental institutions to deal with environmental regulatory issues. Many governments, however, recognize the need to protect their respective nation's health and environment, and are therefore seriously developing legislation that will deal with the unsurmountable environmental management problems. The result is the development of systematic environmental control programs, most of which follow the styles employed by the industrialized countries in Europe, North America, and elsewhere in the world.

4.2 A PARADIGM OF CONTEMPORARY INTERNATIONAL ENVIRONMENTAL LAW

In the past, some polluting behaviors were seen as normal and inevitable consequences of industrialization and national development. As a result, environmental protection in many countries was based on reliance on a nonpunitive, conciliatory style of securing compliance with environmental laws and regulations (UNICRI, 1993). Increasing damage to the environment has, however, changed perceptions of acts of pollution, especially in the industrialized countries. There is now a growing tendency that the developed economies are steadily moving toward criminal sanctions as a means of punishment and deterrence in the arena of environmental offenses (Asante-Duah and Nagy, 1998).

In fact, increasing public concern about the several problems and potentially dangerous situations associated with environmental contamination issues, together with the legal provisions of various legislative instruments and regulatory programs, have all compelled both industry and governmental authorities to carefully formulate responsible environmental management programs. These programs include techniques and strategies needed to provide good waste management methods and technologies, and the development of cost-effective corrective action and risk management programs that will ensure public safety, as well as protect human health, the environment, and public and private properties.

In the formulation of environmental legislative instruments and the design of regulatory systems, it is important to recognize and appreciate the familiar saying that 'pollution knows and respects no geographical boundaries'. In fact, cross-boundary pollution problems have been around for a long time and this scenario is not about to change. Hence, to be truly effective, environmental protection usually must have an international dimension to it. Solutions to cross-boundary environmental issues will therefore tend to require international cooperation. This means that, despite the fact that individual countries will be exercising their sovereign rights in the formulation of environmental legislations for their nations, this should not necessarily happen in isolation; insofar as possible, this should be tied into a global environmental agenda for the international community.

In fact, since solutions to global environmental problems generally require increased international cooperation – both regionally and globally – it should be expected that international environmental law will subjugate national sovereignty to this end. In this vein, it is apparent that the legislation and implementation of most international environmental laws have the tendency to neglect the social factors and the differing economic development status of states – whereas the objective reality is that any global issue is an inseparable and organic whole (L.P. Cheng, in Weiss, 1992). On this basis, it becomes necessary for the international community, in dealing with global environmental issues, to consider the objective conditions of those states that are less developed economically and technologically as an almost separate 'entity' from the industrialized nations. This strategy will, hopefully, help alleviate part of the grim situation associated with the legislation and implementation of international environmental protection plans or laws by several nations/states. In fact, the differing economic and political situations of states may have a dramatic impact on the implementation of international environmental laws in different countries; this fact should be recognized, if successful stories are to be told about existing and new international environmental laws.

In general, all waste management operations and hazardous materials handling or processing activities do constitute a potential source of environmental contamination and therefore may be viewed as a possible source of present or future risks. Effective regulatory and control systems should therefore become an integral part of all waste management programs in particular, and environmental management programs in general; the adoption of such a strategy will help minimize potential effects from the likely consequences of environmental contamination problems.

Ultimately, an established environmental policy that is publicly available and understood at all appropriate levels is fundamental to attaining the requisite environmental management standard. The policy has to indicate commitments to: reduce waste, pollution, and resource consumption; minimize environmental risks and hazard effects; design products with regard to their environmental effects throughout the life cycle; control the effects of raw material processing; and minimize the effects of new developments. Furthermore, a system of environmental management records should be established in order to demonstrate compliance with the requirements of the environmental management system, and to record the extent to which the planned environmental objectives and targets are being met.

SUGGESTED FURTHER READING

Cranor, C.F. 1993. *Regulating Toxic Substances: A Philosophy of Science and the Law.* Oxford University Press, New York.
Dwyer, J.P. 1990. The pathology of symbolic legislation. *Ecology Law Quarterly*, 17: 233–316.
Erickson, R.L. and R.D. Morrison. 1995. *Environmental Reports and Remediation Plans: Forensic and Legal Review.* J. Wiley, New York.
Farmer, A. 1997. *Managing Environmental Pollution.* Routledge, London, UK.
King, J.J. 1995. *The Environmental Dictionary, and Regulatory Cross-Reference*, 3rd edition. J. Wiley, New York.
NAE (National Academy of Engineering). 1993. *Keeping Pace with Science and Engineering: Case Studies in Environmental Regulation.* National Academy Press, Washington, DC.
Rebovich, D.J. 1992. *Dangerous Ground: The World of Hazardous Waste Crime.* Transaction Publishers, London, UK.

REFERENCES

Asante-Duah, D.K. and I.V. Nagy. 1998. *International Trade in Hazardous Waste.* E. & F.N. Spon/Chapman & Hall, London, UK.
Forester, W.S. and J.H. Skinner (eds). 1987. *International Perspectives on Hazardous Waste Management – A Report from the International Solid Wastes and Public Cleansing Association (ISWA) Working Group on Hazardous Wastes.* Academic Press, London, UK.
Garbutt, J. 1995a. *Environmental Law*, 2nd edition. J. Wiley, Chichester, UK.
Garbutt, J. 1995b. *Waste Management Law*, 2nd edition. J. Wiley, Chichester, UK.
Hester, R.E. and R.M. Harrison (eds). 1995. *Waste Treatment and Disposal.* The Royal Society of Chemistry, Herts, UK.
Holmes, G., B.R. Singh, and L. Theodore. 1993. *Handbook of Environmental Management and Technology.* J. Wiley, New York.
Jewell, T. and J. Steele. 1996. UK regulatory reform and the pursuit of 'Sustainable Development': The Environment Act 1995. *Journal of Environmental Law*, 8(2): 283–300.
Lister, C. 1996. *European Union Environmental Law: A Guide for Industry.* J. Wiley, Chichester, UK.
Paustenbach, D.J. (ed.). 1988. *The Risk Assessment of Environmental Hazards: A Textbook of Case Studies.* John Wiley, New York.
UNICRI. 1993. *Environmental Crime, Sanctioning Strategies and Sustainable Development.* United Nations Interregional Crime and Justice Research Institute/Australian Institute of Criminology, UNICRI Publication No. 50, Rome/Canberra.
USEPA (US Environmental Protection Agency). 1974. *Safe Drinking Water Act.* Public Law 93-523.
USEPA. 1985. National Primary Drinking Water Regulations; Volatile Synthetic Organic Chemicals; Final Rule and Proposed Rule, Federal Register, 50: 46830–46901. National Primary Drinking Water Regulations, Synthetic Organic Chemicals, Inorganic Chemicals and Microorganisms; Proposed Rule, *Federal Register*, 50: 56936–47025.
USEPA. 1987. *The New Superfund: What It Is, How It Works.* USEPA, Washington, DC.

USEPA. 1988. *CERCLA Compliance with other Laws Manual (Interim Final)*. Office of Solid Waste and Emergency Response, Washington, DC. EPA/540/G-89/006.

USEPA. 1989. *CERCLA Compliance with Other Laws Manual: Part II – Clean Air Act and Other Environmental Statutes and State Requirements*. EPA/540/G-89/009. OSWER Directive 9234.1-02.

Weiss, E.B. (ed.). 1992. *Environmental Change and International Law: New Challenges and Dimensions*. United Nations University Press, Tokyo, Japan.

PART II
PRINCIPLES OF RISK ASSESSMENT

This part of the book comprises general principles pertaining to risk assessment studies – including a review of the state-of-the-art, the risk assessment process and mechanics, a number of basic concepts used in risk assessment applications, and risk assessment classification systems. It consists of the following specific chapters:

- Chapter 5, *Fundamentals of Hazard, Exposure, and Risk Assessment*, is devoted to some important definitions that are basic to understanding the nature of hazards and risks associated with environmental contamination problems – including a discussion of their applicability to environmental management programs.
- Chapter 6, *The Risk Assessment Process*, describes the generic protocol typically used to complete risk assessments. It also includes a discussion of some contemporary issues pertaining to risk assessment.
- Chapter 7, *Basic Concepts in Risk Assessment Practice*, provides a discussion of various risk assessment concepts finding widespread application in environmental management programs. These include the various measures of risks, risk perception and risk acceptability criteria, conservatism notions, uncertainty issues, and some concepts of probability theory relevant to some aspects of risk assessment.

Chapter Five

Fundamentals of Hazard, Exposure, and Risk Assessment

Hazard is that object with the potential for creating undesirable adverse consequences; exposure is the situation of vulnerability to hazards; and risk is considered to be the probability or likelihood of an adverse effect due to some hazardous situation. In fact, it is the likelihood to harm as a result of exposure to a hazard which distinguishes risk from hazard. For example, a toxic chemical that is hazardous to human health does not constitute a risk unless human receptors/populations are exposed to such a substance – as conceptually illustrated by the Venn diagram representation shown in Figure 5.1. Thus, a complete assessment of potential hazards posed by a substance or an object involves, among several other things, a critical evaluation of available scientific and technical information on the substance or object of concern, as well as the possible modes of exposure. In addition, potential receptors will have to be exposed to the hazards of concern before any risk could be said to exist. In fact, the availability of an adequate and complete information base is a prerequisite to sound hazard, exposure, and risk assessments.

The integrated assessments of hazards, exposures, and risks are indeed a very important contributor to any decision that is aimed at managing hazardous situations. Potential risks are estimated by considering the probability or likelihood of occurrence of harm; the intrinsic harmful features or properties of specified hazards; the population-at-risk (PAR); the exposure scenarios; and the extent of expected harm and potential effects.

5.1 THE NATURE OF HAZARD AND RISK

Hazard is defined as the potential for a substance or situation to cause harm, or to create adverse impacts on populations and/or property. It represents the unassessed loss potential, and may comprise a condition, a situation, or a scenario with the potential for creating undesirable consequences. The degree of hazard will normally be determined by the exposure scenario and the potential effects or responses resulting from any exposures.

There is no universally single accepted definition of risk. Risk may, however, be considered as the probability or likelihood of an adverse effect, or an assessed threat to persons, the environment, and/or property, due to some hazardous situation. It is a measure of the probability and severity of adverse consequences

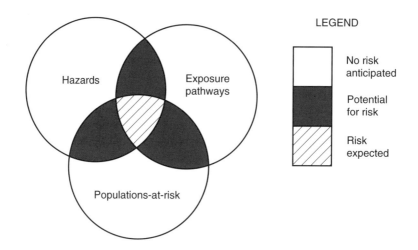

Figure 5.1 When do hazards represent risks?

from an exposure of potential receptors to hazards due to a system failure, and may simply be represented by the measure of the frequency of an event.

Procedures for analyzing hazards and risks may typically comprise several steps (Figure 5.2), consisting of the following general elements:

- Hazard identification and accounting
  - identify hazards (including nature/identity of hazard, location, etc.)
  - identify initiating events (i.e., causes)
  - identify resolutions for hazard
  - define exposure setting

- Vulnerability analysis
  - identify vulnerable zones
  - identify concentration/impact profiles for affected zones
  - determine populations potentially at risk (such as human and ecological populations, and critical facilities)
  - define exposure scenarios

- Consequences/impacts assessment
  - determine risk categories for all identifiable hazards
  - determine probability of adverse outcome (from exposures to hazards)
  - estimate consequences (including severity, uncertainties, etc.).

Some or all of these elements may have to be analysed in a comprehensive manner, depending on the variation and level of detail of the hazard and/or risk analysis that is being performed. The variations typically fall into two broad categories – endangerment assessment (which may be considered as contaminant-based, such as human health and environmental risk assessment associated with

Fundamentals of hazard, exposure, and risk assessment

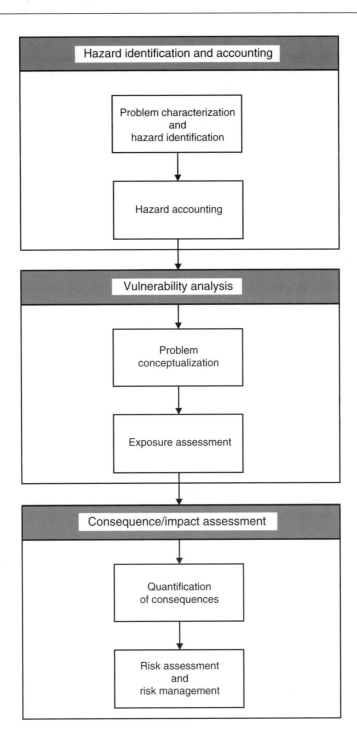

Figure 5.2 Basic steps in the analysis of hazards and risks

chemical contamination), and safety assessment (which is system-failure-based, such as probabilistic risk assessment of hazardous facilities or installations). Ultimately, the final step will consist of developing risk management and/or risk prevention strategies for the problem situation.

5.1.1 Basis for Measuring Risks

Risk represents the assessed loss potential, often estimated by the mathematical expectation of the consequences of an adverse event occurring. It is defined by the product of the two components of the probability of occurrence (p) and the consequence or severity of occurrence (S), viz.:

$$\text{Risk} = p \times S$$

Risk – interpreted as the probability of a harmful event to humans or to the environment that is caused by a chemical, physical, or biological agent – can also be described by the following conceptual relationship:

$$\text{Risk} = [f(\text{I}) \times f(\text{P})] - f(\text{D})$$

where $f(\text{I})$ represents an 'intrinsic risk' factor that is a function of the characteristic nature of the agent or the dangerous properties of the hazard; $f(\text{P})$ is a 'presence' factor that is a function of the quantity of the substance or hazard released into the environment and of all the accumulation and removal methods related to the chemical and physical parameters of the product, as well as to the case-specific parameters typical of the particular environmental setting; and $f(\text{D})$ represents a 'defense' factor that is a function of what society can do in terms of both protection and prevention to minimize the harmful effects of the hazard. Probably, the most important factor in this equation is $f(\text{D})$, which may include both the ordinary defense mechanisms for hazard abatement as well as some legislative measures.

In fact, the level of risk is dependent on the degree of hazard as well as on the amount of safeguards or preventative measures against adverse effects; consequently, risk can also be defined by the following simplistic conceptual relationships:

$$\text{Risk} = \frac{[\text{hazard}]}{[\text{preventative measures}]}$$

or

$$\text{Risk} = f\{\text{hazard, exposure, safeguards}\}$$

where 'preventative measures' or 'safeguards' is considered to be a function of exposure – or rather inversely proportional to the degree of exposure. 'Preventative measures' or 'safeguards' represent the actions that are generally taken to minimize potential exposure of target populations to hazards.

Fundamentals of hazard, exposure, and risk assessment

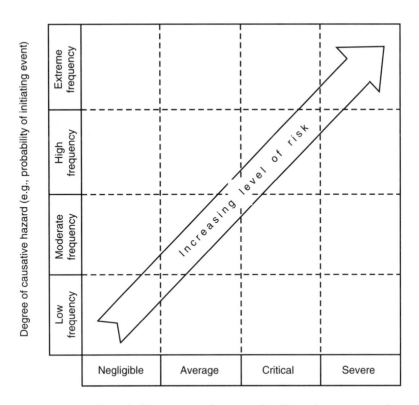

Figure 5.3 Conceptual categories of risk measures

The estimation of risks involves an integration of information on the hazard intensity, exposure frequency, and duration of exposure for all identified exposure routes for the exposed or impacted group(s). Ultimately, the risk measures give an indication of the probability and severity of adverse effects (Figure 5.3), and generally are established with varying degrees of confidence according to the importance of the decision involved. For instance, risk may represent the probability for a chemical to cause adverse impacts to potential receptors as a result of exposures over specified time periods expressed in appropriate terms.

In general, measures used in risk analysis take various forms, depending on the type of problem, degree of resolution appropriate for the situation on hand, and the analysts' preference. It may be expressed in quantitative terms – in which case it could take on values from zero (associated with certainty for no adverse effects) to unity (associated with certainty for adverse effects to occur); in several other cases, risk is only described qualitatively – such as by use of descriptors like 'high', 'moderate', 'low', etc.; or indeed, risk may be described in semi-quantitative/semi-qualitative terms. In any case, the risk qualification or quantification process will normally rely on the use of several measures, parameters and/or tools as reference

> Box 5.1 Typical/common measures, parameters, and/or tools that form the basis for risk qualification or quantification
>
> - Probability distributions (based on probabilistic analyses)
> - Expected values (based on statistical analyses)
> - Economic losses or damages
> - Risk profile diagrams
> - Relative risk (defined by a ratio such as [incidence rate in exposed group]:[incidence rate in nonexposed group])
> - Margin of safety (defined by the ratio of [the highest dose level that does *not* product an adverse effect]:[the anticipated human exposure])
> - Individual lifetime risk (equal to the product of exposure level and severity, e.g., [dose × potency])
> - Population or societal risk (defined by the product of the individual lifetime risk and the population exposed)
> - Frequency–consequence diagrams (also known as F–N curves for fatalities, to define societal risk)
> - Quality of life adjustment (or quality adjusted life expectancy, QALE)
> - Loss of life expectancy (given by the product of individual lifetime risk and the average remaining lifetime)

yardsticks (Box 5.1). Individual lifetime risk is about the most commonly used measure of risk – presented as the probability that the individual will be subjected to an adverse effect from exposure to identified hazards.

5.2 RISK ASSESSMENT DEFINED

Several somewhat differing definitions of risk assessment have been published in the literature by various authors to describe a variety of risk assessment methods and/or protocols (e.g., Asante-Duah, 1990; Bowles et al., 1987; Calabrese and Kostecki, 1992; CEC, 1986; CMA, 1985; Cohrssen and Covello, 1989; Conway, 1982; Cothern, 1993; Covello and Mumpower, 1985; Covello et al., 1986; Crandall and Lave, 1981; Davies, 1996; Glickman and Gough, 1990; Gratt, 1996; Hallenbeck and Cunningham, 1988; JCC, 1986; Kates, 1978; Kolluru et al., 1996; LaGoy, 1994; Lave, 1982; Neely, 1994; NRC, 1982, 1983, 1994; OTA, 1983; Richardson, 1990, 1992; Rowe, 1977; Suter, 1993; Turnberg, 1996; USEPA, 1984; Whyte and Burton, 1980). In a generic sense, risk assessment may be considered to be a systematic process for arriving at estimates of all the significant risk factors or parameters associated with an entire range of 'failure modes' and/or exposure scenarios in connection with some hazard situation(s). In its application to the management of environmental contamination problems, the process encompasses an evaluation of all the significant risk factors associated with all feasible and identifiable exposure scenarios that are the result of contaminant releases into the environment. It involves the characterization of potential adverse consequences or impacts to human and ecological receptors that are potentially at risk from exposure to environmental contaminants.

The risk assessment process seeks to estimate the likelihood of occurrence of adverse effects resulting from exposures of humans and ecological receptors to chemical, physical, and/or biological agents that are present in the environment. It involves the characterization of potential adverse consequences or impacts to humans and ecological receptors following their exposure to environmental, technological, or other hazards. The process consists of a mechanism that utilizes the best available scientific knowledge to establish case-specific responses that will ensure justifiable, cost-efficient and defensible decisions, about hazardous situations. The process concerns the assessment of the importance of all identified risk factors to various stakeholders whose interests are embedded in a candidate problem situation (Petak and Atkisson, 1982). Risk assessment is indeed a powerful tool for developing insights into the relative importance of the various types of exposure scenarios associated with potentially hazardous situations.

5.3 PURPOSE AND ATTRIBUTES OF RISK ASSESSMENT

The conventional paradigm for risk assessment is *'predictive'*, which deals with localized effects of a particular action that could result in adverse effects. However, there also is increasing emphasis on assessments of the effects of environmental hazards associated with existing environmental contamination problems. This assessment of past pollutions, with possible on-going consequences, generally falls under the umbrella of what has been referred to as *'retrospective'* risk assessment (Suter, 1993).

The impetus for a retrospective risk assessment may be a source, observed effects, or evidence of exposure. Source-driven retrospective assessments result from observed pollution that requires elucidation of possible effects (e.g., hazardous waste sites, spills/accidental releases, etc.); effects-driven retrospective assessments result from the observation of apparent effects in the field that require explanation (e.g., fish or bird kills, declining populations of a species, etc.); and exposure-driven retrospective assessments are prompted by evidence of exposure without prior evidence of a source or effects (e.g., the case of a scare over mercury found in the edible portions of dietary fish). In all cases, however, the principal objective of risk assessment is to provide a basis for actions that will minimize the impairment of the environment and/or of public health, welfare, and safety.

In general, risk assessment – which seems to be one of the fastest evolving tools for developing appropriate strategies in relation to environmental management decisions – seeks to answer three basic questions:

- What could potentially go wrong?
- What are the chances for this to happen?
- What are the anticipated consequences, if this should indeed happen?

A complete analysis of risks associated with a given situation or activity will generate answers to these questions. Subsequently, appropriate mitigative activities can be initiated by implementing the necessary corrective action and risk management decisions.

> Box 5.2 An annotation of typical general objectives of a risk assessment
>
> - Determine whether potentially hazardous situations exist – i.e., determine 'baseline' risks and the possible need for corrective action
> - Provide a consistent process for evaluating and documenting public and environmental health threats associated with a potential hazardous situation
> - Estimate the potential threat to public health and/or the environment that is posed by a facility or hazardous situation
> - Estimate potential health risks associated with use of several chemicals and consumer products, to ensure the development and implementation of acceptable public health policies
> - Determine the relative size of different problem situations, in order to facilitate priority setting, where necessary
> - Complete a preliminary scoping in order to identify possible data gaps in the problem evaluation
> - Determine whether there is a need for an immediate response action
> - Identify possible corrective action strategies
> - Provide basis for comparing and choosing between remedial action alternatives
> - Provide a basis for determining levels of chemicals that can remain at a given locale, and still be adequately protective of public health and the environment
> - Provide for the risk management informational needs of property owners and general community

As Whyte and Burton (1980) succintly indicate, a major objective of risk assessment is to help develop risk management decisions that are more systematic, more comprehensive, more accountable and more self-aware of appropriate programs than has often been the case in the past. Ultimately, tasks performed during the risk assessment will help answer the infamous 'how safe is safe enough?' and/or 'how clean is clean enough?' questions. Subsequently, risk management comes in to help address the question of 'what can be done about the prevailing situation?'.

5.3.1 The Purpose

The overall goal in a risk assessment – achievable by fulfilling several general objectives (Box 5.2) – is to identify potential 'system failure modes' and exposure scenarios, intended to facilitate the design of methods to reduce the probability of 'failure' and the attending public health, socio-economic, and environmental consequences of any 'failure' and/or exposure events. Its purpose is to provide, insofar as possible, a complete information set to risk managers, so that the best possible decision can be made concerning a potentially hazardous situation. Ultimately, the risk assessment process provides a framework for developing the risk information necessary to assist decision-making.

Information developed in the risk assessment will typically facilitate decisions about the allocation of resources for safety improvements and hazard/risk reduction, by directing attention and efforts to the features and exposure pathways that dominate the risks. The results of the analysis will generally provide decision-makers with a more justifiable basis for determining risk acceptability, and also aid in choosing between possible corrective measures developed for risk mitigation programs. Indeed, it is imperative to make case-specific risk assessment an integral part of all environmental management programs that are associated with environmental contamination problems.

Oftentimes, the information generated in a risk assessment is used to determine the need for, and the degree of mitigation required for environmental contamination problems. For instance, risk assessment techniques and principles can generally be employed to facilitate the development of effective site characterization and corrective action programs for contaminated lands scheduled for decommissioning. In addition to providing information about the nature and magnitude of potential health and environmental risks associated with a contaminated site problem, risk assessment also provides a basis for judging the need for remediation. Furthermore, risk assessment can be used to compare the risk reductions afforded by different remedial or risk control strategies.

The use of risk assessment techniques in contaminated land cleanup plans in particular, and corrective action programs in general, is becoming increasingly important in several places. The risk assessment serves as a useful tool for evaluating the effectiveness of remedies at contaminated sites and also for establishing cleanup objectives (including the determination of cleanup levels) that will produce efficient, feasible, and cost-effective remedial solutions. Its general purpose is to gather sufficient information that will allow for an adequate and accurate characterization of potential risks associated with a project site. In general, a risk assessment process is utilized to determine whether the level of risk at a contaminated site warrants remediation, and to further project the amount of risk reduction necessary to protect public health and the environment. Subsequently, an appropriate corrective action plan can be developed and implemented for the case site and/or the impacted area.

5.3.2 The Attributes

The risk assessment process will generally utilize the best available scientific knowledge and data to establish case-specific responses to environmental contamination problems. For example, the assessment of health and environmental risks associated with potentially contaminated sites and facilities may contribute, in a significant way, to the processes involved in corrective action planning; in risk mitigation and risk management strategies; and in the overall management of potentially contaminated site problems.

Depending on the scope of the analysis, methods used in estimating risks may be either qualitative or quantitative. Thus, the process may be one of data analysis or modeling, or a combination of the two. In fact, the process of quantifying risks does, by its very nature, give a better understanding of the strengths and weaknesses of the potential hazards being examined. It shows where a given effort can do the most good in modifying a system, in order to improve its safety and efficiency.

The major attributes of risk assessment that are particularly relevant to environmental management programs associated with environmental contamination problems include:

- Identification and ranking of all existing and anticipated potential hazards.
- Explicit consideration of all current and possible future exposure scenarios.
- Qualification and/or quantification of risks associated with the full range of hazard situations, system responses, and exposure scenarios.
- Identification of all significant contributors to the critical pathways, exposure scenarios, and/or total risks.
- Determination of cost-effective risk reduction policies, via the evaluation of risk-based remedial action alternatives and/or the adoption of efficient risk management and risk prevention programs.
- Identification and analysis of sources of uncertainties associated with environmental management programs.

Each attribute will ultimately play an important role in the overall environmental management program. The type and degree of detail of any risk assessment depends on its intended use. Its purpose will shape the data needs, the protocol, the rigor, and related efforts. In general, data generated in a risk assessment are used to determine the need for, and the degree of mitigation that may be required for an environmental contamination problem. Current regulatory requirements are particularly important considerations in the application of risk assessment to environmental contamination problems. In any case, the processes involved in any risk assessment generally require a multidisciplinary approach – covering several areas of expertise in most cases.

It is noteworthy that there are inherent uncertainties associated with risk assessments due to the fact that the risk assessor's knowledge of the causative events and controlling factors usually is limited, and also because the results obtained depend, to a reasonable extent, on the methodology and assumptions used. Furthermore, risk assessment can impose potential delays in the implementation of appropriate corrective measures – albeit the overall gain in program efficiency is likely to more than compensate for any delays.

SUGGESTED FURTHER READING

Cox, S.J. and N.R.S. Tait. 1991. *Reliability, Safety and Risk Management: Integrated Approach*. Butterworth–Heinemann, Oxford, UK.
Kleindorfer, P.R. and H.C. Kunreuther (eds). 1987. *Insuring and Managing Hazardous Risks: From Seveso to Bhopal and Beyond*. Springer-Verlag, Berlin, Germany.
Lu, F.C. 1985. Safety assessments of chemicals with threshold effects. *Regulatory Toxicology and Pharmacology*, 5: 121–132.
Saxena, J. and F. Fisher (eds). 1981. *Hazard Assessment of Chemicals*. Academic Press, New York.
Starr, C., R. Rudman, and C. Whipple. 1976. Philosophical basis for risk analysis. *Annual Review of Energy*, 1: 629–662.
Van Ryzin, J. 1980. Quantitative risk assessment. *Journal of Occupational Medicine*, 22(5): 321–326.

REFERENCES

Asante-Duah, D.K. 1990. Quantitative risk assessment as a decision tool for hazardous waste management. In: *Proceedings of 44th Purdue Industrial Waste Conference* (May, 1989), Lewis Publishers, Chelsea, MI, pp. 111–123.

Bowles, D.S., L.R. Anderson, and T.F. Glover. 1987. Design level risk assessment for dams. *Proceedings of Structural Congress, ASCE*: 210–225.

Calabrese, E.J. and P.T. Kostecki. 1992. *Risk Assessment and Environmental Fate Methodologies*. Lewis Publishers/CRC Press, Boca Raton, FL.

CEC (Commission on the European Communities). 1986. *Risk Assessment for Hazardous Installations*. Pergamon Press, Oxford, UK.

CMA (Chemical Manufacturers Association). 1985. *Risk Analysis in the Chemical Industry*. Government Institutes, Inc., Rockville, MD.

Cohrssen, J.J. and V.T. Covello. 1989. *Risk Analysis: A Guide to Principles and Methods for Analyzing Health and Environmental Risks*. National Technical Information Service, US Dept. of Commerce, Springfield, VA.

Conway, R.A. (ed.). 1982. *Environmental Risk Analysis of Chemicals*. Van Nostrand Reinhold, New York.

Cothern, C.R. (ed.). 1993. *Comparative Environmental Risk Assessment*. Lewis Publishers/ CRC Press, Boca Raton, FL.

Covello, V.T. and J. Mumpower. 1985. Risk analysis and risk management: an historical perspective. *Risk Analysis*, 5: 103–120.

Covello, V.T., J. Menkes and J. Mumpower (eds). 1986. *Risk Evaluation and Management*. Contemporary Issues in Risk Analysis, Vol. 1. Plenum Press, New York.

Crandall, R.W. and B.L. Lave (eds). 1981. *The Scientific Basis of Risk Assessment*. Brookings Institution, Washington, DC.

Davies, J.C. 1996. *Comparing Environmental Risks*. Resources for the Future, Washington, DC.

Glickman, T.S. and M. Gough (eds). 1990. *Readings in Risk*. Resources for the Future, Washington, DC.

Gratt, L.B. 1996. *Air Toxic Risk Assessment and Management: Public Health Risk from Normal Operations*. Van Nostrand Reinhold, New York.

Hallenbeck, W.H. and Cunningham, K.M. 1988. *Quantitative Risk Assessment for Environmental and Occupational Health*. Lewis Publishers, Chelsea, MI.

JCC (J.C. Consltancy Ltd, London). 1986. *Risk Assessment for Hazardous Installations*. Pergamon Press, Oxford, UK.

Kates, R.W. 1978. *Risk Assessment of Environmental Hazard*. SCOPE Report 8, J. Wiley, New York.

Kolluru, R.V., S.M. Bartell, R.M. Pitblado, and R.S. Stricoff (eds). 1996. *Risk Assessment and Management Handbook (for Environmental, Health, and Safety Professionals)*. McGraw-Hill, New York.

LaGoy, PK. 1994. *Risk Assessment, Principles and Applications for Hazardous Waste and Related Sites*. Noyes Data Corp., Park Ridge, NJ.

Lave, L.B. (ed.). 1982. *Quantitative Risk Assessment in Regulation*. The Brooking Institute, Washington, DC.

Neely, W.B. 1994. *Introduction to Chemical Exposure and Risk Assessment*. Lewis Publishers/ CRC Press, Boca Raton, FL.

NRC (National Research Council). 1982. *Risk and Decision-Making: Perspective and Research*. NRC Committee on Risk and Decision-Making. National Academy Press, Washington, DC.

NRC. 1983. *Risk Assessment in the Federal Government: Managing the Process*. National Academy Press, Washington, DC.

NRC. 1994. *Building Consensus Through Risk Assessment and Risk Management*. National Academy Press, Washington, DC.

OTA (Office of Technology Assessment). 1983. *Technologies and Management Strategies for Hazardous Waste Control.* Congress of the US, Office of Technology Assessment, Washington, DC.

Petak, W.J., and A.A. Atkisson. 1982. *Natural Hazard Risk Assessment and Public Policy: Anticipating the Unexpected.* Springer-Verlag, New York.

Richardson, M.L. (ed.). 1990. *Risk Assessment of Chemicals in the Environment.* Royal Society of Chemistry, Cambridge, UK.

Richardson, M.L. (ed.). 1992. *Risk Management of Chemicals.* Royal Society of Chemistry, Cambridge, UK.

Rowe, W.D. 1977. *An Anatomy of Risk.* John Wiley, New York.

Suter II, G.W. 1993. *Ecological Risk Assessment.* Lewis Publishers, Boca Raton, FL.

Turnberg, W.L. 1996. *Biohazardous Waste: Risk Assessment, Policy, and Management.* J. Wiley, New York.

USEPA (US Environmental Protection Agency). 1984. *Risk Assessment and Management: Framework for Decision Making.* Washington, DC. EPA/600/9-85-002.

Whyte, A.V. and I. Burton (eds). 1980. *Environmental Risk Assessment.* SCOPE Report 15, J. Wiley, New York.

Chapter Six

The Risk Assessment Process

Risk assessment – in its application to environmental contamination problems – involves a process used to compile and organize scientific information for environmental management purposes. This is used to help identify potential problems, establish priorities, and provide a basis for regulatory actions. In general, risk assessments may be classified as *retrospective* – focusing on injury after the fact (e.g., nature and level of risks at a given contaminated site), or may be considered as *predictive* – such as evaluating possible future harm to human health or the environment (e.g., risks anticipated if a newly developed food additive is approved for use in consumer food products). In the investigation of environmental contamination problems, the focus of most risk assessments tends to be on a determination of potential risks to human and ecological receptors. Although these represent different types of populations, the mechanics of the evaluation process are similar.

The risk assessment process can be used both to provide a 'baseline' estimate of existing risks attributable to an agent or hazard, and to determine the potential reduction in exposure and risk given various corrective action situations. Several techniques are available for performing risk assessments; most of the techniques are structured around decision analysis procedures, to facilitate comprehensible solutions for even complicated problems. A number of the risk assessment approaches commonly encountered in the literature of environmental management and/or relevant to the management of environmental contamination problems are elaborated further in Part IV of this title.

It is noteworthy that, in general, much of the effort in the development of risk assessment methodologies has been directed at human health risk assessments (as reflected by the differences in the depth of coverage for human health versus ecological risk assessment that can be found in the literature). However, the fundamental components of the risk assessment process for other biological organisms parallel those for human receptors, and can indeed be described in similar terms.

6.1 ELEMENTS OF THE RISK ASSESSMENT PROCESS

Specific forms of risk assessment generally differ considerably in their levels of detail. Most risk assessments, however, share the same general logic – consisting of four basic elements (Figure 6.1). A discussion of these fundamental elements follows, with more detailed elaboration given in Part III of this volume and also

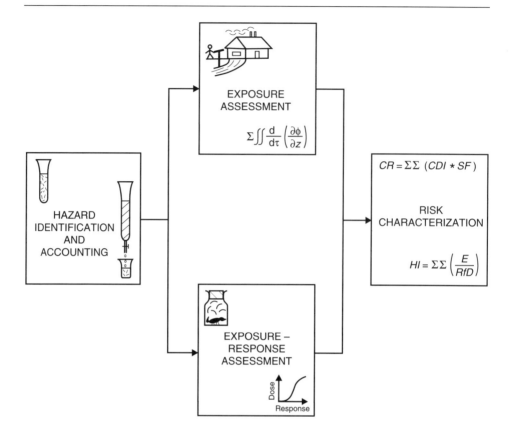

Figure 6.1 Illustrative elements of a risk assessment process

elsewhere in the risk analysis literature (e.g., CAPCOA, 1990; CEC, 1986; CMA, 1985; Cohrssen and Covello, 1989; Conway, 1982; Cothern, 1993; Gheorghe and Nicolet-Monnier, 1995; Hallenbeck and Cunningham, 1988; Huckle, 1991; Kates, 1978; Kolluru et al., 1996; LaGoy, 1994; Lave, 1982; McColl, 1987; McTernan and Kaplan, 1990; Neely, 1994; NRC, 1982, 1983, 1994; OTA, 1983; Paustenbach, 1988; Richardson, 1990; Rowe, 1977; Suter, 1993; USEPA, 1984, 1989a, 1989b; Whyte and Burton, 1980).

6.1.1 Hazard Identification and Accounting

Hazard identification and accounting involves a qualitative assessment of the presence of, and the degree of hazard that an agent could have on potential receptors. It consists of gathering and evaluating data on the types of health effects or diseases that may be produced by a chemical, and the exposure conditions under which environmental damage, injury, or disease will be produced.

In the context of environmental management of potential contamination problems, this may consist of the identification of contaminant sources; a compilation of the lists of all contaminants present at the locale; the identification and selection of the specific chemicals of potential concern (that should become the focus of the risk assessment), based on their specific hazardous properties (such as persistence, bioaccumulative properties, toxicity, and general fate and behavior properties); and a compilation of summary statistics for the key constituents selected for further investigation and evaluation. In identifying the chemicals of potential concern, an attempt is generally made to select all chemicals that could represent the major part (viz., $\geq 95\%$) of the risks associated with site- or facility-related exposures.

6.1.2 Exposure–Response Evaluation

The exposure–response evaluation, or the effects assessment, is the estimation of the relationship between dose or level of exposure to a substance and the incidence and severity of an effect. It considers the types of adverse effects associated with contaminant exposures, the relationship between magnitude of exposure and adverse effects, and related uncertainties (such as the weight-of-evidence of a particular chemical's carcinogenicity in humans).

In the context of environmental contamination problems, this evaluation will generally include a toxicity assessment and/or a dose–response evaluation. The toxicity assessment typically consists of compiling toxicological profiles for the chemicals of potential concern. Dose–response relationships are then used to quantitatively evaluate the toxicity information, and to characterize the relationship between dose of the contaminant administered or received and the incidence of adverse effects on the exposed population. From the quantitative dose–response relationship, appropriate toxicity values can be derived and subsequently used to estimate the incidence of adverse effects occurring in populations at risk for different exposure levels.

6.1.3 Exposure Assessment and Analysis

An exposure assessment is conducted to estimate the magnitude of actual and/or potential receptor exposures to environmental contaminants, the frequency and duration of these exposures, the nature and size of the populations potentially at risk (i.e., the risk group), and the pathways by which the risk group may be exposed. Several physical and chemical characteristics of the chemicals of concern will provide an indication of the critical exposure features. These characteristics can also provide information necessary to determine the chemical's distribution, intake, metabolism, residence time, excretion, magnification, and half-life or breakdown to new chemical compounds.

To complete a typical exposure analysis for an environmental contamination problem, populations potentially at risk are identified, and concentrations of the chemicals of concern are determined in each medium to which potential receptors may be exposed. Finally, using the appropriate case-specific exposure parameter values, the intakes of the chemicals of concern are estimated. The exposure estimates can then be used to determine whether any threats exist based on prevailing exposure conditions for the particular problem situation.

6.1.4 Risk Characterization and Consequence Determination

Risk characterization is the process of estimating the probable incidence of adverse impacts to potential receptors under a set of exposure conditions. Typically, the risk characterization summarizes and then integrates outputs of the exposure and toxicity assessments in order to qualitatively and/or quantitatively define risk levels. This usually will include an elaboration of uncertainties associated with the risk estimates. Exposures resulting in the greatest risk can be identified in this process; mitigative measures can then be selected to address the situation in order of priority, and according to the levels of imminent risks.

An adequate characterization of risks from hazards associated with environmental contamination problems allows risk management and corrective action decisions to be better focused. To the extent feasible, the risk characterization should include the distribution of risk amongst the target populations.

6.2 RISK ASSESSMENT AS A DIAGNOSTIC TOOL

Often risk assessment is considered an integral part of the diagnostic assessment of environmental contamination problems. In general, a risk assessment process is utilized to determine whether the level of risk associated with an environmental contamination problems warrants critical risk management actions, and to further project the amount of risk reduction necessary to protect public health and the environment. In its application to the investigation of environmental contamination problems, the risk assessment process encompasses an evaluation of all the significant risk factors associated with all feasible and identifiable exposure scenarios that are the result of contaminant releases into the environment. It involves the characterization of potential adverse consequences or impacts to human and ecological receptors that are potentially at risk from exposure to environmental contaminants.

Procedures generally used in the risk assessment of environmental contamination problems will comprise the following tasks:

- Identification of the sources of contamination.
- Definition of the contaminant migration pathways.
- Identification of populations potentially at risk.
- Determination of the specific chemicals of potential concern.
- Determination of frequency of potential receptor exposures to environmental contaminants.
- Evaluation of contaminant exposure levels.
- Determination of receptor response to chemical exposures.
- Estimation of impacts or damage resulting from receptor exposures to the chemicals of potential concern.

The risk assessment process 83

Potential risks are estimated by considering the probability or likelihood of occurrence of harm; the intrinsic harmful features or properties of specified hazards; the populations potentially at risk; the exposure scenarios; and the extent of expected harm and potential effects.

In most applications, risk assessment is used to provide a baseline estimate of existing risks that are attributable to a specific agent or hazard; the baseline risk assessment consists of an evaluation of the potential threats to human health and the environment in the absence of any remedial or response action. It can also be used to determine the potential reduction in exposure and risk under various corrective action scenarios, as well as to support remedy selection in mitigative programs. Invariably, risk management and corrective action decisions about environmental contamination problems are made primarily on the basis of potential human health and ecological risks.

6.2.1 Baseline Risk Assessments

Baseline risk assessments involve an analysis of the potential adverse health and environmental effects (current or future) caused by hazardous substance releases into the environment, in the absence of any actions to control or mitigate these releases (i.e., under an assumption of 'no-action'). That is, the baseline risk assessment provides an estimate of the potential risks to human health and the environment resulting from receptor exposure to environmental contaminants in the absence of mitigative actions. Because this type of assessment identifies the primary health and environmental threats associated with an environmental contamination problem, it also provides valuable input to the development and evaluation of alternative risk management and mitigative options. In fact, baseline risk assessments are usually conducted to evaluate the need for, and the extent of corrective action required for an environmental contamination problem. That is, they provide the basis and rationale as to whether or not remedial action is necessary.

In general, a baseline risk assessment contributes to the characterization of environmental contamination problems. It further facilitates the development, evaluation, and selection of appropriate corrective action response alternatives, which may include consideration of a 'no-action' remedial alternative, where appropriate. The results of the baseline risk assessment are generally used to:

- Document the magnitude of risk at a given locale, and the primary causes of that risk.
- Help determine whether any response action is necessary for the problem situation.
- Prioritize the need for remedial action, where several problem situations are involved.
- Provide the basis for quantifying remedial action objectives.
- Develop and modify remedial action goals.
- Support and justify 'no further action' decisions, by documenting the threats posed by the environmental contamination problem, based on expected exposure scenarios.

Baseline risk assessments are case-specific and therefore may vary in both detail and the extent to which qualitative and quantitative analyses are used. The level of effort required to conduct a baseline risk assessment depends largely on the complexity and particular circumstances (such as regulatory systems, criteria, and guidances) associated with the environmental contamination problem.

6.3 RISK ASSESSMENT IN PRACTICE

A number of techniques are available for conducting risk assessments. Invariably, the methods of approach consist of the several basic procedural elements/components shown in Figure 6.2, and further outlined/itemized in Box 6.1. The key issues requiring significant attention in the processes involved will typically relate to the following questions:

- What chemicals pose the greatest risk?
- What are the concentrations of the contaminants of concern at and near the project location?
- Which exposure pathways are the most important?
- Which population groups, if any, face significant risk as a result of possible exposures?
- What is the range of risks to the problem location?
- What are the health and environmental implications for any identifiable range of corrective action alternatives?

As a general guiding principle, risk assessments should be carried out in an iterative fashion – to be adjusted to incorporate new scientific information and regulatory changes – with the ultimate goal of minimizing public health and economic consequences associated with a potentially hazardous situation. An iterative approach would start with relatively inexpensive screening techniques – and then for hazards suspected of exceeding *de minimis* risk, further evaluation is conducted by moving on to more resource-intensive levels of data gathering, model construction, and model application (NRC, 1994). That is, risk assessments should be conducted at different levels – from simple, quick, and inexpensive screening calculations to much more complex and refined efforts at much greater expense – in order to characterize risks under different sets of circumstances.

In general, risk assessment will normally be conducted in an iterative fashion that grows in depth with increasing problem complexity. Consider as an example, a site-specific risk assessment that is used to evaluate/address potential health impacts associated with chemical releases from industrial facilities or hazardous waste sites. A tiered approach is generally recommended in the conduct of such site-specific risk assessments. Usually, this will involve two broad levels of detail – screening and comprehensive. In the screening evaluation, relatively simple models, conservative assumptions, and default generic parameters are used to calculate an upper-bound risk estimate associated with a chemical release from the case facility. No detailed/comprehensive evaluation is warranted if the estimate is below a pre-established reference or target level. On the other hand, if the screening risk estimate is above

The risk assessment process

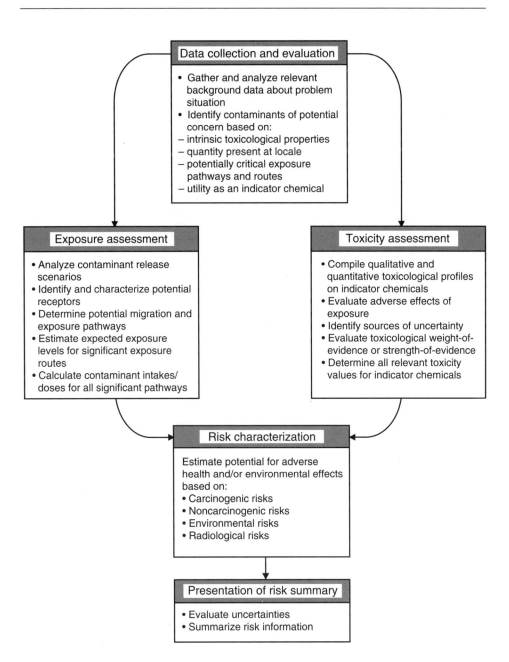

Figure 6.2 Fundamental procedural components of a risk assessment for an environmental contamination problem

> **Box 6.1   Illustrative basic outline for a risk assessment report**
>
> | Section topic | Basic subject matter |
> |---|---|
> | Section topic | Basic subject matter |
> | General overview | • Background information on the case problem or facility
• The risk assessment process
• Purpose and scope of the risk assessment
• The risk assessment technique and method of approach
• Legal and regulatory issues in the risk assessment
• Limits of application for the risk assessment |
> | Data collection | • Environmental areas and media of potential concern
• General case-specific data collection considerations
• Assessment of the data quality objectives
• Identification of uncertainties |
> | Data evaluation | • General case-specific data evaluation considerations
• Identification, quantification, and categorization of contaminants
• Statistical analyses of relevant environmental data
• Screening and selection of the chemicals of potential concern
• Identification of uncertainties |
> | Exposure assessment | • Characterization of the exposure setting (to include the physical setting and populations potentially at risk)
• Identification of the source areas, significant migration pathways, and potentially impacted or receiving media
• Determination of the important environmental fate and transport processes for the chemicals of potential concern, including cross-media transfers
• Determination of the likely and significant receptor exposure pathways
• Development of representative conceptual model(s) for the problem situation
• Development of realistic exposure scenarios (to include both current and potential future possibilities)
• Estimation/modeling of exposure point concentrations for the chemicals of potential concern found in the significant environmental media
• Quantification of exposures (i.e., computation of potential receptor intakes and resultant doses for the applicable exposure scenarios)
• Identification of uncertainties associated with exposure parameters |
> | Toxicity assessment | • Compilation of the relevant toxicological profiles of the chemicals of potential concern
• Determination of the appropriate and relevant toxicity index parameters
• Identification of uncertainties relating to the toxicity information |
> | Risk characterization | • Estimation of the human carcinogenic risks from carcinogens |

	Box 6.1 (continued)
Section topic	Basic subject matter
	• Estimation of the noncarcinogenic effects for systemic toxicants • Estimation of ecological hazard effects • Sensitivity analyses of relevant parameters • Identification and evaluation of uncertainties associated with the risk estimates
Risk summary discussion	• Summarization of risk information • Discussion of all identifiable sources of uncertainties

the reference level, then the more comprehensive/detailed evaluation should be carried out – in order to ascertain the existence of any significant risk. The purpose of such a tiered approach is to optimize the use of resources – viz., a comprehensive/detailed risk assessment is performed only when truly warranted.

Irrespective of the level of detail, however, a well-defined protocol should be used to assess imminent risks as part of an overall policy decision on risk mitigation measures needed by society. A decision on the level (e.g., qualitative, quantitative, or combinations thereof) at which an analysis is carried out will usually be based on the complexity of the situation and the level of risk involved, anticipated, or predicted.

6.4 RISK ASSESSMENT AS AN HOLISTIC TOOL FOR ENVIRONMENTAL MANAGEMENT

Risk assessment is a process used to evaluate the collective demographic, geographic, physical, chemical, biological and related factors associated with environmental contamination problems; this helps to determine and characterize possible risks to public health and the environment. The overall objective of such an assessment is to determine the magnitude and probability of actual or potential harm that the environmental contamination problem poses to human health and the environment. In fact, human populations and ecological receptors are continuously in contact with varying amounts of environmental contaminants present in air, water, soil, and food. Thus, managing the health and environmental risks associated with environmental contaminants usually requires an integrated model to effectively assess the environmental fate and behavior of the constituents of concern, as well as to determine potential human and ecosystem exposures. Ultimately, methods for linking contaminant sources in multiple environmental media (such as air, water, and soil) to human and ecological receptor exposures are often necessary to facilitate the development of sound corrective action and risk management programs.

As an holistic approach to environmental management, risk assessment integrates all relevant environmental and health issues and concerns surrounding a specific problem situation (such as a particular site, facility, or industrial activity) in order to arrive at risk management decisions that are acceptable to all stakeholders (usually comprised of industry, regulatory agencies – representing government, environmental advocacy groups, and the affected or threatened community or individuals). Among other things, the overall strategy should incorporate information that helps to answer the following pertinent questions:

- Why is the project being undertaken?
- How will results and conclusions from the project be used?
- What specific processes and methodologies will be utilized?
- What are the uncertainties and limitations surrounding the study?
- What contingency plans exist for resolving identified problem(s)?

In general, effective risk communication (see Chapter 19) becomes a very important element of the holistic approach to managing environmental contamination problems. Thus, the conveying of risk information derived from a risk assessment should be considered as a very essential integral part of the overall approach.

SUGGESTED FURTHER READING

Covello, V.T. and M.W. Merkhofer. 1993. *Risk Assessment Methods: Approaches for Assessing Health and Environmental Risks*. Plenum Press, New York.
Hertz, D.B. and H. Thomas. 1983. *Risk Analysis and Its Applications*. John Wiley, New York.
Kastenberg, W.E. and H.C. Yeh. 1993. Assessing public exposure to pesticides – contaminated ground water. *Journal of Ground Water*, 31(5): 746–752.
OECD (Organization for Economic Cooperation and Development). 1986. *Report of the OECD Workshop on Practical Approaches to the Assessment of Environmental Exposure*, 14–18 April 1986, Vienna, Austria.
OECD. 1989. *Compendium of Environmental Exposure Assessment Methods for Chemicals*. Environmental Monographs, No. 27, OECD, Paris, France.
Onishi, Y., A.R. Olsen, M.A. Parkhurst, and G. Whelan. 1985. Computer-based environmental exposure and risk assessment methodology for hazardous materials. *Journal of Hazardous Materials*, 10: 389–417.
USEPA (United States Environmental Protection Agency). 1990. *Development of Risk Assessment Methodology for Surface Disposal of Municipal Sludge*. Office of Research and Development, Washington, DC. EPA/600/6-90/001.
Wilson, R. and E.A.C. Crouch. 1987. Risk assessment and comparisons: an introduction. *Science*, 236: 267–270.

REFERENCES

CAPCOA (California Air Pollution Control Officers Association). 1990. *Air Toxics 'Hot Spots' Program. Risk Assessment Guidelines*. CAPCOA, Los Angeles, CA.
CEC (Commission on the European Communities). 1986. *Risk Assessment for Hazardous Installations*. Pergamon Press, Oxford, UK.
CMA (Chemical Manufacturers Association). 1985. *Risk Analysis in the Chemical Industry*. Government Institutes, Inc., Rockville, MD.

Cohrssen, J.J. and V.T. Covello. 1989. *Risk Analysis: A Guide to Principles and Methods for Analyzing Health and Environmental Risks*. National Technical Information Service, US Dept. of Commerce, Springfield, VA.

Conway, R.A. (ed.). 1982. *Environmental Risk Analysis of Chemicals*. Van Nostrand Reinhold, New York.

Cothern, C.R. (ed.). 1993. *Comparative Environmental Risk Assessment*. Lewis Publishers/CRC Press, Boca Raton, FL.

Gheorghe, A.V. and M. Nicolet-Monnier. 1995. *Integrated Regional Risk Assessment*, Volumes I and II. Kluwer Academic Publishers, Dordrecht, The Netherlands.

Hallenbeck, W.H. and K.M. Cunningham. 1988. *Quantitative Risk Assessment for Environmental and Occupational Health*. Lewis Publishers, Chelsea, MI.

Huckle, K.R. 1991. *Risk Assessment – Regulatory Need or Nightmare. Selected Papers*. Shell Centre, London, UK.

Kates, R.W. 1978. *Risk Assessment of Environmental Hazard*. SCOPE Report 8, J. Wiley, New York.

Kolluru, R.V., S.M. Bartell, R.M. Pitblado, and R.S. Stricoff (eds). 1996. *Risk Assessment and Management Handbook (for Environmental, Health, and Safety Professionals)*. McGraw-Hill, New York.

LaGoy, P.K. 1994. *Risk Assessment, Principles and Applications for Hazardous Waste and Related Sites*. Noyes Data Corp., Park Ridge, NJ.

Lave, L.B. (ed.). 1982. *Quantitative Risk Assessment in Regulation*. The Brooking Institute, Washington, DC.

McColl, R.S. (ed.). 1987. *Environmental Health Risks: Assessment and Management*. Institute for Risk Research, University of Waterloo Press, Waterloo, Ontario, Canada.

McTernan, W. and Kaplan, E. 1990. *Risk Assessment for Groundwater Pollution Control*. ASCE Monograph, New York.

Neely, W.B. 1994. *Introduction to Chemical Exposure and Risk Assessment*. Lewis Publishers/CRC Press, Boca Raton, FL.

NRC (National Research Council). 1982. *Risk and Decision-Making: Perspective and Research*. NRC Committee on Risk and Decision-Making. National Academy Press, Washington, DC.

NRC. (1983). *Risk Assessment in the Federal Government: Managing the Process*. National Academy Press, Washington, DC.

NRC (National Research Council). 1994. *Building Consensus Through Risk Assessment and Risk Management*. National Academy Press, Washington, DC.

OTA (Office of Technology Assessment). 1983. *Technologies and Management Strategies for Hazardous Waste Control*. Congress of the US, Office of Technology Assessment, Washington, DC.

Paustenbach, D.J. (ed.). 1988. *The Risk Assessment of Environmental Hazards: A Textbook of Case Studies*. John Wiley, New York.

Richardson, M.L. (ed.). 1990. *Risk Assessment of Chemicals in the Environment*. Royal Society of Chemistry, Cambridge, UK.

Rowe, W.D. 1977. *An Anatomy of Risk*. John Wiley, New York.

Suter II, G.W. 1993. *Ecological Risk Assessment*. Lewis Publishers, Boca Raton, FL.

USEPA (US Environmental Protection Agency). 1984. *Risk Assessment and Management: Framework for Decision Making*. Washington, DC. EPA/600/9-85-002.

USEPA. 1989a. *Risk Assessment Guidance for Superfund. Volume I – Human Health Evaluation Manual (Part A)*. Office of Emergency and Remedial Response, Washington, DC. EPA/540/1-89/002.

USEPA. 1989b. *Risk Assessment Guidance for Superfund. Volume II – Environmental Evaluation Manual*. Office of Emergency and Remedial Response, Washington, DC. EPA/540/1-89/001.

Whyte, A.V. and I. Burton (eds). 1980. *Environmental Risk Assessment*. SCOPE Report 15, J. Wiley, New York.

Chapter Seven

Basic Concepts in Risk Assessment Practice

In order to adequately evaluate the risks due to some hazard situation, several concepts may need to be employed – irrespective of whether they are presented in a qualitative, semi-quantitative, or quantitative manner. Of particular interest, quantitative probability estimates in risk analyses are often based on available statistical data, and to some extent on expert judgment. Consequently, several concepts of probability may be required to analyze the available information that will be utilized in the risk evaluation process. A summary of the notations and theorems pertaining to some probability definitions and concepts commonly used in probabilistic risk analyses are given in Appendix C. A more detailed review can be found in several textbooks of statistics and probability theory (e.g., Berthouex and Brown, 1994; Freund and Walpole, 1987; Larsen and Marx, 1985; Miller and Freund, 1985; Sharp, 1979; Siddall, 1983; Wonnacott and Wonnacott, 1972). Several other fundamental concepts and definitions that will facilitate a better understanding of the risk assessment process and application principles, and that may also affect risk management decisions, are introduced below in this chapter.

7.1 INDIVIDUAL VERSUS GROUP RISKS

In the application of risk assessment to environmental management programs, it often becomes important to distinguish between individual and societal risks, in order that the most appropriate one can be used in the analysis of case-specific problems.

Individual risks are considered to be the frequency at which a given individual could potentially sustain a given level of adverse consequence from the realization or occurrence of specified hazards. Societal risk, on the other hand, relates to the frequency and the number of individuals sustaining some given level of adverse consequence due to the occurrence of specified hazards.

Individual risk estimates are more appropriate in cases where individuals face relatively high risks. However, when individual risks are not inequitably high, then it becomes important during resources allocation to consider possible society-wide risks which might be relatively higher.

At an individual level, the choice of whether or not to accept a risk is primarily a personal decision, whereas on a societal level – where values tend to be in conflict, and decisions often produce 'winners' and 'losers' – the decision to accept or reject a risk is more difficult (Cohrssen and Covello, 1989). In fact, no numerical level of risk will likely receive universal acceptance, but also eliminating all risks is an impossible task – especially for our modern society in which people have become so accustomed to numerous 'hazard-generating' luxuries of life. Indeed, for many activities and technologies of today, some level of risk has to be tolerated in order to gain the benefits of the activity or technology. Consequently, levels of risk that may be considered as tolerable or relatively 'safe enough' should typically be identified/defined – at least on the societal level – to facilitate rational risk management and related decision-making tasks.

7.2 WHAT CONSTITUTES AN 'ACCEPTABLE' RISK?

An important issue in risk assessment is the risk acceptability level, i.e., what level of risk society can allow for a specified hazard situation – also recognizing that the desirable is not always attainable. In any case, with maintenance of public health and safety being crucial, it should be realized that such reasoning as budgetary constraints may not by themselves be justifiable enough reason to perch acceptable levels on the higher side of the risk spectrum. All things considered, it must be acknowledged that risk acceptability usually will have a spatial and temporal variability to it.

7.2.1 The Risk Acceptability Criteria: *de Minimis* versus *de Manifestis* Risks

An important concept in risk management – especially in relation to human health risks – is that there are levels of risk that are so great that they must not be allowed to occur at all costs, and there are other risk levels that are so low that they are not worth bothering with even at insignificant costs; these are known, respectively, as *de manifestis* and *de minimis* levels (Kocher and Hoffman, 1991; Suter, 1993; Travis et al., 1987; Whipple, 1987). Risk levels between these bounds are typically balanced against costs, technical feasibility of mitigative actions, and other socio-economic, political and legal considerations, in order to determine their acceptability or tolerability.

It is noteworthy that the concept of *de manifestis* risk is generally not seen as being controversial because, after all, some hazard effects are clearly unacceptable, whereas the *de minimis* risk concept tends to be controversial in view of the implicit idea that some exposures to and effects of pollutants or hazards are acceptable (Suter, 1993). In any case, it is still desirable to use these types of criteria to eliminate obviously trivial risks from further risk management actions – considering that society cannot completely eliminate or prevent all human and environmental health effects associated with environmental contamination problems. In fact, every social system has a target risk level – whether it is explicitly indicated or not – that represents tolerable limits to danger that the society is prepared to accept in consequence of potential benefits that could accrue from a given activity; this may be represented by the *de minimis* or 'acceptable' risk level.

7.2.1.1 The *de minimis* or 'acceptable' risk

Risk is *de minimis* if the incremental risk produced by an activity is sufficiently small so that there is no incentive to modify the activity (Cohrssen and Covello, 1989; Covello et al., 1986; Fischhoff et al., 1981; Whipple, 1987). These are levels judged to be too insignificant to be of any social concern or to justify use of risk management resources to control them, compared with other beneficial uses for the often limited resources available in practice. Simply stated, the *de minimis* principle assumes that extremely low risks are trivial and need not be controlled. A *de minimis* risk level would therefore represent a cut-off, below which a regulatory agency could simply ignore alleged problems or hazards. Thus, in the general process of establishing 'acceptable' risk levels, it is possible to use *de minimis* levels below which one need not be concerned (Rowe, 1983).

The concept of *de minimis* or acceptable risk is essentially a threshold concept, in that it postulates a threshold of concern below which there would be indifference to changes in the level of risk. In fact, considerable controversy exists concerning the concept of 'acceptable' risk in the risk/decision analysis literature. In practice, acceptable risk is the risk associated with the most acceptable decision – rather than being acceptable in an absolute sense. It has been pointed out (Massmann and Freeze, 1987) that acceptable risk is decided in the political arena and that 'acceptable' risk really means 'politically acceptable' risk. Current regulatory requirements are particularly important considerations in establishing such acceptable risk levels.

In general, the selection of a *de minimis* risk level is contingent upon the nature of the risks, the stakeholders involved, and a host of other contextual variables (such as other risks being compared against). This means that *de minimis* levels will be fuzzy (in that they can never be precisely specified), and relative (in that they will depend on the special circumstances). Also, establishing a *de minimis* risk level is often extremely difficult because people perceive risks differently. Furthermore, the cumulative burden of risks could make a currently insignificant risk become significant in the future. Consequently, stricter *de minimis* standards will usually become necessary in dealing with newly introduced risks affecting the same population groups.

There are several approaches to deriving the *de minimis* risk levels, but that which is selected should be justifiable based on the expected socio-economic, environmental, and public health impacts. A common approach in placing risks in perspective is to list many risks (which are considered similar in nature) along with some quantitative measures of the degree of risk. Typically, risks below the level of one in a million (i.e., 10^{-6} chance of premature death will often be considered insignificant or *de minimis* by regulatory agencies, since this compares favorably with risk levels from 'normal' human activities (e.g., 10^{-3} for smoking a pack of cigarettes/day, or rock climbing, etc.; 10^{-4} for heavy drinking, home accidents, driving motor vehicles, farming; 10^{-5} for truck driving, home fires, skiing, living downstream of a dam, use of contraceptive pills, etc.; 10^{-6} for diagnostic X-rays, fishing, etc.; and 10^{-7} for drinking about 10 liters of diet soda containing saccharin, etc.). In considering a *de minimis* risk level, however, the possibility of multiple *de minimis* exposures with consequential large aggregate risk should not be overlooked. In fact, Whipple (in Paustenbach, 1988) suggests the use of a *de minimis* probability idea that will help develop a generally workable *de minimis* policy.

In summary, *de minimis* is a lower bound on the range of acceptable risk for a given activity. When properly utilized, a *de minimis* risk concept can help prioritize focus of attention with respect to risks in a socially beneficial way. It may define the threshold for regulatory involvement. Ultimately, in determining a target risk level associated with an environmental management situation, a more practical and realistic acceptable risk level ought to be specified. It is only after deciding on an acceptable risk level that an environmental management program can be addressed in a most cost-effective manner.

7.2.2 Consideration of Risk Perception Issues

The perception of risks generally varies amongst individuals and/or groups, and may even change with time. Risk perception may therefore be considered as having both spatial and temporal dimensions. In general, issues relating to risk perception become a very important consideration in environmental management decisions because, in some situations, the perception of a group of people may alter the priorities assigned to the reduction of competing risks. In fact, the differences between risk perception and risk estimation could have crucial consequences for the assessment, management, and communication of risks. The particular risks estimated in a risk assessment may not be consistent with the perceptions or concerns of those individuals most directly affected.

In general, risks which are involuntary (e.g., environmental risks) or 'novel' seem to arouse more concern from the target/affected populations than those that are voluntary or 'routine', and therefore accepted by the affected individuals (van Leeuwen and Hermens, 1995). Thus, 'natural' toxins and contaminants in foods may be considered acceptable (even though they may cause illness) whereas food additives (used in foodstuffs to assist in preservation) may be less acceptable to some people (Richardson, 1986). Perceptions about risk are also influenced by sources of information, styles of presentation, personal background and educational levels, cultural contexts, and the dimensions of a particular risk problem. For instance, cultural explanations for risk management controversies – in regards to the ways people differ in their thinking about risk (or risk acceptability for that matter) – have gained increasing recognition/acceptance over the past decade and a half or so (Earle and Cvetkovich, 1997). In fact, several value judgments become an important component of the consequential decision-making process – with the value judgments involving very complex social processes. A fairly well-established hierarchy of risk 'tolerability' issues has indeed emerged in recent times – including those identified/enumerated in Box 7.1 (Cassidy, 1996; Cohrssen and Covello, 1989; Lowrance, 1976) – that can affect the acceptable risk decision.

> Box 7.1 Key factors affecting the 'tolerability' of risk by individuals and society
>
> - Voluntariness (i.e., voluntary vs. involuntary exposures)
> - Response time (i.e., delayed vs. immediate effects)
> - Source (i.e., natural vs. human-made risks)
> - Controllability (i.e., controllable vs. uncontrollable)
> - Perception of personal control
> - Familiarity with the type of hazard (i.e., old/known vs. new/unknown hazards or risks)
> - Perceptions about potential benefits (i.e., exposure is an essential vs. exposure is a luxury)
> - Nature of hazard and/or consequences (i.e., ordinary vs. catastrophic)
> - Perception of the extent and type of risk
> - Perceptions about comparative risks for other activities
> - Reversibility of effects (i.e., reversible vs. irreversible)
> - Perceptions about available choices (i.e., no alternatives available vs. availability of alternatives)
> - Perceptions about equitability/fairness of risk distribution
> - Continuity of exposure (i.e., occasional vs. continuous)
> - Visual indicators of risk factors or levels (i.e., tangible vs. intangible risks)

7.3 CONSERVATISMS IN RISK ASSESSMENTS

Many of the parameters and assumptions used in hazard, exposure, and risk evaluation studies tend to have high degrees of uncertainties associated with them. Thus, it is common practice for safe design and analysis, to model risks such that risk levels determined for management decisions are preferably over-estimated. Such conservative (also, often cited as 'worst-case', or 'plausible upper-bound') estimates used in risk assessment are based on the premise that pessimism in risk assessment (with resultant high estimates of risks) is more protective of public health and/or the environment.

For example, in most risk assessments, scenarios have often been developed that will reflect the worst possible exposure pattern; this notion of 'worst-case scenario' in the risk assessment refers to the event or series of events resulting in the greatest exposure or potential exposure. Also, quantitative cancer risk assessments are typically expressed as plausible upper bounds rather than estimates of central tendency; but when several plausible upper bounds are added together, the question arises as to whether the overall result is still plausible (Bogen, 1994; Burmaster and Harris, 1993; Cogliano, 1997). In any case, it is believed that, although the overall risk depends on the independence, additivity, synergistic/antagonistic interactions among the carcinogens, and the number of risk estimates, as well as the shapes of the underlying risk distributions, sums of upper bounds still provide useful information about the overall risk.

On the other hand, gross exaggeration of actual risks could lead to poor decisions being made with respect to the often limited resources available for general risk mitigation purposes. Thus, after establishing a so-called worst-case scenario, it is often desirable to also develop and analyse more realistic or 'normal' scenarios, so that the level of risk posed by a hazardous situation can be better bounded, by selecting 'best' or 'most likely' sets of assumptions for the risk assessment. But in deciding on what realistic assumptions are to be used in a risk assessment, it is imperative that the analyst chooses parameters that will, at worst, result in erring on the side of safety.

Lately, a number of investigators have been elaborating on a variety of ways and means of making risk assessments more realistic – rather than the dependence on wholesale compounded conservative assumptions (see, e.g., Anderson and Yuhas, 1996; Burmaster and von Stackelberg, 1991; Cullen, 1994; Maxim, in Paustenbach, 1988). Among other things, there is the need to undertake sensitivity analyses – including the use of multiple assumption sets that reflect a wider spectrum of exposure scenarios. This is important because regulations based on the so-called upper-bound estimate or worst-case scenario may address risks that are almost nonexistent and impractical. Indeed, risk assessments using extremely conservative biases do not provide risk managers with the quality information needed to formulate efficient and cost-effective management strategies. Also, using plausible upper-bound risk estimates or worst-case scenarios may lead to spending scarce and limited resources to regulate or control insignificant risks, whiles more serious risks are probably being ignored.

7.4 RECOGNITION OF UNCERTAINTY AS AN INTEGRAL COMPONENT OF RISK ASSESSMENT

A major difficulty in decision-making resides in the uncertainties of system characteristics for the situation at hand. Uncertainty is the lack of confidence in the estimate of a variable's magnitude or probability of occurrence. Engineering judgment becomes an important factor in problem-solving under uncertainty, and decision analysis provides a means of representing the uncertainties in a manner that allows informed discussion.

The presence of uncertainty means, in general, that the best outcome obtainable from an evaluation and/or analysis cannot be guaranteed. Nonetheless, as has been noted by Bean (1988), decisions ought to be made even in an uncertain setting, otherwise several aspects of environmental management actions could become completely paralyzed. In fact, there are inevitable uncertainties associated with risk estimates, but these uncertainties do not invalidate the use of risk estimates in the decision-making process. However, it is important to identify and define the confidence levels associated with the evaluation.

Depending on the specific level of detail of a risk assessment, the type of uncertainty that dominates at each stage of the analysis can be different (see Chapter 12). In view of the fact that risk assessment constitutes a very important part of the environmental management decision-making process, it is essential that all apparent sources of uncertainty are well documented.

The uncertainty can be characterized via sensitivity analysis and/or probability analysis techniques (USEPA, 1989). The technique selected depends on the availability of input data statistics. Sensitivity analyses require data on the range of values for each exposure factor in the scenario; probabilistic analyses require data on the range and probability function (or distribution) of each exposure factor within the scenario. Through such analyses, uncertainties can be assessed properly, and their effects on given decisions accounted for systematically. In this manner the risk associated with given decision alternatives may be delineated and then appropriate corrective measures taken accordingly.

Uncertainty analysis can indeed be performed qualitatively or quantitatively – with sensitivity analysis often being a useful adjunct to the uncertainty analysis. Sensitivity analysis entails the determination of how rapidly the output of an analysis changes with respect to variations in the input. Thus, in addition to presenting the best estimate, an evaluation will also *provide a range of likely estimates in the form of a sensitivity analysis*. In fact, a sensitivity analysis should generally become an integral part of any risk evaluation process. Further discussion of this topic appears later in Chapter 12 of this volume.

7.5 RISK ASSESSMENT VERSUS RISK MANAGEMENT

Risk assessment is generally conducted to aid risk management decisions, designed to minimize health and environmental risks to society. Whereas risk assessment focuses on evaluating the likelihood of adverse effects, risk management involves the selection of a course of action in response to an identified risk that is based on many other factors (e.g., social, legal, political, or economic) in addition to the risk assessment results. Depending on the problem situation, different degrees of detail may be required for the processes; however, the continuum of acute to chronic hazards and exposures should be fully investigated in a comprehensive assessment, so that the complete spectrum of risks can be defined for subsequent risk management decisions.

In fact, it has long been recognized that nothing is wholly safe or dangerous *per se*, but that the object involved, and the manner and conditions of use determine the degree of hazard or safety. Consequently, it may rightly be concluded that there is no escape from all risk no matter how remote, but that there only are choices among risks (Daniels, 1978). In that sense, risk assessment is usually designed to offer an opportunity to understand a system better by adding an orderliness and completeness to a problem evaluation. It generally embodies the heuristic approach of empirical learning that will provide a 'best knowledge' estimate of the relative importance of risks.

In any case, risk assessment has a usefulness only if it is used properly. The overall risk analysis process must also recognize the fact that hazard perceptions and risk thresholds – which can significantly impact the ultimate risk management decision – tend to be quite different in different parts of the world.

SUGGESTED FURTHER READING

Bromley, D.W. and K. Segerson (eds). 1992. *The Social Response to Environmental Risk: Policy Formulation in an Age of Uncertainty*. Kluwer Academic Publishers, Boston, MA.

CSA (Canadian Standards Association). 1991. *Risk Analysis Requirements and Guidelines*. Canadian Standards Association, Rexdale, Ontario, Canada, CAN/CSA-Q634-91.

DoE (Department of the Environment). 1995. *A Guide to Risk Assessment and Risk Management for Environmental Protection*. UK Department of the Environment, HMSO, London.

Henderson, M. 1987. *Living with Risk: The Choices, The Decisions*. The British Medical Association Guide, J. Wiley, New York.

Johnson, B.B. and V.T. Covello. 1987. *Social and Cultural Construction of Risk: Essays on Risk Selection and Perception*. Kluwer Academic Publishers, Norwell MA.

Lu, F.C. 1988. Acceptable daily intake: inception, evolution, and application. *Regulatory Toxicology and Pharmacology*, 8: 45–60.

NRC (National Research Council). 1996. *Understanding Risk: Informing Decisions in a Democratic Society*. National Academy Press, Washington, DC.

Ruckelshaus, W.D. 1985. Risk, science, and democracy. *Issues in Science and Technology*, Spring 1985: 19–38.

Schrader-Frechette, K.S. 1991. *Risk and Rationality*. University of California Press, Berkeley, CA.

Schwing, R.C. and W.A. Albers, Jr. (eds). 1980. *Societal Risk Assessment: How Safe is Safe Enough?* Plenum Press, New York.

Slovic, P. 1997. Public perception of risk. *Journal of Environmental Health*, 59(9): 22–24.

REFERENCES

Anderson, P.S. and A.I. Yuhas. 1996. Improving risk management by characterizing reality: a benefit of probabilistic risk assessment. *Human and Ecological Risk Assessment*, 2: 55–58.

Bean, M.C. 1988. Speaking of risk. *ASCE Civil Engineering*, 58(2): 59–61.

Berthouex, P.M. and L.C. Brown. 1994. *Statistics for Environmental Engineers*. Lewis Publishers/CRC Press, Boca Raton, FL.

Bogen, K.T. 1994. A note on compounded conservatism. *Risk Analysis*, 14: 379–381.

Burmaster, D.E. and R.H. Harris. 1993. The magnitude of compounding conservatisms in Superfund risk assessments. *Risk Analysis*, 13: 131–134.

Burmaster, D.E. and K. von Stackelberg. 1991. Using Monte Carlo simulations in public health risk assessments: estimating and presenting full distributions of risk. *Journal of Exposure Analysis and Environmental Epidemiology*, 1: 491–512.

Cassidy, K. 1996. Approaches to the risk assessment and control of major industrial chemical and related hazards in the United Kingdom. *International Journal of Environment and Pollution*, 6(4–6): 361–387.

Cogliano, V.J. 1997. Plausible upper bounds: are their sums plausible? *Risk Analysis*, 17(1): 77–84.

Cohrssen, J.J. and V.T. Covello. 1989. *Risk Analysis: A Guide to Principles and Methods for Analyzing Health and Environmental Risks*. National Technical Information Service (NTIS), US Dept. of Commerce, Springfield, VA.

Covello, V.T., J. Menkes and J. Mumpower (eds). 1986. *Risk Evaluation and Management*. Contemporary Issues in Risk Analysis, Vol. 1. Plenum Press, New York.

Cullen, A.C. 1994. Measures of compounding conservatism in probabilistic risk assessment. *Risk Analysis*, 14(4): 389–393.

Daniels, S.L. 1978. Environmental evaluation and regulatory assessment of industrial chemicals. Presented at the *51st Annual Conference, Water Pollution Control Federation* (WPCF), Anaheim, CA.

Earle, T.C. and G. Cvetkovich. 1997. Culture, cosmopolitanism, and risk management. *Risk Analysis*, 17(1): 55–65.

Fischhoff, B., S. Lichtenstein, P. Slovic, S. Derby, and R. Keeney. 1981. *Acceptable Risk*. Cambridge University Press, New York.
Freund, J.E. and R.E. Walpole (eds). 1987. *Mathematical Statistics*. Prentice-Hall, Englewood Cliffs, NJ.
Kocher, D.C. and F.O. Hoffman. 1991. Regulating environmental carcinogens: where do we draw the line? *Environmental Science and Technology*, 25: 1986–1989.
Larsen, R.J. and M.L. Marx. 1985. *An Introduction to Probability and Its Applications*. Prentice-Hall, Englewood Cliffs, NJ.
Lowrance, W.W. 1976. *Of Acceptable Risk: Science and the Determination of Safety*. William Kaufman, Los Altos, CA.
Massmann, J. and R.A. Freeze. 1987. Groundwater contamination from waste management sites: the interaction between risk-based engineering design and regulatory policy 1. Methodology 2. Results. *Water Resources Research*, 23(2): 351–380.
Miller, I. and J.E. Freund. 1985. *Probability and Statistics for Engineers*, 3rd edition. Prentice-Hall, Englewood Cliffs, NJ.
Paustenbach, D.J. (ed.). 1988. *The Risk Assessment of Environmental Hazards: A Textbook of Case Studies*. J. Wiley, New York.
Richardson, M.L. (ed.). 1986. *Toxic Hazard Assessment of Chemicals*. Royal Society of Chemistry, London, UK.
Rowe, W.D. 1983. *Evaluation Methods for Environmental Standards*. CRC Press, Boca Raton, FL.
Sharp, V.F. 1979. *Statistics for the Social Sciences*. Little, Brown & Co., Boston, MA.
Siddall, J.N. 1983. *Probabilistic Engineering Design: Principles and Applications*. Marcel Dekker, New York.
Suter II, G.W. 1993. *Ecological Risk Assessment*. Lewis Publishers, Boca Raton, FL.
Travis, C.C., S.A. Richter, E.A.C. Crouch, R. Wilson, and E.D. Klema. 1987. Cancer risk management. *Environmental Science and Technology*, 21: 415–420.
USEPA (US Environmental Protection Agency). 1989. *Exposure Factors Handbook*. Washington, DC. EPA/600/8-89/043.
van Leeuwen, C.J. and J.L.M. Hermens (eds). 1995. *Risk Assessment of Chemicals: An Introduction*. Kluwer Academic Publishers, Dordrecht, The Netherlands.
Whipple, C. 1987. *De Minimis Risk*. Contemporary Issues in Risk Analysis, Vol. 2. Plenum Press, New York.
Wonnacott, T.H. and R.J. Wonnacott. 1972. *Introductory Statistics*, 2nd edition. J. Wiley, New York.

PART III

PRINCIPAL ELEMENTS OF A RISK ASSESSMENT

This part of the book reviews and develops methods for the analysis of the several practical components of risk assessment, as may be applied to environmental management programs. It comprises the following specific chapters:

- Chapter 8, *Determination of Contaminant Fate and Behavior in the Environment*, describes some of the most important properties, processes, and parameters that affect the fate and transport of environmental contaminants. Also included here are selected environmental fate algorithms for estimating concentrations of chemicals released into the environment.
- Chapter 9, *Hazard Identification, Data Collection, and Data Evaluation*, gives a brief overview of the major sources of hazards that may contribute to a variety of environmental contamination problems and possible risk situations. It further presents strategies for the design of adequate data collection and evaluation programs during the investigation of an environmental contamination problem.
- Chapter 10, *Design of Conceptual Models and Exposure Analysis*, first addresses the design and use of case-specific conceptual models in the evaluation of environmental contamination problem situations. It then goes on to present methodologies for the estimation of potential receptor exposures to chemical contaminants present in various environmental matrices or compartments.
- Chapter 11, *The Toxicology of Environmental Contaminants and Hazard Effects Determination*, consists of a discussion of methods used in the evaluation of the toxicity of a variety of chemical constituents associated with environmental contamination problems.
- Chapter 12, *Risk Characterization and Uncertainty Analysis*, alludes to the major risk estimation methods that are usually employed in the evaluation of a variety of environmental contamination problems. It further identifies several relevant methods for dealing with the inevitable uncertainties that surround the risk evaluation process employed in environmental management programs.

Chapter Eight

Determination of Contaminant Fate and Behavior in the Environment

Once a chemical has been determined to present a potential health or environmental hazard, then the first concern is one of the likelihood for and degree of exposure. The fate of chemical substances released into the environment forms a very important basis for evaluating the exposure of potential receptors to toxic chemicals.

Contaminants released into the environments are controlled by a complex set of processes – consisting of transport, transformation, degradation and decay, cross-media transfers, and/or biological uptake and bioaccumulation. Environmental fate and transport analyses offer a way to assess the movement of chemicals between environmental compartments, further to the prediction of the long-term fate of such chemicals in the environment.

In general, as pollutants are released into various environmental media, several factors contribute to their migration from one environmental matrix into another, or their phase change from one physical state into another. The processes and phenomena that affect the fate and transport of contaminants should therefore be recognized as an important part of any environmental management program that is designed to address an environmental contamination problem. The relevant phenomena involved in the fate and behavior of environmental contaminants, together with the important factors affecting the processes involved are annotated in this chapter.

8.1 THE CROSS-MEDIA TRANSFER OF CONTAMINANTS BETWEEN ENVIRONMENTAL COMPARTMENTS

Chemicals present in one environmental matrix may be affected by several complex processes and phenomena, facilitating transfers into other media. The potential for cross-media transfers of pollutants from the soil medium into other media is particularly significant. Contaminated soil is indeed the main source of chemical repository for most environmental pollutants. Consequently, soils often become the principal focus of attention in the investigation of environmental contamination problems. This is reasonable because soils at such locales not only serve as a medium of exposure to potential receptors, but also serve as a long-term reservoir for contaminants to be released into other media. In fact, the soil media is by no means an inert repository, since there is an active interchange of chemicals between soils and water, air, and biota (Figure 8.1) (Asante-Duah, 1996; Brooks et al., 1995), with the main driving forces in contaminant transport in soils being advective and diffusive in nature (Yong et al., 1992).

The movement of contaminants through soils is generally very complex, with some pollutants moving rapidly while others move rather slowly from one environmental matrix into another. In general, the affinity that contaminants have for soil affects their mobility by retarding transport. For instance, hydrophobic or cationic contaminants that are migrating in solution are subject to retardation effects. The hydrophobicity of a contaminant can indeed greatly affect its fate, which explains some of the different rates of contaminant migration that occur in the subsurface environment. Also, the phenomenon of adsorption is a major reason why the sediment zones of surface water systems may become highly contaminated with specific organic and inorganic chemicals. On the other hand, chemical constituents having a moderate to high degree of mobility can leach from soils into groundwater; volatile constituents may contribute to subsurface gas in the vadose zone, and also possible releases into the atmosphere. Conversely, it is possible for cross-media transport of constituents from other media into soils to take place; for example, chemical constituents may be transported into the soil matrix via deposition of suspended particulates from the atmosphere, and also through releases of subsurface gas.

In general, the distribution of organic chemicals among environmental compartments can be defined in terms of such simple equilibrium expressions (Swann and Eschenroeder, 1983) as illustrated in Figure 8.2 – where the K_w, K_{oc} and *BCF* symbols refer to partitioning coefficients that are elaborated further in Section 8.2. The general assumption is that all environmental compartments are well mixed in order to achieve equilibrium between them.

The partitioning of inorganic chemicals is somewhat different from organic constituents. Metals generally exhibit relatively low mobilities in soils (Evans, 1989). Also, relatively insignificant partitioning would be expected among environmental compartments for metals. Rather, the inorganics will tend to adsorb onto soils (which may become airborne or be transported by surface erosion) and sediments (that may be transported in water).

8.1.1 Phase Distribution of Environmental Contaminants

Environmental contamination may exist in different physical states, and may generally be present in a variety of environmental matrices. For instance, when fluids that are immiscible with water, called nonaqueous phase liquids (NAPLs), are released at a site and then enter the subsurface environment, they tend to exist as distinct fluids that flow separately from the water phase. Fluids less dense than water, or light NAPLs (LNAPLs) migrate downward through the vadose (unsaturated) zone, but tend to form lenses that 'float' on top of an aquifer upon reaching the water table. Typical examples of LNAPLs include hydrocarbon fuels such as gasoline, heating oil, kerosene, jet fuel, and aviation gas; such LNAPLs will pool and spread as a floating free product layer on top of the water table if they are

Determination of contaminant fate and behavior in the environment 105

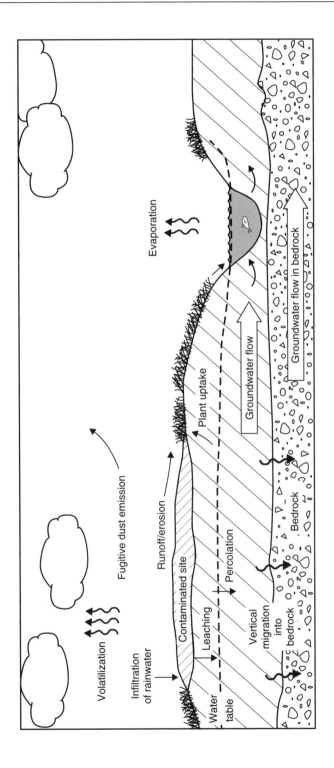

Figure 8.1 Illustrative sketch of the cross-media transfers of environmental contaminants. (Source: Asante-Duah, 1996)

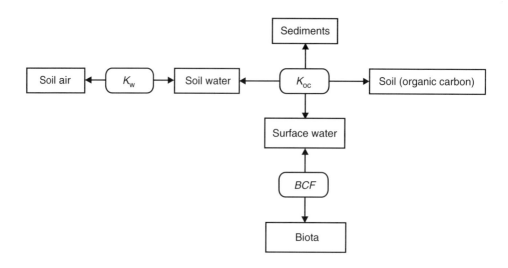

Figure 8.2 Major components of the partitioning of contaminants between environmental compartments

released to the subsurface in sufficient quantities. Denser-than-water NAPLs, or dense NAPLs (DNAPLs) will tend to 'sink' into the aquifer. Typical examples of DNAPLs include chlorinated hydrocarbons (e.g., trichloroethylene [TCE], tetrachloroethene [PERC], chlorophenols, chlorobenzenes, and PCBs), coal tar wastes, creosote-based wood-treating oils, some pesticides, etc.; such DNAPLs can pass across the water table and may be found at greater depths within the saturated zone of a groundwater aquifer. Contact with groundwater or infiltrating recharge water causes some of the chemical constituents of NAPLs to dissolve, resulting in aquifer contamination, further to any trail of contamination left in the soil matrix. In general, contamination present at such a contaminated site may typically show up in the following phases:

- Adsorbed contamination (onto solid phase matter or soils).
- Vapor phase contamination (present in the vadose zone due to volatilization into soil gas).
- Dissolved contamination (in water, present in both the unsaturated and saturated soil zones).
- Free product (as residual and mobile immiscible fluids, e.g., as LNAPL floating on the surface of the water table, or as DNAPL that sinks deep into the groundwater zone, or as NAPL persisting in the soil pore spaces).

This contamination system that exist in the soils, water, vapor phase, or the NAPLs tends towards a state of equilibrium, such that the chemical potential or fugacity is equal in all the phases that co-exist in the environment. Changes in equilibrium between the phases can be occurring on a continuing basis as a result of several extraneous factors that affect the concentration gradients.

8.2 IMPORTANT FATE AND TRANSPORT PROPERTIES, PROCESSES, AND PARAMETERS OF ENVIRONMENTAL CONTAMINANTS

As pollutants are released into various environmental media, several factors contribute to their migration and transport. For example, in the groundwater system, the solutes in the porous media will move with the mean velocity of the solvent by advective mechanism. In addition, other mechanisms governing the spread of contaminants include hydraulic dispersion and molecular diffusion. Furthermore, the transport and concentration of the solute(s) are affected by reversible ion exchange with soil grains; the chemical degeneration with other constituents; fluid compression and expansion; and in the case of radioactive materials, by the radioactive decay.

Examination of a contaminant's physical and chemical properties will often allow an estimation of its degree of environmental partitioning, migration and/or attenuation. Several important physical and chemical properties, processes, and parameters affecting the environmental fate and/or cross-media transfers of environmental contaminants are briefly annotated below, with further detailed discussions offered in the literature elsewhere (e.g., CDHS, 1986; Evans, 1989; Hemond and Fechner, 1994; Lyman et al., 1990; Samiullah, 1990; Swann and Eschenroeder, 1983; USEPA, 1985a, 1989e; Yong et al., 1992).

8.2.1 Physical State

Contaminants present in the environment may exist in any or all of three major physical states – viz.: solid, liquid, and vapor states (e.g., solids adsorbed onto soils, free product or dissolved contaminants, and vapor phase in the soil vadose zones of a contaminated site). Contaminants in the solid phase are generally less susceptible to release and migration than the fluids. However, certain processes (such as leaching, erosion and/or runoff, and physical transport of chemical constituents) can act as significant release mechanisms, irrespective of the physical state of a contaminant.

8.2.2 Water Solubility

The solubility of a chemical in water is the maximum amount of the chemical that will dissolve in pure water at a specified temperature. Solubility is an important factor affecting a chemical constituent's release and subsequent migration and fate in the surface water and groundwater environments. In fact, among the various parameters affecting the fate and transport of organic chemicals in the environment, water solubility is one of the most important, especially with regard to hydrophilic compounds.

Typically, solubility affects mobility, leachability, availability for biodegradation, and the ultimate fate of a given constituent. In general, highly soluble chemicals are easily and quickly distributed by the hydrologic system. Such chemicals tend to have relatively low adsorption coefficients for soils and sediments, and also relatively low bioconcentration factors in aquatic biota. Furthermore, they tend to be more readily biodegradable. Substances which are more soluble are more likely to desorb from soils and less likely to volatilize from water.

In combination with vapor pressure, water solubility yields a chemical's Henry's Law constant – which determines whether or not the chemical will volatilize from water into air. In combination with the chemical's solubility in fats (obtained from the octanol–water partition coefficient), water solubility predicts whether or not a chemical will tend to concentrate in living organisms and also whether a soil contaminant will remain bound to soil or will leach from soil to contaminate groundwater or surface water bodies.

8.2.3 Diffusion

Diffusivity describes the movement of a molecule in a liquid or gas medium as a result of differences in concentrations. Diffusive processes create mass spreading due to molecular diffusion, in response to concentration gradients. The higher the diffusivity, the more likely a chemical is to move in response to concentration gradients. Thus, diffusion coefficients are used to describe the movement of a molecule in a liquid or gas medium as a result of differences in concentration; they can also be used to calculate the dispersive component of chemical transport.

8.2.4 Dispersion

Dispersive processes create mass mixing due to system heterogeneities (e.g., velocity variations). Consequently, for example, as a pulse of contaminant plume migrates through a soil matrix, the peaks in concentration are decreased by spreading. Dispersion is indeed an important attenuation mechanism that results in the dilution of a contaminant; the degree of spreading or dilution is proportional to the size of the dispersion coefficients.

8.2.5 Volatilization

Volatilization is the process by which a chemical compound evaporates from one environmental compartment into the vapor phase. The volatilization of chemicals is indeed a very important mass-transfer process. The transfer process from the source (e.g., water body, sediments, or soils) to the atmosphere is dependent on the physical and chemical properties of the compound in question, the presence of other pollutants, the physical properties of the source media, and the atmospheric conditions.

Knowledge of volatilization rates is important in the determination of the amount of chemicals entering the atmosphere and the change of pollutant concentrations in the source media. Volatility is therefore considered a very important parameter for chemical hazard assessments. Several important measures of a chemical's volatility or volatilization rate are enumerated below.

- *Henry's Law constant.* Henry's Law constant (H) provides a measure of the extent of chemical partitioning between air and water at equilibrium. It indicates the relative tendency of a constituent to volatilize from aqueous solution into the atmosphere, based on the competition between its vapor pressure and water solubility. This parameter is important to determining the potential for cross-media transport into air.

    Contaminants with low Henry's Law constant values will tend to favor the aqueous phase and will therefore volatilize into the atmosphere more slowly than constituents with high values. As a general guideline: H values in the range of 10^{-7} to 10^{-5} (atm-m^3/mol) represent low volatilization; H between 10^{-5} and 10^{-3} (atm-m^3/mol) means volatilization is not rapid but possibly significant; and $H > 10^{-3}$ (atm-m^3/mol) implies volatilization is rapid. The variation in H between chemicals is indeed extensive.

- *Vapor pressure.* Vapor pressure is the pressure exerted by a chemical vapor in equilibrium with its solid or liquid form at any given temperature. It is a relative measure of the volatility of a chemical in its pure state, and is an important determinant of the rate of volatilization. The vapor pressure of a chemical can be used to calculate the rate of volatilization of a pure substance from a surface, or to estimate a Henry's Law constant for chemicals with low water solubility.

    In general, the higher the vapor pressure, the more volatile a chemical compound and therefore the more likely the chemical is to exist in significant quantities in a gaseous state. Thus, constituents with high vapor pressure are more likely to migrate from soil and groundwater, to be transported into air.

- *Boiling point.* Boiling point (BP) is the temperature at which the vapor pressure of a liquid is equal to the atmospheric pressure on the liquid. At this temperature, a substance transforms from the liquid into a vapor phase.

    Besides being an indicator of the physical state of a chemical, the BP also provides an indication of its volatility. Other physical properties, such as critical temperature and latent heat (or enthalpy) of vaporization may be predicted by use of a chemical's normal BP as an input.

8.2.6 Partitioning and the Partition Coefficients

The partitioning of a chemical between several phases within a variety of environmental matrices is considered a very important fate and behavior property for contaminant migration in the environment. The partition coefficient is a measure of the distribution of a given compound in two phases and is expressed as a concentration ratio. Several important measures of the partitioning phenomena are enumerated below.

- *Water/air partition coefficient.* The water/air partition coefficient (K_w) relates the distribution of a chemical between water and air. It consists of an expression that is equivalent to the reciprocal of Henry's Law constant (H), i.e.,

$$K_w = \frac{C_{water}}{C_{air}} = \frac{1}{H}$$

  where C_{air} is the concentration of the chemical in air (expressed in units of μg/L) and C_{water} is the concentration of the chemical in water (in μg/L).

- *Octanol/water partition coefficient.* The octanol/water partition coefficient (K_{ow}) is defined as the ratio of a chemical's concentration in the octanol phase (organic) to its concentration in the aqueous phase of a two-phase octanol/water system, represented by:

$$K_{ow} = \frac{\text{concentration in octanol phase}}{\text{concentration in aqueous phase}}$$

  This dimensionless parameter provides a measure of the extent of chemical partitioning between water and octanol at equilibrium. It has become a particularly important parameter in studies of the environmental fate of organic chemicals.

  K_{ow} can be used to predict the magnitude of an organic constituent's tendency to partition between the aqueous and organic phases of a two-phase system, such as surface water and aquatic organisms. The higher the value of K_{ow}, the greater the tendency of an organic constituent to adsorb to soil or waste matrices containing appreciable organic carbon or to accumulate in biota. It has been found to be related to water solubility, soil/sediment adsorption coefficients, and bioaccumulation factors for aquatic life. In fact, high K_{ow} values are generally indicative of a chemical's ability to accumulate in fatty tissues and therefore bioaccumulate in the foodchain. It is also a key variable in the estimation of skin permeability for chemical constituents.

  In general, chemicals with low K_{ow} (<10) values may be considered relatively hydrophilic, whereas those with high K_{ow} (>10 000) values are very hydrophobic. Thus, the greater the K_{ow}, the more likely a chemical is to partition to octanol than to remain in water. The hydrophilic chemicals tend to have high water solubilities, small soil or sediment adsorption coefficients, and small bioaccumulation factors for aquatic life.

- *Organic carbon adsorption coefficient.* The sorption characteristics of a chemical may be normalized to obtain a sorption constant based on organic carbon which is essentially independent of any soil material. The organic carbon adsorption coefficient (K_{oc}) provides a measure of the extent of partitioning of a chemical constituent between soil or sediment organic carbon and water at equilibrium. Also called the organic carbon partition coefficient, K_{oc} is a measure of the tendency for organics to be adsorbed by soil and sediment, and is expressed by:

$$K_{oc} \text{ [mL/g]} = \frac{\mu g \text{ chemical adsorbed per g weight of soil or sediment organic carbon}}{\mu g \text{ chemical dissolved per mL of water}}$$

The extent to which an organic constituent partitions between the solid and solution phases of a saturated or unsaturated soil, or between runoff water and sediment, is determined by the physical and chemical properties of both the constituent and the soil (or sediment). The K_{oc} is chemical-specific and largely independent of the soil or sediment properties. The tendency of a constituent to be adsorbed to soil is, however, dependent on its properties and also on the organic carbon content of the soil or sediment.

Values of K_{oc} typically range from 1 to 10^7; the higher the K_{oc}, the more likely a chemical is to bind to soil or sediment than to remain in water. That is, constituents with a high K_{oc} have a tendency to partition to the soil or sediment. In fact, this value is also a measure of the hydrophobicity of a chemical; the more highly sorbed, the more hydrophobic (or the less hydrophilic) a substance.

- *Soil–water partition coefficient.* The mobility of contaminants in soil depends not only on properties related to the physical structure of the soil, but also on the extent to which the soil material will retain, or adsorb, the pollutant constituents. The extent to which a constituent is adsorbed depends on the physico-chemical properties of the chemical constituent and of the soil. The sorptive capacity must therefore be determined with reference to a particular constituent and soil pair.

The soil–water partition coefficient (K_d), also called the soil/water distribution coefficient, is generally used to quantify soil sorption. K_d is the ratio of the adsorbed contaminant concentration to the dissolved concentration at equilibrium, and for most environmental concentrations it can be approximated by:

$$K_d \text{ [mL/g]} = \frac{\text{concentration of adsorbed chemical in soil } (\mu g \text{ chemical per g soil})}{\text{concentration of chemical in solution in water } (\mu g \text{ chemical per mL water})}$$

K_d provides a soil- or sediment-specific measure of the extent of chemical partitioning between soil or sediment and water, unadjusted for dependence on organic carbon. On this basis, K_d describes the sorptive capacity of the soil and allows estimation of the concentration in one medium, given the concentration in the adjoining medium. For hydrophobic contaminants: $K_d = f_{oc} K_{oc}$, where f_{oc} is the fraction of organic carbon in the soil.

In general, the higher the value of K_d, the less mobile is a contaminant; this is because, for large values of K_d, most of the chemical remains stationary and attached to soil particles due to the high degree of sorption. Thus, the higher the K_d the more likely a chemical is to bind to soil or sediment than to remain

in water. Invariably, the distribution of a chemical between water and adjoining soil or sediment may be described by this equilibrium expression that relates the amount of chemical sorbed to soil or sediment to the amount in water at equilibrium.
- *Bioconcentration factor.* The bioconcentration factor (*BCF*) is the ratio of the concentration of a chemical constituent in an organism or whole body (e.g., a fish) or specific tissue (e.g., fat) to the concentration in its surrounding medium (e.g., water) at equilibrium, given by:

$$BCF = \frac{(\text{concentration in biota})}{(\text{concentration in surrounding medium})}$$

$$= \frac{[(\mu g \text{ chemical per g biota such as fish})]}{[(\mu g \text{ chemical per mL medium such as water})]}$$

The *BCF* indicates the degree to which a chemical residue may accumulate in aquatic organisms, coincident with ambient concentrations of the chemical in water; it is a measure of the tendency of a chemical in water to accumulate in the tissue of an organism. In this regard, the concentration of the chemical in the edible portion of the organism's tissue can be estimated by multiplying the concentration of the chemical in surface water by the fish or biota *BCF* for that chemical. Thus, the average concentration in fish or biota is given by:

$$C_{\text{fish-biota}} (\mu g/kg) = C_{\text{water}} (\mu g/L) \times BCF$$

where C_{water} is the concentration in water. This parameter is indeed an important determinant for human exposure to chemicals via ingestion of aquatic foods. The partitioning of a chemical between water and biota (e.g., fish) also gives a measure of the hydrophobicity of the chemical.

Values of *BCF* typically range from 1 to over 10^6. In general, constituents exhibiting a *BCF* greater than unity are potentially bioaccumulative, but those exhibiting a *BCF* greater than 100 cause the greatest concern (USEPA, 1987). Ranges of *BCF*s for various constituents and organisms can be used to predict the potential for bioaccumulation, and therefore to determine whether sampling of the biota is a necessary part of an environmental characterization program. The accumulation of chemicals in aquatic organisms is indeed of increased concern as a significant source of environmental and health hazard.

8.2.7 Sorption and the Retardation Factor

Sorption, which collectively accounts for both adsorption and absorption, is the partitioning of a chemical constituent between the solution and solid phases. In this partitioning process, molecules of the dissolved constituents leave the liquid phase and attach to the solid phase; this partitioning continues until a state of equilibrium is reached. The practical result of the partitioning process is a phenomenon called retardation.

Retardation is the chemical-specific, dynamic process of adsorption to, and desorption from solid materials. It is typically characterized by a parameter called the retardation factor or retardation coefficient. In the assessment of the environmental fate and transport properties of chemical contaminants, reversible equilibrium and controlled sorption may be simulated by the use of the retardation factor or coefficient.

- *Retardation factor.* The retardation factor, R_f, is defined as the ratio of (C_{mobile} + C_{sorbed}) to C_{mobile}, where C_{mobile} and C_{sorbed} are the mobile and sorbed chemical concentrations, respectively. Thus,

$$R_f = 1 + \frac{C_{sorbed}}{C_{mobile}}$$

R_f can be calculated for a contaminant as a function of the chemical's soil–water partition coefficient (K_d), and also the bulk density (β) and porosity (n) of the medium through which the contaminant is moving. Typically, the retardation factors are calculated for linear sorption, in accordance with the following relationship:

$$R_f = 1 + \frac{\beta K_d}{n} = \left[1 + \frac{\beta K_{oc} f_{oc}}{n}\right]$$

where $K_d = K_{oc} \times f_{oc}$, and f_{oc} is the organic carbon fraction.

The velocity of a contaminant is one of the most important variables in any groundwater quality modeling study. Sorption affects the solute seepage velocity through retardation, which is a function of R_f. Estimating R_f is therefore very important if solute transport is to be adequately represented. In the aquifer system, the retardation factor gives a measure of how fast a compound moves relative to groundwater (Hemond and Fechner, 1994; Nyer, 1993). Defined in terms of groundwater and solute concentrations, therefore,

$$R_f = \frac{\text{groundwater velocity } [\nu]}{\text{solute velocity } [\nu^*]}$$

For example, a retardation factor of two indicates that the specific compound is traveling at one-half the groundwater flow rate, and a retardation factor of five means that a plume of the dissolved compound will advance only one-fifth as fast as the groundwater parcel. This will usually become a very important parameter in the design of groundwater remediation systems. In particular, sorption (and retardation) can have major effects on pump-and-treat cleanup times and volumes of water to be removed from a contaminated aquifer system.

- *Sorption.* Under equilibrium conditions, a sorbing solute will partition between the liquid and solid phases according to the value of R_f. The fraction of the total contaminant mass contained in an aquifer which is dissolved in the solution phase, $F_{\text{dissolved}}$, and the sorbed fraction, F_{sorbed}, can be calculated as follows:

$$F_{\text{dissolved}} = \frac{1}{R_f}$$

$$F_{\text{sorbed}} = 1 - \left[\frac{1}{R_f}\right]$$

In general, if a compound is strongly adsorbed, then it also means this particular compound will be highly retarded.

8.2.8 Degradation

Degradation, whether biological, physical or chemical, is often reported in the literature as a half-life, which is generally measured in days. It is usually expressed as the time it takes for one-half of a given quantity of a compound to be degraded. Several important measures of the degradation phenomena are enumerated below.

- *Chemical half-lives.* Half-lives are used as measures of persistence, since they indicate how long a chemical will remain in various environmental media; long half-lives (e.g., greater than a month or a year) are characteristic of persistent constituents. Media-specific half-lives provide a relative measure of the persistence of a chemical in a given medium, although actual values can vary greatly depending on case-specific conditions. For example, the absence of certain microorganisms at a site, or the number of microorganisms, can influence the rate of biodegradation, and therefore, the half-life for specific compounds. As such, half-life values should be used only as a general indication of a chemical's persistence in the environment. In general, however, the higher the half-life value, the more persistent a chemical is likely to be.
- *Biodegradation.* Biodegradation is one of the most important environmental processes affecting the breakdown of organic compounds. It results from the enzyme-catalyzed transformation of organic constituents, primarily by microorganisms. As a result of biodegradation, the ultimate fate of a constituent introduced into several environmental systems (e.g., soil, water, etc.) may be any compound other than the parent compound that was originally released into the environment. Biodegradation potential should therefore be carefully evaluated in the design of environmental monitoring programs – in particular for contaminated site assessment programs. It is noteworthy that biological degradation may also initiate other chemical reactions, such as oxygen depletion in microbial degradation processes, creating anaerobic conditions and the initiation of redox-potential-related reactions.

- *Photolysis.* Photolysis (or photodegradation) can be an important dissipative mechanism for specific chemical constituents in the environment. Similar to biodegradation, photolysis may cause the ultimate fate of a constituent introduced into an environmental system (e.g., surface water, soil, etc.) to be different from the constituent originally released. Hence, photodegradation potential should be carefully evaluated in designing sampling and analysis, as well as environmental monitoring programs.
- *Chemical degradation.* Similar to photodegradation and biodegradation, chemical degradation – primarily through hydrolysis and oxidation/reduction (redox) reactions – can also act to change chemical constituent species from what the parent compound used to be when it was first introduced into the environment. For instance, oxidation may occur as a result of chemical oxidants being formed during photochemical processes in natural waters. Similarly, reduction of constituents may take place in some surface water environments (primarily those with low oxygen levels). Hydrolysis of organics usually results in the introduction of a hydroxyl group (–OH) into a constituent structure; hydrated metal ions (particularly those with a valence ≥ 3) tend to form ions in aqueous solution, thereby enhancing species solubility.

8.3 THE MODELING OF CONTAMINANT MIGRATION IN A CONTAMINANT FATE AND TRANSPORT ASSESSMENT

Contaminants entering the environment tend to be partitioned or distributed across various environmental compartments. A good prediction of contaminant concentrations in the various environmental media is essential to adequately characterize environmental contamination problems, the results of which can also be used to support risk assessment and environmental management decisions.

Mathematical algorithms are typically used to predict the potential for contaminants to migrate from an environmental compartment into potential receptor locations or environmental compliance boundaries. For example, relevant exposure point concentrations associated with an environmental contamination problem can be so determined for the potentially affected populations under the stipulated set of exposure scenarios. If the transport of compounds associated with this situation is under steady-state conditions, monitoring data are generally adequate to determine potential exposure concentrations. If there are no data available, or if conditions are transient (such as pertains to a migrating plume in groundwater), models are better used to predict exposure concentrations. Many factors – including the fate and transport properties of the chemicals of concern – must be considered in the model selection process. In any case, *in lieu* of an established trend in historical data indicating the contrary, a potential environmental contamination problem may be considered to be in steady-state with its surroundings.

Overall, mathematical models often serve as valuable tools for evaluating the behavior and fate of chemical constituents in various environmental media. The transport and fate of contaminants can be predicted through the use of various methods – ranging from simple mass balance and analytical procedures to multi-dimensional numerical solution of coupled differential equations. Several models of possible interest in environmental management programs are listed below in Section 8.3.3; a broader selection of environmental models, together with model selection criteria and limitations, are discussed elsewhere in the literature of exposure modeling (e.g., CCME, 1994; CDHS, 1986; Clark, 1996; Feenstra et al., 1991; Ghadiri and Rose, 1992; Gordon, 1985; Haith, 1980; Honeycutt and Schabacker, 1994; Johnson and Ettinger, 1991; Jury et al., 1984; Mulkey, 1984; NRC, 1989; Schnoor, 1996; USEPA, 1985b, 1987, 1988a, 1988b; Williams et al., 1996). The effective use of models in contaminant fate and behavior assessment depends greatly on the selection of models most suitable for this purpose.

In general, the appropriateness of a particular model depends on the characteristics of the particular problem. Thus, the screening of models should be tied to the project goals. Indeed, the wrong choice of models could result in the generation of false information, with consequential negative impacts on any decision made thereof. Ultimately, the choice of appropriate fate and transport models that will give reasonable indications of the contaminant behavior will help produce a realistic conceptual representation of the problem; this is important to the characterization of any environmental contamination problem, which in turn is a prerequisite to developing reliable environmental management policies and strategies.

8.3.1 Utility and Application of Environmental Models

Environmental models are typically designed to serve a variety of purposes, most importantly the following (Schnoor, 1996):

- To gain better understanding of the fate and transport of chemicals existing in, or to be introduced into, the environment.
- To determine the temporal and spatial distributions of chemical exposure concentrations at potential receptor locations.
- To predict future consequences of exposure under various chemical loading or release conditions, exposure scenarios, or management action alternatives.

One of the major benefits associated with the use of mathematical models in environmental management programs relates to the fact that environmental concentrations useful for exposure assessment and risk characterization can be estimated for several locations and time-periods of interest. Since field data frequently are limited and insufficient to accurately and completely characterize an environmental contamination problem, models can be particularly useful for studying spatial and temporal variabilities, together with potential uncertainties. In addition, sensitivity analyses can be performed by varying specific parameters and then using models to explore the ramifications (as reflected by changes in the model outputs). Models can indeed be used for several purposes in the study of environmental contamination problems.

In general, models usually simulate the response of a simplified version of a complex system. As such, their results are imperfect. Nonetheless, when used in a technically responsible manner, they can provide a very useful basis for making technically sound decisions about an environmental contamination problem. They are particularly useful where several alternative scenarios are to be compared. In such cases, all the alternatives are compared on a similar basis; thus, whereas the numerical results of any single alternative may not be exact, the comparative results of showing that one alternative is superior to others will usually be valid. Ultimately, the effective use of models in corrective action assessment and risk management programs depends greatly on the selection of models most suitable for this purpose.

8.3.2 Model Selection

Numerous model classification systems with different complexities exist in practice – broadly categorized as analytical or numerical models, depending on the degree of mathematical sophistication involved in their formulation. Analytical models are models with simplifying underlying assumptions, often sufficient and appropriate for well-defined systems for which extensive data are available, and/or for which the limiting assumptions are valid. Whereas analytical models may suffice for some situations, numerical models (with more stringent underlying assumptions) may be required for more complex configurations and complicated systems.

The choice of which model to use for specific applications is subject to numerous factors. Thus, simply choosing a more complicated model over a simple one will not necessarily ensure a better solution in all situations. In fact, since a model is a mathematical representation of a complex system, some degree of mathematical simplification usually must be made about the system being modeled. Data limitations must be weighted appropriately, since it usually is not possible to obtain all of the input parameters due to the complexity (e.g., anisotropy and non-homogeneity) of natural systems.

Ultimately, the type of model selected will be dependent on the overall goal of the the assessment, the complexity of the problem, the type of contaminants of concern, the nature of the impacted and threatened media that are being considered in the investigation, and the type of corrective actions being considered. General guidance for the effective selection of models in corrective action assessments and risk management decisions is provided in the literature elsewhere (e.g., CCME, 1994; CDHS, 1990; Clark, 1996; Cowherd et al., 1985; DOE, 1987; NRC, 1989; Schnoor, 1996; USEPA, 1987, 1988a, 1988b; Yong et al., 1992; Zirschy and Harris, 1986). In several environmental assessment situations, a 'ballpark' or 'order-of-magnitude' (i.e., a rough approximation) estimate of the contaminant behavior and fate is usually all that is required for most analyses – in which case simple analytical models usually will suffice.

8.3.3 Selected Environmental Models Potentially Applicable to Risk Assessment Studies and Environmental Management Programs

Table 8.1 consists of selected models that may be applied to some aspect of risk assessment and environmental management problems. The choice of one particular

Table 8.1 Listing of selected environmental models potentially applicable to risk assessment and environmental management problems

Environmental compartment	Model name	Model type and description	Model uses/applications	Sources of model information and/or developer
Air	BOXMOD (The BOX MODel)	BOXMOD is an interactive steady-state, simple atmospheric areal source box model for screening chemicals. It is applicable to regions containing many diffuse emission sources within its boundaries, such as in an urban area	BOXMOD calculates a single annual average concentration applicable to the entire region based on a uniform areal emission rate. It is used for detailed screening assessments of contaminated areas	US EPA's Office of Air Quality Planning and Standards (OAQPS), Research Triangle Park, NC
Air	ISCLT (Industrial Source Complex Long-Term model)	ISCLT is a long-term sector-averaged environmental model that uses statistical wind summaries to calculate annual ground-level concentrations or dispersion values	ISCLT is an air model that calculates annual ground-level concentrations or deposition values, and then estimates risk and exposure level using these values. It is used for modeling long-term air exposures associated with point and areal sources of air emissions	US EPA's Office of Air Quality Planning and Standards (OAQPS), Research Triangle Park, NC
Air	ISCST (Industrial Source Complex Short-Term model)	ISCST is a short-term model that uses a finite line source approach to model areal sources. Each square areal source is modeled as a single line segment oriented normal to the wind direction. The model does not accurately account for source-receptor geometry	The ISCST algorithm is used to model short-term air exposures associated with air emissions from an areal source	US EPA's Office of Air Quality Planning and Standards (OAQPS), Research Triangle Park, NC

Air	FDM (Fugitive Dust Model)	FDM is a computerized analytical air quality model specifically designed for computing concentration and deposition impacts from fugitive dust sources. The sources may be point, line or areal; it contains no plume rise algorithm FDM accounts for deposition losses as well as pollutant dispersion. The model is generally based on Gaussian plume formulation for computing concentrations, with improved gradient-transfer deposition algorithm	FDM models both short-term and long-term average particulate emissions from surface mining and similar sources A primary use of FDM is for the computation of concentrations and deposition rates resulting from emission sources such as hazardous waste sites where fugitive dust is a concern. Concentration and deposition are computed at all user-selected receptor locations	Support Center for Regulatory Air Models, Office of Air Quality Planning and Standards (OAQPS), US EPA, Research Triangle Park, NC
Air	OML (Operationelle Meteorologiske Luftkvalitetsmodeller)	OML is a modern Gaussian plume model, based on boundary layer scaling instead of relying on Pasquill stability classification. It describes dispersion of a passive, or possibly buoyant, gas from a number of sources. The model can be used for both high and low sources, but is not suitable for complex terrain conditions The model requires information on emission and meteorology on an hourly basis. It then computes a time-series of concentrations at user-specified receptor points, from which statistics are extracted and presented to the user	The OML model is a modern Gaussian plume model intended to be used for distances up to about 20 km from the source. The source is typically one or more stacks, and possibly also areal sources Typically, the OML model is applied for the purpose of making regulatory decisions; for instance, it is a recommended model for environmental impact assessments when new industrial sources are being planned	National Environmental Research Institute, DK-4000 Roskilde, Denmark

continues overleaf

Table 8.1 (continued)

Environmental compartment	Model name	Model type and description	Model uses/applications	Sources of model information and/or developer
Air	MLTT (The Model for a Long range and Transboundary Transport)	MLTT is a 3-dimensional model that calculates concentrations of pollutants in the atmosphere and their deposition on the surface. MLTT is designed to address multi-pollutant emission sources for multiple polluting substances with varying properties. The model incorporates advective transport, turbulence and diffusion, and deposition schemes	Intended fields of application for MLTT include the forecasting and monitoring of a large-scale distribution for polluting substances in the atmosphere and their deposition on the surface. Specific areas of application of this model include emergency planning, scientific research, and air quality assessments	State Institute for Applied Ecology–Russia, 117463 Moscow, Russia
Air	CemoS/Air	CemoS/Air is a box-model for calculating the stationary chemical concentration in air caused by an areal emission source. The model simulation consists of calculating the atmospheric transport of a chemical following an areal emission. It assumes that atmospheric and meteorological conditions are constant	CemoS/Air can be used to estimate the exposure concentration of environmental chemicals. Besides the stationary chemical concentration, the model can be used to estimate chemical deposition, photodegradation, advection, and annual inhalation dose.	UFIS/Institute for Environmental Systems Research, University of Osnabrueck, Germany

Air	CemoS/Plume	CemoS/Plume is a 'plume-model' for calculating the stationary chemical concentration in air caused by a point emission source The model simulation consists of calculating the atmospheric transport of a chemical following point emission. It assumes that atmospheric and meteorological conditions are constant. Advective transport by wind as well as dilution by the atmosphere is taken into account, but dispersion in the wind direction is neglected and so are the elimination of the chemical and any reaction with the ground	CemoS/Plume can be used to estimate the deposition to ground and potential inhalation dose of environmental chemicals from stationary chemical concentrations. It is possible to determine the concentration of a chemical at a given point or along the main wind direction	UFIS/Institute for Environmental Systems Research, University of Osnabrueck, Germany

continues overleaf

Table 8.1 (continued)

Environmental compartment	Model name	Model type and description	Model uses/applications	Sources of model information and/or developer
Air	DREAM (the Danish Rimpuff and Eulerian Accidental release Model)	DREAM is a comprehensive, 3-dimensional tracer model that is based on a combination of a Lagrangian short-scale puff model and an Eulerian long-range transport model. The Lagrangian model is used in the area near the source to describe the transport, dispersion, and deposition in the initial phase of the release and the Eulerian model is used for long-range transport calculations in the model domain (which covers the whole of Europe) Advanced 2-D and 3-D visualization and animation techniques are important tools for the development of the model, for the validation of the model results, and for a better understanding of the different processes studied by the model	DREAM has been developed for studying transport, dispersion, and deposition of air pollution caused by a single but strong source – such as from a major accident	National Environmental Research Institute (NERI), Department of Atmospheric Environment, DK-4000 Roskilde, Denmark

Determination of contaminant fate and behavior in the environment 123

Surface water	EXAMS-II (EXposure Analysis Modeling System)	EXAMS is a steady-state contaminant fate and transport numerical, finite-difference model, offering 1, 2, or 3-dimensional compartmental solutions in surface water bodies. It is based on a series of equations which account for interactions between the canonical aquatic environment into which a chemical is released, the chemistry of a given chemical, and the toxicant loading quantities. It includes process models of the physical, chemical, and biological phenomena governing the transport and fate of compounds	EXAMS simulates the fate of organic chemicals in surface water bodies (i.e., rivers, lakes, reservoirs, estuaries). It can estimate the time-varying and/or steady-state concentrations of the chemical in the water body in various phases (dissolved, sediment, sorbed, biosorbed). The model has been designed to evaluate the consequences of longer-term, primarily time-averaged chemical loadings that ultimately result in trace-level contamination of aquatic systems. It is suitable for modeling synthetic organic chemicals for freshwater, nontidal aquatic systems	Environmental Research Laboratory, Office of Research and Development, USEPA, Athens, GA
Surface water	REACHSCA (REACH SCAn model)	REACHSCA is a simple dilution model used to estimate steady-state chemical concentration in surface water bodies due to continuous loading from a single discharging facility	REACHSCA is used to estimate steady-state chemical concentration in surface water bodies (mainly river reaches)	Office of Toxic Substances, USEPA, Washington, DC

continues overleaf

Table 8.1 (*continued*)

Environmental compartment	Model name	Model type and description	Model uses/applications	Sources of model information and/or developer
Surface water	WTRISK (Waterborne Toxic RISK assessment model)	WTRISK consists of a framework that employs a risk assessment methodology in which mathematical models simulating chemical fate and transport processes can be linked to determine pollutant concentrations in all appropriate environmental media. These predicted concentrations are then used as input values for modeling nearby population exposures and potential health risks	WTRISK provides a flexible framework for the risk assessment of toxic substances in surface water. It aids the estimation of the source terms or quantities of toxic substances released into the environment	EPRI (Electric Power Research Institute), Palo Alto, CA
Surface water	SARAH	SARAH is a semi-analytical steady-state, 1-dimensional surface water exposure assessment model. It models a contaminated leachate plume feeding a downgradient surface water body (stream or river). Bioaccumulation in fish, degradation, sorption, dilution, and volatilization are included	SARAH is a surface water backcalculation procedure that is used to model a contaminated leachate plume feeding a downgradient stream or river	Environmental Research Laboratory, USEPA, College Station Road, Athens, GA

Surface water	PDM (Probabilistic Dilution Model)	PDM models exceedances of specified concentration levels in streams. The model estimates are based on statistical distribution of daily volume flow, and on solution of mass balance dilution evaluation	PDM estimates the percentage of time that a given concentration level may be exceeded in receiving surface water bodies	Office of Toxic Substances, Exposure Evaluation Division, USEPA, Washington, DC
Surface water	CemoS/Water	CemoS/Water is a steady-state model for the estimation of the pollution of a river by a chemical, caused by a single continuous pollution source like the discharge of sewage into the river The model assumes stationary conditions for the water flow and the material flow. The concentration profile of a chemical in a river, caused by a single release source, is described by an analytical equation derived from the mass balance for the river. Hence, all elimination processes are described by an aggregated first-order degradation rate. The sorption of the chemical on suspended material as well as the concentrations in sediment and in biota are calculated from the concentration in the water with the help of partition coefficients. Advection of the chemical and elimination processes due to degradation, sedimentation, and volatilization are taken into consideration	CemoS/Water can be used to calculate the concentration profile and the mass balance of the chemical in a river, as well as the concentrations in the sediment and in the biota living in the river (e.g., fishes)	UFIS/Institute for Environmental Systems Research, University of Osnabrueck, Germany

continues overleaf

Table 8.1 (continued)

Environmental compartment	Model name	Model type and description	Model uses/applications	Sources of model information and/or developer
Groundwater	SOLUTE (SOLUTE transport model)	SOLUTE is a basic program package of analytical models for solute transport in groundwater. The package includes several subprograms, including UNITS (for conversion of units); ERFC (to calculate error functions and complementary error functions); ONED1 (for solute transport in 1-D); WMPLUME and SLUG (for solute transport in 2-dimensions); RADIAL and LTIRD (for 2-dimensional radial flow); and PLUME3D and SLUG3D (for 3-dimensional transport)	SOLUTE is used in solute transport modeling to estimate receptor exposure concentration distributions	International Ground Water Modeling Center (IGWMC) – Europe, TNO Institute of Applied Geoscience, Delft, The Netherlands, Institute for Ground-water Research and Education, Colorado School of Mines, Golden, CO
Groundwater	AT123D (Analytical Transient 1-2-3 Dimensional simulation model)	AT123D is a saturated zone analytical transient 1-2-3-dimensional simulation model that is used to simulate chemical movement and waste transport in the aquifer system. It predicts spread of a contaminant plume through groundwater, and can handle constant as well as time-varying chemical release to groundwater. The model takes account of both adsorption and degradation. Output concentrations are time-varying	AT123D is an environmental model that predicts spread of a contaminant plume (chemical, thermal, or radioactive) through groundwater (saturated zone) and estimates the chemical concentration within groundwater at positions on a user-specified 3-dimensional grid	Oak Ridge National Laboratory (ORNL), Environmental Sciences Division, TN

Groundwater	MYGRT (Migration of solutes in the subsurface environment)	MYGRT is a 1-, 2-dimensional fate model that provides a method for computing the fate of reacting or nonreacting inorganic chemicals released to the groundwater environment. The simulation is based on quasi-analytical solutions to the conservation of mass equations, including advection, dispersion, and retardation. Simulation of both continuous and finite-duration solute releases are possible MYGRT gives results that are 'ballpark' approximations to the dynamic and complex problem of solute migration predictions	MYGRT allows users to calculate concentration of solute at an elapsed time from the release time. Calculations can also be made to estimate time taken for a chemical to travel the distances of interest	EPRI (Electric Power Research Institute), Palo Alto, CA
Groundwater	VHS (Vertical and Horizontal Spread model)	VHS is a steady-state, analytical model, used to simulate the dispersion of contaminants and calculate contaminant concentrations at a receptor point or well directly downgradient of a waste disposal area	VHS is a groundwater dilution model used for predicting steady-state contaminant concentrations at receptor locations	USEPA, Research Triangle Park, NC

continues overleaf

Table 8.1 (*continued*)

Environmental compartment	Model name	Model type and description	Model uses/applications	Sources of model information and/or developer
Groundwater	MOC (Method-of-Characteristics model for solute transport)	MOC is a finite-difference computer model that is applicable to 1- or 2-dimensional problems involving steady-state or transient flow. The model couples the groundwater flow equation with the solute transport equation MOC is based on a rectangular, block-centered, finite-difference grid. The method of characteristics is used in the model to solve the solute transport equation. By coupling the flow equation with the solute-transport equation, the model can be applied to both steady-state and transient flow problems MOC computes changes in concentration over time caused by the processes of convective transport, hydrodynamic dispersion, and mixing (or dilution) from fluid sources. The model assumes that solute is nonreactive and that gradients of fluid density, viscosity, and temperature do not affect the velocity distribution; aquifer may be heterogeneous and/or anisotropic	MOC is used for calculating changes in the concentration of dissolved (nonreactive) chemical species in flowing groundwater, by the model simulating solute transport in flowing groundwater. The purpose of the simulation is to compute the concentration of a dissolved chemical species in an aquifer at any specified place and time	US Geological Survey (USGS), Water Resources Department, Reston, VA The Holcomb Research Institute, Indianapolis, IN

Soil vadose (unsaturated) zone	CMLS (Chemical Movement in Layered Soils) model	CMLS is an analytical model developed as a management tool to describe the fate and transport of pesticides in layered soils, and to estimate the amount of chemical at a certain position at a particular time

CMLS assumes movement of the chemical in liquid phase only, and allows for a finite source. Volatilization is not considered; dispersion and diffusion of the chemical is ignored; and degradation is defined as a first-order process | The CMLS model is typically used to estimate the time for a chemical entering the unsaturated zone to reach a certain depth | Office of Research and Development, USEPA, Robert S. Kerr Environmental Lab., Ada, OK |
| Soil vadose (unsaturated) zone | VIP (Vadose zone Interactive Process) model | VIP is a 1-dimensional, numerical (finite-difference) fate and transport model, designed for simulating the movement of compounds in the unsaturated zone, that is the result of land application of oily wastes

VIP considers dual soil zones (a plow zone and a treatment zone), and considers the source to be infinite | VIP is a useful model for predicting contaminant transport of residual constituents in settings similar to a land treatment area with respect to contaminant transport

VIP models chemical migration in a uniform unsaturated zone as a result of land application | Office of Research and Development, US EPA, Robert S. Kerr Environmental Lab., Ada, OK |

continues overleaf

Table 8.1 (continued)

Environmental compartment	Model name	Model type and description	Model uses/applications	Sources of model information and/or developer
Soil vadose (unsaturated) zone	VLEACH (Vadose zone LEACHing model)	VLEACH is a 1-dimensional, finite-difference model developed to simulate the transport of contaminants displaying linear partitioning behavior through the vadose zone to the water table by aqueous advection and diffusion. Linear equilibrium partitioning is used to determine chemical concentrations between the aqueous, gaseous, and adsorbed phases, and a finite source can be considered	VLEACH provides conservative estimates of contaminant migration in soil. Multiple layers are modelled, and water flow is assumed to be steady-state	Environmental Research Lab., Office of Research and Development, USEPA, Athens, GA
Soil vadose (unsaturated) zone	RITZ (Regulatory and Investigative Treatment Zone model)	RITZ is a steady-state, 1-dimensional, unsaturated zone analytical model for pollutant transport. It can account for soil solution, volatilization and atmospheric losses, and biological degradation of chemicals. It considers the effect of an oil phase on pollutant transport	RITZ is useful in predicting contaminant transport of residual constituents in settings similar to a land treatment area with respect to contaminant transport	Office of Environmental Processes and Effects Research, Oak Ridge, TN

| Soil vadose (unsaturated) zone | SESOIL (SEasonal SOIL compartment model) | The model was designed to predict fate of contaminants in a land treatment scenario and considers downward movement of chemicals. Output includes mass transport to groundwater, that can become input to a groundwater dilution model

SESOIL is a seasonal soil compartment model that estimates the rate of vertical chemical transport and transformation in the soil column in terms of mass and concentration distributions among the soil, water, and air phases in the unsaturated soil zone. It is designed for long-term environmental fate simulations of pollutants in the vadose zone

SESOIL is a 1-dimensional unsaturated zone model for both organic and inorganic chemicals | Estimates the rate of vertical chemical transport and transformation in the soil column. Also has capability to simulate contaminant transport in washload at soil surface and volatilization rates to the atmosphere

SESOIL can be used for assessing risks from landfill disposal; accidental leaks or spills onto land; agricultural applications; leaking underground storage tanks; and deposition from the atmosphere | Office of Toxic Substances, Exposure Evaluation Division, USEPA, Washington, DC |
|---|---|---|---|---|
| Soil vadose (unsaturated) zone | SWAG (Simulated Waste Access to Groundwater) | SWAG is a 3-compartment analytical computer model for organic pollutant transport, that considers transformations in the soil–geological matrix | SWAG predicts organic pollutant transport into groundwater | Office of Environmental Processes and Effects Research, Oak Ridge, TN |

continues overleaf

Table 8.1 (continued)

Environmental compartment	Model name	Model type and description	Model uses/applications	Sources of model information and/or developer
Multimedia (air, water, and soil)	ENPART (ENvironmental PARTitioning model)	ENPART uses simple physical–chemical data to estimate equilibrium concentration ratios of a chemical between the environmental compartments of air, water, and soil. It is an approximate method intended to identify chemicals which may require further testing. ENPART is a fugacity-based model that estimates the steady-state equilibrium or dynamic partitioning of organic chemicals among environmental compartments. It serves more as a screening tool	ENPART is a multi-media model that estimates equilibrium concentration ratios of a chemical between the environmental compartments of air, water, and soil. It also provides a second level of concentration and mass partitioning called dynamic partitioning. Subsequently, potential exposures to the chemical in each environmental compartment are estimated	Office of Toxic Substances, Exposure Evaluation Division, USEPA, Washington, DC
Multimedia (air, water, and soil)	POSSM (PCB On-Site Spill Model)	POSSM is an exposure assessment methodology. It consists of a chemical transport and fate model capable of considering all of the key processes controlling chemical losses from a spill site (including volatilization, leaching to groundwater, and chemical washoff from the land surface due to runoff/erosion)	POSSM provides a quantitative framework for estimating general public exposure levels associated with spills from utility electrical equipment. The methodology was developed primarily for PCBs, but is applicable to a wide range of organic chemicals	EPRI (Electric Power Research Institute), Palo Alto, CA

POSSM contains several relevant subprograms, including PTDIS (air model), RIVLAK (surface water model), GROUND (groundwater model), and EXPOSE (exposure intake model)

On-site environmental concentrations can be estimated with POSSM; off-site concentrations can be estimated with one of the relatively simple transport and fate models for air (PTDIS), surface water (RIVLAK), and groundwater (GROUND) that are also incorporated into the methodology

Given estimates of on-site and/or off-site concentrations and the characteristics and activity patterns of the receptors of concern, inhalation, ingestion, and dermal exposure levels can be calculated using the sub-model EXPOSE

Table 8.1 (continued)

Environmental compartment	Model name	Model type and description	Model uses/applications	Sources of model information and/or developer
Multimedia (air, water, and soil)	INPOSSM (INteractive PCB On-Site Spill Model)	INPOSSM is an interactive PCB exposure assessment model. It includes a chemical transport and fate model capable of considering key processes controlling chemical losses from a PCB or organic chemical spill site, including volatilization, leaching to groundwater, and chemical washoff from the land surface due to runoff/erosion	INPOSSM uses a Monte Carlo simulation model in evaluating the variability of predicted chemical losses, which then forms a basis for the exposure assessment	EPRI (Electric Power Research Institute), Palo Alto, CA
Multimedia (air, water, and soil)	MCPOSSM (Monte Carlo PCB On-Site Spill Model)	MCPOSSM is a chemical spill exposure assessment methodology, providing a quantitative framework for estimating uncertainties of chemical levels associated with spills. The core of the methodology is the PCB On-Site Spill Model (POSSM), and a Monte Carlo chemical transport and fate model capable of considering key processes controlling chemical losses from a spill site, including volatilization, leaching to groundwater, and chemical washoff from the land surface due to runoff/erosion	MCPOSSM provides a distribution of concentrations over time and probabilities of exceeding specified levels (i.e., probability of a worst-case level), which then forms a basis for the exposure assessment	EPRI (Electric Power Research Institute), Palo Alto, CA

Multimedia (air, water, soil, biota)	SMCM (Spatial–Multimedia-Compartmental-Model)	SMCM describes the fate of chemicals in a conventional air–water–soil–sediment system under steady- or unsteady-state conditions. It is based on a modeling approach that estimates the multimedia partitioning of organic pollutants in local environments SMCM consists of coupled partial and ordinary differential equations that are solved simultaneously by finite-difference method and using operator-splitting techniques. It is a hybrid transport and fate model that makes use of both uniform and nonuniform 1-dimensional compartments. It has the capability of simulating a variety of pollutant transport phenomena	SMCM is designed to predict the multimedia concentrations of organic chemicals in the environment. It is used for the analysis of the multimedia distribution of organic pollutants in the environment	The National Center for Intermedia Transport, UCLA, Los Angeles, CA

continues overleaf

Table 8.1 (*continued*)

Environmental compartment	Model name	Model type and description	Model uses/applications	Sources of model information and/or developer
Multimedia (within soil zones)	CemoS/Soil	CemoS/Soil is a dynamic model used to simulate the transport and the accumulation of a chemical in a soil column The basis of the model is a 1-dimensional advection–diffusion/dispersion equation. This equation describes the dynamic behavior of a chemical under steady-state water transport regime by neglecting the transient water flow and changing water content in soil over time. CemoS/Soil uses three analytical solutions of this equation, found for specific boundary conditions. The velocity of the vertical transport of the chemical is estimated from the Darcy water flux (calculated from the water balance at the soil surface assuming steady-state water flow)	CemoS/Soil can be used to calculate the concentration profile of a chemical in soil, as well as the fractions of the chemical in the soil-water, the soil-air and the soil-matrix. Further outputs are the velocity of the vertical transport of the chemical and the total diffusion/dispersion coefficient of the chemical in the soil The model can be used in three specific situations: single input of a chemical at the soil surface; transport from a contaminated soil layer into deeper layers; and transport and residence in case of a continuous injection	UFIS/Institute for Environmental Systems Research, University of Osnabrueck, Germany

Footnote: Regardless of how much environmental monitoring data is available, it is almost always desirable to have: an estimate of chemical concentrations under different sets of conditions; results for a future chemical loading scenario; a predicted 'hindcast' or reconstructed history of chemical releases; and/or estimates at alternate locations where field data do not exist. This is when environmental models usually come in handy. But there is a word of caution – best illustrated with the analogous view held by society about 'social' models, as expressed by Kaplan (Kaplan, 1964 – cited in Aris, 1994), that: '*Models are undeniably beautiful, and a man may justly be proud to be seen in their company. But they may have their hidden vices. The question is, after all, not only whether they are good to look at, but whether we can live happily with them.*' This view does indeed compare very well with the underlying principles in the selection and use of environmental models – calling for the careful choice of such models to support environmental management programs. Most importantly, it should be recognized that a given mathematical model that performs extremely well under one set of circumstances may not necessarily be appropriate for other similar or comparable situations for a variety of reasons.

model over another – some of which are propietary or a registered trademark – will generally be problem-specific; *the mention of any particular model in this title does not necessarily constitute an endorsement of such product as the most preferred*, since each has its own merits and limitations. Also, it should be noted that this listing is by no means complete and exhaustive. The on-line service provided by the Internet – the most widely used international network communication service – may be used to obtain additional information on a variety of environmental models that could prove useful to case-specific applications.

8.4 SELECTED ENVIRONMENTAL FATE ALGORITHMS FOR ESTIMATING CONTAMINANT CONCENTRATIONS

Once contaminants are released into the environment, the pollutants may be transported into various media and environmental matrices. A variety of mathematical algorithms and models are usually employed to support the determination of contaminant fate and transport in the environment; the results are then used in estimating the consequential exposures and risks to potential receptors that may result from the release and behaviors of an environmental contaminant.

In general, releases from potential contamination sources can cause human exposures to contaminants in a variety of ways, such as the following:

- Direct inhalation of airborne vapors and also respirable particulates.
- Deposition of airborne contaminants onto soils, leading to human exposure via dermal absorption or ingestion.
- Ingestion of food products that have been contaminated as a result of deposition onto crops or pasture lands, and introduction into the human foodchain.
- Ingestion of contaminated dairy and meat products from animals consuming contaminated crops or waters.
- Deposition of airborne contaminants on waterways, uptake through aquatic organisms, and eventual human consumption.
- Leaching and runoff into water resources, and consequential human exposures to contaminated waters.

Some simple example models and equations that may be employed in the estimation of contaminant concentrations in air, soil, water, and food products are presented below – with much emphasis given to the air medium, since this represents a very important 'transfer' medium and yet is often lacking in actual environmental field sampling data.

8.4.1 Estimation of Contaminant Concentrations in Air

Air pollution presents one of the greatest risks to human health and the environment globally. The long list of health problems caused or aggravated by air pollution includes respiratory problems, cancer, and eye irritations; environmental problems range from damage to crops and vegetation to increased acidity of lakes which renders them uninhabitable for fish and other aquatic life (Holmes et al., 1993). Of particular interest, air emissions from contaminated sites often are a major source of human exposure to toxic or hazardous substances. Contaminated sites can therefore pose significant risks to public health as a result of possible airborne release of soil particulate matter laden with toxic chemicals, and/or volatile emissions. In fact, even very low level air emissions could pose significant threats to exposed individuals, especially if toxic or carcinogenic contaminants are involved. Furthermore, remedial actions – especially ones involving excavation – may create much higher emissions than baseline or undisturbed conditions. Consequently, there is increased attention on the assessment of risks associated with air toxic releases.

Air emissions of critical concern relate to volatile organic chemicals (VOCs), semi-VOCs, particulate matter, and other chemicals associated with wind-borne particulates such as metals, PCBs, dioxins, etc. Volatile chemicals may be released into the gaseous phase from such sources as landfills, surface impoundments, contaminated surface waters, open/ruptured tanks or containers, etc. Also, there is the potential for subsurface gas movements into underground structures such as pipes and basements, and eventually into indoor air. Additionally, toxic chemicals adsorbed to soils may be transported to the ambient air as particulate matter or fugitive dust.

In general, chemical concentration in air – represented by the ground-level concentration (GLC) – is a function of the source emission rate and the dilution factor at the points of interest. This is generally estimated by using the following type of mathematical relationship:

$$GLC = ER \times DF$$

where GLC is the ground-level concentration ($\mu g/m^3$); ER is the pollutant emission rate (g/s); and DF is a dilution factor, provided by dispersion modeling ($\mu g/m^3/g/s$).

8.4.1.1 Classification of air emission sources

Air emissions may be classified as coming from either point or area sources. Point sources include vents (e.g., landfill gas vents) and stacks (e.g., incinerator and air stripper releases); area sources are generally associated with ground-level emissions (e.g., from landfills, lagoons, and contaminated surface areas).

Area sources are released at ground level and disperse there, with less influence of winds and turbulence; point sources, generally, come from a stack, and are emitted with an upward velocity, often at a height significantly above ground level. Thus, point sources are more readily diluted by mixing and diffusion, further to being at greater heights, so that ground-level concentrations are reduced. This means that area source emissions may be up to 100 times as hazardous on a mass per time basis as point sources and stacks. This scenario demonstrates the importance of adequately describing the source type in the assessment of potential air impacts associated with environmental contamination problems.

Air contaminant emissions may also be grouped into two major categories – gas phase emissions and particulate matter emissions. The emission mechanisms associated with gas phase and particulate matter releases are quite different. Gas phase emissions primarily involve organic compounds but may also include certain metals (such as mercury); these emissions may be released through several mechanisms such as volatilization, biodegradation, photo-decomposition, hydrolysis, and combustion (USEPA, 1989d). Particulate matter emissions can be released through wind erosion, mechanical disturbances, and combustion. For airborne particulates, the particle size distribution plays an important role in inhalation exposures. Large particles tend to settle out of the air more rapidly than small particles, but may be important in terms of noninhalation exposures. Very small particles (2.5–10 μm diameter) are considered to be respirable and thus pose a greater health hazard than the larger particles.

In addition to gas phase and particulate emissions, aerosols are an added mechanism – particularly for combustion processes. Aerosol processes can lead to chemical transformations, especially on surfaces of particulates. They are important because the usual control devices (i.e., baghouses, precipitators, etc.) do not remove aerosols with any degree of efficiency.

8.4.1.2 Determination of air contaminant emission potential

Once released to the ambient air, a contaminant is subject to simultaneous transport and diffusion processes in the atmosphere; these conditions are significantly affected by meteorological, topographical, and source factors. Additional fundamental atmospheric processes (other than atmospheric transport and diffusion) that affect airborne contaminants include transformation, deposition, and depletion. The extent to which all these atmospheric processes act on the contaminant of concern determines the magnitude, composition, and duration of the release; the route of human exposure; and the impact of the release on the environment and public health.

Air emissions may pose potential public health risks due to possible airborne particulate matter laden with toxic chemicals, and/or volatile emissions from volatile or semi-volatile chemical species generated from the source. The most important chemical parameters to consider in the evaluation of volatile air emissions are the vapor pressure and the Henry's Law constant. Vapor pressure is a useful screening indicator of the potential for a chemical to volatilize from the media in which it exists. The Henry's Law constant is particularly important in estimating the tendency of a chemical to volatilize from a surface impoundment or water; it also indicates the tendency for a chemical to partition between, for instance, the soil and gas phase from soil water in the vadose zone or groundwater.

Several site-specific factors (such as site integrity, presence of cracks or fissures, soil organic content, soil moisture, microbial activity, ambient temperature, and the site area) influence volatilization, emission rates, and ambient chemical concentrations at contaminated sites and vicinities. Persistent chemicals which readily adsorb to soil (i.e., high K_d or K_{oc} values) are resistant to biodegradation, are not readily volatilized, and are most likely to remain in surface soils; such contaminated sites with surface contamination may generate contaminated airborne particulates of significant concern. In particular, particles of <10 μm diameter are usually considered respirable and subject to air quality control and standards. Details of the exposure assessment process and the assumptions involved, as well as guidelines and criteria for assessing the potential for particulate and volatile emissions from contaminated sites, can be found in the literature relating to air toxics programs (e.g., CAPCOA, 1990; CDHS, 1986).

8.4.1.3 Dispersion modeling and the estimation of air emissions

Methods for estimating air emissions from contaminated sources may consist of emissions measurement (direct and indirect); air monitoring (supplemented by modeling, as appropriate); and predictive emissions modeling. General and specific protocols for estimating emission levels for contaminants from several sources are available in the literature (e.g., CAPCOA, 1990; CDHS, 1986; Mackay and Leinonen, 1975; Mackay and Yeun, 1983; Thibodeaux and Hwang, 1982; USEPA, 1989a, 1989b, 1989c, 1990a, 1990b). In all situations, case-specific data should be used whenever possible, in order to increase the accuracy of the emission rate estimates. In fact, the combined approach of environmental fate analysis and field monitoring should provide an efficient and cost-effective strategy for investigating the impacts of air pathways on potential receptors, given a variety of meteorological conditions.

Atmospheric dispersion modeling has indeed become an integral part of the planning and decision-making process in the assessment of public health and environmental impacts from environmental contamination problems. It is an approach that can be used to provide contaminant concentrations at potential receptor locations of interest based on emission rates and meteorological data.

A number of general assumptions are normally made in the assessment of contaminant releases into the atmosphere, including the following:

- air dispersion and particulate deposition modeling of emissions adequately represent the fate and transport of chemical emission to ground level;
- the composition of emission products found at ground level is identical to the composition found at source, but concentrations are different;
- the potential receptors are exposed to the maximum annual average ground-level concentrations from the emission sources for 24 h/day, throughout a 70-year lifetime – a conservative assumption; and
- there are no losses of chemicals through transformation and other processes (such as biodegradation or photodegradation) – a conservative assumption.

Naturally, the accuracy of the model predictions depends on the accuracy and representativeness of the input data. In general, model input data will include emissions and release parameters, meteorological data, and receptor locations. Existing air monitoring data (if any) for the case area can be used in designing a receptor grid and in selecting indicator chemicals to be modeled. This can also provide insight to background concentrations. Table 8.1 listed selected models that may be utilized in an air pathway exposure analysis (APEA); this list which is by no means complete and exhaustive indicates the range and variations in the available models.

There are many levels of complexity implicit in the models used in air pathway assessments; a description of the mathematical structure for such models is presented elsewhere (e.g., USEPA, 1992). Some selected screening-level air emission models are discussed below for illustrative purposes; these include typical algorithms for both volatile and nonvolatile emissions. All nonvolatile compounds are generally considered to be bound onto particulates. For the purposes of a screening evaluation, a volatile substance may be defined as any chemical having a vapor pressure greater than 1×10^{-3} mmHg or a Henry's Law constant greater than 1×10^{-5} atm-m^3/mol (DTSC, 1994). Thus, chemicals with values less than or equal to these are generally considered as nonvolatile compounds.

8.4.1.4 Screening-level estimation of airborne dust concentrations

In the estimation of potential risks from fugitive dust inhalation, an estimate of respirable (<10 μm aerodynamic diameter, denoted by the symbol PM-10 or PM_{10}) fraction and concentrations are required. The amount of nonrespirable (>10 μm aerodynamic diameter) concentrations may also be needed to estimate deposition of wind-blown emissions which will eventually reach potential receptors via other routes such as ingestion and dermal exposures.

Air models for fugitive dust emission and dispersion can be used to estimate the applicable exposure point concentrations of respirable particulates from contamination sources (e.g., CAPCOA, 1989; CDHS, 1986; DOE, 1987; USEPA, 1988b, 1989d). In such models, fugitive dust dispersion concentrations evaluated are typically represented by a three-dimensional Gaussian distribution of particulate emissions from the source.

In general, the inhalation of contaminants that are adsorbed onto particulate matter depends upon the concentration of suspended particulates (dust), the fraction of the dust that is respirable (generally defined as particles with an aerodynamic diameter of 10 μm or less), and the fraction of the dust that derives from the contaminated source. For the screening analysis of particulate inhalation, the chemical concentration in air may be calculated according to the following equation:

$$C_a = C_s \times PM_{10} \times CF \times f$$

where C_a = concentration in air (μg/m^3); C_s = contaminant concentration in soil (mg/kg); PM_{10} = airborne concentration of respirable dust (less than 10 μm in diameter), usually assumed to be 50 μg/m^3; CF = conversion factor, equal to 10^{-6} kg/mg; and f = fraction of the dust that derives from the contaminated source.

Typically, a screening level assumption is made that, for nonVOCs, particulate contamination levels are directly proportional to the maximum soil concentrations. The particulate concentration used in this type of characterization may be set at 50 μg/m^3 – typical of a value that is attributable to the US National Ambient Air Quality Standard for annual average respirable portion (PM_{10}) of suspended particulate matter (USEPA, 1993). By using the above assumptions, the nonVOC concentration in air is given as:

$$C_a = C_s \times (5 \times 10^{-5})$$

It is noteworthy that this estimation procedure is not applicable to a site that is particularly dusty – as would be expected in situations where the air quality standard for suspended particulate matter is routinely exceeded (i.e., $PM_{10} > 50 \ \mu g/m^3$).

8.4.1.5 Screening level estimation of airborne vapor concentrations

The most important chemical parameters to consider in the evaluation of volatile air emissions are the vapor pressure and the Henry's Law constant. Vapor pressure is a useful screening indicator of the potential for a chemical to volatilize from the media in which it exists. The Henry's Law constant is particularly important in estimating the tendency of a chemical to volatilize from a surface impoundment or water; it also indicates the tendency of a chemical to partition between the soil and gas phase from soil water in the vadose zone or groundwater.

A vaporization model may be used to calculate flux from volatiles present in soils into the overlying air zone. Typically, the following general equation will be used to estimate the average emission over a residential lot (USEPA, 1988b, 1990a, 1992):

$$E_i = \frac{2AD_e P_a K_{as} C_i \times CF}{(3.14 \alpha T)^{1/2}}$$

where:

- E_i = average emission rate of contaminant i over a residential lot during the exposure interval (mg/s)
- A = area of contamination (cm²), with a typical default value of 4.84×10^6 cm²
- D_e = effective diffusivity of compound (cm²/s)
  = $D_i (P_a^{3.33}/P_t^2)$
- D_i = diffusivity in air for compound i (cm²/s)
- P_t = total soil porosity (dimensionless)
  = $1 - (\beta/\rho)$
- β = soil bulk density (g/cm³), with typical default value of 1.5 g/cm³
- ρ = particle density (g/cm³), with typical default value of 2.65 g/cm³
- P_a = air-filled soil porosity (dimensionless)
  = $P_t - \theta_m \beta$
- θ_m = soil moisture content (cm³/g), with typical default value of 0.1 cm³/g
- K_{as} = soil/air partition coefficient (g/cm³)
  = $(H_c/K_d) \times CF2 = (H_c/K_d) \times 41$
- H_c = Henry's Law constant (atm-m³/mol)
- K_d = soil/water partition coefficient (cm³ water/g soil, or L/kg)
- $CF2$ = 41, a conversion factor that changes H_c into a dimensionless form
- C_i = bulk soil concentration of contaminant i (i.e., chemical concentration in soil, mg/kg $\times 10^{-6}$ kg/mg) (g/g soil)

CF = 10^3 mg/g, a conversion factor
α = conversion factor composed of several quantities defined above

$$\alpha = \frac{D_e \times P_a}{\{P_a + [\rho(1 - P_a)/K_{as}]\}}$$

T = exposure interval (s), with a typical default value of 30 years = 9.5×10^8 seconds.

The equation for estimating emission rates of VOCs can be reduced to the following form:

$$E_i = \frac{1.6 \times 10^5 \times D_i \times \frac{H_c}{K_d} \times C_i}{\sqrt{\left[D_i \times \frac{0.023}{\left\{0.284 + \left[0.046 \times \frac{K_d}{H_c}\right]\right\}}\right]}}$$

It is noteworthy that this equation is valid only in situations when no 'free product' is expected to exist in the soil vadose zone. Analytical procedures for checking the saturation concentrations, intended to help determine whether free products exist, is available in the literature (e.g., DTSC, 1994; USEPA, 1988b, 1990a, 1992). Under circumstances when the soil contaminant concentration is greater than the calculated saturation concentration for the contaminant (implying the presence of free product), a more sophisticated evaluation scheme should be adopted.

The potential air concentration of chemicals in the breathing zone as a result of volatilization of chemicals through the soil surface is calculated above each discrete area of concern. A simple box model (Hwang and Falco, 1986; USEPA, 1988b, 1990a, 1992) can be used to provide an estimate of ambient air concentrations using the total emission rate calculated above. The length dimensions of the hypothetical box within which mixing will occur is usually based on the minimum dimensions of a residential lot in the applicable locality/region (Hadley and Sedman, 1990). Consequently, a screening-level estimate of the ambient air concentration is given by:

$$C_a = \frac{E_i}{(LS \times V \times MH)}$$

where C_a = ambient air concentration (mg/m^3); E_i = total emission rate (mg/s); LS = length dimension perpendicular to wind (m), with a typical default value of 22 m based on the length dimension of a square residential lot with area 484 m^2 (Hadley and Sedman, 1990; USEPA, 1992); V = average wind speed within mixing zone (m/s), with typical default value of 2.25 m/s (Hadley and Sedman, 1990; USEPA, 1992); MH = mixing height (m), with typical default value of 2 m, or the height of the average breathing zone for an adult (Hadley and Sedman, 1990; USEPA, 1992).

By using the above-indicated default values, the ambient air concentration can be estimated by:

$$C_a = \frac{E_i}{99}$$

where E_i would already have been estimated as shown previously above.

8.4.1.6 Screening-level estimation of volatilization to shower air

An often encountered scenario in human health risk assessments is the volatilization of contaminants from contaminated water into shower air during a bathing/showering activity. A simple/common model that may be used to derive contaminant concentration in air from concentration in domestic water consists of a very simple box model of volatilization; air concentration is derived from volatile emission rate by treating the shower as a fixed volume with perfect mixing and no outside air exchange, so that air concentration increases linearly with time. Thus, the following equation can be used to determine the average air concentration in the bathroom during a shower activity (generally for chemicals with a Henry's Law constant of $\geq 2 \times 10^{-7}$ atm-m^3/mol only) (HRI, 1995):

$$C_{sha} = \frac{[C_w \times f \times F_w \times t]}{2 \times [V \times 1000 \mu g/mg]}$$

where C_{sha} is the average air concentration in the bathroom during a shower activity (mg/m^3); C_w is the concentration of contaminant in the tap water (μg/L); f is the fraction of contaminant volatilized (unitless); F_w is the water flow rate in the shower (L/h); t is the shower duration (hours); and V is the bathroom volume (m^3).

The following equation can be used to determine the average air concentration in the bathroom after a shower activity (generally for chemicals with a Henry's Law constant of $\geq 2 \times 10^{-7}$ atm-m^3/mol only) (HRI, 1995):

$$C_{sha2} = \frac{[C_w \times f \times F_w \times t]}{[V \times 1000 \mu g/mg]}$$

In these simplified representations, the models assume the following: there is no air exchange in the shower – this assumption tends to over-estimate contaminant concentration in bathroom air; there is perfect mixing within the bathroom (i.e., the contaminant concentration is equally dispersed throughout the volume of the bathroom) – this assumption tends to under-estimate contaminant concentration in the shower air; the emission rate from water is independent of instantaneous air concentration; and the contaminant concentration in the bathroom air is determined by the amount of contaminants emitted into the box (i.e., $[C_w \times f \times F_w \times t]$) divided by the volume of the bathroom (V) (HRI, 1995).

8.4.1.7 Estimation of household air contamination due to volatilization from water

Chemical concentrations in household indoor air due to contamination in domestic water may be estimated for volatile chemicals (generally for chemicals with a

Henry's Law constant of $\geq 2 \times 10^{-7}$ atm-m^3/mol only), in accordance with the following relationship (HRI, 1995):

$$C_{ha} = \frac{[C_w \times WFH \times f]}{[HV \times ER \times MC \times 1000 \mu g/mg]}$$

where C_{ha} is the chemical concentration in air (mg/m^3); C_w is the concentration of contaminant in the tap water (μg/L); WFH is the water flow through the house (L/day); f is the fraction of contaminant volatilized (unitless); HV is the house volume (m^3/house); ER is the air exchange rate (house/day); and MC is the mixing coefficient (unitless).

8.4.2 Estimation of Contaminant Concentrations in Soils

The average chemical concentration in soil (C_s) is a function of the release/deposition (Q), accumulation period (τ), chemical-specific soil half-life (T), mixing depth (d), and soil bulk density (γ), expressed in accordance with the following functional relationship:

$$C_s = f(Q, \tau, T, d, \gamma)$$

The application of such estimation procedures becomes necessary when, for example, it is required to determine the vertical migration characteristics of environmental contaminants released into the soil medium.

8.4.3 Estimation of Contaminant Concentrations in Water

Historically, surface waters were among the first environmental media to receive widespread attention with regard to environmental pollution problems. This attention was due in part to the high visibility and extensive public usage of surface waters, as well because of their historical use as waste receptors (Hemond and Fechner, 1994). But it must be recognized that groundwater resources are just about as vulnerable to environmental contamination. It is worth mentioning the fact that groundwater is extensively used by public water supply systems in several places around the world; thus, it is important to give very close attention to the seemingly 'hidden' groundwater pollution problems.

In certain circumstances designed to address groundwater contamination problems, it becomes necessary to estimate the leachate concentrations of chemical contaminants. For organic chemicals, soil-water concentrations can be calculated as follows (Feenstra et al., 1991):

$$C_{s-w} = \frac{\rho C_s}{[\theta_w + \rho K_d]} = \frac{\rho C_s}{[\theta_w + \rho K_{oc} f_{oc}]}$$

where C_{s-w} is the soil-water concentration (µg/L); ρ is the bulk density (kg/L); C_s is the soil concentration (mg/kg); θ_w is the porosity; K_d is the partition coefficient (L/kg); K_{oc} is the organic carbon partition coefficient; and f_{oc} is the percent organic carbon content of soil.

Subsequently, a simple water-balance equation can be used to calculate a dilution factor to account for reduction of soil leachate concentration from mixing in an aquifer. In general, as soil leachate moves through soil and groundwater, contaminant concentrations are attenuated by adsorption and degradation; in the aquifer, dilution by clean groundwater further reduces concentrations before contaminants reach receptor points (e.g., drinking water wells). This reduction in concentration can be expressed by a dilution-attenuation factor (*DAF*), defined as the ratio of soil leachate concentration to receptor point concentration. The lowest possible *DAF* is 1 – corresponding to the situation where there is no dilution or attenuation of a contaminant (i.e., when the concentration at the receptor location equals the soil leachate concentration). On the other hand, high *DAF* values correspond to a large reduction in contaminant concentration from the contaminated soil to the receptor well. Assuming for the sake of simplicity that only one of the dilution-attenuation processes is important – viz., contaminant dilution in groundwater – then a simple mixing zone equation derived from a water-balance relationship is used to calculate a site-specific dilution factor, as follows (USEPA, 1996a, 1996b):

$$DF = 1 + \left[\frac{K \times i \times d}{I \times L}\right]$$

where *DF* is the dilution factor (unitless); K is the aquifer hydraulic conductivity (m/year); i is the hydraulic gradient (m/m); I is the infiltration rate (m/year); d is the mixing zone depth (m); and L is the source length parallel to groundwater flow (m). The mixing-zone depth is estimated by the following equation (USEPA, 1996a, 1996b):

$$d = [\sqrt{(0.0112 \times L^2)}] + d_a \left[1 - \exp\left(\frac{-LI}{Kid_a}\right)\right]$$

where d_a is the aquifer thickness (m). It is noteworthy that the mixing-zone depth should not exceed aquifer thickness; thus, the aquifer thickness should be considered as the upper limit for mixing-zone depth.

8.4.4 Estimation of Contaminant Concentrations in Vegetation

The average concentration in and on vegetation (C_v) is a function of direct deposition and root translocation or uptake from exposed soil, estimated according to the following functional relationship:

Determination of contaminant fate and behavior in the environment 147

$$C_v = C_{dep} + C_{trans} + C_{air}$$

where C_v is the average concentration in and on specific types of vegetation (μg/kg); C_{dep} is the concentration due to direct deposition (μg/kg); C_{trans} is the concentration due to translocation (μg/kg); and C_{air} is the concentration due to uptake from exposed soil (μg/kg).

8.4.5 Estimation of Contaminant Concentrations in Animal Products

The average concentration in animal products depends on which routes of exposure exist for the animals. Animal exposure routes include inhalation, soil ingestion, ingestion of contaminated feed and pasture, ingestion of contaminated water, and dermal contacts. This is estimated in accordance with the following relationship:

$$C_{fa} = (INH + ING_{water} + ING_{feed} + ING_{pasture/grazing} + ING_{soil} + DER) \times F_i$$

where:

C_{fa}	=	average concentration in farm animals and their products (μg/kg)
INH	=	dose through inhalation (μg/day)
	=	$RR \times GLC$
RR	=	inhalation rate for animal (m³/day)
GLC	=	ground-level concentration (μg/m³)
ING_{water}	=	dose through water ingestion (μg/day)
	=	$WIR \times \%SW \times C_w$
WIR	=	water ingestion rate for animal (kg/day)
$\%SW$	=	% water ingested from a contaminated source
C_w	=	average concentration in water (μg/kg)
ING_{feed}	=	dose through feed ingestion (μg/day)
	=	$(1 - \%G) \times FI \times L \times C_v$
$\%G$	=	% diet provided by grazing
FI	=	feed ingestion rate (kg/day)
L	=	% of feed other than pasture locally grown
C_v	=	concentration in feed (μg/kg)
$ING_{pasture/grazing}$	=	dose through pasture/grazing (μg/day)
	=	$\%G \times C_v \times FI$
$\%G$	=	% diet provided by grazing
C_v	=	concentration in pasture/grazing material (μg/day)
FI	=	feed ingestion rate (kg/day)
ING_{soil}	=	dose through soil ingestion (μg/kg)
	=	$SIR \times C_s$
SIR	=	soil ingestion rate for animal (kg/day)
C_s	=	average soil concentration (μg/kg)
DER	=	dose through skin absorption from dermal contacts (μg/day)
F_i	=	transfer coefficient of contaminant from diet to animal product (day/kg).

8.4.5.1 Contaminant bioconcentration in meat and dairy products

The tendency of certain chemicals to become concentrated in animal tissues, relative to their concentrations in the ambient environment, in many cases can be attributed to the fact that the chemicals are lipophilic (i.e., they are more soluble in fat than in water) (HRI, 1995). Thus, these chemicals accumulate in the fatty portion of animal tissue. Accordingly, the concentration of such a chemical in animal tissue (or other animal products for that matter) may be considered to reflect the chemical's inherent bioconcentration factor (BCF). The bioconcentration of chemicals in meat is primarily dependent on the partitioning of chemical compounds to fat deposits (HRI, 1995). Thus,

$$C_x = BCF \times F \times C_w$$

where C_x is the chemical concentration in animal tissue or dairy product; BCF is the chemical-specific bioconcentration factor for tissue fat – indicating the tendency of the chemical to accumulate in fat; F is the fat content of the tissue or dairy product; and C_w is the chemical concentration in water (HRI, 1995; USEPA, 1986).

8.4.6 Estimation of Contaminant Concentrations in Fish Products

Fish tissue contaminant concentrations may be predicted from water concentrations using chemical-specific BCFs, which predict the accumulation of contaminants in the lipids of the fish. The average chemical concentration in fish is based on the concentration in water and a bioconcentration factor, estimated by the following relationship (HRI, 1995):

$$C_f = C_w \times BCF \times 1000$$

where C_f is the concentration in fish (μg/kg), C_w is the concentration in water (mg/L), and BCF is the bioconcentration factor.

If fish tissue concentrations are predicted from sediment concentrations, a two-step process is used; first, sediment concentration is used to calculate water concentrations, then water concentrations are used to predict fish tissue concentrations. The former is carried out in accordance with the following equation:

$$C_{water} = \frac{C_{sediment}}{[K_{oc} \times OC \times DN]}$$

where C_{water} is the concentration of the chemical in water; $C_{sediment}$ is the concentration of the chemical in sediment; K_{oc} is the chemical-specific organic carbon partition coefficient; OC is the organic carbon content of the sediment; and DN is the sediment density (relative to water density).

8.5 FACTORS AFFECTING CONTAMINANT FATE AND TRANSPORT IN THE ENVIRONMENT

Environmental contamination can be transported far away from its primary source(s) of origination via a variety of natural processes (such as erosion and leaching), resulting in the possible birth of new environmental contamination problems. On the other hand, some natural processes work to lessen or attenuate contaminant concentrations in the environment through mechanisms of natural attenuation (such as dispersion/dilution, sorption and retardation, photolysis, and biodegradation).

In general, the physical and chemical characteristics of constituents present in the environment determine the fate and transport properties of the contaminants, and thus their degree of migration through the environment. Some of the particularly important constituent properties affecting the fate and transport of contaminants in the environment include the following:

- Solubility in water (which relates to leaching, partitioning, and mobility in the environment).
- Partitioning coefficients (relating to cross-media transfers, bioaccumulation potential, and sorption by organic matter).
- Vapor pressure and Henry's Law constant (relating to atmospheric mobility and the rate of vaporization or volatilization).
- Degradation/half-life (relating to the degradation of contaminants and the resulting transformation products).
- Retardation factor (which relates to the sorptivity and mobility of the constituent within the solid–fluid media).

These parameters were discussed earlier in Section 8.2. Further details and additional parameters of possible interest are discussed elsewhere in the literature (e.g., Devinny et al., 1990; Evans, 1989; Hemond and Fechner, 1994; Lindsay, 1979; Lyman et al., 1990; Mahmood and Sims, 1986; Mansour, 1993; Neely, 1980; Swann and Eschenroeder, 1983; Thibodeaux, 1979, 1996; USEPA, 1989e; Yong et al., 1992).

Several characteristics of the physical environment (such as amount of ambient moisture, humidity levels, temperatures, and wind speed; geologic, hydrologic, pedologic, and watershed characteristics; topographic features of the impacted location and its vicinity; vegetative cover of problem location and surrounding area; and land-use characteristics) may also influence the environmental fate of the chemicals of concern. Other factors such as initial contaminant concentration in the impacted media, and media pH may additionally affect the release of a chemical constituent from the environmental matrix in which it is found.

In general, the degree of chemical migration from an environmental compartment depends on both the physical and chemical characteristics of the individual constituents, and also on the physical, chemical, and biological characteristics of the affected media. For example, physical characteristics of the contaminants such as solubility and volatility influence the rate at which chemicals leach into groundwater or escape into the atmosphere. The characteristics of the environmental setting (such as the geologic or hydrogeologic features) also affect the rate of contaminant migration. In addition, under various environmental conditions, some chemicals will readily degrade to substances of relatively low toxicity, while other chemicals may undergo complex reactions to become more toxic than the parent chemical constituent.

All other factors being equal, the extent and rate of contaminant movement are a function of the physical containment of the chemical constituents or the contaminated zone. A classical illustration pertains to the fact that a low permeability cap over a contaminated site will minimize water percolation from the surface and therefore minimize leaching of chemicals into an underlying aquifer.

Environmental fate and transport analysis and modeling is typically used to assess the movement of chemicals between environmental compartments. For instance, simple mathematical models can be used to guide the decisions involved in estimating the potential spread of contaminant plumes; where applicable, monitoring equipment can then be located in areas expected to have elevated contaminant concentrations and/or in areas considered upgradient and downgradient of a contaminant plume.

SUGGESTED FURTHER READING

Cohen, Y. 1986. Organic pollutant transport. *Environmental Science and Technology*, 20(6): 538–545.
Mitchell, J.K. 1993. *Fundamentals of Soil Behavior*, 2nd edition. J. Wiley, New York.
Prager, J.C. (1995). *Environmental Contaminant Reference Databook*. Van Nostrand Reinhold, New York.
Seip, H.M. and A.B. Heiberg (eds). 1989. *Risk Management of Chemicals in the Environment*. NATO, Challenges of Modern Society, Vol. 12. Plenum Press, New York.
USEPA (US Environmental Protection Agency). 1985. *Modeling Remedial Actions at Uncontrolled Hazardous Waste Sites*. Office of Emergency and Remedial Response, Washington, DC. EPA/540/2-85/001.
Wilson, D.J. 1995. *Modeling of In Situ Techniques for Treatment of Contaminated Soils*. Technomic Publishing Co., Lancaster, PA.

REFERENCES

Aris, R. 1994. *Mathematical Modelling Techniques*. Dover Publications, Inc., New York.
Asante-Duah, D.K. 1996. *Managing Contaminated Sites: Problem Diagnosis and Development of Site Restoration*. J. Wiley, Chichester, UK.
Brooks, S.M. et al. 1995. *Environmental Medicine*. Mosby, Mosby-Year Book, Inc., St Louis, MO.
CAPCOA (California Air Pollution Control Officers Association). 1989. *Air Toxics Assessment Manual*. California Air Pollution Control Officers Association, Draft Manual, August 1987 (amended, 1989). Los Angeles, CA.
CAPCOA. 1990. *Air Toxics 'Hot Spots' Program. Risk Assessment Guidelines*. CAPCOA, Los Angeles, CA.
CCME (Canadian Council of Ministers of the Environment). 1994. *Subsurface Assessment*

Handbook for Contaminated Sites. CCME, The National Contaminated Sites Remediation Program (NCSRP), Report No. CCME-EPC-NCSRP-48E (March, 1994), Ottawa, Ontario, Canada.

CDHS (California Department of Health Services). 1986. *The California Site Mitigation Decision Tree Manual*. CDHS, Toxic Substances Control Division, Sacremento, CA.

CDHS. 1990. *Scientific and Technical Standards for Hazardous Waste Sites*. Prepared by CDHS, Toxic Substances Control Program, Technical Services Branch, Sacremento, CA.

Clark, M. 1996. *Transport Modeling for Environmental Engineers and Scientists*. J. Wiley, New York.

Cowherd, C.M., G.E. Muleski, P.J. Engelhart, and D.A. Gillette. 1985. *Rapid Assessment of Exposure to Particulate Emissions from Surface Contamination Sites*. Prepared for US EPA, Office of Health and Environmental Assessment, Washington, DC, EPA/600/8-85/002.

Devinny, J.S., L.G. Everett, J.C.S. Lu, and R.L. Stollar. 1990. *Subsurface Migration of Hazardous Wastes*. Van Nostrand Reinhold, New York.

DOE (US Department of Energy). 1987. *The Remedial Action Priority System (RAPS): Mathematical Formulations*. US Dept. of Energy, Office of Environment, Safety and Health, Washington, DC.

DTSC (Department of Toxic Substances Control). 1994. *Preliminary Endangerment Assessment Guidance Manual (A Guidance Manual for Evaluating Hazardous Substance Release Sites)*. California Environmental Protection Agency, DTSC, Sacramento, CA.

Evans, L.J. 1989. Chemistry of metal retention by soils. *Environmental Science and Technology*, 23(9), 1047–1056.

Feenstra, S., D.M. Mackay, and J.A. Cherry. 1991. A method for assessing residual NAPL based on organic chemical concentrations in soil samples. *Ground Water Monitoring Review*, Spring Issue: 189–197.

Ghadiri, H. and C.W. Rose (eds). 1992. *Modeling Chemical Transport in Soils: Natural and Applied Contaminants*. CRC Press/Lewis Publishers, Boca Raton, FL.

Gordon, S.I. 1985. *Computer Models in Environmental Planning*. Van Nostrand Reinhold, New York.

Hadley, P.W. and R.M. Sedman. 1990. A health-based approach for sampling shallow soils at hazardous waste sites using the $AAL_{soil\ contact}$ criterion. *Environmental Health Perspectives*, 18: 203–207.

Haith, D.A. 1980. A mathematical model for estimating pesticide losses in runoff. *Journal of Environmental Quality*, 9(3): 428–433.

Hemond, H.F. and E.J. Fechner. 1994. *Chemical Fate and Transport in the Environment*. Academic Press, San Diego, CA.

Holmes, G., B.R. Singh, and L. Theodore. 1993. *Handbook of Environmental Management and Technology*. J. Wiley, New York.

Honeycutt, R.C. and D.J. Schabacker (eds). 1994. *Mechanisms of Pesticide Movement into Groundwater*. Lewis Publishers/CRC Press, Boca Raton, FL.

HRI (Hampshire Research Institute). 1995. *Risk*Assistant for Windows*. HRI, Alexandria, VA.

Hwang, S.T. and J.W. Falco. 1986. Estimation of multimedia exposures related to hazardous waste facilities. In *Pollutants in a Multimedia Environment*. Y. Cohen (ed.). Plenum Press, New York, pp. 229–264.

Johnson, P.C. and R.A. Ettinger. 1991. A heuristic model for predicting the intrusion rate of contaminant vapours into buildings. *Environmental Science and Technology*, 25(8): 1445–1452.

Jury, W.A., W.J. Farmer, and W.F. Spencer. 1984. Behavior assessment model for trace organics in soil: II. Chemical classification and parameter sensitivity. *Journal of Environmental Quality*, 13(4): 567–572.

Kaplan, A. 1964. *The Conduct of Enquiry: Methodology for Behavioral Science*. Chandler Publishing Co., San Francisco, CA.

Lindsay, W.L. 1979. *Chemical Equilibria in Soils*. Wiley-Interscience, New York.

Lyman, W.J., W.F. Reehl and D.H. Rosenblatt. 1990. *Handbook of Chemical Property Estimation Methods: Environmental Behavior of Organic Compounds.* American Chemical Society, Washington, DC.

Mackay, D. and P.J. Leinonen. 1975. Rate of evaporation of low-solubility contaminants from water bodies. *Environmental Science and Technology,* 9: 1178–1180.

Mackay, D. and A.T.K. Yeun. 1983. Mass transfer coefficient correlations for volatilization of organic solutes from water. *Environmental Science and Technology,* 17: 211–217.

Mahmood, R.J. and R.C. Sims. 1986. Mobility of organics in land treatment systems. *Journal of Environmental Engineering,* 112(2): 236–245.

Mansour, M. (ed.). 1993. *Fate and Prediction of Environmental Chemicals in Soils, Plants, and Aquatic Systems.* Lewis Publishers/CRC Press, Boca Raton, FL.

Mulkey, L.A. 1984. Multimedia fate and transport models: an overview. *Journal of Toxicology – Clinical Toxicology,* 21(1–2): 65–95.

Neely, W.B. 1980. *Chemicals in the Environment (Distribution, Transport, Fate, Analysis).* Marcel Dekker, New York.

NRC (National Research Council). 1989. *Ground Water Models: Scientific and Regulatory Applications.* National Academy Press, New York.

Nyer, E.K. 1993. *Practical Techniques for Groundwater and Soil Remediation.* Lewis Publishers, Boca Raton, FL.

Samiullah, Y. 1990. *Prediction of the Environmental Fate of Chemicals.* Elsevier Applied Science (in association with BP), London, UK.

Schnoor, J.L. 1996. *Environmental Modeling: Fate and Transport of Pollutants in Water, Air and Soil.* J. Wiley, New York.

Swann, R.L. and A. Eschenroeder (eds). 1983. *Fate of Chemicals in the Environment.* ACS Symposium Series 225, American Chemical Society, Washington, DC.

Thibodeaux, L.J. 1979. *Chemodynamics: Environmental Movement of Chemicals in Air, Water and Soil.* J. Wiley, New York.

Thibodeaux, L.J. 1996. *Environmental Chemodynamics,* 2nd edition. J. Wiley, New York.

Thibodeaux, L.J. and S.T. Hwang. 1982. Landfarming of petroleum wastes – modeling the air emission problem. *Environmental Progress,* 1: 42–46.

USEPA (US Environmental Protection Agency). 1985a. *Chemical, Physical, and Biological Properties of Compounds Present at Hazardous Waste Sites.* Report prepared by Clement Associates for the USEPA, September 1985.

USEPA. 1985b. *Rapid Assessment of Exposure to Particulate Emissions from Surface Contamination Sites.* Office of Health and Environmental Assessment, Washington, DC. EPA/600/8-85/002, NTIS PB85-192219.

USEPA. 1986. *Methods for Assessing Exposure to Chemical Substances, Volume 8: Methods for Assessing Environmental Pathways of Food Contamination.* Office of Toxic Substances, Washington, DC. EPA/560/5-85-008.

USEPA. 1987. *RCRA Facility Investigation (RFI) Guidance.* Washington, DC. EPA/530/SW-87/001.

USEPA. 1988a. *GEO-EAS (Geostatistical Environmental Assessment Software) User's Guide.* Environmental Monitoring Systems Laboratory, Office of R&D, Las Vegas, NV. EPA/600/4-88/033a.

USEPA. 1988b. *Superfund Exposure Assessment Manual.* OSWER Directive 9285.5-1. USEPA, Office of Remedial Response, Washington, DC. EPA/40/1-88/001.

USEPA. 1989a. *Application of Air Pathway Analyses for Superfund Activities.* Air/Superfund National Technical Guidance Study Series. Procedures for Conducting Air Pathway Analyses for Superfund Applications, Volume I. Interim Final. Office of Air Quality Planning and Standards, Research Triangle Park, NC. EPA/450/1-89-001.

USEPA. 1989b. *Estimation of Air Emissions from Cleanup Activities at Superfund Sites.* Air/Superfund National Technical Guidance Study Series, Volume III. Interim Final. Office of Air Quality Planning and Standards, Research Triangle Park, NC. EPA/450/1-89-003.

USEPA. 1989c. *Procedures for Conducting Air Pathway Analyses for Superfund Applications.* Volume IV – Procedures for Dispersion Modeling and Air Monitoring for Superfund Air

Pathway Analysis. Air/Superfund National Technical Guidance Study Series. Interim Final. Office of Air Quality Planning and Standards, Research Triangle Park, NC. EPA/450/1-89-004.

USEPA. 1989d. *Review and Evaluation of Area Source Dispersion Algorithms for Emission Sources at Superfund Sites.* Office of Air Quality Planning and Standards, Research Triangle Park, NC. EPA/450/4-89-020.

USEPA. 1989e. *Risk Assessment Guidance for Superfund. Volume I – Human Health Evaluation Manual (Part A).* Office of Emergency and Remedial Response, Washington, DC. EPA/540/1-89/002.

USEPA. 1990a. *Estimation of Baseline Air Emissions at Superfund Sites.* Air/Superfund National Technical Guidance Study Series. Procedures for Conducting Air Pathway Analyses for Superfund Applications, Volume II. Office of Air Quality Planning and Standards, Research Triangle Park, NC. EPA/450/1-89-002a.

USEPA. 1990b. *Air/Superfund National Technical Guidance Study Series. Development of Example Procedures for Evaluating the Air Impacts of Soil Excavation Associated with Superfund Remedial Actions.* Office of Air Quality Planning and Standards, Research Triangle Park, NC. EPA/450/4-90-014.

USEPA. 1992. *Guideline for Predictive Baseline Emissions Estimation Procedures for Superfund Sites.* Interim Final. Office of Health and Environmental Assessment Air/Superfund National Technical Guidance Study Series, Cincinatti, Ohio. EPA/450/I-92-002.

USEPA. 1993. *National Ambient Air Quality Standard for Particulate Matter.* Federal Register: 40 CFR, Part 50.6.

USEPA. 1996a. *Soil Screening Guidance: Technical Background Document.* Office of Emergency and Remedial Response, Washington, DC. EPA/540/R-95/128.

USEPA. 1996b. *Soil Screening Guidance: User's Guide.* Office of Emergency and Remedial Response, Washington, DC. EPA/540/R-96/018.

Williams, D.R., J.C. Paslawski, and G.M. Richardson. 1996. Development of a screening relationship to describe migration of contaminant vapors into buildings. *Journal of Soil Contamination*, 5(2): 141–156.

Yong, R.N., A.M.O. Mohamed, and B.P. Warkentin. 1992. *Principles of Contaminant Transport in Soils.* Developments in Geotechnical Engineering, 73, Elsevier Scientific Publishers B.V., Amsterdam, The Netherlands.

Zirschy, J.H. and D.J. Harris. 1986. Geostatistical analysis of hazardous waste site data. *ASCE Journal of Environmental Engineering*, 112(4).

Chapter Nine

Hazard Identification, Data Collection, and Data Evaluation

The first issue in any attempt to conduct a risk assessment relates to answering the seemingly straightforward question – 'does a hazard exist?'. The hazard identification component of a risk assessment for environmental contamination problems involves establishing the presence of an environmental contaminant or stressor that could potentially cause an adverse effect in potential receptors. The process involves a consideration of the major sources of hazards that may contribute to a variety of environmental contamination problems and possible risk situations.

Once the presence of potential environmental contaminants has been established, environmental samples will usually be gathered and analyzed for the likely chemicals of concern in the appropriate media of interest. The relevant data should help identify the chemicals present at a given locale that are to be the focus of the risk assessment process. Effective analytical protocols in the sampling and laboratory procedures are generally required, to help minimize uncertainties associated with the data collection and evaluation aspects of the risk assessment.

9.1 SOURCES OF ENVIRONMENTAL HAZARDS

The major sources of environmental hazards associated with environmental contamination problems generally pertain to the handling and management of hazardous materials (Box 9.1). Chemical constituents from the variety of sources may be released by several different mechanisms, resulting in the contamination of various environmental media or matrices (Table 9.1). As a consequence, there is a corresponding variability in the range and type of hazards and risks that may be anticipated from different environmental contamination problems.

Mostly, qualitative information on potential sources and likely consequences of hazard releases is all that is required during this early stage – the hazard identification phase – of the risk assessment process. To add a greater level of sophistication in the hazard identification process, quantitative techniques may be incorporated in order to determine the likelihood of an actual release situation occurring. The quantitative methods may consist of the use of environmental modeling techniques (see Chapter 8) to determine contaminant release potentials.

> Box 9.1 Major sources of hazards associated with environmental contamination problems
>
> - Landfills
> - Waste tailings
> - Waste piles
> - Surface impoundments
> - Materials stockpiles
> - Above-ground storage tanks
> - Underground storage tanks
> - Land application of wastewaters
> - Pipelines for hazardous materials
> - Spills from loading and unloading of hazardous materials
> - Spills from hazardous materials transport accidents
> - Irrigation practices
> - Pesticide, herbicide, and fertilizer applications
> - Urban runoff
> - Mining and mine drainage
> - Injection wells for hazardous wastes
> - Treatment system and incinerator emissions

9.2 HAZARD ACCOUNTING AND ENVIRONMENTAL INVESTIGATION

When it is suspected that some degree of hazard exists at a particular locale, it becomes necessary to further investigate the situation and to fully take account of the prevailing or anticipated hazards. This may be accomplished by the use of a well-designed environmental investigation program.

Environmental investigations consist of the planned and managed sequence of activities carried out to determine the nature and distribution of contaminants associated with potential environmental contamination problems. The activities involved usually are comprised of the following tasks (BSI, 1988):

- identification of the principal hazards;
- design of sampling and analysis programs;
- collection and analysis of environmental samples; and
- recording or reporting of laboratory results for further evaluation.

In order to get the most out of an environmental investigation, it must be conducted in a systematic manner. Systematic methods help focus the purpose, the required level of detail, and the several topics of interest – such as physical characteristics of the problem locale, likely contaminants, extent and severity of possible contamination, effects of contaminants on populations potentially at risk, probability of harm to human health and the environment, and possible hazards during risk management and corrective action activities (Cairney, 1993).

Table 9.1 Major causes and mechanisms of contaminant releases into various environmental media

Representative contaminant source	Typical release causes and mechanisms	Primary receiving/impacted media
Surface impoundments (e.g., lagoons, ponds, pits)	Loading/unloading activities Overtopping dikes and surface runoff Seepage and infiltration/percolation Fugitive dust generation Volatilization	Air Soils and sediments Surface water Groundwater
Waste management units (e.g., landfill, land treatment unit, and waste pile)	Migration of releases outside unit's runoff collection and containment system Migration of releases outside the containment area from loading and unloading operations Seepage and infiltration Leachate migration Fugitive dust generation Volatilization	Air Soils and sediments Surface water Groundwater Subsurface gas (in soil pores, vents, and cracks migrating through soil)
Waste management zones (e.g., container storage area and storage tanks)	Migration of runoff outside containment area Loading/unloading area spills Leaking drums, leaks through tank shells, and leakage from cracked or corroded tanks Releases from overflows Leakage from coupling/uncoupling operations	Air Soils and sediments Surface water Groundwater Subsurface gas (in soil pores, vents, and cracks migrating through soil)
Waste treatment plants/facilities	Effluent discharge to surface and groundwater resources	Surface water (by dissolution, dispersion, transport, etc.) Sediments (from adsorbed chemicals) Groundwater
Incinerators	Routine releases from waste handing/preparation activities Leakage due to mechanical failure Stack emissions	Air Foliage (from particulate deposition and atmospheric fallout) Soils (from particulate deposition and atmospheric washout) Surface water (from particulate deposition and atmospheric washout)

continues overleaf

Table 9.1 (*continued*)

Representative contaminant source	Typical release causes and mechanisms	Primary receiving/impacted media
Injection wells	Leakage from waste handling operations at the well head	Groundwater (by dissolution, diffusion, dispersion, etc.) Surface water (from groundwater recharge)

9.2.1 Development of Data Quality Objectives

Data quality objectives (DQOs) are statements that specify the data needed to support environmental management decisions. They are described to establish the desired degree of data reliability, the specific data requirements and considerations, and an assessment of the data applications as determined by the overall study objective(s). The DQO represents the full set of qualitative and quantitative constraints needed to specify the level of uncertainty that an analyst can accept when making a decision based on a particular set of data.

The DQO process consists of a planning tool that enables an investigator to specify the quality of the data required to support the objectives of a given study. In the course of the investigation of potential environmental contamination problems, DQOs are typically used as qualitative and quantitative statements that specify the quality of data required to support risk assessment and environmental management programs. These are determined based on the end uses of the data to be collected.

DQOs are indeed an important aspect of the quality assurance requirements in the entire environmental management process. Despite the fact that the DQO process is considered flexible and iterative, it generally follows a well-defined sequence of stages that will allow effective and efficient data management. In any event, it is apparent that, as the quantity and quality of data increases, the risk of making a wrong decision generally decreases (Figure 9.1) (USEPA 1987a, 1987b). Consequently, the DQO process should be carefully developed to allow for a responsible program that will produce adequate quantity and quality of information required for environmental management decisions. In fact, there are several benefits to establishing adequate DQOs for environmental management programs, including the following (CCME, 1993):

- The DQO process ensures that the data generated are of known quality.
- All projects have some inherent degree of uncertainty, and DQOs help data users plan for uncertainty. By establishing DQOs, data users evaluate the consequences of uncertainty and specify constraints on the level of uncertainty that can be tolerated in the expected study results. Thus, the likelihood of an incorrect decision is gaged *a priori*.

Hazard identification, data collection, and data evaluation 159

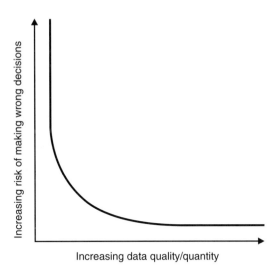

Figure 9.1 Relationship between decision risk and data quality/quantity

- The DQO process facilitates communication among data users, data collectors, managers, and other technical staff before time and money are spent collecting data. This will result in increased cost-efficiency in the data collection program.
- The DQO process provides a logistical structure for study planning that is iterative and that encourages the data users to narrow many vague objectives to one or a few critical questions.
- The structure of the DQO process provides a convenient way to document activities and decisions that can prove useful in litigation or administrative procedures.
- The DQO process establishes quantitative criteria for use as a cut-off line, as to when to stop collecting environmental samples.

Overall, the DQO process results in a well thought-out sampling and analysis plan. Consequently, DQOs should preferably be established prior to data collection activities, to ensure that data collected are sufficient and of adequate quality for their intended uses. The DQO should indeed be integrated with development of the sampling and analysis plan, and should be revised as needed – based on the results of each data collection activity.

9.2.2 Design of an Environmental Investigation Program

Often, environmental investigation activities are designed and implemented in accordance with several regulatory and legal requirements of the geographical region or area in which a potential environmental contamination problem is located. In a typical investigation, the influence of the responsible regulatory agencies may affect the several operational elements necessary to complete the

Table 9.2 Major tasks and important elements of an environmental investigation program

Task	Elements
Problem definition	• Define project objectives (including the level of detail and topics of interest) • Determine data quality objectives (DQOs)
Preliminary evaluation	• Collect and analyze existing information (i.e., review available background information, previous reports, etc.) • Conduct visual inspection of the locale (i.e., reconnaissance surveys) • Construct preliminary conceptual model of the problem situation
Sampling design	• Identify information required to refine conceptual model of the problem situation • Identify constraints and limitations (e.g., accessibility, regulatory controls, utilities, and financial limitations) • Define sampling, analysis, and interpretation strategies • Determine possible exploratory techniques and testing program
Implementation of sampling and analysis plans	• Conduct necessary and appropriate exploratory work, and perform appropriate testing at the problem location • Carry out sampling activities • Compile record of investigation logs, photographs, and sample details • Perform laboratory analyses
Data evaluation	• Compile a database of relevant information at the locale • Carry out logical analysis of environmental data • Refine conceptual model for the problem situation
Interpretation of results	• Enumerate implications of environmental investigation results • Prepare a report on findings

environmental investigation program (Asante-Duah, 1996). Irrespective of whichever regulatory authority is involved, however, the basic environmental investigation strategy adopted for environmental contamination problems typically will comprise the specific tasks and general elements summarized in Table 9.2 (Cairney, 1993; USEPA, 1988b). Ultimately, the data derived from the environmental investigation may be used to perform a risk assessment, that becomes a very important element in the environmental management decision.

The general types of environmental data and information required in the investigation of potential environmental contamination problems relate to contaminant identities; contaminant concentrations in the key sources and media of interest; characteristics of sources and contaminant release potential; and characteristics of the physical and environmental setting that can affect the fate, transport, and persistence of the contaminants (USEPA, 1989a). In fact, because of the inherent variability in the materials and the diversity of processes used in industrial activities, it is not unexpected to find a wide variety of contaminants associated with an environmental contamination problem. Thus, the design and implementation of a substantive data collection and evaluation program is vital to the effective management of environmental contamination problems. A file of detailed background information on the critical contaminants of potential concern should be compiled as part of the environmental investigation program.

In general, a phased approach to environmental sampling encourages the identification of key data requirements in the environmental investigation process. This ensures that the data collection effort is always directed at providing adequate information that meets the data quantity and quality requirements of the study. As a basic understanding of the particular problem situation is achieved, subsequent data collection efforts focus on identifying and filling in any remaining data gaps from a previous phase of the environmental investigation. Any additionally acquired data should be such as to further improve the understanding of the problem situation, and also consolidate information necessary to manage the environmental contamination problem in an effective manner. In this way, the overall environmental investigation effort can be continually re-scoped to minimize the collection of unnecessary data and to maximize the quality of data acquired.

The analysis of previously acquired and newly generated data serves to provide an initial basis to understanding the nature and extent of contamination, which in turn aids in the design of appropriate risk assessment and environmental management programs. Consequently, at any reasonable stage of an environmental investigation, all available site information should be compiled and analyzed to develop a conceptual model for the problem situation. This representation should incorporate contaminant sources and 'sinks', the nature and behavior of the environmental contaminants, migration pathways, the affected environmental matrices, and potential receptors.

Overall, the data gathering process should provide a logical, objective, and quantitative balance between the time and resources available for collecting the data and the quality of data, based on the intended use of such data.

9.3 DATA COLLECTION AND ANALYSIS

The investigation of a potential environmental contamination problem must provide information on all contaminants known, suspected, or believed to be present at a given locale. In fact, the investigation should cover all compounds for which the history of local activities, current visible signs of pollution, or public concerns suggest the possibility of contamination by such substances. In addition to establishing the concentration of contaminants at a locale, an environmental investigation should be designed to provide an indication of the naturally occurring or anthropogenic background levels of the target contaminants in the local environment.

Ultimately, several chemical-specific factors (such as toxicity or potency, concentration, mobility, persistence, bioaccumulative or bioconcentration potential, synergistic or antagonistic effects, potentiation or neutralizing effects, frequency of detection, and naturally-occurring or anthropogenic background thresholds) are used to screen and select the specific target contaminants that will become the focus of a detailed risk assessment and environmental management process. Typically, the selected target chemicals are those environmental contaminants that are generally the most mobile and persistent; consequently, they reflect the likelihood of contamination at the problem location.

9.3.1 Data Collection and Analysis Strategies

Traditionally, the characterization of environmental contamination has been accomplished by taking several environmental samples, sealing them in sample containers, and shipping them for laboratory analyses. When the analytes of interest are volatile organic compounds (VOCs), the sample may be extracted with solvent upon arrival at the laboratory and the extract subsequently analyzed in a gas chromatograph (GC) or a gas chromatograph/mass spectrometer (GC/MS). Whereas these procedures tend to accommodate broad-spectrum analysis and low levels of detection, they are also known to suffer severe limitations with respect to precision and accuracy. The failings of this approach are the result of both the heterogeneity of the sampled medium and the ease with which VOCs vaporize and escape from the sample during the activities preceding the analysis. Consequently, environmental sampling and analysis can grossly misrepresent VOC concentrations in affected media, by biasing them to the low end. In some cases, the under-estimation can be so significant as to seriously affect the results of an environmental investigation effort, and ultimately the development of a risk management and corrective action plan for the problem situation. It is noteworthy, however, that whereas individual environmental samples may under-estimate VOC levels, a suite of samples from across an impacted area will reliably indicate if a contamination problem exists and is likely to indeed characterize the nature of the plume adequately.

Data are generally collected at several stages of the environmental investigation, with initial data collection efforts usually limited to developing a general understanding of the problem situation. In areas where the contamination source is known, the sampling strategy should be targeted around that source. Normally sampling points should be located at regular distances along lines radiating from the contaminant source. Provisions should also be made in the investigation to collect additional samples of small, isolated pockets of material which are visually suspect. A variety of data collection and analysis protocols exist in the literature (e.g., Boulding, 1994; Byrnes, 1994; CCME, 1993, 1994; Csuros, 1994; Garrett, 1988; Hadley and Sedman, 1990; Keith, 1992; Millette and Hays, 1994; O'Shay and Hoddinott, 1994; Schulin et al., 1993; Thompson, 1992; USEPA, 1982, 1985a, 1985b; Wilson et al., 1995) that may be employed or adopted in the environmental investigation program.

Further to collecting and analyzing samples from the impacted area, background (or control site) samples are typically collected and evaluated to determine the possibility of a potential environmental contamination problem contributing to off-site contamination levels in the vicinity of the problem site. The background samples would not have been significantly influenced by contamination from the project site. However, these samples are obtained from an environmental matrix that has similar basic characteristics to those of the matrix at the project site, in order to provide a justifiable basis for comparison.

Control locations, considered not to have been impacted by an environmental contamination problem, are indeed important to understanding the significance of environmental sampling and monitoring data. Such locations should generally have similar characteristics (i.e., identical in their physical and environmental settings) to the potentially contaminated area or locale under investigation. Background samples collected at control locations or sites are typically used to demonstrate whether or not a project location or site is truly contaminated. The background sampling results allow a technically valid scientific comparison to be made between environmental samples (suspected of containing chemical contaminants associated with the problem location) and control location or site samples (possibly containing only naturally low or anthropogenic levels of the same chemicals).

9.3.2 Background Sampling Considerations

Background sampling is conducted to distinguish site-related contamination from naturally occurring or other nonsite-related levels of select constituents. Anthropogenic levels (which are concentrations of chemicals that are present in the environment due to human-made, nonsite sources, such as industry and automobiles), rather than naturally occurring levels, are preferably used as a basis for evaluating background sampling data.

There are two types of background or control locations/sites – local and area – whose differentiation is based primarily on the closeness of the control site to the environmental sampling or project site. Local control sites are usually adjacent or very near the potentially impacted project sites, whereas an area control site is in the same general area or region as the project site, *but not adjacent to it*. In selecting and working with either type of control site, the following general principles should be observed insofar as possible (CCME, 1993; Keith, 1991):

- Control sites/locations generally should be upwind, upstream, and/or upgradient from the environmental sampling site.
- When possible, control site/location samples should be taken first to avoid possible cross-contamination from the environmental sampling site.
- Travel between control sites/locations and environmental sampling areas should be minimized because of potential cross-contamination caused by humans, equipment, and/or vehicles.

In general, local control sites are preferable to area control sites because they are physically closer. However, when a suitable local control site cannot be found, an area control site will generally provide for the requisite background sampling information.

Ultimately, sampling information from the control sites is used to establish background thresholds. The background threshold is meant to give an indication of the level of contamination in the environment that may not necessarily be attributed to the potential environmental contamination problem under investigation. This serves to provide a reference 'point-of-departure' that can be used to determine the magnitude of contamination in other environmental samples obtained from a contaminated locale. Ideally, background samples (or control site samples) are preferably collected near the time and place of the environmental samples of interest.

9.3.2.1 Factors affecting background sampling design

To satisfy acceptable criteria required of background samples, the following requirements should be carefully incorporated into the design of background sampling programs (Asante-Duah, 1996):

- *Significance of matrix effects on environmental contaminant levels.* Unless background samples are collected and analyzed under the same conditions as the environmental test samples, the presence and/or levels of the analytes of interest and the effects of the matrix on their analysis cannot be known or estimated with any acceptable degree of certainty. Therefore, background samples of each significantly different matrix must always be collected when different types of matrices are involved, such as various types of water, sediments, and soils in or near a sampling site area. For example, it has been observed in a number of investigations that the analysis of data from different soil types in the same background area reveal levels of select inorganic constituents that are over twice as high in silt/clay as in sand (e.g., LaGoy and Schulz, 1993). Thus, it is important to give adequate consideration to the effects of natural variations in soil composition when one is designing a field sampling program.
- *Number of background samples.* A minimum of three background samples per medium will usually be collected, although more may be desired especially in complex environmental settings. In general, if the natural variability of a particular constituent present in the environment is relatively large, the sampling plan should reflect this case-specific characteristic.
- *Background sampling locations.* In typical sampling programs, background air samples would consist of upwind air samples and, perhaps, different height samples; background soil samples would be collected near a site in areas upwind and upslope of the site; background groundwater samples generally come from upgradient well locations, in relation to groundwater flow direction(s) at the impacted area; and background surface water and sediment samples may be collected under both high and low flow conditions at upstream locations, and insofar as practicable, sample collection from nearby lakes and wetlands should comprise shallow and deep samples (when sufficient water depth allows), to account for such differences potentially resulting from stratification or incomplete mixing.

More detailed background sampling considerations and strategies for the various environmental media of general interest can be found in the literature elsewhere (e.g., Keith, 1991; Lesage and Jackson, 1992; USEPA, 1988b, 1989b).

9.3.2.2 Evaluation of background sampling data

The statistical evaluation of background sampling results will typically comprise a determination of whether or not there is a difference between contaminant average concentrations in the background areas and the chemical concentrations at the problem location – and especially if indeed the concentrations are higher at the problem location. Broadly speaking, however, a statistically significant difference between background samples and local contamination should not, by itself, be a cause for alarm nor should it trigger off a remedial action; further evaluations (such as conducting a risk assessment) will ascertain the significance of the contamination. In cases where corrective action is required, the established background thresholds may be used as part of the basis for setting realistic remedial action goals. The ability to estimate background or threshold concentrations can indeed become a critical factor in the formulation of reasonable remedial action objectives associated with potential environmental contamination problems.

It is noteworthy that, even at background concentration levels, a given locale may still be posing significant risks to human health and the environment. Whereas remedial action may not be required under such circumstances, it is important to evaluate and document this situation so that at least some risk management or institutional control measures can be implemented to protect populations potentially at risk in this area.

9.3.3 Evaluation of Quality Control Samples

Most environmental sampling and analysis procedures offer numerous opportunities for sample contamination from a variety of sources (Keith, 1988). To be able to address and account for possible errors arising from 'foreign' sources, quality control (QC) samples are typically included in the sampling and analytical schemes. The QC samples are analytical quality control samples analyzed in the same manner as the environmental samples, and are subsequently used in the measurement of any contamination that may have been introduced into a sample along its life cycle from the field (i.e., point of collection) to the laboratory (i.e., place of analysis). To prevent the inclusion of 'foreign' constituents in the characterization of environmental contamination and/or in a risk assessment, the concentrations of the chemicals detected in 'control' samples must be compared with concentrations of the same chemicals detected in the environmental samples from the impacted media.

In general, QC samples containing common laboratory contaminants are evaluated differently from QC samples which contain chemicals that are not common laboratory contaminants. Thus, if the QC samples contains detectable levels of known common laboratory contaminants (e.g., acetone, 2-butanone [methyl ethyl ketone], methylene chloride, toluene, and the phthalate esters), then the environmental sample results may be considered as positive only if the concentrations in the sample exceed approximately ten times the maximum amount detected in any QC sample (DTSC, 1994; USEPA, 1989b, 1990). For QC samples containing detectable levels of one or more organic or inorganic chemicals that are not considered to be common laboratory contaminants, environmental sample results may be considered as positive only if the concentration of the chemical in the environmental sample exceeds approximately five times the maximum amount detected in any QC sample (DTSC, 1994; USEPA, 1989b, 1990).

Invariably, QC samples become an essential component of any environmental investigation program. This is because firm conclusions cannot be drawn from the environmental investigation unless adequate controls have been included as part of the sampling and analytical protocols (Keith, 1988).

Environmental QC samples can indeed be a very important reference datum for the evaluation of environmental sampling data. The analysis of environmental QC samples provides a way to determine whether contamination has been introduced into a sample set either in the field while the samples were being collected and transported to the laboratory, or in the laboratory during sample preparation and analysis.

9.4 EVALUATION OF ENVIRONMENTAL SAMPLING DATA

Over the years, extensive technical literature has been developed regarding the 'best' probability distribution to utilize in different scientific applications. Of the many statistical distributions available, the Gaussian (or normal) distribution has been widely utilized to describe environmental data. However, there is considerable support for the use of the lognormal distribution in describing environmental data; the use of lognormal statistics for the data set X_1, X_2, X_3, ..., X_n requires that the logarithmic transform of these data (i.e., $\ln[X_1]$, $\ln[X_2]$, $\ln[X_3]$, ..., $\ln[X_n]$) can be expected to be normally distributed. Consequently, chemical concentration data in the environment have been described by the lognormal distribution, rather than by a normal distribution (Gilbert, 1987; Leidel and Busch, 1985; Rappaport and Selvin, 1987; Saltzman, 1997). In fact, the use of a normal distribution (whose central tendency is measured by the arithmetic mean) to describe environmental contaminant distribution, rather than lognormal statistics (whose central tendency is defined by the geometric mean), will often result in significant over-estimation, and may be overly conservative – albeit some investigators may argue otherwise (e.g., Parkhurst, 1998). Parkhurst (1998) argues that geometric means are biased low and do not represent components of mass balances properly, *whereas* arithmetic means are unbiased, easier to calculate and understand, scientifically more meaningful for concentration data, and more protective of public health; however, he also concedes to the nonuniversality of this school of thought. All these arguments and counter-arguments only go to reinforce the fact that no one particular parameter or distribution may be appropriate for every situation, and that care must be exercised in the choice of statistical methods for data manipulation.

More sophisticated methods, such as geostatistical techniques that account for spatial variations in concentrations may also be employed for estimating the average environmental concentrations (e.g., USEPA, 1988a; Zirschy and Harris, 1986). For example, a technique called block kriging is frequently used to estimate soil chemical concentrations in sections of contaminated sites in which only sparse sampling data exist. In this case, the site is divided into blocks (or grids), and concentrations are determined within blocks by using interpolation procedures that incorporate sampling data in the vicinity of the block. The sampling data are weighted in proportion to the distance of the sampling location from the block.

9.4.1 General Approach to the Statistical Analysis of Environmental Sampling Data

Statistical procedures used for the evaluation of environmental data can indeed significantly affect the conclusions of a given environmental characterization and risk assessment program. Consider, for instance, the use of a normal distribution (whose central tendency is measured by the arithmetic mean) to describe environmental contaminant distribution, rather than lognormal statistics (whose central tendency is defined by the geometric mean); the former will often result in significant over-estimation of contamination levels. Appropriate statistical methods should therefore be utilized in the evaluation of environmental sampling data (e.g., in relation to the choice of proper averaging techniques). Furthermore, contamination levels and exposures may have temporal variations; the dynamic nature of such parameters should, insofar as possible, be incorporated in the evaluation of the environmental data.

In general, statistical procedures used in the evaluation of environmental data should reflect the character of the underlying distribution of the data set. The appropriateness of any distribution assumed or used for a given data set should preferably be checked prior to its application; this can be accomplished by using some goodness-of-fit methods (see, e.g., Cressie, 1994; Freund and Walpole, 1987; Gilbert, 1987; Miller and Freund, 1985; Sharp, 1979; Wonnacott and Wonnacott, 1972).

Ultimately, the process/approach used to estimate a potential receptor's exposure point concentration (EPC) to a chemical constituent associated with an environmental contamination problem will comprise: determining the distribution of the environmental sample data, and fitting the appropriate distribution to the data (e.g., normal, lognormal, etc.); and developing the basic statistics for the sample data – to include calculation of such relevant statistical parameters such as the upper 95% confidence limit (UCL_{95}). Subsequently, the EPC will usually be defined as the minimum of either the UCL or the maximum sample data value, conceptually represented as follows:

$$EPC = \min (UCL_{95} \text{ or max value})$$

The *EPC* so obtained, which may be significantly different from the field-measured concentration, represents the true contaminant level at the location of the potential receptor. This value will therefore be used to calculate the chemical intake/dose by the populations potentially at risk.

The choice of statistical parameters for environmental characterization programs is indeed critical to the risk management and/or corrective action decisions about an environmental contamination problem. Several of the available statistical methods and procedures finding widespread use in environmental management programs can be found in the literature on statistics (e.g., Berthouex and Brown, 1994; Cressie, 1994; Freund and Walpole, 1987; Gibbons, 1994; Gilbert, 1987; Hipel, 1988; Miller and Freund, 1985; Ott, 1995; Sharp, 1979; Wonnacott and Wonnacott, 1972; Zirschy and Harris, 1986). Some commonly used methods of approach that find general application in the evaluation of environmental data are briefly discussed below.

9.4.1.1 Parametric versus nonparametric statistics

There are several statistical techniques available for analyzing data that are not dependent on the assumption that the data follow any particular statistical distribution. These distribution-free methods, referred to as *nonparametric* statistical tests, have fewer and less stringent assumptions. Conversely, several assumptions have to be met before one can use a *parametric* test. In fact, whenever the set of requisite assumptions are met, it is preferred to use a parametric test, because it is more powerful than the nonparametric test. However, to reduce the number of underlying assumptions required to test a hypothesis (such as the presence of specific trends in a data set), nonparametric tests are typically employed. Nonparametric techniques are generally selected when the sample sizes are small and the statistical assumptions of normality and homogeneity of variance are tenuous.

In general, nonparametric tests are usually adopted for use in environmental impact assessments because the statistical characteristics of the often messy environmental data make it difficult, or even unwise, to use many of the available parametric methods. It is noteworthy, however, that nonparametric tests tend to ignore the magnitude of the observations in favor of the relative values or ranks of the data. Consequently, as Hipel (1988) notes, a given nonparametric test with few underlying assumptions that is designed, for instance, to test for the presence of a trend may only provide a 'yes' or 'no' answer as to whether or not a trend may be contained in the data. The output from the nonparametric test may not give an indication of the type or magnitude of the trend. To have a more powerful test about what is occurring, many assumptions must be made, and as more and more assumptions are formulated, a nonparametric test begins to look more like a parametric test (Hipel, 1988).

9.4.1.2 Hypothesis testing

A typical and common example of statistical applications in environmental management programs relates to the use of the analysis-of-variance (ANOVA) model in hypotheses testing of environmental variables. ANOVA is a method for partitioning the total variation in a set of data into the different sources of variation that are present. ANOVA models are also used to analyze the effects of the independent variable(s) on the dependent variable(s).

The application of ANOVA to environmental impact evaluations will generally result in a summary table that provides a convenient form for presenting information contained in the environmental data sets. For instance, ANOVA can be used to study data from groundwater monitoring wells as an integral part of a corrective action program. In the context of groundwater monitoring, wells or groups of wells represent the independent variables and the groundwater contaminant concentrations represent the dependent variable. The ANOVA will help determine whether different wells (or group of wells) have significantly different concentrations of the contaminants of concern, by comparing site or compliance monitoring wells with background wells. The contrasts of interest may involve a comparison between the mean concentration of the background wells and the mean concentration of each compliance well.

9.4.1.3 Selection of statistical averaging techniques

The selection of appropriate methods of approach to averaging a set of environmental sampling data can have profound effects on the resulting concentration, especially for data sets of sampling results that are not normally distributed. Consequently, reasonable discretion should be exercised in the selection of an averaging technique during the analysis of environmental sampling data. For example, when dealing with lognormally distributed data, geometric means are often used as a measure of central tendency, to ensure that a few very high values do not exert excessive influence on the characterization of the distribution. If, however, high concentrations do indeed represent 'hot-spots' in a spatial distribution, then using the geometric mean would inappropriately discount the contribution of these high contaminant concentrations present in the environment. This is particularly true if the spatial pattern indicates that areas of high concentration are in close proximity to compliance boundaries or points of exposure to populations potentially at risk.

The geometric mean has indeed been extensively used as an averaging parameter in the past. Its principal advantage is in minimizing the effects of 'outlier' values (i.e., a few values that are much higher or lower than the general range of sample values). Its corresponding disadvantage is that discounting these values may be inappropriate, when they represent true variations in concentrations from one part of a contaminated area to another (such as a 'hot-spot' versus a 'cold-spot'). As a measure of central tendency, the geometric mean is most appropriate if sample data are lognormally distributed, without an obvious spatial pattern.

The arithmetic mean – commonly used when referring to an 'average' – is more sensitive to a small number of extreme values or a single 'outlier' compared to the geometric mean. Its corresponding advantage is that true high concentrations will not be inappropriately discounted. With limited sampling data, however, this may not necessarily provide a conservative enough estimate of environmental contamination.

Table 9.3 Environmental sampling data used to illustrate the effects of statistical averaging techniques on concentration predictions

Sampling event	Concentration of benzene in water (μg/L)	
	Original 'raw' data, X	Log-transformed data, $Y = \ln(X)$
1	0.049	−3.016
2	0.056	−2.882
3	0.085	−2.465
4	1.200	0.182
5	0.810	−0.211
6	0.056	−2.882
7	0.049	−3.016
8	0.048	−3.037
9	0.062	−2.781
10	0.039	−3.244
11	0.045	−3.101
12	0.056	−2.882

In fact, none of the above measures, in themselves, may be appropriate in the face of limited and variable sampling data. Current applications tend to favor the use of an upper confidence limit (UCL) on the average concentration. Even so, if the UCL exceeds the maximum detected value, then the latter is used as the source term. In situations where there is a discernible spatial pattern to contaminant concentration data, standard approaches to data aggregation and analysis may usually be inadequate or even inappropriate.

To demonstrate the possible effects of the choice of statistical distributions and/or averaging techniques on the analysis of environmental data, consider a case involving the estimation of the mean, standard deviation, and confidence limits of monthly groundwater sampling data from a contaminated site. In order to compare the selected statistical parameters based on the assumption that this data is normally distributed versus an alternative assumption that the data is lognormally distributed, the several statistical manipulations enumerated below are carried out on the 'raw' and log-transformed data for the concentrations of benzene in the groundwater samples shown in Table 9.3.

1. Calculate the following statistical parameters for the 'raw' data: mean, standard deviation, and 95% confidence limits (see standard statistics textbooks for details of procedures involved). The arithmetic mean, standard deviation, and 95% confidence limits (95% CL) for a set of n values are defined, respectively, as follows:

$$X_m = \frac{\sum_{i=1}^{n} X_i}{n}, \quad SD_x = \sqrt{\frac{\sum_{i=1}^{n}(X_i - X_m)^2}{n-1}} \quad \text{and} \quad CL_x = X_m \pm \frac{ts}{\sqrt{n}}$$

where t is the value of the Student t-distribution (refer to standard statistical texts) for the desired confidence level and degrees of freedom, $(n-1)$, and s is an estimate of the standard deviation from the mean (X_m). Thus,

$X_m = 0.213$ µg/L
$SD_x = 0.379$ µg/L
$CL_x = 0.213 \pm 0.241$ (i.e., $-0.028 \leq CI_x \leq 0.454$) and $UCL_x = 0.454$ µg/L

where X_m = arithmetic mean of 'raw' data; SD_x = standard deviation of 'raw' data; CI_x = 95% confidence interval (95% CI) of 'raw' data; and UCL_x = 95% upper confidence level (95% UCL) of 'raw' data.

Note that, the development of 95% confidence limits for the untransformed data gives a confidence interval of $0.213 \pm 0.109t = 0.213 \pm 0.241$ (where $t = 2.20$, obtained from the Student t-distribution for $[n-1] = 12-1 = 11$ degrees of freedom), indicating the possibility for a nonzero probability of a negative concentration value.

2. Calculate the following statistical parameters for the log-transformed data: mean, standard deviation, and 95% confidence limits (see standard statistics textbooks for details of procedures involved). The geometric mean, standard deviation, and 95% confidence limits (95% CL) for a set of n values are defined, respectively, as follows:

$$X_{gm} = \text{antilog}\left\{\frac{\sum_{i=1}^{n} \log X_i}{n}\right\}, \quad SD_x = \sqrt{\left[\frac{\sum_{i=1}^{n}(X_i - X_{gm})^2}{n-1}\right]} \quad \text{and} \quad CL_x = X_{gm} \pm \frac{ts}{\sqrt{n}}$$

where t is the value of the Student t-distribution (refer to standard statistical texts) for the desired confidence level and degrees of freedom, $(n-1)$, and s is an estimate of the standard deviation of the mean (X_{gm}). Thus,

$Y_{a-mean} = -2.445$
$SD_y = 1.154$
$CL_y = -2.445 \pm 0.733$ (i.e., a confidence interval from -3.178 to -1.712)

where Y_{a-mean} = arithmetic mean of log-transformed data; SD_y = standard deviation of log-transformed data; and CI_y = 95% confidence interval (95% CI) of log-transformed data.

The development of a 95% confidence limit for the log-transformed data gives a confidence interval of $-2.445 \pm 0.333t = -2.445 \pm 0.733$ (where $t = 2.20$, obtained from the Student t-distribution for $(n-1) = 12-1 = 11$ degrees of freedom).

Transforming the average of the Y values back into arithmetic values yields a geometric mean value, $X_{gm} = e^{-2.445} = 0.087$. Furthermore, transforming the confidence limits of the log-transformed values back into the arithmetic realm yields a 95% confidence interval of 0.042 to 0.180 µg/L, consisting of positive concentration values only. That is,

$X_{gm} = 0.087$ μg/L
$SD_x = 3.171$ μg/L
$0.042 \leq CI_x \leq 0.180$ μg/L and $UCL_x = 0.180$ μg/L

where X_{gm} = geometric mean for the 'raw' data; SD_x = standard deviation of the 'raw' data (assuming lognormal distribution); CI_x = 95% confidence interval (95% *CI*) for the 'raw' data (assuming lognormal distribution); and UCL_x = 95% upper confidence level (95% *UCL*) for the 'raw' data (assuming lognormal distribution).

It is apparent that the arithmetic mean, $X_m = 0.213$ μg/L, is substantially larger than the geometric mean of $X_{gm} = 0.087$ μg/L. The reason for this is that two large sample values in the data set (i.e., sampling events no. 4 and no. 5 in Table 9.3) tend to strongly bias the arithmetic mean; the logarithmic transform acts to suppress the extreme values. A similar observation can be made for the 95% upper confidence level (*UCL*) of the normally and lognormally distributed data sets. In general, however, the 95% *UCL* is a preferred statistical parameter to use in the evaluation of environmental data rather than the mean values, irrespective of the type of underlying distribution.

The results from this example analysis illustrate the potential effects that could result from the choice of one distribution type over another, and also the implications of selecting specific statistical parameters in the evaluation of environmental sampling data. In general, the use of arithmetic or geometric mean values for estimating average concentrations would tend to bias the concentration estimates.

9.4.2 Treatment of Censored Data in Environmental Samples

Often, in a given set of environmental samples, certain chemicals will be reliably quantified in some, but not all, of the samples. Environmental data sets may therefore contain observations which are below the instrument or method detection limit, or its corresponding quantitation limit; such data are often referred to as 'censored data' (or 'nondetects' [*NDs*]). This situation may reflect the fact that, either the chemical is truly absent at this location at the time the sample was collected, or the chemical is present but at a concentration below the quantitation limits of the analytical method that was employed in the sample analysis.

Invariably, all laboratory analytical techniques have detection and quantitation limits below which only 'less than' values may be reported; the reporting of such values provides a degree of quantification for the censored data. In such situations, a decision has to be made as to how to treat such *NDs* and associated 'proxy' concentrations. The appropriate procedure, which depends on the general pattern of detection of the chemical in the overall sampling events, may consist of the following determinations (HRI, 1995):

- If a chemical is rarely detected in any medium, and there is little to no reason to expect the chemical to be associated with the environmental contamination problem, then it may be appropriate to exclude it from further analysis. This is particularly true if the chemical is a common laboratory contaminant.

- If a chemical is rarely detected in a specific medium, and there is no reason to expect the chemical to be significantly associated with the environmental contamination problem, then it may be appropriate to exclude it from further analysis of that particular medium.
- If the pattern of a chemical's concentration in samples suggests that it was confined to well-defined 'hot-spots' at the time of sampling, then the potential for contaminant migration should be considered. Consequently, it will be important to include such chemical in the analysis of both the source and receiving media.

In fact, it is customary to assign nonzero values to all environmental sampling data reported as NDs. This is important because, even at or near their detection limits, certain chemical constituents may be of considerable importance in the characterization of an environmental contamination problem. However, uncertainty about the actual values below the detection or quantitation limit can bias or preclude subsequent statistical analyses.

Censored data do indeed create great uncertainties in the data analysis required of the environmental characterization process; such data should therefore be handled in an appropriate manner, as elaborated below.

9.4.2.1 Derivation and use of 'proxy' concentrations

'Proxy' concentrations are usually employed when a chemical is not detected in a specific medium. A variety of approaches are offered in the literature for deriving and using proxy values in environmental data analyses, including the following (HRI, 1995; USEPA, 1989a):

- *Set the sample concentration to zero.* This involves very strong assumptions, and it can rarely be justified that the chemical is not present in the environmental samples.
- *Drop the sample with the nondetect for the particular chemical from further analysis.* This will have the same effect on the data analysis as assigning a concentration that is the average of concentrations found in samples where the chemical was detected.
- *Set the proxy sample concentration to the sample quantitation limit (SQL).* For *ND*s, setting the sample concentration to a proxy concentration equal to the *SQL* (which is a quantifiable number used in practice to define the analytical detection limit) makes the fewest assumptions and tends to be conservative, since the *SQL* represents an upper-bound on the concentration of a *ND*. This approach recognizes that the true distribution of concentrations represented by the *ND*s is unknown.
- *Set the proxy sample concentration to one-half the SQL.* For *ND*s, setting the sample concentration to a proxy concentration equal to one-half the *SQL* assumes that, regardless of the distribution of concentrations above the *SQL*, the distribution of concentrations below the *SQL* is symmetrical.

The common practice involves specifying the sample-specific quantitation limit for the chemical. Notwithstanding the above procedures of using 'proxy' concentrations, re-sampling should always be viewed as the preferred approach to resolving uncertainties that surround *ND* results from environmental samples. Under such circumstances, if the initially reported data represent a problem in sample collection or analytical methods rather than a true failure to detect, then the identified problem may be rectified (e.g., by the use of more sensitive analytical protocols) before critical decisions are made based on the earlier results.

9.4.2.2 Statistical evaluation of 'nondetect' values

The favored approach in the calculation of the applicable statistical values during the evaluation of data containing *ND*s involves the use of a value of one-half of the *SQL*. This approach assumes that the samples are equally likely to have any value between the detection limit and zero, and can be described by a normal distribution. However, when the sample values above the *ND* level are lognormally distributed, it generally may be assumed that the *ND* values are also lognormally distributed. The best estimate of the *ND* values for a lognormally distributed data set is the reported *SQL* divided by the square root of two (i.e., $SQL/\sqrt{2} = SQL/1.414$) (CDHS, 1990; USEPA, 1989a).

In general, during the analysis of environmental sampling data that contains some *ND*s, a fraction of the *SQL* is usually assumed (as a proxy or estimated concentration) for nondetectable levels, instead of assuming a value of zero or neglecting such values. This procedure is generally used, provided there is at least one detected value from the analytical results, and/or if there is reason to believe that the chemical is possibly present in the sample at a concentration below the *SQL*. This approach conservatively assumes that some level of the chemical could be present (eventhough an ND has been recorded) and arbitrarily sets that level at the appropriate percentage of the *SQL* (i.e., *SQL*/2 if the data set is assumed to be normally distributed or *SQL*/1.414 for a lognormally distributed data set). In fact, in some situations the *SQL* value itself may be used if there is strong reason to believe that the chemical concentration is closer to this value, rather than to a fraction of the *SQL*. Where it is apparent that serious biases could result, more sophisticated analytical and evaluation methods may be warranted.

9.5 SELECTING THE CHEMICALS OF POTENTIAL CONCERN

The quantitative evaluation of all chemical constituents found at a contaminated facility may be considered as the most thorough approach for assessing potential risks posed by exposures to such chemicals. This may become necessary if the following criteria for identifying chemicals associated with facility releases into the environment are fulfilled for all the chemicals of interest:

- Positively detected in at least one sample.
- Detected at levels significantly elevated above levels of the same chemicals found in associated blank samples.

Hazard identification, data collection, and data evaluation 175

Box 9.2 Typical important considerations in the screening of environmental contaminants for chemicals of potential concern

- Historical local use information (i.e., historical data concerning chemicals, waste processes, etc. associated with site/facility activities often provide important information concerning the types of, and possible sources for, contaminant releases into the environment)
- Status as a known human carcinogen versus probable or possible carcinogen
- Status as a known human developmental and reproductive toxin
- Mobility, persistence and bioaccumulation, and ecological effects
- Nature of possible transformation products of the site contaminants
- Inherent toxicity of chemical/contaminant
- Concentration–toxicity score, reflecting concentration levels in combination with degree of toxicity. (The chemical score is represented by a risk factor, calculated as the product of the chemical concentration and toxicity value; the ratio of the risk factor for each chemical to the total risk factor approximates the relative risk for each chemical – giving a basis for inclusion or exclusion as a COC)
- Frequency of detection in affected media. (Chemicals that are infrequently detected may be artifacts in the data due to sampling, analytical, or other problems, and therefore may not be related to operations or activities at the project location)
- Status and condition as an essential element – defined as an essential human nutrient, and toxic only at elevated doses, i.e., much higher than those that could be associated with contact at the project site/facility (e.g., Ca or Na generally do not pose a significantly greater risk to health and the environment, but As or Cr may pose a significantly greater risk to health and the environment)

- Detected at levels significantly elevated above naturally occurring or anthropogenic background levels.
- Only tentatively identified, but may be historically associated with the facility or site.
- Transformation (daughter) products of chemicals demonstrated to be present.

However, the list of possible environmental chemicals associated with some complex facilities may be so huge as to necessitate reducing the size to a manageable number. Under such latter circumstances, an 'indicator chemical' approach may be used to provide a short-list of chemicals of potential concern (COCs) that is carried through a risk assessment process (Box 9.2).

In general, chemicals accounting for at least 95% of the risk are to be considered in a comprehensive risk assessment. A typical process flow in the COC selection is shown in Figure 9.2. Details of the processes involved in the selection of the indicator chemicals or COCs for a risk assessment are provided in the literature elsewhere (e.g., CDHS, 1990; OSA, 1992; USEPA, 1989a). It is noteworthy that, in some marginal situations, the indicator chemical approach requires such a significant expenditure of time and effort to implement as to render it as *not* cost-effective; under such circumstances, it probably becomes prudent to carry all contaminants through as COCs for the risk assessment.

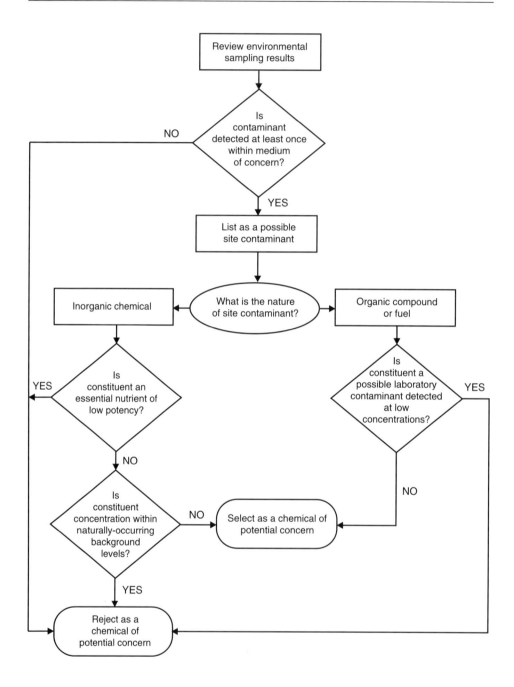

Figure 9.2 Typical process flow for the screening of site contaminants for chemicals of potential concern. (Source: Asante-Duah, 1996)

SUGGESTED FURTHER READING

Aswathanarayana, U. 1995. *Geoenvironment: An Introduction*. A.A. Balkema Publishers, Rotterdam, The Netherlands.
Cothern, C.R. and N.P. Ross (eds). 1994. *Environmental Statistics, Assessment, and Forecasting*. Lewis Publishers/CRC Press, Boca Raton, FL.
Daniel, D.E. (ed.) 1993. *Geotechnical Practice for Waste Disposal*. Chapman & Hall, London, UK.
Hasan, S.E. 1996. *Geology and Hazardous Waste Management*. Prentice-Hall, Englewood Cliffs, NJ.
Kabata-Pendias, A. and H. Pendias. 1984. *Trace Elements in Soils and Plants*. CRC Press, Boca Raton, FL.
Ramamoorthy, S. and E. Baddaloo. 1991. *Evaluation of Environmental Data for Regulatory and Impact Assessment*. Studies in Environmental Science 41, Elsevier Science Publishers B.V., Amsterdam, The Netherlands.
Reeve, R.N. 1994. *Environmental Analysis*. J. Wiley, New York.
Shacklette, H.T. and J.G. Boerngen. 1984. *Element Concentrations in Soils and Other Surficial Materials of the Conterminous United States*. USGS Professional Paper No. 1270.

REFERENCES

Asante-Duah, D.K. 1996. *Managing Contaminated Sites: Problem Diagnosis and Development of Site Restoration*. J. Wiley, Chichester, UK.
Berthouex, P.M. and L.C. Brown. 1994. *Statistics for Environmental Engineers*. Lewis Publishers/CRC Press, Boca Raton, FL.
Boulding, J.R. 1994. *Description and Sampling of Contaminated Soils (A Field Manual)*, 2nd edition. Lewis Publishers/CRC Press, Boca Raton, FL.
BSI (British Standards Institution). 1988. *Draft for Development, DD175: 1988 Code of Practice for the Identification of Potentially Contaminated Land and its Investigation*. BSI, London, UK.
Byrnes, M.E. 1994. *Field Sampling Methods for Remedial Investigations*. Lewis Publishers/CRC Press, Boca Raton, FL.
Cairney, T. (ed.). 1993. *Contaminated Land (Problems and Solutions)*. Blackie Academic & Professional, Glasgow/Chapman & Hall, London/Lewis Publishers, Boca Raton, FL.
CCME (Canadian Council of Ministers of the Environment). 1993. *Guidance Manual on Sampling, Analysis, and Data Management for Contaminated Sites*. Volume I: Main Report (Report CCME EPC-NCS62E), and Volume II: Analytical Method Summaries (Report CCME EPC-NCS66E). The National Contaminated Sites Remediation Program, Winnipeg, Manitoba, December, 1993.
CCME. 1994. *Subsurface Assessment Handbook for Contaminated Sites*. CCME, The National Contaminated Sites Remediation Program (NCSRP), Report No. CCME-EPC-NCSRP-48E (March, 1994), Ottawa, Ontario, Canada.
CDHS (California Department of Health Services). 1990. *Scientific and Technical Standards for Hazardous Waste Sites*. Prepared by the California Department of Health Services, Toxic Substances Control Program, Technical Services Branch, Sacramento, CA.
Cressie, N.A. 1994. *Statistics for Spatial Data*, revised edition. J. Wiley, New York.
Csuros, M. (1994). *Environmental Sampling and Analysis for Technicians*. Lewis Publishers/CRC Press, Boca Raton, Florida.
DTSC (Department of Toxic Substances Control). 1994. *Preliminary Endangerment Assessment Guidance Manual (A Guidance Manual for Evaluating Hazardous Substance Release Sites)*. California Environmental Protection Agency, DTSC, Sacramento, CA.

Freund, J.E. and R.E. Walpole (eds). 1987. *Mathematical Statistics*. Prentice-Hall, Englewood Cliffs, NJ.

Garrett, P. 1988. *How to Sample Groundwater and Soils*. National Water Well Association (NWWA), Dublin, OH.

Gibbons, R.D. 1994. *Statistical Methods for Groundwater Monitoring*. J. Wiley, New York.

Gilbert, R.O. 1987. *Statistical Methods for Environmental Pollution Monitoring*. Van Nostrand Reinhold, New York.

Hadley, P.W. and R.M. Sedman. 1990. A health-based approach for sampling shallow soils at hazardous waste sites using the $AAL_{soil\ contact}$ criterion. *Environmental Health Perspectives*, 18: 203–207.

Hipel, K.W. 1988. Nonparametric approaches to environmental impact assessment. *Water Resources Bulletin*, AWRA, 24(3): 487–491.

HRI (Hampshire Research Institute). 1995. *Risk*Assistant for Windows*. HRI, Alexandria, VA.

Keith, L.H. (ed.). 1988. *Principles of Environmental Sampling*. American Chemical Society (ACS), Washington, DC.

Keith, L.H. 1991. *Environmental Sampling and Analysis – A Practical Guide*. Lewis Publishers, Boca Raton, FL.

Keith, L.H. (ed.). 1992. *Compilation of E.P.A.'s Sampling and Analysis Methods*. Lewis Publishers/CRC Press, Boca Raton, FL.

LaGoy, P.K. and C.O. Schulz. 1993. Background sampling: an example of the need for reasonableness in risk assessment. *Risk Analysis*, 13(5): 483–484.

Leidel, N. and K.A. Busch. 1985. Statistical design and data analysis requirements. In: *Patty's Industrial Hygiene and Toxicology*, Vol. IIIa, 2nd edition, J. Wiley, New York.

Lesage, S. and R.E. Jackson (eds). 1992. *Groundwater Contamination and Analysis at Hazardous Waste Sites*. Marcel Dekker, New York.

Miller, I. and J.E. Freund. 1985. *Probability and Statistics for Engineers*, 3rd edition. Prentice-Hall, Englewood Cliffs, NJ.

Millette, J.R. and S.M. Hays. 1994. *Settled Asbestos Dust Sampling and Analysis*. Lewis Publishers/CRC Press, Boca Raton, FL.

OSA (Office of Scientific Affairs). 1992. *Supplemental Guidance for Human Health Multimedia Risk Assessments of Hazardous Waste Sites and Permitted Facilities*. Cal EPA, DTSC, Sacramento, CA.

O'Shay, T.A. and K.B. Hoddinott (eds). 1994. *Analysis of Soils Contaminated with Petroleum Constituents*. ASTM Publication, STP 1221, ASTM, Philadelphia, PA.

Ott, W.R. 1995. *Environmental Statistics and Data Analysis*. Lewis Publishers/CRC Press, Boca Raton, FL.

Parkhurst, D.F. 1998. Arithmetic versus geometric means for environmental concentration data. *Environmental Science and Technology*, 32(3): 92A–98A.

Rappaport, S.M. and J. Selvin. 1987. A method for evaluating the mean exposure from a lognormal distribution. *Journal of American Industrial Hygiene Association*, 48: 374–379.

Saltzman, B.E. 1997. Health risk assessment of fluctuating concentrations using lognormal models. *Journal of the Air and Waste Management Association*, 47: 1152–1160.

Schulin, R., A. Desaules, R. Webster, and B. Von Steiger (eds). 1993. *Soil Monitoring: Early Detection and Surveying of Soil Contamination and Degradation*. Birkhäuser Verlag, Basel, Switzerland.

Sharp, V.F. 1979. *Statistics for the Social Sciences*. Little, Brown & Co., Boston, MA.

Thompson, S.K. 1992. *Sampling*. J. Wiley, New York.

USEPA (US Environmental Protection Agency). 1982. *Test Methods for Evaluating Solid Waste: Physical/Chemical Methods*, 1st edition, SW-846. USEPA, Washington, DC.

USEPA. 1985a. *Characterization of Hazardous Waste Sites – A Methods Manual, Volume 1: Site Investigations*. Environmental Monitoring Systems Laboratory, Las Vegas, NV. EPA/600/4-84/075.

USEPA. 1985b. *Practical Guide to Ground-Water Sampling*. Robert S. Kerr Environmental

Research Lab., Office of Research and Development, USEPA, Ada, OK. EPA/600/2-85/104.
USEPA. 1987a. *Data Quality Objectives for Remedial Response Activities*. Office of Emergency and Remedial Response, Washington, DC. EPA/540/G-87/003.
USEPA. 1987b. *Data Quality Objectives for Remedial Response Activities: Example Scenario*. USEPA, Washington, DC. EPA/540/G-87/004.
USEPA. 1988a. *GEO-EAS (Geostatistical Environmental Assessment Software) User's Guide*. Environmental Monitoring Systems Laboratory, Office of R&D, Las Vegas, NV. EPA/600/4-88/033a.
USEPA. 1988b. *Guidance for Conducting Remedial Investigations and Feasibility Studies Under CERCLA*. Office of Emergency and Remedial Response, Washington, DC. EPA/540/G-89/004. OSWER Directive 9355.3-01.
USEPA. 1989a. *Risk Assessment Guidance for Superfund. Volume I – Human Health Evaluation Manual (Part A)*. Office of Emergency and Remedial Response, Washington, DC. EPA/540/1-89/002.
USEPA. 1989b. *Soil Sampling Quality Assurance User's Guide*. 2nd edition. Experimental Monitoring Support Laboratory (EMSL), Las Vegas, NV. EPA/600/8-89/046.
USEPA. 1990. *Guidance for Data Useability in Risk Assessment*, Interim Final. Office of Emergency and Remedial Response, Washington, DC. EPA/540/G-90/008.
Wilson, L.G., L.G. Everett, and S.J. Cullen (eds). 1995. *Handbook of Vadose Zone Characterization and Monitoring*. Lewis Publishers/CRC Press, Boca Raton, FL.
Wonnacott, T.H. and R.J. Wonnacott. 1972. *Introductory Statistics*, 2nd edition. J. Wiley, New York.
Zirschy, J.H. and D.J. Harris. 1986. Geostatistical analysis of hazardous waste site data. *ASCE Journal of Environmental Engineering*, 112(4).

Chapter Ten

Design of Conceptual Models and Exposure Analysis

Conceptual models provide a systematic and structured framework for characterizing possible threats posed by potential environmental contamination problems. Typically, the conceptual model aids in organizing and analyzing the basic information relevant to the environmental management situation. Thus, irrespective of the nature of environmental contamination problem, the design of an adequate conceptual model is a generally recommended and vital part of the environmental management program.

An exposure assessment is generally conducted to estimate the magnitude of actual and/or potential receptor exposures to chemical constituents, the frequency and duration of these exposures, and the pathways by which populations are potentially exposed to chemicals released into the environment. The exposure estimates are then used to determine whether any threats exist based on existing exposure conditions associated with an environmental contamination problem.

10.1 DESIGN OF CONCEPTUAL MODELS FOR ENVIRONMENTAL CONTAMINATION PROBLEMS

Conceptual models generally establish a hypothesis about possible contaminant sources, contaminant fate and transport, and possible pathways of exposure to the populations potentially at risk (Figure 10.1). In all cases, the conceptual model helps to identify and document known and suspected sources of contamination, types of contaminants and affected media, known and potential migration pathways, potential exposure pathways and routes, target receiving media, and known or potential human and ecological receptors. Such information can be used to develop a conceptual understanding of the environmental contamination problem, so that potential risks to human health and the environment can be evaluated more completely.

In general, the complexity and degree of sophistication of a conceptual model usually is consistent with the complexity of the particular problem and the amount of data available. For instance, in a typical scenario in which there is a source release, contaminants may be transported from the contaminated media by several processes into other environmental compartments (as depicted by the diagrammatic representation shown in Figure 10.2). In this type of situation, precipitation may infiltrate into soils at a contaminated site and leach contaminants from the soil as it

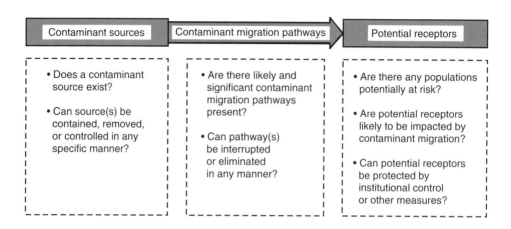

Figure 10.1 Conceptual elements of an environmental contamination problem

migrates through the contaminated material and the unsaturated soil zone. Infiltrating water may continue its downward migration until it encounters the water table at the top of the saturated zone; the mobilized contaminants may be diluted by the available groundwater flow. Once a contaminant enters the groundwater system, it is possible for it to be transported to a discharge point or to a water supply well location. There also is the possibility of continued downward migration of contaminants into a bedrock aquifer system. Contaminants may additionally be carried by surface runoff into surface water bodies. Air releases of particulates and vapors present additional migration pathways for the contaminants. Consequently, all these types of scenarios will typically be evaluated as part of an environmental characterization process for a contaminated site problem.

The development of an adequate conceptual model is indeed a very important aspect of the technical evaluation scheme necessary for the successful completion of most environmental characterization programs. It integrates various physical and environmental setting information, and then provides a basis for human health and ecological risk assessments. The conceptual model is also relevant to the development and evaluation of corrective action programs for a variety of environmental contamination problems.

10.1.1 Elements of a Conceptual Site Model

Typically a conceptual site model (CSM) is developed from available field sampling data, historical records, aerial photographs, and hydrogeologic information about a project site. Once synthesized, this information may be presented in several different forms such as map views (showing sources, pathways, receptors, and the distribution of contamination), cross-sectional views (illustrating hydrogeological sectional components), and/or tabular forms (summarizing and comparing contaminant concentrations against standards such as background thresholds, and/or regulatory or risk-

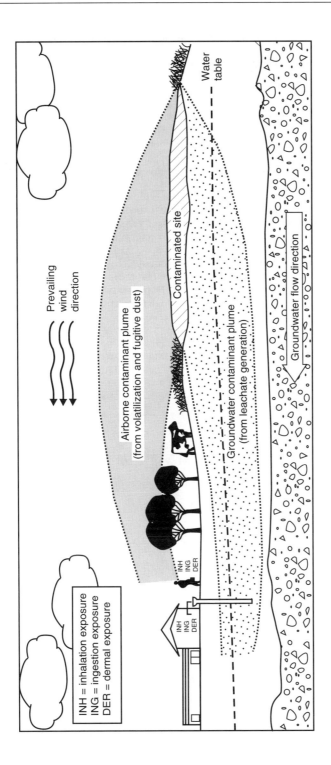

Figure 10.2 A diagrammatic conceptual representation of a contaminated site problem

> **Box 10.1 Major considerations in the design of a CSM**
>
> - Determine the spatial distribution of contaminants
> - Analyze site geology and hydrogeology
> - Determine the extent to which contaminant sources can be adequately identified and characterized
> - Determine the likelihood of releases if the contaminants remain on-site
> - Determine the extent to which natural and artificial barriers currently contain contaminants, and the adequacy of such barriers
> - Identify potential migration pathways
> - Determine the extent to which site contaminants have migrated or are expected to migrate from their source(s)
> - Estimate the contaminant release rates into specific environmental media over time
> - Provide guidance for evaluating the transport, transformation, and fate of contaminants in the environment following their release
> - Identify off-site areas affected by contaminant migration
> - Determine contaminant amount, concentration, hazardous nature, and environmental fate properties of the constituents present in affected areas
> - Determine the human, ecological, and welfare resources potentially at risk
> - Identify the routes of contaminant exposure to populations potentially at risk
> - Determine the likelihood of human and ecological receptors coming into contact with the contaminants of concern
> - Assess the likelihood of contaminant migration posing a threat to public health, welfare, or the environment
> - Provide guidance for calculating and integrating exposures to all populations affected by the various exposure scenarios associated with the contaminated site
> - Determine the extent to which contamination levels could exceed relevant regulatory standards, in relation to public health or environmental standards and criteria

based criteria). In general, a typical CSM will incorporate the following basic elements:

- Identification of site contaminants and determination of their physical/chemical properties.
- Characterization of the source(s) of contamination and site conditions.
- Delineation of potential migration pathways.
- Identification and characterization of all populations and resources that are potentially at risk
- Determination of the nature of inter-connections between contaminant sources, contaminant migration pathways, and potential receptors.

Several considerations and evaluations are indeed important to the design of a realistic and truly representative CSM that will meet the overall goals of a risk assessment and environmental management program (Box 10.1). Site history and preliminary assessment or site inspection data generally are very useful sources of information for developing preliminary CSMs. Subsequently, the CSM should be appropriately modified if the acquisition of additional data and new information necessitates a re-design.

> Primary and Secondary Sources → Migration & Exposure Pathways → Receptor Exposures

Figure 10.3 Exposure scenario evaluation flowchart

10.2 DEVELOPMENT OF EXPOSURE SCENARIOS

An exposure scenario is a description of the activity that brings a population into contact with a contaminated environmental medium. This representation incorporates contaminant sources and 'sinks', the nature and behavior of the contaminants, migration pathways, the affected environmental matrices, and potential receptors (Figure 10.3). Exposure scenarios are developed, based on the movement of contaminants in various environmental compartments. Several tasks are usually undertaken to facilitate the development of complete and realistic exposure scenarios; the critical tasks include the following:

- Determine the sources of contamination.
- Identify the specific constituents of concern.
- Identify the affected environmental media.
- Delineate contaminant migration pathways.
- Identify potential receptors.
- Determine potential exposure routes.
- Construct a representative conceptual model for the problem situation.
- Delineate likely and significant migration and exposure pathways.

On this basis, a realistic set of exposure scenarios can be developed for an environmental management program. In fact, the conceptual model facilitates an assessment of the nature and extent of contamination, and also helps determine the potential impacts from such contamination. Consequently, in as early a stage as possible during an environmental investigation, all available site information should be compiled and analyzed to develop a conceptual model for the problem situation.

The exposure scenario associated with a given hazardous situation may be well defined if the exposure is known to have already occurred. In most cases associated with the investigation of potential environmental contamination problems, however, decisions may have to be made about exposures that may not yet have occurred – in which case hypothetical exposure scenarios are generally developed to facilitate the problem solution. Ultimately, the exposure scenarios developed for a given environmental contamination problem can be used to support an evaluation of the risks posed by the situation, as well as facilitate the development of appropriate environmental management decisions.

10.2.1 The Nature and Spectrum of Exposure Scenarios

A wide variety of *potential* exposure patterns can generally be anticipated from environmental contamination problems. A select list of typical and commonly encountered exposure scenarios in relation to environmental contamination problems will include the following (Asante-Duah, 1996; HRI, 1995):

- Inhalation exposures
    - Indoor air, resulting from potential receptor exposure to contaminants (both volatile constituents and fugitive dust) in indoor ambient air.
    - Indoor air, resulting from potential receptor exposure to volatile chemicals in domestic water that may volatilize inside a house (e.g., during hot water showering) and contaminate indoor air.
    - Outdoor air, resulting from potential receptor exposure to contaminants (both volatile constituents and fugitive dust) in outdoor ambient air.
    - Outdoor air, resulting from potential receptor exposure to volatile chemicals in irrigation water, or other surface water bodies, that may volatilize and contaminate outdoor air.

- Ingestion exposures
    - Drinking water, resulting from potential receptor oral exposure to contaminants in domestic water used for drinking or cooking.
    - Swimming, resulting from potential receptor exposure (via incidental ingestion) to contaminants in surface water bodies.
    - Incidental soil ingestion, resulting from potential receptor exposure to contaminants in dust and soils.
    - Crop consumption, resulting from potential receptor exposure to contaminated foods (such as vegetables and fruits produced in household gardens that used contaminated soils, groundwater, or irrigation water in the cultivation process).
    - Dairy and meat consumption, resulting from potential receptor exposure to contaminated foods (such as locally grown livestock that may be contaminated from domestic water supplies, or from feeding on contaminated crops, or from contaminated air and soils).
    - Receptor exposure through the foodchain due to consumption of game and livestock. Typically, game or livestock may have been exposed by ingestion and/or contacts to contaminated materials. Humans can subsequently be exposed by consuming food that has become contaminated as a result of bioaccumulation through the foodchain.
    - Seafood consumption, resulting from potential receptor exposure to contaminated foods (such as fish and shellfish harvested from contaminated waters, or that have been exposed to contaminated sediments, and consequently have bioaccumulated toxic levels of chemicals in the edible portions).
    - Inter-receptor transfers, such as ingestion of mammalian breast milk containing chemicals absorbed by the feeding mother.

- Dermal exposures
    - Showering, resulting from potential receptor exposure (via skin absorption) to contaminants in domestic water supply.

- Swimming, resulting from potential receptor exposure (via skin absorption) to contaminants in surface water bodies.
- Direct soils contact, resulting from potential receptor exposure to contaminants in outdoor soils.

These types of exposure scenarios will typically be evaluated as part of the environmental characterization process designed to address an environmental contamination problem. It should be emphasized, however, that this listing is by no means complete, since new exposure scenarios are always possible for case-specific situations – albeit this demonstrates the multiplicity and inter-connections of numerous pathways through which populations may be exposed to environmental contaminants. In fact, the listed exposure scenarios may not all be relevant for every environmental contamination problem; on the other hand, a number of other exposure scenarios may have to be evaluated for the local conditions. In any case, once the complete set of potential exposure scenarios have been fully determined, the range of critical exposure pathways can be identified. This information can then be used to support the design of cost-effective environmental management programs.

In general, if numerous potential exposure scenarios exist, or if a complex exposure scenario has to be evaluated, it usually is helpful to use an event-tree model to clarify potential outcomes and/or consequences. The event-tree concept, as exemplified by Figure 10.4, does indeed offer an efficient way to develop exposure scenarios. By using such an approach, the various exposure contingencies can be identified and organized in a systematic manner. Once developed, priorities can be established to help focus the available effort on the aspects of greatest concern. Table 10.1 illustrates an alternative analytical protocol for developing the set of exposure scenarios; it is noteworthy, from the elementary similarities, that this representation is analogous to the event tree structure shown in Figure 10.4.

10.3 THE EXPOSURE ASSESSMENT PROCESS

The exposure assessment process is used to estimate the rates at which chemicals are absorbed by organisms. It generally involves several characterization and evaluation efforts – including a determination of the contaminant distributions leading from release sources to the locations of likely exposure; the identification of significant migration and exposure pathways; the identification of potential receptors, or the populations potentially at risk; the development of conceptual model(s) and exposure scenarios (including a determination of current and future exposure patterns, and the analysis of environmental fate and persistence); the estimation/modeling of exposure point concentrations for the critical pathways and environmental media; and the estimation of chemical intakes for all potential receptors and significant pathways of concern (Box 10.2).

Exposure pathways are one of the most important elements of the exposure assessment process, consisting of the routes that contaminants follow to reach potential receptors. In fact, failure to identify and address any significant exposure

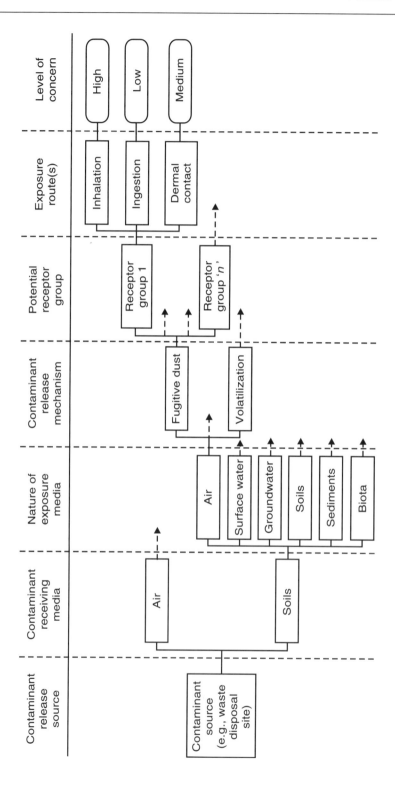

Figure 10.4 Diagrammatic representation of the development of exposure scenarios using an event tree

Table 10.1 Tabular illustration for the development of exposure scenarios

Exposure medium of concern	Potential exposure routes	Examples of typical (secondary) contaminant sources
Air	• Inhalation • Dermal absorption • Ingestion of food products	• Mother's milk • Poultry, meat, and eggs diet • Dairy products
Subsurface gas	• Inhalation	• Mother's milk • Poultry, meat, and eggs diet • Dairy products
Soil	• Soil ingestion • Crop ingestion • Dermal contact • Inhalation of particulates	• Food crops (with plant uptake) • Poultry, meat, and eggs diet • Dairy products
Groundwater	• Inhalation of volatiles • Water ingestion • Crop ingestion • Dermal absorption	• Food crops (with plant uptake from irrigation water) • Poultry, meat, and eggs diet (from use of water in feed) • Dairy products (from use of water in feed)
Surface water	• Inhalation of volatiles • Ingestion of water • Ingestion of contaminated biota • Dermal absorption	• Fish ingestion • Food plants (with plant uptake from irrigation water) • Poultry, meat, and eggs diet (from use of water in feed) • Dairy products (from use of water in feed)
Aquatic and terrestrial biota	• Ingestion within the foodchain	• Fish ingestion

---

Box 10.2 General procedural elements of an integrated exposure assessment process

- *Multimedia contaminant release analysis* (to include characterization of physical setting and monitoring/direct measurement data and modeling estimates; the results provide the basis for evaluating the potential for contaminant transport, transformation, and environmental fate)
- *Contaminant transport and fate analysis* (to include identification of migration and exposure pathways, and determination of contaminant distribution and concentrations; this describes the extent and magnitude of environmental contamination)
- *Exposed population analysis* (to include evaluation of populations contacting chemicals released into the environment; this involves the identification, enumeration, and characterization of those population segments likely to be exposed)
- *Integrated exposure analysis* (to include development of exposure estimates for the selected exposure scenarios)
- *Uncertainty analysis* (to consist of identification of any uncertainties involved, and an evaluation of their separate and cumulative impact on the assessment results)

pathway may seriously detract from the usefulness of any risk assessment, since a complete pathway must be present for receptor exposure to occur. An exposure pathway is considered complete only if all of the following elements are present:

- contaminant source(s);
- mechanism(s) of contaminant release into the environment;
- contaminant migration pathway(s) and exposure route(s); and
- receptor exposure in the affected media.

Exposure pathways are determined by integrating information from an initial environmental characterization with knowledge about potentially exposed populations and their likely behavior. The significance of migration pathways is evaluated on the basis of whether the contaminant migration could cause significant adverse exposures and impacts. The exposure routes (which may consist of inhalation, ingestion, and/or dermal contacts) and duration of exposure (that may be short-term [acute] or long-term [chronic]) will significantly influence the level of impacts on the affected receptors.

Finally, the assessment of human or ecological receptor exposure to environmental contaminants requires translating environmental concentrations into quantitative estimates of the amount of chemical that contacts an individual selected at random from the population-at-risk (PAR) (CDHS, 1990; OSA, 1992). The PAR refers to the receptors who do or plausibly could inhabit or traverse the location that is nearest to the source of contamination. Contact is expressed by the amount of material per unit body weight (mg/kg-day) that enters the lungs (for an inhalation exposure), enters the gastrointestinal tract (for an ingestion exposure), or crosses the stratum corneum (for a dermal contact exposure). This quantity is used as a basis for projecting the incidence of health or ecological detriment within the population. To accomplish this task, several important exposure parameters and/or information will typically need to be acquired (Table 10.2).

10.3.1 Multimedia and Multipathway Exposure Modeling

Multimedia mathematical models are often used to predict the potential for contaminant migration from a contaminant source to potential receptors, using pathways analyses concepts. The common and general types of modeling practice used in the exposure assessment relate to atmospheric, surface water, groundwater, multimedia, and foodchain models. Typically, several modeling scenarios will be simulated and evaluated using the appropriate models for an environmental contamination problem; for example, the study of a contaminated site problem may require the modeling of infiltration of rainwater, erosion/surface runoff release of chemicals, emission of particulates and vapors, chemical fate and transport through the unsaturated zone, chemical transport through the aquifer system, and/or mixing of groundwater with surface water (Figure 10.5). In any case, due to the heterogeneity in environmental compartments and natural systems, models used for exposure assessments should be adequately tested, and sensitivity runs should be

Table 10.2 Typical exposure parameters and information necessary for the estimation of potential receptor exposures

Exposure route	Relevant exposure parameters/data
Inhalation	• Airborne contaminant concentrations (e.g., resulting from showering, bathing, and other uses of contaminated water; or from dust inhalation) • Variation in air concentrations over time • Amount of contaminated air breathed • Fraction of inhaled contaminant absorbed through lungs • Breathing rate • Exposure duration and frequency • Exposure averaging time • Average receptor body weight
Ingestion	• Concentration of contaminant in consumed material (viz: water, food, soils, etc.) • Amount of contaminated material ingested each day (i.e., water ingestion rate; food intake rate; soil ingestion rate; etc.) • Fraction of ingested contaminant absorbed through wall of gastrointestinal tract • Exposure duration and frequency • Exposure averaging time • Average receptor body weight
Dermal (skin) absorption	• Concentration of contaminant in contacted material (viz: water, soils, etc.) • Amount of daily skin contact (i.e., dermal contact with soil; dermal contact with water; etc.) • Fraction of contaminant absorbed through skin during contact period • Period of time spent in contact with contaminated material • Average contact rate • Receptor's contacting body surface area • Exposure duration and frequency • Exposure averaging time • Average receptor body weight

carried out to help determine the most sensitive and/or critical parameters considered in the evaluation.

Ultimately, populations potentially at risk are defined, and concentrations of the chemicals of concern are determined in each medium to which potential receptors may be exposed. Then, using the appropriate case-specific exposure parameter values, the intakes of the chemicals of concern can be estimated. The evaluation could concern past or current exposures, or exposures anticipated in the future. In fact, the exposure assessment also involves describing the nature and size of the population exposed to a substance (i.e., the risk group, which refers to the actual or hypothetical exposed population) and the magnitude and duration of their exposure.

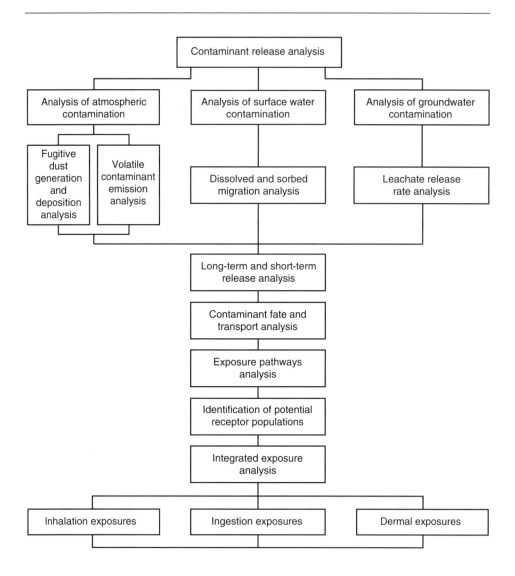

Figure 10.5 Example exposure scenario – (incorporating major migration and exposure pathways for contaminant releases into the environment) – that is typically associated with an integrated exposure assessment process

In general, several characteristics of the chemicals of concern will provide an indication of the critical features of exposure, as well as information necessary to determine the chemical's distribution, uptake, residence time, magnification, and breakdown to new chemical compounds (Hallenbeck and Cunningham, 1988). The physical and chemical characteristics of the chemicals can also affect the intake, distribution, half-life, metabolism, and excretion of such chemicals by potential receptors.

10.3.2 Chemical Intake versus Dose

Once exposure point concentrations in all media of interest have been estimated, the intakes and/or doses to potentially exposed populations can be determined. Intake is defined as the amount of chemical coming into contact with the receptor's visible exterior body (e.g., skin and openings into the body such as mouth and nostrils) or with the abstracted/conceptual exchange boundaries (such as the skin, lungs, or gastrointestinal tract); dose is the amount of chemical absorbed by the body into the bloodstream. The absorbed dose differs significantly from the externally applied dose (called exposure or intake).

For each exposure pathway considered in the exposure assessment, an intake per event is developed. This value quantifies the amount of a chemical contacted during each exposure event – where 'event' may have different meanings depending on the nature of exposure scenario being considered (e.g., each day's inhalation of an air contaminant constitutes one inhalation exposure event). The quantity of a chemical absorbed into the bloodstream per event – represented by the dose – is calculated by further considering pertinent physiological parameters (such as gastrointestinal absorption rates). When the systemic absorption from an intake is unknown, or cannot be estimated by a defensible scientific argument, intake and dose are considered to be the same (i.e., 100% absorption into the bloodstream from contact is assumed). This approach provides a conservative estimate of the actual exposures. In general, conservative estimates assume that the potential receptor is always in the same location, exposed to the same ambient concentration, and that there is 100% absorption upon exposure. These assumptions hardly represent any real-life situation. In fact, lower exposures will be expected due to the fact that potential receptors will generally be exposed to lower or even near-zero levels of the chemicals of concern for the period of time spent outside the impacted areas.

Intakes and doses are normally calculated in the same step of the exposure assessment, whereby the former multiplied by an absorption factor yields the latter value.

10.3.3 Chronic versus Subchronic Exposures

Event-based intake values are converted to final intake values by multiplying the intake per event by the frequency of exposure events, over the time-frame being considered. Chronic daily intake (CDI), which measures long-term (chronic) exposures, is based on the number of events that are assumed to occur within an assumed lifetime for potential receptors; subchronic daily intake (SDI), which represents projected receptor exposures over a short-term period, considers only a portion of a lifetime (USEPA, 1989b). These are calculated by multiplying the average or the reasonably maximum exposure (RME) media concentrations – usually represented by the 95% UCL – by the appropriate receptor exposure and body weight factors.

SDIs are generally used to evaluate subchronic noncarcinogenic effects, whereas CDIs are used to evaluate both carcinogenic risks and chronic noncarcinogenic

> Box 10.3 General equation for estimating potential receptor exposures to environmental contaminants
>
> $$EXP = \frac{(C_{\text{medium}} \times CR \times CF \times FI \times ABS_f \times EF \times ED)}{(BW \times AT)}$$
>
> where:
> EXP = Intake (i.e., the amount of chemical at the exchange boundary), adjusted for absorption (mg/kg-day)
> C_{medium} = Average or reasonably maximum exposure concentration of chemical contacted by potential receptor over the exposure period from the medium of concern (e.g., $\mu g/m^3$ [air]; or $\mu g/L$ [water]; or mg/kg [soil])
> CR = Contact rate, i.e., the amount of contaminated medium contacted per unit time or event (e.g., inhalation rate in m^3/day [air]; or ingestion rate in mg/day [soil], or L/day [water])
> CF = Conversion factor (10^{-6} kg/mg for solid media, or 1.00 for fluid media)
> FI = Fraction of intake from contaminated source (dimensionless)
> ABS_f = Bioavailability or absorption factor (%)
> EF = Exposure frequency (days/year)
> ED = Exposure duration (years)
> BW = Body weight, i.e., the average body weight over the exposure period (kg)
> AT = Averaging time (period over which exposure is averaged – days)
>   = $ED \times 365$ days/year, for noncarcinogenic effects of human exposure
>   = $LT \times 365$ days/year = 70 years $\times$ 365 days/year, for carcinogenic effects of human exposure (assuming an average lifetime, LT, of 70 years)

effects. The short-term exposures can result when, for instance, a particular activity is performed for a limited number of years or when a chemical with a short half-life degrades to negligible concentrations within several months.

10.4 A GENERIC EXPOSURE ESTIMATION MODEL

The methods by which each type of exposure is estimated are well documented in the literature of exposure assessment (e.g., CAPCOA, 1990; DOE, 1987; USEPA, 1988, 1989a, 1989b, 1989c). Typically, potential receptor exposures to environmental contaminants are conservatively estimated according to the generic equation shown in Box 10.3. The concentration in the various media may be obtained from field measurements, estimated by simple mass balance analyses or other appropriate contaminant transport models, or may be determined from equilibrium and partitioning relations. The various exposure parameters may be derived on a case-specific basis, or they may be compiled from regulatory guidance manuals and documents, and related literature (e.g., Binder et al., 1986; Calabrese et al., 1989; CAPCOA, 1990; DTSC, 1994; Finley et al., 1994; Hrudey et al., 1996; LaGoy, 1987; Lepow et al., 1974, 1975; OSA, 1992; Sedman, 1989; Smith, 1987; Stanek and Calabrese, 1990; Travis and Arms, 1988; USEPA, 1987, 1989a, 1989b, 1991, 1992; van Wijnen et al., 1990); these parameters are usually based on information relating to the maximum exposure level resulting from specified categories of receptor activity and/or exposures.

The methods by which each specific type of exposure is estimated, including the relevant exposure estimation equations for specific major routes of exposure, are further discussed later in Part IV of this volume, together with an illustration of the computational steps involved in the calculation of receptor intakes and doses.

10.4.1 Receptor Age Adjustments to Human Exposure Factors

In a refined and comprehensive evaluation, it is generally recommended to incorporate age adjustment factors in the exposure assessment, where appropriate. Age adjustments, necessary when receptor exposures occur from childhood through the adult life, are meant to account for the transitioning of a potential receptor from childhood (requiring one set of intake assumptions and exposure parameters) to adulthood (that requires a different set of intake assumptions and exposure parameters). In particular, in the processes involved in human exposure assessments, it becomes apparent that contact rates tend to be different for children and adults. Consequently, carcinogenic risks may preferably be calculated using age-adjusted factors (Box 10.4). For the sake of simplicity, such age adjustments are usually not made part of most screening-level computational processes.

The use of age-adjusted factors is especially important in specific situations – such as those involving human soil ingestion exposures, which are typically higher during childhood and decrease with age. Further details on the development of age-adjusted factors are provided elsewhere in the literature (e.g., DTSC, 1994; OSA, 1992; USEPA, 1989b).

10.4.2 Incorporating Contaminant Degradation into Exposure Calculations

When certain chemical compounds undergo degradation, potentially more toxic daughter products result (such as is the case when trichloroethylene (TCE) biodegrades to produce vinyl chloride). On the other hand, there are situations where the end-products of degradation are less toxic than the parent compounds. Since receptor exposures could be occurring over long time periods, a more valid approach in exposure modeling will be to take contaminant degradation (or indeed other transformation processes) into consideration during an exposure assessment. Under such circumstances, if significant degradation is likely to occur, then exposure calculations become much more complicated. In that case, contaminant concentrations at release sources are calculated at frequent and short time intervals, and then summed over the exposure period.

To illustrate the concept of incorporating contaminant degradation into exposure assessment, assume first-order kinetics for a hypothetical problem. An approximation of the degradation effects for this scenario can be obtained by multiplying contaminant concentrations by a degradation factor, *DGF*, that is defined by:

> **Box 10.4 Age-adjustment factors to human exposure calculations**
>
> - Ingestion (mg-year/kg-day or L-year/kg-day)
>
> $$INGf_{adj} = \frac{(MIR_c \times ED_c)}{BW_c} + \frac{(MIR_a \times [ED - ED_c])}{BW_a}$$
>
> - Dermal contact (mg-year/kg-day)
>
> $$DERf_{adj} = \frac{(AF \times SA_c \times ED_c)}{BW_c} + \frac{(AF \times SA_a \times [ED - ED_c])}{BW_a}$$
>
> - Inhalation (m³-year/kg-day)
>
> $$INHf_{adj} = \frac{(IRA_c \times ED_c)}{BW_c} + \frac{(IRA_a \times [ED - ED_c])}{BW_a}$$
>
> where:
> $INGf_{adj}$ = Age-adjusted ingestion factor (mg-yr/kg-day)
> $DERf_{adj}$ = Age-adjusted dermal contact factor (mg-yr/kg-day)
> $INHf_{adj}$ = Age-adjusted inhalation factor (m³-yr/kg-day)
> MIR_c = Material ingestion rate – child (mg/day or L/day)
> MIR_a = Material ingestion rate – adult (mg/day or L/day)
> AF = Material adherence factor (mg/cm²)
> SA_c = Child's exposed surface area (cm²)
> SA_a = Adult's exposed surface area (cm²)
> IRA_c = Inhalation rate – child (m³/day)
> IRA_a = Inhalation rate – adult (m³/day)
> ED = Total exposure duration (years)
> ED_c = Exposure duration – child (years)
> ED_a = Exposure duration – adult (years)
> BW_c = Body weight – child, i.e., the average child body weight over the exposure period (kg)
> BW_a = Body weight – adult, i.e., the average adult body weight over the exposure period (kg)

$$DGF = \frac{(1 - e^{-kt})}{kt}$$

where k is a chemical-specific degradation rate constant (days⁻¹) and t is the time period over which exposure occurs (days). For a first-order decaying substance, k is estimated from the following relationship:

$$T_{1/2}[\text{days}] = \frac{0.693}{k} \quad \text{or} \quad k[\text{days}^{-1}] = \frac{0.693}{T_{1/2}}$$

where $T_{1/2}$ is the chemical half-life, which is the time after which the mass of a given substance will be one-half its initial value.

The degradation factor is generally ignored in most exposure calculations. This is especially justifiable if the degradation product is of potentially equal toxicity and of comparable amounts. Although it cannot always be proven that the daughter products result in receptor exposures of comparable levels to the parent compound, the *DGF* is nevertheless ignored in most screening-level exposure assessments.

10.4.3 Averaging Exposure Estimates

In a more realistic exposure assessment, it may be more appropriate and less conservative to estimate exposure to a specific population subgroup over an exposure duration of less than a lifetime, as illustrated by the exposure combinations presented below.

10.4.3.1 *Averaging exposure over population age groups – when environmental concentrations are constant in time*

For situations where environmental concentrations are assumed constant over time but for which exposure is to be averaged over population age groups, the chronic daily exposure may be estimated using the following model form (CDHS, 1990; OSA, 1992):

$$CDI = \left[\frac{1}{\sum_{a=1}^{NG} AT_a} \right] \times \left[\sum_{a=1}^{NG} \left(\frac{CR}{BW} \right)_a \times EF_a \times ED_a \right] \times C_m$$

where $(CR/BW)_a$ is the contact rate per unit body weight, averaged over the age group a; EF_a is the exposure frequency of the exposed population in the age group/category a; ED_a is the exposure duration for the exposed population in the age group/category a; AT_a is the averaging time for the age group a; C_m is the concentration in the environmental medium contacted; and NG is the number of age groups used to represent the whole population.

10.4.3.2 *Averaging exposure over time within a population group – when environmental concentrations vary in time*

For some chemical compounds present in the environment, the assumption that concentrations remain constant in time can result in significant over-estimation of risks. Consequently, a model that accounts for time-varying concentrations may be employed in the exposure estimation process.

Where contaminant concentrations in the source medium varies with time – such as for cases where there are chemicals volatilizing from a contaminated site, or being transformed by degradational processes – then exposures or chronic daily intakes for the exposed population may be estimated using the following general model form (CDHS, 1990; OSA, 1992):

$$CDI = \frac{(CR_m \times EF \times ED)}{(BW \times AT)} \times \int_{t=0}^{ED} C_m(t)\,dt$$

where CR_m is the contact rate in medium m; EF is the exposure frequency of the exposed population; ED is the exposure duration for the exposed population; AT is the averaging time for the population group; and $C_m(t)$ is the time-varying concentration in the environmental medium contacted.

It should be noted, however, that when one contaminant species is transformed such that its concentration decreases in time, then all decay products must be identified; exposure to all toxic decay products must then be modeled and accounted for – recognizing also that the concentrations of decay products could actually be increasing with time.

10.4.3.3 Averaging exposure over population age subgroups – when environmental concentrations vary with time

In some situations involving time-varying environmental concentrations, it may be decided to estimate the exposure to specific population subgroups over an exposure duration of less than a lifetime, and then to use these age subgroups to calculate the lifetime equivalent exposure to an individual drawn at random from the population. Under such circumstances, the following model form can be employed in the exposure estimation (CDHS, 1990; OSA, 1992):

$$CDI = \left[\frac{1}{\sum_{a=1}^{NG} AT_a}\right] \times \left[\sum_{a=1}^{NG}\left(\frac{CR}{BW}\right)_a \times EF_a \times \int_{t=0}^{ED_a} C_m(t)\,dt\right]$$

where $(CR/BW)_a$ is the contact rate per unit body weight, averaged over the age group a; EF_a is the exposure frequency of the exposed population in the age group/category a; ED_a is the exposure duration for the exposed population in the age group/category a; AT_a is the averaging time for the age group a; $C_m(t)$ is the time-varying concentration in the environmental medium contacted; and NG is the number of age groups used to represent the whole population.

10.5 UTILITY OF THE EXPOSURE CHARACTERIZATION

Chemical contaminants entering the environment tend to be partitioned or distributed across various environmental media and biota – with the distribution of chemicals entering the environmental compartments being the result of a number of complex processes. In any case, the potential hazards and/or risks associated with the individual chemicals is very much dependent on the extent of multimedia exposures to potential receptors. Thus, a good prediction of chemical concentrations in the various compartmental media *together with* a carefully executed exposure assessment is essential for the completion of a credible risk assessment.

The information contained in a conceptual model is developed during the various stages of an environmental characterization process, and also from controlled field and laboratory experiments that may be conducted in studies pertaining to the potential environmental contamination problem. As environmental investigations progress, the conceptual model may be revised as necessary and used to direct the next iteration of characterization activities. The updated or finalized conceptual model is then used to develop realistic exposure scenarios for the project, which then form an important basis for completing an effectual and credible risk assessment or environmental management program.

Several techniques may indeed be used for the exposure assessment, including: modeling of anticipated future exposures; environmental monitoring of current exposures; and biological monitoring to determine past exposures. The physicochemical properties of the contaminants of potential concern and the impacted media are important considerations in the exposure modeling. In general, a variety of exposure models and conservative but realistic assumptions regarding contaminant migration and equilibrium partitioning are used to facilitate the exposure quantification process.

SUGGESTED FURTHER READING

Clausing, O., A.B. Brunekreef and J.H. van Wijnen. 1987. A method for estimating soil ingestion by children. *International Archives of Occupational and Environmental Health*, 59: 73–82.
Goldstein, B.D. 1989. The maximally exposed individual: an inappropriate basis for public health decision making. *Environmental Forum*, 6: 13–16.
Keenan, R.E., B.L. Finley and P.S. Price. 1994. Exposure assessment: then, now and quantum leaps in the future. *Risk Analysis*, 14(3): 225–231.
Macari, E.J., J.D. Frost, and L.F. Pumarada (eds). 1995. *Geo-Environmental Issues Facing the Americas*. ASCE Geotechnical Special Publication No. 47, American Society of Civil Engineers, New York.
Miller, D.W. (ed.). 1980. *Waste Disposal Effects on Ground Water*. Premier Press, Berkeley, CA.
NRC (National Research Council). 1988. *Hazardous Waste Site Management: Water Quality Issues*. National Academy Press, Washington, DC.
Raloff, J. 1996. Tap water's toxic touch and vapors. *Science News*, 149(6): 84.
Reed, S.C., R.W. Crites, and E.J. Middlebrooks. 1995. *Natural Systems for Waste Management and Treatment*, 2nd edition. McGraw-Hill, New York.
Wiens, J.A. and K.R. Parker. 1995. Analyzing the effects of accidental environmental impacts: approaches and assumptions. *Ecological Applications*, 5(4): 1069–1083.

REFERENCES

Asante-Duah, D.K. 1996. *Managing Contaminated Sites: Problem Diagnosis and Development of Site Restoration.* J. Wiley, Chichester, UK.

Binder, S., D. Sokal and D. Maughan. 1986. Estimating the amount of soil ingested by young children through tracer elements. *Archives of Environmental Health*, 41: 341–345.

Calabrese, E.J., R. Barnes, E.J. Stanek III, H. Pastides, C.E. Gilbert, P. Veneman, X. Wang, A. Lasztity, and P.T. Kostecki. 1989. How much soil do young children ingest: an epidemiologic study. *Regulatory Toxicology and Pharmacology*, 10: 123–137.

CAPCOA (California Air Pollution Control Officers Association). 1990. *Air Toxics 'Hot Spots' Program. Risk Assessment Guidelines.* CAPCOA, Los Angeles, CA.

CDHS (California Department of Health Services). 1990. *Scientific and Technical Standards for Hazardous Waste Sites.* Prepared by the California Department of Health Services, Toxic Substances Control Program, Technical Services Branch, Sacramento, CA.

DOE (US Department of Energy). 1987. *The Remedial Action Priority System (RAPS): Mathematical Formulations.* US Dept. of Energy, Office of Environment, Safety and Health, Washington, DC.

DTSC (Department of Toxic Substances Control). 1994. *Preliminary Endangerment Assessment Guidance Manual (A Guidance Manual for Evaluating Hazardous Substance Release Sites).* California Environmental Protection Agency, DTSC, Sacramento, CA.

Finley, B., D. Proctor, et al. 1994. Recommended distributions for exposure factors frequently used in health risk assessment. *Risk Analysis*, 14: 533–553.

Hallenbeck, W.H. and Cunningham, K.M. 1988. *Quantitative Risk Assessment for Environmental and Occupational Health.* Lewis Publishers, Chelsea, MI.

HRI (Hampshire Research Institute). 1995. *Risk*Assistant for Windows.* HRI, Alexandria, VA.

Hrudey, S.E., W. Chen, and C.G. Rousseaux. 1996. *Bioavailability in Environmental Risk Assessment.* Lewis Publishers/CRC Press, Boca Raton, FL.

LaGoy, P.K. 1987. Estimated soil ingestion rates for use in risk assessment. *Risk Analysis*, 7(3): 355–359.

Lepow, M.L., M. Bruckman, L. Robino, S. Markowitz, M. Gillette, and J. Kapish, 1974. Role of airborne lead in increased body burden of lead in Hartford children. *Environmental Health Perspectives*, 6: 99–101.

Lepow, M.L., L. Bruckman, M. Gillette, S. Markowitz, R. Robino, and J. Kapish. 1975. Investigations into sources of lead in the environment of urban children. *Environmental Research*, 10: 415–426.

OSA (Office of Scientific Affairs). 1992. *Supplemental Guidance for Human Health Multimedia Risk Assessments of Hazardous Waste Sites and Permitted Facilities.* Cal EPA, DTSC, Sacramento, CA.

Sedman, R.M. 1989. The development of applied action levels for soil contact: a scenario for the exposure of humans to soil in a residential setting. *Environmental Health Perspectives*, 79: 291–313.

Smith, A.H. 1987. Infant exposure assessment for breast milk dioxins and furans derived from waste incineration emissions. *Risk Analysis*, 7: 347–353.

Stanek, E.J. and E.J. Calabrese. 1990. A guide to interpreting soil ingestion studies. *Regulatory Toxicology and Pharmacology*, 13: 263–292.

Travis, C.C. and A.D. Arms. 1988. Bioconcentration of organics in beef, milk, and vegetation. *Environmental Science & Technology*, 22: 271–274.

USEPA (US Environmental Protection Agency). 1987. *RCRA Facility Investigation (RFI) Guidance.* Washington, DC. EPA/530/SW-87/001.

USEPA. 1988. *Superfund Exposure Assessment Manual.* USEPA, Office of Remedial Response, Washington, DC. EPA/540/1-88/001, OSWER Directive 9285.5-1.

USEPA. 1989a. *Exposure Factors Handbook.* Office of Health and Environmental Assessment, Washington, DC. EPA/600/8-89/043.

USEPA. 1989b. *Risk Assessment Guidance for Superfund. Volume I – Human Health Evaluation Manual (Part A)*. Office of Emergency and Remedial Response, Washington, DC. EPA/540/1-89/002.

USEPA. 1989c. *Exposure Assessment Methods Handbook*. Office of Health and Environmental Assessment, USEPA, OH.

USEPA. 1991. *Risk Assessment Guidance for Superfund, Volume I: Human Health Evaluation Manual. Supplemental Guidance. 'Standard Default Exposure Factors' (Interim Final)*. March, 1991. Office of Emergency and Remedial Response, Washington, DC. OSWER Directive: 9285.6-03.

USEPA. 1992. *Dermal Exposure Assessment: Principles and Applications*. Office of Research and Development, Washington, DC. EPA/600/8-91/011B.

Van Wijnen, J.H., P. Clausing, and B. Brunekreef. 1990. Estimated soil ingestion by children. *Environmental Research*, 51: 147–162.

Chapter Eleven

The Toxicology of Environmental Contaminants and Hazard Effects Determination

Toxicology is the study of how specific chemical substances cause injury/undesirable effects to living cells and/or whole organisms. It generally consists of studies conducted to determine the following:

- how easily the chemical enters the organism;
- how the chemical behaves in the organism;
- how rapidly the chemical is removed from the organism;
- what cells are affected by the chemical; and
- what cell functions are impaired.

This chapter discusses the fundamental analytical procedures that are generally employed in the evaluation of the hazard effects or toxicity of various chemical constituents present in the environment.

11.1 IDENTIFICATION OF TOXIC SUBSTANCES

The toxic characteristics of a substance are usually categorized according to the organs or systems they affect (e.g., kidney, liver, nervous system, etc.) or the disease they cause (e.g., birth defects, cancer, etc.). Typically, the identification of toxic substances begins with the retrieval of a variety of pertinent information that is available on the suspected agent (Box 11.1) (CDHS, 1986; Smith, 1992).

Toxicity tests may reveal that a substance produces a wide variety of adverse effects on different organs or systems of the body, or that the range of effects is narrow. Some effects may occur only at high dosages, and only the most sensitive indicators of a substance's toxicity may become manifest at lower doses (USEPA, 1985b).

11.1.1 Manifestations of Toxicity

Toxic responses, regardless of the organ or system in which they occur, can be of several types (USEPA, 1985b). For some, the *severity* of the injury increases as the dose increases. One of the goals of toxicity studies is to determine the 'no-observed-effect level' (*NOEL*) – which is the dose at which no toxic effect is seen in an

> **Box 11.1 Typical information requirements for the identification of chemical toxicity**
>
> - Physical and chemical properties
> - Routes of exposure
> - Metabolic and pharmacokinetic properties
> - Structure–activity relationships
> - Toxicological effects
> - Short-term tests
> - Long-term animal tests
> - Human epidemiologic studies
> - Clinical data

organism; this becomes an important input in the development of toxicity parameters for use in risk assessments.

For other cases, the severity of an effect may not increase with dose, but the *incidence* of the effects will increase with increasing dose. This type of response is properly characterized as probabilistic – since increasing the dose increases the probability (i.e., risk) that the abnormality or alteration will develop in the exposed population.

Often with toxic effects – including cancer – both the severity and the incidence increase as the level of exposure is raised. The increase in severity is a result of increased damage at higher doses, whereas the increase in incidence is a result of differences in individual sensitivity. Furthermore, the site at which a substance acts (e.g., kidney, liver, etc.) may change as the dose changes.

In general, as the duration of exposure increases, both the *NOEL* and the doses at which effects appear decrease; in some cases, new effects not apparent upon exposure of short duration become manifest.

Toxic responses also vary in their degree of *reversibility*. In some cases, an effect will disappear almost immediately following cessation of exposure, whereas at the other extreme exposures will result in a permanent injury. Most toxic responses tend to fall somewhere between these extremes.

Seriousness is yet another characteristic of a toxic response. Certain types of toxic damage are clearly adverse and are a definite threat to health, whereas other types of effects may not be of obvious health significance.

Finally, it is noteworthy that potential receptor populations (especially humans) tend to be exposed to mixtures of chemicals, as opposed to the single toxic agent scenario often presented in hazard evaluations. Consequently, several results may occur from chemical mixtures – including additive, synergistic, and antagonistic effects – that may have to be addressed, even if only qualitatively.

11.1.2 Dose–Response Relationships

The dose–response relationship is about the most fundamental concept in toxicology. The relationship between the degree of exposure to a chemical (viz., the

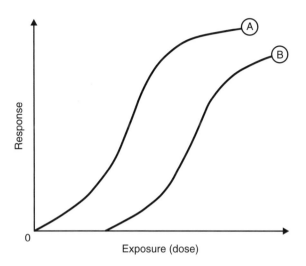

Figure 11.1 Schematic representation of exposure–response relationships: Illustration of dose–response relationship for (A) = nonthreshold chemicals and (B) = threshold chemicals

dose) and the magnitude of chemical-induced effects (viz., the response) is described by a dose–response curve – often referred to by 'stressor–response profile' in the characterization of ecological effects. Dose–response curves fall into two general categories/groups (Figure 11.1): those in which no response is observed until some minimum (i.e., threshold) dose is reached, and those in which no threshold is apparent – meaning that response is expected for any dose, no matter how small. For some chemicals, a very small dose causes no observable effects whereas a higher dose will result in some toxicity, and still higher doses cause even greater toxicity – up to the point of fatality; such chemicals are called *threshold chemicals* (curve B in Figure 11.1). For other chemicals, such as most carcinogens, the threshold concept may not be applicable, in which case no minimum level is required to induce adverse and overt toxicity effects (curve A in Figure 11.1).

The most important part of the dose–response curve with a threshold chemical is the dose at which significant effects first begin to occur (Figure 11.2). The highest dose which does not produce an observable adverse effect is the 'no-observed-adverse-effect level' (*NOAEL*), and the lowest dose that produces an observable adverse effect is the 'lowest-observed-adverse-effect level' (*LOAEL*). For non-threshold chemicals, the dose–response curve behaves differently, in that there is no dose that is free of risk. When one is dealing specifically with the characterization of ecological effects, the dose–response curves (also, stressor–response profiles) may be represented as functional relationships between the amounts of a chemical substance and its morbidity/lethality (Figure 11.3).

In general, several important variables (Box 11.2) determine the characteristics of dose–response relationships, and which parameters should be given careful consideration in performing toxicity tests and when interpreting toxicity data (USEPA, 1985a, 1985b).

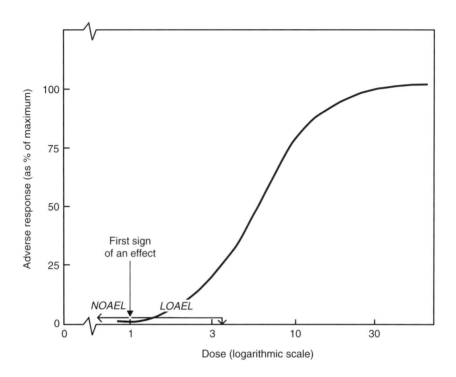

Figure 11.2 Illustrative relationship between 'toxicological endpoints' for a typical dose–response curve for threshold chemicals

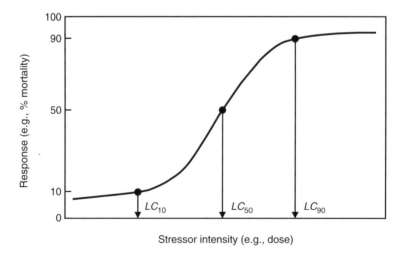

Figure 11.3 A schematic of a stressor–response curve, showing effect levels corresponding to 10% (LC_{10}), 50% (LC_{50}) and 90% (LC_{90}) lethalities or mortalities

> Box 11.2   Important parameters/factors to consider in toxicity assessments
>
> - *Route of exposure.* The toxicity of some chemicals depends on whether the route of exposure is by inhalation, ingestion, or dermal contact. Also, there may be local responses at the absorption site (i.e., lungs, gastrointestinal tract, and skin).
> - *Duration/frequency of exposure.* The toxicity of many chemicals depends not only on dose (i.e., the amount of chemical contacted or absorbed each day) but also on the length of exposure (i.e., number of days, weeks, or years).
> - *Test species characteristics.* Differences among species with respect to absorption, excretion, or metabolism of chemicals, as well as several other factors (such as genetic susceptibility) should be carefully evaluated in the choice of appropriate animal test species.
> - *Individual characteristics.* Individual members of a population (especially humans) are not identical and usually do not respond identically to equal exposures to a chemical. It is therefore important to identify any subgroups that may be more sensitive to a chemical than the general population.
> - *Toxicological endpoints.* This refers to the nature of toxic effects. The endpoints represent the changes detected in test animals, which become an index of the chemical's toxicity. Some commonly measured endpoints are carcinogenicity, hepatotoxicity (i.e., liver toxicity), mutagenicity, neurotoxicity, renal toxicity, reproductive toxicity, teratogenicity, etc. One of the most important parts of any toxicity study is the selection of the best endpoint to monitor – usually the most sensitive with respect to dose–response changes, the severity of effects, and whether an effect is reversible or irreversible.

11.2   CATEGORIZATION OF HUMAN TOXIC EFFECTS

For the purpose of human health risk assessment, chemicals are usually categorized into carcinogenic and noncarcinogenic groups. Chemicals that give rise to toxic endpoints other than cancer and gene mutations are often referred to as 'systemic toxicants' because of their effects on the function of various organ systems; the toxic endpoints are referred to as 'noncancer or systemic toxicity'. Most chemicals that produce noncancer toxicity do not cause a similar degree of toxicity in all organs, but usually demonstrate major toxicity to one or two organs; these are referred to as the target organs of toxicity for the chemicals (Klaassen et al., 1986; USEPA, 1989). In addition, chemicals that cause cancer and gene mutations also commonly evoke other toxic effects (i.e., systemic toxicity).

Noncarcinogens generally are believed to operate by 'threshold' mechanisms, i.e., the manifestation of systemic effects requires a threshold level of exposure or dose to be exceeded during a continuous exposure episode. Thus, noncancer or systemic toxicity is generally treated as if there is an identifiable exposure threshold below which there are no observable adverse effects. In fact, chronic noncarcinogenic health effects are assumed to exhibit a threshold level – i.e., continuous exposure to levels below the threshold produce no adverse or noticeable health effects. This characteristic distinguishes systemic endpoints from carcinogenic and mutagenic endpoints, which are often treated as 'nonthreshold' processes. That is, the threshold concept and principle is not applicable for carcinogens, since it is believed that no thresholds exist for this group.

11.2.1 Basis for the 'Threshold' versus 'Nonthreshold' Concepts

For many noncarcinogenic effects, protective mechanisms are believed to exist in the mammalian physiological system that must be overcome before the adverse effect of a chemical constituent is manifested. Consequently, a range of exposures exists from zero to some finite value – called the 'threshold' – that can be tolerated by the exposed organism with essentially no likelihood of adverse effects.

Carcinogenesis, unlike many noncarcinogenic health effects, is generally thought to be a phenomenon for which risk evaluation based on presumption of a threshold is inappropriate (USEPA, 1989). Thus, it is assumed that, for carcinogens, any finite exposure could result in a clinical state of disease. This hypothesized mechanism for carcinogenesis is referred to as 'nonthreshold' because there is believed to be essentially no level of exposure to such a chemical that does not pose a finite probability, however small, of generating a carcinogenic response.

It is noteworthy, however, that among some professional groups, there is the belief that certain carcinogens require a threshold exposure level to be exceeded to provoke carcinogenic effects (e.g., Wilson, 1996, 1997). In fact, opinion among regulatory scientists seems to have recently been returning to the ancient presumption that at least some cancer-causing substances induce effects through a threshold process (Wilson, 1997) – implying that for such substances there exists a finite level of exposure or dose at which no finite response is indicated.

11.2.2 Determination of Chemical Carcinogenicity

An important issue in chemical carcinogenesis relates to initiators and promoters. A *promoter* is defined as an agent which results in an increase in cancer induction when it is administered some time after a receptor has been exposed to an *initiator*. A *co-carcinogen* differs from a promoter only in that it is administered at the same time as the initiator. It is believed that initiators, co-carcinogens, and promoters do not usually induce tumors when administered separately. *Complete carcinogens* act as both initiator and promoter (OSTP, 1985). Even so, most regulatory agencies do not usually distinguish between initiators and promoters, because it is very difficult to confirm that a given chemical acts by promotion alone (OSHA, 1980; OSTP, 1985; USEPA, 1984).

A chemical's potential for human carcinogenicity is inferred from the available information relevant to the potential carcinogenicity of the chemical and from judgments as to the quality of the available studies. In general, carcinogens may be categorized into the following identifiable groupings (IARC, 1982; Theiss, 1983; USDHS, 1989):

- 'Known human carcinogens', defined as those chemicals for which there is sufficient evidence of carcinogenicity from studies in humans to indicate a causal relationship between exposure to the agent and human cancer.
- 'Reasonably anticipated to be carcinogens', referring to those chemical substances for which there is limited evidence for carcinogenicity in humans and/or

sufficient evidence of carcinogenicity in experimental animals. Sufficient evidence in animals is demonstrated by positive carcinogenicity findings in multiple strains and species of animals, in multiple experiments, or to an unusual degree with regard to incidence, site or type of tumor, or age of onset.
- 'Sufficient evidence' and 'limited evidence' of carcinogenicity, used in the criteria for judging the adequacy of available data for identifying carcinogens, refer only to the amount and adequacy of the available evidence and not to the potency of carcinogenic effect on the mechanisms involved.

Evidence of possible carcinogenicity in humans comes primarily from epidemiological studies and long-term animal exposure studies at high doses which have subsequently been extrapolated to humans. Results from these studies are supplemented with information from short-term tests, pharmacokinetic studies, comparative metabolism studies, molecular structure–activity relationships, and other relevant information sources.

11.2.3 Carcinogen Classification Systems

Carcinogenic chemicals are generally classified into several categories, depending on the 'weight-of-evidence' or 'strength-of-evidence' available on the particular chemical's carcinogenicity (Hallenbeck and Cunningham, 1988; Huckle, 1991; IARC, 1982; USDHS, 1989; USEPA, 1986). A chemical's potential for human carcinogenicity is inferred from the available information relevant to the potential carcinogenicity of the chemical, and from judgments as to the quality of the available studies.

The two carcinogenicity evaluation philosophies – one based on 'weight-of-evidence' and the other on 'strength-of-evidence' – seem to have found common acceptance and usage. Systems that employ the weight-of-evidence evaluations consider and balance the negative indicators of carcinogenicity with those showing carcinogenic activity (Box 11.3); schemes using the strength-of-evidence evaluations consider combined strengths of all positive animal tests (human epidemiology studies and genotoxicity) to rank a chemical without evaluating negative studies, nor considering potency or mechanism (Huckle, 1991). Other varying carcinogen classification schemes do indeed exist globally within various regulatory and legislative groups.

11.2.3.1 Weight-of-evidence classification

A weight-of-evidence approach is used by the US EPA to classify the likelihood that an agent in question is a human carcinogen. This is a classification system for characterizing the extent to which available data indicate that an agent is a human carcinogen (or for some other toxic effects such as developmental toxicity). A three-stage procedure is utilized, as follows:

- Stage 1 – the evidence is characterized separately for human studies and for animal studies.

> Box 11.3 Summary of pertinent/comparative factors affecting the weight-of-evidence for human carcinogens
>
> *Factors increasing weight-of-evidence*
>
> - Evidence of human causality
> - Evidence of animal effects relevant to humans
> - Coherent inferences
> - Comparable metabolism and toxikinetics between species
> - Mode of action comparable across species
>
> *Factors decreasing weight-of-evidence*
>
> - No evidence or relevant data showing human causality
> - No evidence or data on relevance of animal effects to humans
> - Conflicting data
> - Metabolism and toxikinetics between species not comparable
> - Mode of action *not* comparable across species

- Stage 2 – the human and animal evidence are integrated into a presumptive overall classification.
- Stage 3 – the provisional classification is modified (i.e., adjusted upwards or downwards), based on analysis of the supporting evidence.

The result is that chemicals are placed into one of five categories, in accordance with the USEPA Carcinogen Assessment Group (CAG) weight-of-evidence categories for potential carcinogens. Proposed guidelines for the classification of the weight-of-evidence for human carcinogenicity have been published by the USEPA (USEPA, 1984); these guidelines are adaptations from those of the International Agency for Research on Cancer (IARC, 1984, 1987, 1988), and consist of the categorization of the weight-of-evidence into the five groups – viz., Groups A–E (Box 11.4) – further discussed below.

- *Group A – Human carcinogen.* For this group, there is sufficient evidence from epidemiologic studies to support a causal association between exposure to the agent and human cancer. The following three criteria must be satisfied before a causal association can be inferred between exposure and cancer in humans (Hallenbeck and Cunningham, 1988):

  – no identified bias which could explain the association;
  – possibility of confounding factors (i.e., variables other than chemical exposure level which can affect the incidence or degree of the parameter being measured) has been considered and ruled out as explaining the association; and
  – association is unlikely to be due to chance.

  This group is used only when there is sufficient evidence from epidemiologic studies to support a causal association between exposure to the agents and cancer.

- *Group B – Probable human carcinogen.* This group includes agents for which the weight-of-evidence of human carcinogenicity based on epidemiologic studies is

> Box 11.4   USEPA's weight-of-evidence classification system for potential carcinogens
>
> | USEPA Group | Reference Category |
> |---|---|
> | A | Human carcinogen (i.e., known human carcinogen) |
> | B | Probable human carcinogen:
　B1　indicates limited human evidence
　B2　indicates sufficient evidence in animals and inadequate or no evidence in humans |
> | C | Possible human carcinogen |
> | D | Not classifiable as to human carcinogenicity |
> | E | No evidence of carcinogenicity in humans
(or, evidence of noncarcinogenicity for humans) |

'limited' and also includes agents for which the weight-of-evidence of carcinogenicity based on animal studies is 'sufficient'. The category consists of agents for which the evidence of human carcinogenicity from epidemiologic studies ranges from almost sufficient to inadequate.

This group is divided into two subgroups, reflecting higher (Group B1) and lower (Group B2) degrees of evidence. Usually, category B1 is reserved for agents for which there is limited evidence of carcinogenicity to humans from epidemiologic studies; limited evidence of carcinogenicity indicates that a causal interpretation is credible but that alternative explanations such as chance, bias, or confounding could not be excluded. Inadequate evidence indicates that one of the following two conditions prevailed (Hallenbeck and Cunningham, 1988):

- there were few pertinent data; or
- the available studies, while showing evidence of association, did not exclude chance, bias, or confounding.

When there are inadequate data for humans, it is reasonable to regard agents for which there is sufficient evidence of carcinogenicity in animals as if they presented a carcinogenic risk to humans. Therefore, agents for which there is 'sufficient' evidence from animal studies and for which there is 'inadequate' evidence from human (epidemiological) studies or 'no data' from epidemiologic studies would usually result in a classification of B2 (CDHS, 1986; Hallenbeck and Cunningham, 1988; USEPA, 1986).

- *Group C – Possible human carcinogen.* This group is used for agents with limited evidence of carcinogenicity in animals in the absence of human data. Limited evidence means that the data suggest a carcinogenic effect, but are limited for the following reasons (Hallenbeck and Cunningham, 1988):

  - the studies involve a single species, strain, or experiment; or
  - the experiments are restricted by inadequate dosage levels, inadequate duration of exposure to the agent, inadequate period of follow-up, poor survival, too few animals, or inadequate reporting; or
  - an increase in the incidence of benign tumors only.

Group C classification relies on a wide variety of evidence, including the following (Hallenbeck and Cunningham, 1988; USEPA, 1986): definitive malignant tumor response in a single well conducted experiment that does not meet conditions for 'sufficient' evidence; tumor response of marginal statistical significance in studies having inadequate design or reporting; benign but not malignant tumors, with an agent showing no response in a variety of short-term tests for mutagenicity; and responses of marginal statistical significance in a tissue known to have a high and variable background rate.

- *Group D – Not classifiable as to human carcinogenicity.* This group is generally used for agents with inadequate animal evidence of carcinogenicity and also inadequate evidence from human (epidemiological) studies. Inadequate evidence means that, because of major qualitative or quantitative limitations, the studies cannot be interpreted as showing either the presence or absence of a carcinogenic effect.
- *Group E – No evidence of carcinogenicity in humans.* This group is used for agents for which there is evidence of noncarcinogenicity for humans, together with no evidence of carcinogenicity in at least two adequate animal tests in different species, or no evidence in both adequate animal and human (epidemiological) studies. The designation of an agent as being in this group is based on the available evidence and should not be interpreted as a definitive conclusion that the agent will not be a carcinogen under any circumstances.

11.2.3.2 Strength-of-evidence classification

The IARC bases its classification on the strength-of-evidence philosophy. The corresponding IARC classification system, comparable or equivalent to the USEPA description presented above, is shown in Box 11.5 – and further discussed below.

- *Group 1 – Known human carcinogen.* This group is generally used for agents with sufficient evidence from human (epidemiological) studies as to human carcinogenicity.
- *Group 2A – Probable human carcinogen.* This group is generally used for agents for which there is sufficient animal evidence, evidence of human carcinogenicity, or at least limited evidence from human (epidemiological) studies. These are probably carcinogenic to humans, with (usually) at least limited human evidence.
- *Group 2B – Possible human carcinogen.* This group is generally used for agents for which there is sufficient animal evidence and inadequate evidence from human (epidemiological) studies, or there is limited evidence from human (epidemiological) studies in the absence of sufficient animal evidence. These are probably carcinogenic to humans, but (usually) have no human evidence.
- *Group 3 – Not classifiable.* This group is generally used for agents for which there is inadequate animal evidence and inadequate evidence from human (epidemiological) studies. There is sufficient evidence of carcinogenicity in experimental animals.

> Box 11.5 IARC's strength-of-evidence classification system for potential carcinogens
>
> | IARC Group | Category |
> |---|---|
> | 1 | Human carcinogen (i.e., known human carcinogen) |
> | 2 | Probable or possible human carcinogen:
 2A indicates limited human evidence (i.e., probable)
 2B indicates sufficient evidence in animals and inadequate or no evidence in humans (i.e., possible) |
> | 3 | Not classifiable as to human carcinogenicity |
> | 4 | No evidence of carcinogenicity in humans |

- *Group 4 – Noncarcinogenic to humans.* This group is generally used for agents for which there is evidence for lack of carcinogenicity. They are probably not carcinogens.

11.3 EVALUATION OF CHEMICAL TOXICITY

A toxicity assessment is conducted as part of a risk assessment, in order to qualitatively and quantitatively determine the potential for adverse human and ecological health effects to result from exposure to environmental contamination problems. This involves an evaluation of the types of adverse effects associated with chemical exposures, the relationship between the magnitude of exposure and adverse effects, and related uncertainties such as the weight-of-evidence of a particular chemical's carcinogenicity in humans.

The quantitative portion of the toxicity assessment entails identifying the relevant toxicity indices against which exposure point intakes and doses can be compared during the risk characterization stage of the overall assessment. Such assessment may include a consideration of experimental studies that uses animal data for extrapolation to humans, as well as epidemiological studies. The qualitative aspect of the assessment provides summaries of the adverse health effects, typical environmental levels or background concentrations, toxicokinetics, toxicodynamics, and ecotoxicology associated with each chemical of potential concern.

A comprehensive toxicity assessment for chemicals associated with environmental contamination problems is generally accomplished in two steps: hazard effects assessment and dose–response assessment. These steps are briefly discussed below, and in more detail elsewhere in the literature (e.g., Casarett and Doull, 1975; Klaassen et al., 1986; USEPA, 1989).

11.3.1 Hazard Effects Assessment

Hazard effects assessment is the process used to determine whether exposure to an agent can cause an increase in the incidence of an adverse health effect (e.g., cancer, birth defects, etc.); it includes a characterization of the nature and strength of the evidence of causation. The process involves gathering and evaluating data on the types of health injury or disease that may be produced by a chemical and on the conditions of exposure under which injury or disease is produced. Hazard effects assessment may also involve characterizing the behavior of a chemical within the receptor's body and the interactions it undergoes with organs, cells, or even parts of cells. Data of the latter types may be of value in answering the ultimate question of whether the forms of toxicity known to be produced by a substance in one population group or in experimental settings, are also likely to be replicated in real-life situations.

Methods commonly used for assessing the hazardous nature of substances include (Lave, 1982; NRC, 1991; Talbot and Craun, 1995):

- case clusters;
- structural toxicology;
- laboratory study of simple test systems;
- long-term animal bioassays; and
- human (epidemiological) studies.

Case clusters are based on the identification of an abnormal pattern of disease. This procedure tends to be more powerful in identifying hazards especially when the resulting condition is extremely rare; when the health condition is more common in the general population, the method is not very powerful. Since the population at risk is essentially never known in detail, the case cluster method necessarily yields no conclusive evidence, only rather vague suspicions.

Structural toxicology involves searching for similarities in chemical structure that might identify carcinogens. The close association between mutagens and carcinogens leads to a general presumption that mutagenic substances are also carcinogenic.

Animal bioassays are laboratory experimentations, generally with rodents; statistical models are used to extrapolate from animal bioassays to humans.

Epidemiology is a more scientific, systematic form of case cluster analysis with an attempt to control for confounding factors in the experimental design or statistical analysis.

In practice toxicity studies are generally conducted to identify the nature of health damage produced by a substance and the range of doses over which damage is produced (Box 11.6) (USEPA, 1985b). The assessment of the toxicity of a chemical substance involves identification of the adverse effects which the chemical causes and systematic study of how these effects depend upon dose, route and duration of exposure, and test organisms. This information is typically derived from studies falling into one of the general protocols/categories annotated below (Cohrssen and Covello, 1989; Moeller, 1997; USEPA, 1985a, 1985b).

- *Laboratory animal studies*, which evaluate the toxicity of a chemical with special reference and/or the ultimate goal of predicting the toxicity in humans. Testing protocols in animals are designed to identify the principal adverse effects of a

> **Box 11.6 Summary reasons for conducting toxicity studies**
>
> - To identify the specific organs or systems of the body that may be damaged by a substance
> - To identify specific abnormalities or diseases (such as cancer, birth defects, nervous disorders, or behavioral problems) that a substance may produce
> - To establish the conditions of exposure and dose that give rise to specific forms of damage or disease
> - To identify the specific nature and course of the injury or disease produced by a substance
> - To identify the biological processes that underlie the production of observable damage or disease

chemical as a function of dose, route of exposure, species and sex of test animals, and duration of exposure.

- *Clinical case studies in humans*, in which there are case-by-case investigations of the symptoms and diseases in humans who are exposed to a toxic substance at doses high enough to call for medical attention. Exposures may be accidental (e.g., a farmer applying pesticide without proper protection) or, in rare cases, intentional (e.g., suicide or homicide cases). Tragically, this sort of direct toxicological observation is especially valuable in characterizing toxic responses of clinical significance in humans – far better than extrapolations from laboratory animals to humans.
- *Epidemiologic studies*, which seek to determine whether a correlation exists between chemical exposure and frequency of disease or health problems in large groups of human populations. It involves the examination of persons who have been inadvertently exposed to one or more chemical agents. The major advantages of epidemiological studies are that they are based on large numbers of humans and exposure levels are usually subclinical. Thus, the data are directly relevant, with no need to extrapolate from animal data or to make projections from a small number of humans exposed to a high dose of the chemical, as for clinical studies.
- *Ecotoxicological studies*, which assess the toxic effects of chemical substances on indigenous aquatic and terrestrial plants and animals. The presence of toxic substances in the environment may adversely affect the abundance; species composition and diversity; stability; productivity; and physiological conditions of indigenous fish, plant, and wildlife populations. The studies involved are usually similar in design and objectives to studies involving laboratory animals; the main difference is that species employed in ecotoxicological tests are selected to be representative of indigenous fish, plants, and wildlife, whereas laboratory animals are intended to serve as models for humans.

The overall purpose of the hazard effects assessment is to review and evaluate data pertinent to answering two questions – whether an agent may pose a hazard to potential receptors, and under what circumstances an identified hazard may be manifested.

11.3.2 Dose–Response Assessment

Dose–response assessment is the process of quantitatively evaluating toxicity information and characterizing the relationship between the dose of the contaminant administered or received (i.e., exposure to an agent) and the incidence of adverse health effects in the exposed populations. The process consists of estimating the potency of the specific compounds by use of dose–response relationships. For example, in the case of carcinogens, this involves estimating the probability that an individual exposed to a given amount of chemical will contract cancer due to that exposure; potency estimates may be given as 'unit risk factor' (expressed in mg/m^3 or ppb) or as 'potency slopes' (in units of $[mg/kg-day]^{-1}$). Data are generally derived from animal studies or, less frequently, from studies in exposed populations.

The dose–response assessment first addresses the relationship of dose to the degree of response observed in an experiment or human study. When environmental exposures are outside the range of observations, extrapolations are necessary in order to estimate or characterize the dose relationship. The extrapolations will typically be made from: high to low doses, animal to human responses, and/or one route of exposure to another.

11.3.2.1 The nature of dose–response extrapolation models

Three major classes of mathematical extrapolation models are often used for relating dose and response in the subexperimental dose range, viz.:

- tolerance distribution models – including probit, logit, and Weibull;
- mechanistic models – including one-hit, multi-hit, and multistage; and
- time-to-occurrence models – including lognormal and Weibull.

Indeed, other independent models – such as linear, quadratic, and linear-cum-quadratic – may also be employed for this purpose. The details of these are beyond the scope of this discussion, but are elaborated elsewhere in the literature (e.g., Brown, 1978; CDHS, 1986; Crump, 1981; Crump and Howe, 1984; Gaylor and Kodell, 1980; Gaylor and Shapiro, 1979; Hogan, 1983; Krewski and Van Ryzin, 1981). In any case, the primary models used to extrapolate from nonthreshold effects associated with carcinogenic responses that are observed at high doses to responses at low doses include the following (Jolley and Wang, 1993):

- *Linearized multistage model*, which assumes that there are multiple stages to cancer; it fits curve to the experimental data, and is linear from the upper confidence level to zero.
- *One-hit model*, which assumes there is a single stage for cancer and that one molecular or radiation interaction induces malignant change; it is a very conservative model, and corresponds to the simplest mechanistic model of carcinogenesis.
- *Multi-hit model*, which assumes several interactions are needed before a cell becomes transformed; it is the least conservative model.

- *Probit model*, which assumes probit (lognormal) distribution for tolerance of exposed population; it is appropriate for acute toxicity, but questionable for cancer.
- *Physiologically-based pharmacokinetic (PB-PK) models*, which incorporate pharmacokinetic and mechanistic data into the extrapolation; they possess data-rich requirements with great promise for extensive utilization in the future, as more biological data becomes available.

It is noteworthy that most of the techniques used to compensate for toxicity assessment uncertainties (such as the use of large safety factors, conservative assumptions, and extrapolation models) are designed to err on the side of safety. For these reasons, many regulatory agencies tend to use the so-called linearized multistage model for conservatism. In general, however, the choice of model is determined by its consistency with the current understanding of the mechanisms of carcinogenesis.

In fact, several models have been proposed for the quantitative extrapolations of carcinogenic effects to low dose levels. However, among these models, the USEPA, for instance, recommends a linearized multistage model (USEPA, 1986). The linearized multistage model conservatively assumes linearity at low doses. Alternative models that are generally less conservative do exist which do not assume a linear relationship.

There is often no sound basis, in a biological sense, for choosing one model over another. When applied to the same data, the various models can produce a wide range of risk estimates. The model recommended by the USEPA produces among the highest estimates of risk, and thus provides a greater margin of protection for human health. Moreover, this model does not provide a 'best estimate' or point estimate of risk, but rather an upper-bound probability that the actual risk will be less than the predicted risk 95% of the time. But, given that no single model will apply for all chemicals, it is important to identify risk on a case-by-case basis. In fact, Huckle (1991) suggests a presentation of the best estimate of risk (or range, with an added margin of safety) from two or three appropriate models, or a single value based on 'weight-of-evidence', rather than using simply the linearized multistage model. Exceptions may occur, however, for cases of poorly studied chemicals.

11.3.2.2 Dose–response quantification

In general, the risks of a substance cannot be ascertained with any degree of confidence unless dose–response relationships are quantified, even if the substance is known to be toxic. Dose–response relationships are generally used to determine what dose of a particular chemical causes specific levels of toxic effects to potential receptors. In fact, there may be many different dose–response relationships for a substance if it produces different toxic effects under different conditions of exposure. In any case, the response of a toxicant depends on the mechanism of its action; for the simplest scenario, the response, R, is directly proportional to its concentration, $[C]$, so that:

$$R = k[C]$$

where k is a rate constant. This would be the case for a pollutant that metabolizes rapidly but, even so, the response and the value of the rate constant would tend to differ for different risk groups of individuals and for unique exposures. For instance, if the toxicant accumulates in the body, the response is defined by:

$$R = k[C]t^n$$

where t is the time and n is a constant. For cumulative exposures, the response would generally increase with time. Thus, the cumulative effect may show as linear until a threshold is reached, after which secondary effects begin to affect and enhance the responses. The cumulative effect may be related to what is referred to as the 'body burden' (BB). The body burden is determined by the relative rates of absorption (ABS), storage (STR), elimination (ELM), and biotransformation (BTF), according to the following relationship (Meyer, 1983):

$$BB = ABS + STR - ELM - BTF$$

Each of the factors involved in the quantification of the body burden is dependent on a number of biological and physiochemical factors.

In fact, the response of an individual to a given dose cannot be truly quantitatively predicted since it depends on many extraneous factors, such as general health and diet of individual receptors or the PAR. Nonetheless, from the quantitative dose–response relationship, toxicity values can be derived and used to estimate the incidence of adverse effects occurring in potential receptors at different exposure levels.

It is noteworthy that the fundamental principles underlying the dose–response assessment for carcinogenic chemicals remain arguable – especially in relation to the tenet that there is some degree of carcinogenic risk associated with every potential carcinogen, no matter how small the dose. The speculation and/or belief that chemically induced cancer is a nonthreshold process/phenomenon may be false, but represents a conservative default policy necessary to ensure adequate protection of human health, albeit this shortcoming should be kept in perspective in consequential policy decisions about the assessment.

11.4 UTILITY OF THE HAZARD EFFECTS ASSESSMENT

A hazard effects evaluation is generally conducted as part of a risk assessment, in order to qualitatively and/or quantitatively assess the potential for adverse effects from receptor exposure to environmental contaminants. For most environmental contamination problems, this usually will comprise toxicological evaluations, the ultimate goal of which is to derive reliable estimates of the amount of chemical exposure which may be considered 'tolerable' (or 'acceptable' or 'reasonably safe') for humans or other organisms.

Typically, risk assessments rely heavily on existing toxicity information developed for specific chemicals; a summary listing of such toxicological parameters is shown in Appendix D for selected environmental chemicals. Where toxicity information does not exist at all, a decision may be made to estimate toxicological data from that of similar compounds (with respect to molecular weight and structure–activity). Structure–activity analysis is a technique which can be applied to derive an estimate for the toxicity of a chemical when direct experimental or observational data are lacking.

SUGGESTED FURTHER READING

Freese, E. 1973. Thresholds in toxic, teratogenic, mutagenic, and carcinogenic effects. *Environmental Health Perspectives*, 6: 171–178.

Furst, A. 1990. Yes, but is it a human carcinogen? *Journal of the American College of Toxicology*, 9: 1–18.

Hayes, A.W. (ed.). 1982. *Principles and Methods of Toxicology*. Raven Press, New York.

Hughes, W.W. 1996. *Essentials of Environmental Toxicology: The Effects of Environmentally Hazardous Substances on Human Health*. Taylor & Francis Publishers, London, UK.

IARC (International Agency for Research on Cancer) (1972–1985). *IARC Monographs on the Evaluation of the Carcinogenic Risk of Chemicals to Man. (Multivolume work.)* IARC, World Health Organization, Geneva.

Klaassen, C.D. (ed.). 1996. *Casarett and Doull's Toxicology: The Basic Science of Poisons*, 5th edition. McGraw-Hill, New York.

Lu, F.C. 1985. *Basic Toxicology*. Hemisphere, Washington, DC.

Merck. 1989. *The Merck Index: An Encyclopedia of Chemicals, Drugs and Biologicals*. 11th (Centennial) edition. Merck & Co., Inc., Rockway, NJ.

Sax, N.I. 1979. *Dangerous Properties of Industrial Materials*, 5th edition. Van Nostrand Reinhold, New York.

Sax, N.I. and R.J. Lewis, Sr. 1987. *Hawley's Condensed Chemical Dictionary*. Van Nostrand Reinhold, New York.

Sitnig, M. 1985. *Handbook of Toxic and Hazardous Chemicals and Carcinogens*. Noyes Data Corp., Park Ridge, NJ.

Stenesh, J. 1989. *Dictionary of Biochemistry and Molecular Biology*. J. Wiley, New York.

Weisman, J. 1996. AMS adds realism to chemical risk assessment. *Science*, 271(5247): 286–287.

REFERENCES

Brown, C. 1978. Statistical aspects of extrapolation of dichotomous dose response data. *Journal National Cancer Institute*, 60: 101–108.

Casarett, L.J. and J. Doull. 1975. *Toxicology: The Basic Science of Poisons*. Macmillan Publishing Co., New York.

CDHS (California Department of Health Services). 1986. *The California Site Mitigation Decision Tree Manual*. CDHS, Toxic Substances Control Division, Sacramento, CA.

Cohrssen, J.J. and V.T. Covello. 1989. *Risk Analysis: A Guide to Principles and Methods for Analyzing Health and Environmental Risks*. National Technical Information Service, US Dept. of Commerce, Springfield, VA.

Crump, K.S. 1981. An improved procedure for low-dose carcinogenic risk assessment from animal data. *Journal of Environmental Toxicology*, 5: 339–346.

Crump, K.S. and R.B. Howe. 1984. The multistage model with time-dependent dose pattern: applications of carcinogenic risk assessment. *Risk Analysis*, 4: 163–176.

Gaylor, D.W. and R.L. Kodell. (1980). Linear interpolation algorithm for low dose risk assessment of toxic substances. *Journal of Environmental Pathology, Toxicology and Oncology*, 4: 305–312.

Gaylor, D.W. and R.E. Shapiro. 1979. Extrapolation and risk estimation for carcinogenesis. In: *Advances in Modern Toxicology Volume 1, New Concepts in Safety Evaluation Part 2*, M.A. Mehlman, R.E. Shapiro, and H. Blumenthal (eds). Hemisphere, New York, pp. 65–85.

Hallenbeck, W.H. and K.M. Cunningham. 1988. *Quantitative Risk Assessment for Environmental and Occupational Health*. Lewis Publishers, Chelsea, MI.

Hogan, M.D. 1983. Extrapolation of animal carcinogenicity data: limitations and pitfalls. *Environ. Health Perspectives*, 47: 333–337.

Huckle, K.R. 1991. *Risk Assessment – Regulatory Need or Nightmare?* Shell Publications, Shell Centre, London, UK.

IARC (International Agency for Research on Cancer). 1982. *IARC Monographs on the Evaluation of the Carcinogenic Risk of Chemicals to Humans. Chemicals, Industrial Processes and Industries Associated with Cancer in Humans*. Supplement 4 (292 pp.) IARC, Lyon, France.

IARC. 1984. *IARC Monographs on the Evaluation of the Carcinogenic Risk of Chemicals to Humans*, Vol. 33, World Health Organization, Lyon, France.

IARC. 1987. *IARC Monographs on the Evaluation of Carcinogenic Risks to Humans. Overall Evaluations of Carcinogenicity*. Supplement 7 (440 pp.). IARC, Lyon, France.

IARC. 1988. *IARC Monographs on the Evaluation of Carcinogenic Risks to Humans*, Vol. 43. (pp. 15–32). World Health Organization, Lyon, France.

Jolley, R.L. and R.G.M. Wang (eds). 1993. *Effective and Safe Waste Management: Interfacing Sciences and Engineering with Monitoring and Risk Analysis*. Lewis Publishers, Boca Raton, FL.

Klaassen, C.D., M.O. Amdur, and J. Doull (eds). 1986. *Casarett and Doull's Toxicology: The Basic Science of Poisons*, 3rd edition. Macmillan Publishing Company, New York.

Krewski, D. and J. Van Ryzin. 1981. Dose response models for quantal response toxicity data. In: *Statistics and Related Topics*, M. Csorgo, D.A. Dawson, J.N.K. Rao, and A.K.E. Saleh (eds). North Holland, New York, pp. 201–231.

Lave, L.B. (ed.). 1982. *Quantitative Risk Assessment in Regulation*. The Brooking Institute, Washington, DC.

Meyer, C.R. 1983. Liver dysfunction in residents exposed to leachate from a toxic waste dump. *Environmental Health Perspectives*, 48: 9–13.

Moeller, D.W. 1997. *Environmental Health*, Revised edition. Harvard University Press, Cambridge, MA.

NRC (National Research Council). 1991. *Environmental Epidemiology (Public Health and Hazardous Wastes)*. National Academy Press, Washington, DC.

OSHA (Occupational Safety and Health Administration). 1980. Identification, classification, and regulation of potential occupational carcinogens. *Federal Register*, 45: 5002–5296.

OSTP (Office of Science and Technology Policy). 1985. Chemical carcinogens: a review of the science and its associated principles. *Federal Register*, 50: 10372–10442.

Smith, R.P. 1992. *A Primer of Environmental Toxicology*. Lea & Febiger, Philadelphia, PA.

Talbot, E.O. and G.F. Craun (eds). 1995. *Introduction to Environmental Epidemiology*. Lewis Publishers/CRC Press, Boca Raton, FL.

Theiss, J.C. 1983. The ranking of chemicals for carcinogenic potency. *Regulatory Toxicology and Pharmacology*, 3: 320–328.

USDHS (US Department of Health and Human Services). 1989. *Public Health Service. Fifth Annual Report on Carcinogens*. Summary. US Government Printing Office, Washington, DC.

USEPA (US Environmental Protection Agency). 1984. Proposed guidelines for carcinogen, mutagenicity, and developmental toxicant risk assessment. *Federal Register*, 49: 46294–46331.

USEPA. 1985a. *Principles of Risk Assessment: A Nontechnical Review*. Office of Policy Analysis, Washington, DC.

USEPA. 1985b. *Toxicology Handbook*. Office of Waste Programs Enforcement, Washington, DC.

USEPA. 1986. Guidelines for carcinogen risk assessment. *Federal Register*, 51(185): 33992-34003, CFR 2984, September 24, 1986.

USEPA. 1989. *Risk Assessment Guidance for Superfund. Volume I – Human Health Evaluation Manual (Part A)*. Office of Emergency and Remedial Response, Washington, DC. EPA/540/1-89/002.

Wilson, J.D. 1996. *Threshold for Carcinogens: A Review of the Relevant Science and its Implications for Regulatory Policy*. Discussion Paper 96-21, Resources for the Future, Washington, DC.

Wilson, J.D. 1997. So carcinogens have thresholds: how do we decide what exposure levels should be considered safe? *Risk Analysis*, 17(1): 1–3.

Chapter Twelve

Risk Characterization and Uncertainty Analysis

Risk characterization consists of estimating the probable incidence of adverse impacts to potential receptors under various exposure conditions that are associated with a hazard situation. It involves an integration of the hazard effects and exposure assessments in order to arrive at an estimate of risk to the exposed population.

Depending on the nature of populations potentially at risk from an environmental contamination problem, different types of risk measures or parameters may be employed in the risk characterization process. Typically, the cancer risk estimates and hazard quotient/hazard index estimates are used to define potential risks to human health (see Chapter 13); the ecological quotient (similar to the human health hazard quotient) estimates are used to define risks to potential ecological receptors (see Chapter 14); and the risk costs associated with a 'pathway probability' concept is employed in probabilistic risk characterizations (see Chapter 15). The general details of the various risk characterization models and measures finding application in commonly encountered environmental contamination problems are discussed in Part IV of this volume.

12.1 RISK PRESENTATION AND SUMMARIZATION

Risk values are often stated simply as a number – such as is expressed by the risk probability of occurrence of additional cases of cancer (e.g., a cancer risk of 1×10^{-6}, reflecting the estimated number of excess cancer cases in a population), or by the hazard index of noncancer health effects like neurotoxicity or birth defects (e.g., a hazard index of 1, reflecting the degree of harm from a given level of exposure). One of the most important points to remember in all cases of risk presentation, however, is that the numbers by themselves may not tell the whole story. For instance, despite the fact that the numerical values may be identical, a human cancer risk value of 10^{-6} for an 'average exposed person' (e.g., someone exposed via food products) is not to be taken to be the same as a cancer risk of 10^{-6} for a 'maximally/most exposed individual' (e.g., someone exposed from living in a highly contaminated area). In fact, omission of the qualifier terms – 'average' or 'maximally/most exposed' – could mean an incomplete description of the true risk scenarios, and could result in poor risk management strategies and/or a failure in risk communication tasks. Thus, it is very important to know and to recognize such apparently subtle differences in the risk summarization – or indeed throughout the risk characterization process. In fact, this reflects on a notion that any qualitative aspect of a risk characterization (which may also include an explicit recognition of all assumptions, uncertainties, etc.) may be as important as its quantitative component (i.e., the risk numbers). The qualitative considerations are indeed essential to making a judgment about the reliability of the calculated risk numbers, and therefore the confidence to attach to the characterization of potential risks.

12.1.1 Graphical Presentation of the Risk Summary Information

Several graphical representations may be employed in presenting a summary of the requisite risk information that has been developed from the risk characterization efforts. Examples of such graphical forms include the following:

- *Pie charts*, such as shown in Figure 12.1 to illustrate the hazard index contributions from different environmental contamination sources.
- *Horizontal bar charts*, such as shown in Figure 12.2 to illustrate the hazard index contributions associated with different exposure routes and receptor groups.
- *Vertical bar charts*, such as shown in Figure 12.3 to illustrate the hazard index and cancer risk contributions from different chemical contaminants.
- *Variety of relational plots*, such as shown in Figures 12.4 through 12.6 to illustrate various graphical relationships used to characterize risk associated with environmental contamination problems.

This listing is by no means complete; other novel representations that may consist of variations or convolutions of the above may indeed be found to be more appropriate and/or useful for some case-specific situations.

12.1.2 Presenting and Managing Uncertain Risks

Inevitably, some degree of uncertainty remains in quantitative risk estimates in virtually all fields of applied risk analysis. A carefully executed analysis of uncertainties therefore plays a very important role in all risk assessments. On the other hand, either or both of a comprehensive qualitative analysis and rigorous quantitative analysis of uncertainties will be of little value if such analysis results are not clearly presented for effective use in the decision-making process. A number of approaches have been suggested by investigators like Cox and Ricci (in Paustenbach, 1988) for presenting risk analysis results to decision-makers, including the following:

- Risk assessment results should be presented in a sufficiently disaggregated form (showing risks for different subgroups) so that key uncertainties and heterogeneities are not lost in the aggregation.

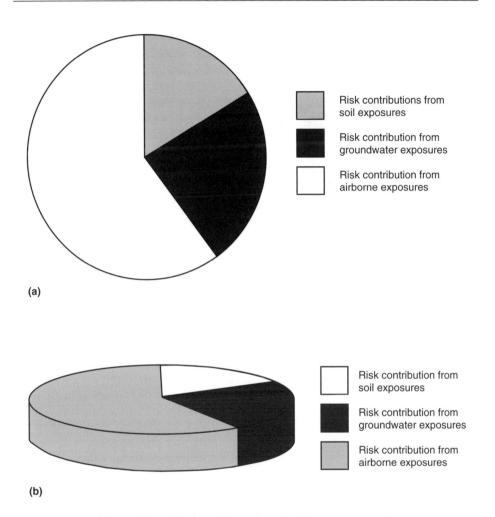

Figure 12.1 Pie chart illustration of risk summary results

- Confidence bands around the predictions of statistical models can be useful, but uncertainties about the assumptions of the model itself should also be presented.
- Both individual (e.g., for the typical and most threatened individuals in the population) and population/group risks should be presented, so that the equity of the distribution of individual risks in the population can be appreciated and taken into account.
- Any uncertainties, heterogeneities, or correlations across individual risks should be identified.
- Population risks can be described at the 'micro' level (viz: in terms of frequency distribution of individual risks), or at the 'macro' level (viz: using decision-analytic models, in terms of attributes such as equivalent number of life-years).

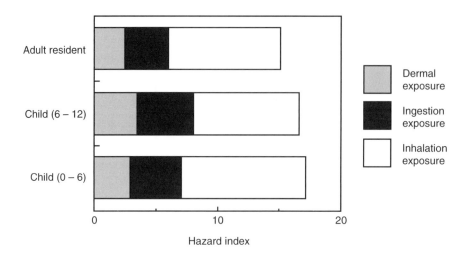

Figure 12.2   Horizontal bar chart illustration of risk summary results

Furthermore, sensitivity analyses for key assumptions are recommended for use extensively. These serve to identify the sensitivity of the calculated result to the various input assumptions – and thus identify key uncertainties – as well as to bracket potential risks so that policy-makers can make more informed choices.

In fact, several factors may still contribute to the over- or under-estimation of risks. For example, in human health risk assessments, some factors – such as lack of potency data for some carcinogenic chemicals; risk contributions from compounds formed in environmental media, such as transformation products, but that are not quantified; and the fact that all risks are assumed to be additive, although certain combinations of exposure may have synergistic (i.e., greater than additive) effects – will typically under-estimate health impacts associated with chemicals evaluated in the assessment. On the other hand, another set of factors – such as the fact that many unit risk and potency factors are often considered plausible upper-bound estimates of carcinogenic potency, when indeed the true potency of the chemical could be considerably lower; exposure estimates are often very conservative; and possible antagonistic effects, for chemicals whose combined presence reduce toxic impacts, are not accounted for – may cause an analysis to over-estimate risks.

In general, the results of deterministic risk assessments should be interpreted with caution, and never construed as absolute measures of risk. Even so, such point estimates of risk may still be useful in a qualitative sense for ranking different programs or issues. Probabilistic methods, however, must be encouraged as the logical evolution of risk assessment, and should be accompanied by the development of risk management methods that can utilize the richness of information provided by Monte Carlo assessments and other similar techniques (Zemba et al., 1996). In fact, the danger of mis-characterizing high-end, central tendency, and other exposure levels can only be alleviated by the development of full probabilistic analyses.

Risk characterization and uncertainty analysis

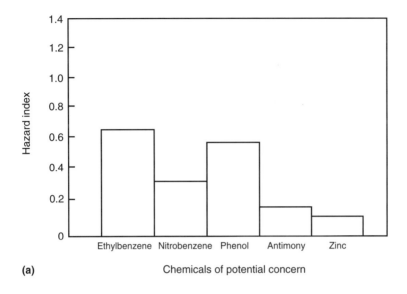

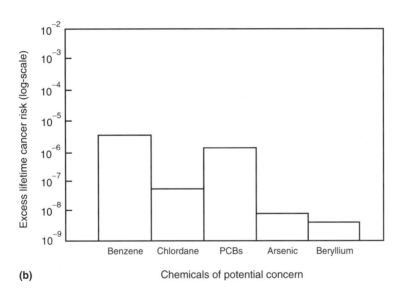

Figure 12.3 Vertical bar chart illustration of risk summary results: (a) Illustrative presentation of the relative contribution of individual chemicals to overall hazard index estimates associated with a hypothetical public water supply system. (b) Illustrative presentation of the relative contribution of individual chemicals to overall cancer risk estimates associated with a hypothetical public water supply system (semi-log plot)

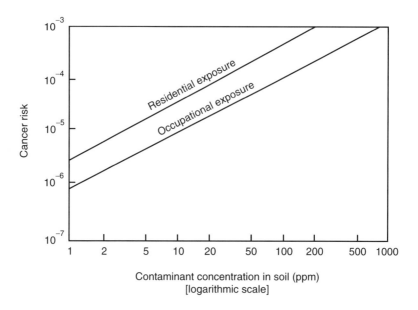

Figure 12.4 Illustrative sketch of the effects of choice of exposure scenarios on dose and risk estimates

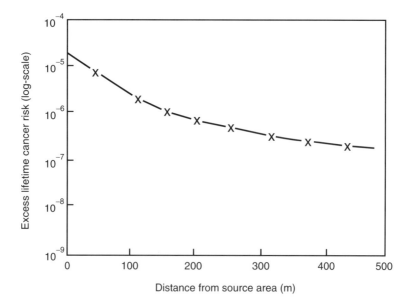

Figure 12.5 Illustrative sketch of the variation of estimated cancer risks with distance from contaminant source: a semi-log plot of cancer risk estimates from receptor exposures to benzene in groundwater at several different locations downgradient of a release source

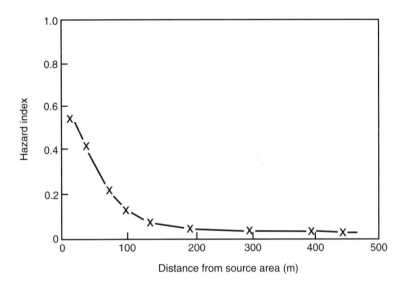

Figure 12.6 Illustrative sketch of the variation of estimated hazard index with distance from contaminant source: an arithmetic-scale plot of hazard index estimates from receptor exposures to ethylbenzene in groundwater at several different locations downgradient of a release source

12.2 UNCERTAINTY AND VARIABILITY ISSUES IN RISK ASSESSMENT

Risk assessments tend to be highly uncertain as well as highly variable. *Variability* (or *stochasticity*) arises from true heterogeneity in characteristics such as dose–response differences within a population or differences in body weight or differences in rates of food and water intakes/ingestion or differences in contaminant levels in the environment; whereas *uncertainty* represents a lack of knowledge about factors such as adverse effects or contaminant levels, which may be reduced with additional study or investigations.

In general, some parameters used in risk assessments may reflect both variability and uncertainty under different sets of circumstances or conditions. However, insofar as possible, stochastic variability and knowledge uncertainty should be segregated in the evaluation processes employed during the risk assessment. Indeed, in principle, uncertainty can be reduced by obtaining more information while variability is irreducible.

12.2.1 The Need for Uncertainty and Variability Analyses

A number of factors – directly or indirectly related to uncertainties and variabilities – may indeed cause an analysis to either under-estimate or over-estimate true risks associated with an environmental contamination problem. For instance, it is always possible that a chemical whose toxic properties have not been thoroughly tested may be more toxic than originally believed or anticipated; a chemical not tested for carcinogenicity or teratogenicity may in fact display those effects. Furthermore, a limitation of analysis for selected 'indicator chemicals' may have some limiting (even if insignificant) effects. In any event, all risk estimates involve some degree of uncertainty, especially because of the inability of the risk assessor to quantify all the information that must indeed be used. Uncertainty analysis should therefore become an integral part of all risk assessments, regardless of the scope or level of detail. It is indeed prudent and essential to the credibility of the risk assessment, to describe the relevant uncertainties in as great a detail as possible.

The degree to which variability and uncertainty are addressed in a given study depends largely on the scope of the risk assessment and the resources available. For the study of variability, stochastic models are used as the more realistic representations of reality, rather than the use of deterministic models. In any case, as a guiding principle, the discussion of uncertainty and variability should, ideally, reflect the type and complexity of the risk assessment – with corresponding levels of efforts for the risk assessment and the analysis or discussion of uncertainty and variability.

12.3 TYPES AND NATURE OF UNCERTAINTY

The uncertainties that typically arise in risk assessments can be of three general types: uncertainties in parameter values (e.g., use of incomplete or biased values); uncertainties in parameter modeling (e.g., issue of model adequacy/inadequacy); and uncertainties in the degree of completeness (e.g., representativeness of evaluation scenarios) – further discussed below.

- *Parameter uncertainties* arise from the need to estimate parameter values from limited or inadequate data. Such uncertainties are inherent because the available data are usually incomplete, and the analyst must make inferences from a state of incomplete knowledge. Examples of uncertainties in parameter values relate to such issues as incomplete or biased data; applicability of available data to the particular case (i.e., generic versus case-specific data); etc.
- *Modeling uncertainties* stem from inadequacies in the various models used to evaluate hazards, exposures, and consequences, and also from the deficiencies of the models in representing reality. Examples of uncertainties in modeling relate to such issues as model adequacy; whether uncertainty is introduced by the mathematical or numerical approximations that are made for convenience; use of models outside its range of validity; etc.
- *Completeness/scenario uncertainties* relate to the inability of the analyst to evaluate exhaustively all contributions to risk. They refer to the problem of assessing what may have been omitted in the analysis. Examples of uncertainties in the degree of completeness may relate to such questions as to whether the analyses have been taken to sufficient depth; whether all important hazard sources and exposure possibilities have been addressed; etc.

For all practical purposes, these uncertainties are propagated through the analysis. To the extent possible, and as a complement to uncertainty analysis, a sensitivity analysis provides insight into the possible range of results. Sensitivity analysis entails the determination of how rapidly the output of an analysis changes with respect to variations in the input. Sensitivity studies do not usually incorporate the error range or uncertainty of the input – thus serving as a distinguishing element from uncertainty analyses.

Depending on the specific aspect or component of the risk assessment being performed, the type of uncertainty that dominates at each stage of the analysis can be different. Each type of uncertainty can be characterized either qualitatively or quantitatively. Various levels of uncertainty analysis can therefore be characterized by the degree to which each type of uncertainty is quantitatively analyzed.

12.3.1 Common Sources of Uncertainty in Endangerment Assessments

Considerable uncertainty is inherent in the human and ecological risk assessment process. In particular, uncertainties arise due to the use of several assumptions and inferences necessary to complete a risk assessment. For instance, human health risk assessments usually involve extrapolations and inferences to predict the occurrence of adverse health effects under certain conditions of exposure to chemicals in the environment, based on knowledge of the adverse effects that occur under a different set of exposure conditions (e.g., different dose levels and/or species). Because of these types of extrapolations and projections, there is considerable uncertainty in the resulting conclusions due in part to the several assumptions that are part of the process mechanics.

In fact, because of the various limitations and uncertainties, the results of a risk assessment cannot be considered as an absolutely accurate determination of risks. Commonly encountered limitations and uncertainties of considerable significance in relation to several components of the risk assessment process are enumerated below (see, e.g., Calabrese, 1984; Clewell and Andersen, 1985; Dourson and Stara, 1983; USEPA, 1989b).

- *Uncertainties in general extrapolations relevant to toxicity information.* Whereas some chemicals have been studied extensively under a variety of exposure conditions in several species (including humans), others may have only limited investigations done on them. This latter group will tend to have inherent limitations in toxicity data (arising for several reasons). Also, because data that specifically identify the hazards to humans as a result of their exposure to various chemicals of concern under the conditions of likely human exposure may not exist, it becomes necessary to infer such hazard effects by extrapolating from data obtained under different exposure conditions, usually in experimental animals. This introduces three major types of uncertainties – those related to extrapolating from one species to another (i.e., uncertainties in interspecies extrapolation), those relating to extrapolation from a high-dose region curve to a low-dose region (i.e., uncertainties in intraspecies extrapolation), and those

related to extrapolating from one set of exposure conditions to another (i.e., uncertainties due to differences in exposure conditions).
- *Uncertainties from quantitative extrapolations and adjustments in dose–response evaluation.* Experimental studies to determine the carcinogenic effects due to low exposure levels often encountered in the environment generally are not feasible. This is because such effects are not readily apparent in the relatively short time-frame over which it is usually possible to conduct such a study. Consequently, various mathematical models are used to extrapolate from the high doses used in animal studies to the doses encountered in exposure to ambient environmental concentrations. Extrapolating from a high dose (of animal studies) to a low dose (for human effects) introduces a level of uncertainty which could be significantly large, and which may have to be addressed. For instance, in human health risk assessments, no-observed-adverse-effect levels (*NOAELs*) and cancer potency slope factors (SFs) from animal studies are usually divided by a factor of 10 to account for extrapolation from animals to humans and by an additional factor of 10 to account for variability in human responses (see Chapter 13). Given the recognized differences among species in their responses to toxic insult, and between strains of the same species, it is apparent that additional uncertainties will be introduced when these types of quantitative extrapolations and adjustments are made in the dose–response evaluation.
- *Uncertainty associated with the toxicity of chemical mixtures.* The effects of combining two chemicals may be synergistic (effect when outcome of combining two chemicals is greater than the sum of the inputs), antagonistic (effect when the outcome is less than the sum of the two inputs), or under potentiation (i.e., when one chemical has no toxic effect but combined with another chemical that is toxic, produces a much more toxic effect). Indeed, chemicals present in a mixture can interact to yield a new chemical or one can interfere with the absorption, distribution, metabolism, or excretion of another. Notwithstanding all these possible scenarios, risk assessments often assume toxicity to be additive – resulting in an important source of uncertainty.
- *Limitations in model form.* Exposure scenarios and contaminant transport models usually are a major contributor of uncertainty to risk assessments. Apart from general model imperfections, environmental transport models usually oversimplify reality, contributing one form of uncertainty or another. Also, the natural variability in environmental and exposure-related parameters causes variability in exposure factors, and therefore in exposure estimates developed on this basis. This therefore begs the question of how close to reality the model function and output are likely to be.
- *Consideration of 'background' exposures.* For the most part, risk assessment methods used in practice tend to ignore background exposures; instead, the process considers only incremental risk estimates for the exposed populations. Consequently, such risk estimates do not address what constitutes the true health risks to the public – of which background exposures could be contributing in a very significant way.
- *Representativeness of sampling data.* Uncertainties arise from random and systematic errors in the type of measurement and sampling techniques often

used in environmental characterization activities. For instance, professional judgment (based on engineering and scientific assumptions) is frequently used for sampling design and also to make decisions on how to correct for data gaps – albeit this process has some inherent uncertaintes associated with it.

In general, uncertainties are difficult to quantify or, at best, the quantification of uncertainty is itself uncertain. Thus, the risk levels generated in a risk assessment are useful only as a yardstick and decision-making tool for the prioritization of problem situations, rather than being construed as actual expected rates of disease, or adversarial impacts in exposed populations. They are used only as an estimate of risks, based on current level of knowledge coupled with several assumptions. Quantitative descriptions of uncertainty, which could take into account random and systematic sources of uncertainty in potency, exposure, intakes, etc., would help present the spectrum of possible true values of risk estimates, together with the probability (or likelihood) associated with each point in the spectrum.

12.4 ANALYSIS OF UNCERTAINTIES AND VARIABILITY

An uncertainty analysis consists of the process that translates uncertainties about models, variables, and input data and the random variability in measured parameters into uncertainties in output variables (Calabrese and Kostecki, 1991; Finkel, 1990; Iman and Helton, 1988). The analysis of uncertainties will typically involve the following fundamental elements:

- evaluation of uncertainties in the input of each of the relevant tasks;
- propagation of input uncertainties through each task;
- combination/convolution of the uncertainties in the output from the various tasks; and
- display and interpretation of the uncertainties in the final results.

The goal of an analysis of uncertainties is to provide decision-makers with the complete spectrum of information concerning the quality of an assessment – including the potential variability in the estimated parameters, the major data gaps, and the effect that such data gaps have on the accuracy and reasonableness of the estimates that are developed (Borgen, 1990; Covello et al., 1987; Cox and Ricci, 1992; Finkel and Evans, 1987; Helton, 1993; Hoffmann and Hammonds, 1992; Morgan and Henrion, 1991; USEPA, 1989a). Analysis and presentation of the uncertainties allow analysts or decision-makers to better evaluate the assessment results in the context of other factors being considered. This, in turn, will generally result in a more sound and open decision-making process.

An uncertainty analysis can be performed qualitatively or quantitatively. Whether qualitative or quantitative in nature, the analysis considers uncertainties in the database, uncertainties arising from assumptions in modeling, and the completeness of the analysis.

12.4.1 Qualitative Analysis of Uncertainties

The qualitative analysis of uncertainties typically involves a determination of the general quality and reasonableness of the risk assessment data, parameters, and results. Qualitative analysis is paramount to screening, preliminary, and intermediate-level assessments (USEPA, 1989a).

As part of the qualitative analysis, the cause of uncertainty is initially determined. The basic cause of uncertainty is a lack of knowledge on the part of the analyst because of inadequate, or even nonexistent, experimental and operational data on processes and parameters. The specific causes of uncertainty can be categorized as follows (USEPA, 1989a):

- Measurement error (resulting from measurement techniques employed in the study).
- Sampling error (arising from the degree of representativeness of sampled data to actual population).
- Natural variability.
- Model limitations (reflecting on how close to reality models employed are).
- Application and quality of generic or indirect empirical data.
- Professional/expert judgment (reflecting on the possible unreliability of scientific assumptions that may have been revoked/used).

In general, once the causes of the uncertainties have been identified, the impact that these uncertainties have on the assessment results would then have to be determined. Insofar as possible, measures to minimize the impacts of such uncertainties on the results should be elaborated. In any case, the explicit presentation of the qualitative analysis results will transmit the level of confidence in the results to the decision-maker, facilitating the implementation of appropriate environmental management actions.

12.4.2 Quantitative Analysis of Uncertainties

In addition to a qualitative analysis, detailed assessments may also require quantitative uncertainty analysis techniques. The quantitative analysis of uncertainties, often employed in detailed assessments, usually will proceed via sensitivity analysis and/or probabilistic analysis (e.g., Monte Carlo simulation techniques). The technique of choice depends on the availability of input data statistics. The approach will generally allow for a deviation from the conservative and rather unrealistic approach of making point estimates of risks, that has 'traditionally' been used in risk assessment programs.

In general, quantitative analysis of uncertainty becomes very important and necessary when prior risk screening calculations indicate a potential problem, or when mitigation may result in high costs, or when it is necessary to establish the relative importance of contaminants and exposure pathways. On the other hand, if calculated contaminant doses/intakes or risks are most obviously small and/or if the consequence of a 'wrong' prediction/decision based on the calculated risk is negligible, then quantitative analysis of uncertainty may not be necessary or a worthwhile effort.

12.4.2.1 Sensitivity analysis

Sensitivity analysis, often a useful adjunct to uncertainty analysis, comprises the processes that examine the relative change or response of output variables caused by variation of the input variables and parameters (Calabrese and Kostecki, 1991; Iman and Helton, 1988). It is a technique that tests the sensitivity of an output variable to the possible variation in the input variables of a given model. Performance of sensitivity testing requires data on the range of values for each relevant model parameter.

The purpose of sensitivity analysis is to identify the influential input variables and develop bounds on the model output. When computing the sensitivity with respect to a given input variable, all other input variables are held fixed at their nominal values. By identifying the influential/critical input variables, more resources can be directed to reduce their uncertainties and hence reduce the output uncertainty.

12.4.2.2 Probabilistic analysis: the application of Monte Carlo simulation techniques

Typically, probabilistic analysis techniques are employed/used to quantify uncertainties in risk assessment (e.g., Burmaster, 1996; Finley and Paustenbach, 1994; Finley et al., 1994a, 1994b; Lee and Kissel, 1995; Lee et al., 1995; Macintosh et al., 1994; Power and McCarty, 1996; Richardson, 1996; Smith et al., 1992). The probabilistic analysis may be applied to the evaluation of risks, so that uncertainties are accounted for systematically. Probabilistic risk analyses may indeed serve several purposes, including being used to: propagate uncertainty in the estimate of exposure dose and risk; properly prioritize resources for risk reduction activities; and simulate stochastic variability among individuals in a population.

Probabilistic analyses require data on the range and probability function (or distribution) of each model parameter. In fact, a central part of probabilistic risk analyses is the selection of probability distributions for the uncertain input variables (Haas, 1997; Hamed and Bedient, 1997). Thus, it is generally recommended to undertake a formal selection among various distributional families, along with a formal goodness-of-fit test, in order to obtain the most suitable family appropriate for characterizing the case data set.

The favored probabilistic approach for assessing uncertainty is via Monte Carlo simulation (e.g., McKone, 1994; McKone and Borgen, 1991; Price et al., 1996; Smith, 1994; Thompson et al., 1992). Monte Carlo simulation is a statistical technique by which a quantity is calculated repeatedly, using randomly selected/generated scenarios for each calculation cycle – and typically presenting the results

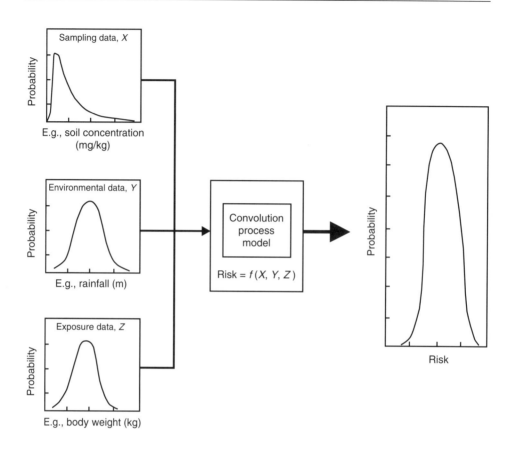

Figure 12.7 Conceptual illustration of the Monte Carlo simulation procedure

in simple graphs and tables. The results from the simulation process approximate the full range of possible outcomes, and the likelihood of each.

The Monte Carlo simulation involves assigning a joint probability distribution to the input variables; the procedure yields a concomitant distribution that is strictly a consequence of the assumed distributions of the model inputs and the assumed functional form of the model (Figure 12.7). Several considerations may be important in the selection of appropriate probability distribution to represent the relevant input parameters (Box 12.1) (Finley et al., 1994a; USEPA, 1989a). Unless specific information on the relationships between the relevant parameters is available, values for the required input parameters will normally be assumed to be independent.

Monte Carlo simulations can indeed be used to develop numerical estimates of uncertainties that allow efficient ways to extend risk assessment methods to the estimation of point values as well as distributions of risks posed by environmental contamination problems. In using Monte Carlo techniques, most or all input

> Box 12.1 Important considerations in the selection of appropriate probability distribution in a Monte Carlo simulation
>
> - *A uniform distribution* would be used to represent a factor/parameter when nothing is known about the factor except its finite range. The use of a uniform distribution assumes that all possible values within the range are equally likely
> - *A triangular distribution* would be used if the range of the parameter and its mode are known
> - *A beta distribution* (scaled to the desired range) may be most appropriate if the parameter has a finite range of possible values and a smooth probability function is desired
> - *A gamma, lognormal, or Weibull distribution* may be an appropriate choice if the parameter only assumes positive values. The gamma distribution is probably the most flexible, especially because its probability function can assume a variety of shapes by varying its parameters, and it is mathematically tractable
> - *A normal distribution* may be an appropriate choice if the parameter has an unrestricted range of possible values and is symmetrically distributed around its mode

variables to the risk assessment models become random variables with known or estimated probability density functions (pdfs). Within this framework, a variable can take a range of values with a known probability.

In general, when Monte Carlo simulation is applied to risk assessment, the risk presentation appears as frequency or probability distribution graphs – as illustrated by Figure 12.8 – from which the mean, median, variance, and/or percentile levels/values can be extracted.

12.5 UTILITY OF A RISK ANALYSIS

During risk characterization, chemical-specific toxicity information is compared against both field-measured and estimated contaminant exposure levels (and in some cases, those levels predicted through fate and transport modeling) in order to determine whether contaminant concentrations associated with an environmental contamination problem are of significant concern.

Risk characterization is indeed the final step in the risk assessment process, and becomes the first input into risk management programs. Thus, it serves as a bridge between risk assessment and risk management, and is therefore a key step in the decision-making process developed to address environmental contamination problems.

Through probabilistic modeling and analyses, uncertainties associated with the risk evaluation process can be assessed properly and their effects on a given decision accounted for systematically. In this manner the risks associated with given decisions may be delineated and then appropriate corrective measures taken accordingly.

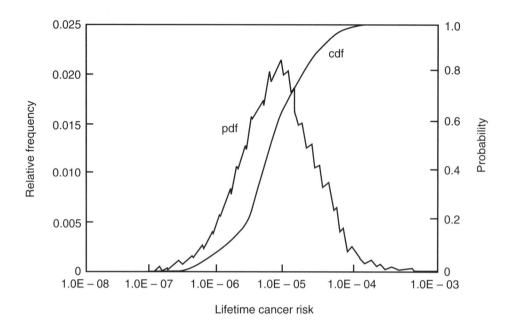

Figure 12.8 An illustrative sketch of a plot from a Monte Carlo simulation analysis (showing probability density function [pdf] and cumulative distribution function [cdf] for lifetime cancer risks from a contaminated site)

SUGGESTED FURTHER READING

Batchelor, B., J. Valdés, and V. Araganth. 1998. Stochastic risk assessment of sites contaminated by hazardous wastes. *Journal of Environmental Engineering*, 124(4): 380–388.

Finkel, A.M. 1995. Toward less misleading comparisons of uncertain risks: the example of aflatoxin and algar. *Environmental Health Perspectives*, 103: 376–385.

Hawley, J.K. 1985. Assessment of health risks from exposure to contaminated soil. *Risk Analysis*, 5(4): 289–302.

Hertwich, E.G., W.S. Pease, and T.E. McKone. 1998. Evaluating toxic impact assessment methods: what works best. *Environmental Science and Technology*, 32(5): 138A–144A.

Hoel, D.G., D.W. Gaylor, R.L. Kirschstein, U. Saffiotti, and M.A. Schneiderman. 1975. Estimation of risks of irreversible, delayed toxicity. *Journal of Toxicology and Environmental Health*, 1: 133–151.

REFERENCES

Borgen, K.T. 1990. *Uncertainty in Environmental Health Risk Assessment*. Garland Publishing, New York.

Burmaster, D.E. 1996. Benefits and costs of using probabilistic techniques in human health risk assessments – with emphasis on site-specific risk assessments. *Human and Ecological Risk Assessment*, 2: 35–43.

Calabrese, E.J. 1984. *Principles of Animal Extrapolation.* J. Wiley, New York.
Calabrese, E.J. and P.T. Kostecki (eds). 1991. *Hydrocarbon Contaminated Soils*, Volume 1. Lewis Publishers, Chelsea, MI.
Clewell, H.J. and M.E. Andersen. 1985. Risk assessment extrapolations and physiological modeling. *Toxicology and Industrial Health*, 1: 111–132.
Covello, V.T., et al. 1987. *Uncertainty in Risk Assessment, Risk Management, and Decision Making.* Advances in Risk Analysis, Vol. 4. Plenum Press, New York.
Cox, L.A. and P.F. Ricci. 1992. Dealing with uncertainty – from health risk assessment to environmental decision making. *Journal of Environmental Engineering – ASCE*, 118(2): 77–94.
Dourson, M.L. and J.F. Stara. 1983. Regulatory history and experimental support of uncertainty (safety) factors. *Regulatory Toxicology and Pharmacology*, 3: 224–238.
Finkel, A. 1990. *Confronting Uncertainty in Risk Management.* Resources for the Future, Washington, DC.
Finkel, A.M. and J.S. Evans. 1987. Evaluating the benefits of uncertainty reduction in environmental health risk management. *Journal of the Air Pollution Control Association*, 37: 1164–1171.
Finley, B. and D.P. Paustenbach. 1994. The benefits of probabilistic exposure assessment: three case studies involving contaminated air, water, and soil. *Risk Analysis*, 14: 53–73.
Finley, B., D. Proctor, et al. 1994a. Recommended distributions for exposure factors frequently used in health risk assessment. *Risk Analysis*, 14: 533–553.
Finley, B., P.K. Scott, and D.A. Mayhall. 1994b. Development of a standard soil-to-skin adherence probability density function for use in Monte Carlo analyses of dermal exposures. *Risk Analysis*, 14: 555–569.
Haas, C.N. 1997. Importance of distributional form in characterizing inputs to Monte Carlo risk assessments. *Risk Analysis*, 17(1): 107–113.
Hamed, M.M. and P.B. Bedient. 1997. On the effect of probability distributions of input variables in public health risk assessment. *Risk Analysis*, 17(1): 97–105.
Helton, J.C. 1993. Risk, uncertainty in risk, and the EPA release limits for radioactive waste disposal. *Nuclear Technology*, 101: 18–39.
Hoffmann, F.O. and J.S. Hammonds. 1992. *An Introductory Guide to Uncertainty Analysis in Environmental and Health Risk Assessment.* Environmental Sciences Division, Oak Ridge National Lab., TN, ESD Publication 3920. Prepared for the US Dept. of Energy, Washington, DC.
Iman, R.L. and J.C. Helton. 1988. An investigation of uncertainty and sensitivity analysis techniques for computer models. *Risk Analysis*, 8: 71–90.
Lee, R.C. and J.C. Kissel. 1995. Probabilistic prediction of exposures to arsenic contaminated residential soil. *Environmental Geochemistry and Health*, 17: 159–168.
Lee, R.C., J.R. Fricke, W.E. Wright, and W. Haerer. 1995. Development of a probabilistic blood lead prediction model. *Environmental Geochemistry and Health*, 17: 169–181.
Macintosh, D.L., G.W. Suter II, and F.O. Hoffman. 1994. Uses of probabilistic exposure models in ecological risk assessments of contaminated sites. *Risk Analysis*, 14: 405–419.
McKone, T.E. 1994. Uncertainty and variability in human exposures to soil contaminants through home-grown food: a Monte Carlo assessment. *Risk Analysis*, 14: 449–463.
McKone, T.E. and K.T. Bogen. 1991. Predicting the uncertainties in risk assessment. *Environmental Science and Technology*, 25: 1674–1681.
Morgan, M.G. and M. Henrion. 1991. *Uncertainty: A Guide to Dealing with Uncertainty in Quantitative Risk and Policy Analysis.* Oxford University Press, Oxford, UK.
Paustenbach, D.J. (ed.). 1988. *The Risk Assessment of Environmental Hazards: A Textbook of Case Studies.* J. Wiley, New York.
Power, M. and L.S. McCarty. 1996. Probabilistic risk assessment: betting on its future. *Human and Ecological Risk Assessment*, 2: 30–34.
Price, P.S., C.L. Curry, et al. 1996. Monte Carlo modeling of time-dependent exposures using a microexposure event approach. *Risk Analysis*, 16(3): 339–348.

Richardson, G.M. 1996. Deterministic versus probabilistic risk assessment: strengths and weaknesses in a regulatory context. *Human and Ecological Risk Assessment*, 2: 44–54.

Smith A.E., P.B. Ryan, and J.S. Evans. 1992. The effect of neglecting correlations when propagating uncertainty and estimating population distribution of risk. *Risk Analysis*, 12: 457–474.

Smith, R.L. 1994. Use of Monte Carlo simulation for human exposure assessment at a Superfund site. *Risk Analysis*, 14: 433–439.

Thompson, K.M., D.E. Burmaster, and A.C. Crouch. 1992. Monte Carlo techniques for quantitative uncertainty analysis in public health risk assessments. *Risk Analysis*, 12: 53–63.

USEPA (US Environmental Protection Agency). 1989a. *Exposure Factors Handbook*. Office of Health and Environmental Assessment, Washington, DC. EPA/600/8-89/043.

USEPA. 1989b. *Risk Assessment Guidance for Superfund. Volume I – Human Health Evaluation Manual (Part A)*. Office of Emergency and Remedial Response, Washington, DC. EPA/540/1-89/002.

Zemba, S.G., L.G. Green, E.A.C. Crouch, and R.R. Lester. 1996. Quantitative risk assessment of stack emissions from municipal waste combustors. *Journal of Hazardous Materials*, 47: 229–275.

PART IV

RISK ASSESSMENT TECHNIQUES AND METHODS OF APPROACH

This part of the book highlights specific methods typically employed in the evaluation of risks associated with environmental problems, and used to facilitate environmental management decisions. This includes a general overview of the state-of-the-art, the major risk assessment tools usually employed in environmental management programs, the principles and methods for evaluating human health and ecological risks, and the use of probabilistic risk assessment to evaluate aspects of technological risks relevant to environmental management problems. The elaboration also identifies typical/potential applications of the risk assessment methods and techniques. This part comprises the following specific chapters:

- Chapter 13, *Human Health Risk Assessments*, is devoted to the principles and procedures for completing human health risk assessments, as part of an environmental management program.
- Chapter 14, *Ecological Risk Assessments*, deals with the general considerations and methods of approach necessary for the completion of risk assessments pertaining to ecological receptors that are potentially affected by an environmental contamination problem.
- Chapter 15, *Probabilistic Risk Assessments*, discusses a number of analytical tools/techniques generally available for the determination of risk probabilities associated with environmental management facilities and/or containment systems.

Chapter Thirteen

Human Health Risk Assessments

Human health risk assessment is defined as the characterization of the potential adverse health effects associated with human exposures to environmental hazards (NRC, 1983). In a typical health risk assessment process, the extent to which potential human receptors have been, or could be exposed to chemical(s) associated with an environmental contamination problem is determined. The extent of exposure is then considered in relation to the type and degree of hazard posed by the chemical(s), thereby permitting an estimate to be made of the present or future health risks to the populations-at-risk.

Human health risk assessment techniques can be employed to better develop responsible environmental management programs. The scope of applications for the health risk assessment methodology may vary greatly – with some common specific applications and uses identified in Chapter 20. In fact, it is almost imperative to make human health risk assessment an integral part of all environmental management programs, except that the level of detail will be case-specific, ranging from qualitative through semi-quantitative to detailed quantitative analyses.

This chapter elaborates the major components of the human health risk assessment methodology, as may be applied to the evaluation of an environmental contamination problem.

13.1 THE HEALTH RISK ASSESSMENT METHODOLOGY

Figure 13.1 shows the basic components and steps typically involved in a comprehensive human health risk assessment designed for use in environmental management programs.

Invariably, all environmental management programs start with a hazard identification/data collection and data evaluation phase. The data evaluation aspect of a human health risk assessment consists of an identification and analysis of the chemicals associated with an environmental contamination problem that should become the focus of the environmental management program. In this process, an attempt is generally made to select all chemicals that could represent the major part of the risks associated with case-related exposures; typically, this will consist of all constituents contributing ≥95% of the overall risks. Chemicals are screened based on such parameters as toxicity, carcinogenicity, concentrations of the detected chemicals, and the frequency of detection in the sampled matrix.

The exposure assessment phase of the human health risk assessment is used to estimate the rates at which chemicals are absorbed by potential receptors. Since

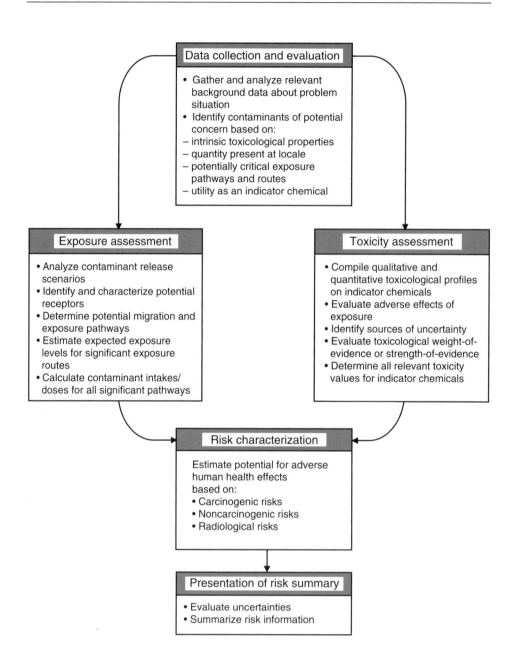

Figure 13.1 Components of the human health risk assessment process

most potential receptors tend to be exposed to chemicals from a variety of sources and/or in different environmental media, an evaluation of the relative contributions of each medium and/or source to total chemical intake could be critical in a multipathway exposure analysis. In fact, the accuracy with which exposures are characterized could be a major determinant of the ultimate validity of the risk assessment.

The quantitative evaluation of toxicological effects consists of a compilation of toxicological profiles (including the intrinsic toxicological properties of the chemicals of concern, which may include their acute, subchronic, chronic, carcinogenic, and/or reproductive effects) and the determination of appropriate toxicity indices (see Appendix D).

Finally, the risk characterization consists of estimating the probable incidence of adverse impacts to potential receptors under various exposure conditions. It involves an integration of the toxicity and exposure assessments, resulting in a quantitative estimation of the actual and potential risks and/or hazards due to exposure to each key chemical constituent, and also the possible additive effects of exposure to mixtures of the chemicals of potential concern.

Several key aspects of the human health risk assessment methodology are presented in the following sections, further to earlier discussions in Part III of this volume; additional details can be found elsewhere in the literature (e.g., Hoddinott, 1992; Huckle, 1991; NRC, 1983; Patton, 1993; Paustenbach, 1988; Ricci, 1985; Ricci and Rowe, 1985; USEPA, 1984a, 1984b, 1985a, 1986a, 1986b, 1986c, 1986d, 1987a, 1987b, 1989d, 1991, 1992; Van Leeuwen and Hermens, 1995).

13.2 POTENTIAL RECEPTOR EXPOSURES TO ENVIRONMENTAL CONTAMINANTS

The analysis of potential human receptor exposures to environmental contaminants often involves several complex issues related to various contaminant migration and exposure pathways. In any event, potential receptors may ultimately become exposed to a variety of environmental contaminants via several different exposure routes – represented primarily by the inhalation, ingestion, and dermal exposure routes (Figure 13.2).

In general, once the concentrations of pollutants are determined for the various environmental matrices (viz: air, soil, water, plants, and animal products), these values usually are used to estimate potential human receptor exposures. Exposures are evaluated by calculating the average daily dose (ADD) and/or the lifetime average daily dose ($LADD$).

The carcinogenic effects (and sometimes the chronic noncarcinogenic effects) associated with an environmental contamination problem involve estimating the $LADD$; for noncarcinogenic effects, the ADD is usually used. The ADD differs from the $LADD$ in that the former is not averaged over a lifetime; rather, it is the average daily dose pertaining to the days of exposure. The maximum daily dose (MDD) will typically be used in estimating acute or subchronic exposures.

246 Risk assessment in environmental management

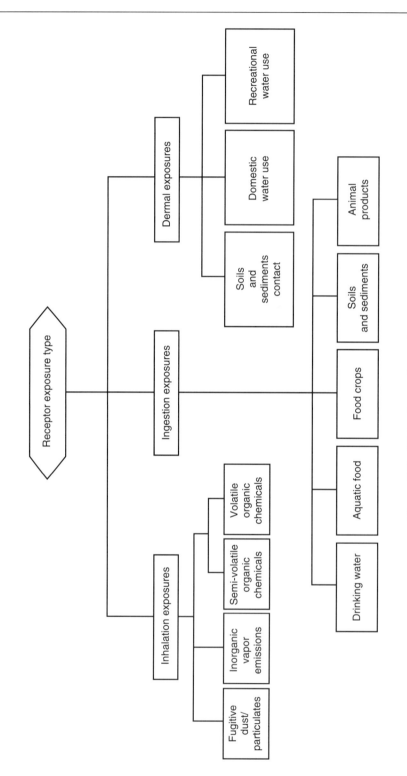

Figure 13.2 Major types of receptor exposures

The major routes of potential receptor exposures to chemical constituents associated with environmental contamination problems are discussed below – including the algorithms/equations used for the estimation of the respective potential receptor exposures; these algorithms and related ones are elaborated in greater detail elsewhere in the literature (e.g., CAPCOA, 1990; CDHS, 1986; DTSC, 1994; McKone, 1989; McKone and Daniels, 1991; NRC, 1991a, 1991b; USEPA, 1986c, 1988, 1989a, 1989b, 1991).

13.2.1 Potential Receptor Inhalation Exposures

Two major types of inhalation exposures are generally considered in the investigation of potential environmental contamination problems (see Figure 13.2) – broadly categorized into the inhalation of airborne particulates from fugitive dust, in which all individuals within approximately 80 km ($\approx$50 miles) radius of a contaminant source are potentially impacted; and the inhalation of volatile compounds (i.e., airborne, vapor-phase chemicals).

Potential inhalation intakes may be estimated based on the length of exposure, the inhalation rate of the exposed individual, the concentration of contaminant in the inhaled air, and the amount retained in the lungs – conservatively represented by the following relationship:

$$\text{Inhalation exposure (mg/kg-day)} = \frac{[GLC \times RR \times CF]}{BW}$$

where GLC is the ground-level concentration (μg/m^3); RR is the respiration rate (m^3/day); CF is a conversion factor (= 1 mg/1000 μg = 1.0E-03 mg/μg); and BW is the body weight (kg).

Potential receptor inhalation exposures specific to contaminated particulates from wind-borne fugitive dust releases, and to volatile compounds from airborne vapor-phase emissions are annotated below.

- *Receptor inhalation exposure to particulates from contaminated fugitive dust.* Box 13.1 shows the algorithm for calculating potential receptor intakes resulting from the inhalation of wind-borne fugitive dust (CAPCOA, 1990; DTSC, 1994; USEPA, 1988, 1989a, 1989b). The contaminant concentration in air, C_a, is defined by the ground-level concentration (GLC) – represented by the respirable (*PM*-10) particles – expressed in μg/m^3.
- *Receptor inhalation exposure to volatile compounds.* Box 13.2 shows the algorithm for calculating potential receptor intakes resulting from the inhalation of airborne vapor-phase chemicals (CAPCOA, 1990; DTSC, 1994; USEPA, 1988, 1989a, 1989b). The vapor-phase contaminant concentration in air is assumed to be in equilibrium with the concentration in the release source.

It is noteworthy that showering generally represents a system that promotes release of VOCs from water, due to high turbulence, high surface area, and

Box 13.1 Equation for estimating inhalation exposure to contaminants in fugitive dust

$$INH_a = \frac{(C_a \times IR \times RR \times ABS_s \times ET \times EF \times ED)}{(BW \times AT)}$$

where:
- INH_a = Inhalation intake (mg/kg-day)
- C_a = Chemical concentration of airborne particulates (defined by the ground-level concentration (GLC), and represented by the respirable, PM-10 particles) (mg/m³)
- IR = Inhalation rate (m³/h)
- RR = Retention rate of inhaled air (%)
- ABS_s = Percent of chemical absorbed into the bloodstream (%)
- ET = Exposure time (h/day)
- EF = Exposure frequency (days/year)
- ED = Exposure duration (years)
- BW = Body weight (kg)
- AT = Averaging time (period over which exposure is averaged – days)
  = $ED \times 365$ days/year, for noncarcinogenic effects
  = 70 years × 365 days/year, for carcinogenic effects

Box 13.2 Equation for estimating inhalation exposure to vapor-phase contaminants

$$INH_{av} = \frac{(C_{av} \times IR \times RR \times ABS_s \times ET \times EF \times ED)}{(BW \times AT)}$$

where:
- INH_{av} = Inhalation intake (mg/kg-day)
- C_{av} = Chemical concentration in air (the vapor-phase contaminant concentration in air is assumed to be in equilibrium with the concentration in the release source) (mg/m³)
- IR = Inhalation rate (m³/h)
- RR = Retention rate of inhaled air (%)
- ABS_s = Percent of chemical absorbed into the bloodstream (%)
- ET = Exposure time (h/day)
- EF = Exposure frequency (days/year)
- ED = Exposure duration (years)
- BW = Body weight (kg)
- AT = Averaging time (period over which exposure is averaged – days)
  = $ED \times 365$ days/year, for noncarcinogenic effects
  = 70 years × 365 days/year, for carcinogenic effects

Box 13.2A Equation for estimating inhalation exposure to vapor-phase contaminants during showering activity

$$INH = \left[C_w \times FV \times \left\{\frac{ET_1}{(VS \times 2)} + \frac{ET_2}{VB}\right\}\right] \times \frac{(IR \times RR \times VW \times ABS_s \times EF \times ED)}{(BW \times AT)}$$

$$= [ACB_{sh}] \times \frac{(IR \times RR \times VW \times ABS_s \times EF \times ED)}{(BW \times AT)}$$

where:
- INH = Inhalation intake while showering (mg/kg-day)
- C_w = Concentration of contaminant in water – adjusted for water treatment purification factor, T_f, which is the fraction remaining after treatment (i.e., $C_w = C_w$-source $\times T_f$) (mg/L)
- FV = Fraction of contaminant volatilized (unitless)
- ET_1 = Length of exposure in shower (h/day)
- ET_2 = Length of additional exposure in enclosed bathroom (h/day)
- VS = Volume of shower stall (m^3)
- VB = Volume of bathroom (m^3)
- IR = Breathing/inhalation rate (m^3/h)
- RR = Retention rate of inhaled air (%)
- VW = Volume of water used in shower (L)
  = water flow rate (F_w [L/h]) × shower duration (h)
- ABS_s = Percent of chemical absorbed into the bloodstream (%)
- EF = Exposure frequency (days/year)
- ED = Exposure duration (years)
- BW = Body weight (kg)
- AT = Averaging time (period over which exposure is averaged – days)
- ACB_{sh} = Average air concentration in bathroom during a shower activity

$$= \left[C_w \times FV \times \left\{\frac{ET_1}{(VS \times 2)} + \frac{ET_2}{VB}\right\}\right]$$

Note: The concentration of contaminants in water may further be adjusted for environmental degradation by multiplying by a factor of e^{-kt}, where k (in days^{-1}) is the environmental degradation constant of the chemical, and t (in days) is the average time of transit through the water distribution system; this yields a new C_w value equal to $(C_w)(e^{-kt})$ to be used for the intakes computation.

small droplets of water involved. In fact, recent studies have shown that risks from inhalation while showering can be comparable to – if not greater than – risks from drinking contaminated water (McKone, 1987). Thus, this exposure scenario represents a particularly important one to evaluate in a human health risk assessment, as appropriate. In this case, the concentration of the contaminants in the shower air is assumed to be in equilibrium with the concentration in the water (DOE, 1987). In the case of volatile compounds released while bathing, the exposure relationship may be defined by the specific equation shown in Box 13.2A (USEPA, 1988, 1989a, 1989b). Other assumptions used in this model include the following: there is no air exchange in the shower (this assumption tending to over-estimate the concentration of contaminants in the air in the bathroom); there is perfect mixing within the bathroom (this assumption tending to under-estimate the concentration of contaminants in the air in the shower); and the emission rate from water is independent of instantaneous air concentration.

13.2.2 Potential Receptor Ingestion Exposures

The major types of ingestion exposures that could affect environmental management decisions consist of the oral intake of contaminated soils, food products, and waters (see Figure 13.2). In general, exposure through ingestion is a function of the concentration of the contaminant in the material ingested (e.g., soil, water, or food products such as crops, and dairy/beef), the gastrointestinal absorption of the pollutant in solid or fluid matrix, and the amount ingested – conservatively estimated as follows:

$$\text{Water ingestion exposure (mg/kg-day)} = \frac{(CW \times WIR \times GI)}{BW}$$

$$\text{Soil ingestion exposure (mg/kg-day)} = \frac{(CS \times SIR \times GI)}{BW}$$

$$\text{Crop ingestion exposure (mg/kg-day)} = \frac{(CS \times RUF \times CIR \times GI)}{BW}$$

$$\text{Dairy and beef products ingestion exposure (mg/kg-day)} = \frac{(CD \times FIR \times GI)}{BW}$$

where CW is the chemical concentration in water (mg/L); WIR is the water consumption rate (L/day); CS is the chemical concentration in soil (mg/kg); SIR is the soil consumption rate (kg/day); RUF is the root uptake factor; CIR is the crop consumption rate (kg/day); CD is the concentration of chemical in diet (mg/kg) – for grazing animals, the concentration of chemicals in tissue, CT, is $CT = BCF \times F \times CD$, where BCF is the bioconcentration factor (fat basis) for the organism, expressed as {mg/kg fat}/{mg/kg of diet}, and F is the fat content of tissues (in kg fat/kg tissue); FIR is the food (meat and dairy) consumption (kg/day); GI is the gastrointestinal absorption factor; and BW is the body weight (kg).

The total dose received by the potential receptors from chemical ingestions will, in general, be dependent on the absorption of the chemical across the gastrointestinal (GI) lining. The scientific literature provides some estimates of such absorption factors for various chemical substances. For chemicals without published absorption values and for which absorption factors are not implicitly

> Box 13.3 Equation for estimating ingestion exposure to contaminated water used for culinary purposes
>
> $$ING_{dw} = \frac{(C_w \times WIR \times FI \times ABS_s \times EF \times ED)}{(BW \times AT)}$$
>
> where:
> ING_{dw} = Ingestion intake, adjusted for absorption (mg/kg-day)
> C_w = Chemical concentration in drinking water (mg/L)
> WIR = Average ingestion rate (L/day)
> FI = Fraction ingested from contaminated source (unitless)
> ABS_s = Bioavailability/gastrointestinal (GI) absorption factor (%)
> EF = Exposure frequency (days/year)
> ED = Exposure duration (years)
> BW = Body weight (kg)
> AT = Averaging time (period over which exposure is averaged – days)

accounted for in toxicological parameters, absorption may conservatively be assumed to be 100%.

Potential receptor ingestion exposures through the oral intake of contaminated waters, through the consumption of contaminated food products, and through the incidental ingestion of contaminated soils/sediments are annotated below.

- *Receptor exposure through ingestion of drinking water.* In general, exposure to contaminants via the ingestion of contaminated fluids or solids may be estimated using the algorithm shown in Box 13.3 (CAPCOA, 1990; DTSC, 1994; USEPA, 1988, 1989a, 1989b). This consists of the applicable relationship for estimating the exposure intake that occurs through the ingestion of drinking water.

  As a special situation, *receptor exposure through incidental ingestion of water during swimming activities* (as a result of the ingestion of contaminated surface water during recreational activities) may be estimated by using the algorithm shown in Box 13.3A.

- *Receptor exposure through ingestion of food.* Exposure from the ingestion of food can occur via the ingestion of plant products, fish, animal products, and mother's milk. The general algorithm for estimating the exposure intake through the ingestion of foods is shown in Box 13.4 – with corresponding relationships defined below for specific types of food products.

  – *Ingestion of plant products.* Exposure through the ingestion of plant products, ING_p, is a function of the type of plant, the gastrointestinal absorption factor, and the fraction of plants ingested that are affected by the environmental pollutants of concern. The calculation is done for each plant type according to the algorithm presented in Box 13.4A (CAPCOA, 1990; USEPA, 1989a).

  – *Bioaccumulation and ingestion of seafood.* Exposure from the ingestion of contaminated fish (e.g., coming from contaminated surface water bodies)

Box 13.3A Equation for estimating incidental ingestion exposure to contaminated surface water during recreational activities

$$ING_r = \frac{(CW \times CR \times ABS_s \times ET \times EF \times ED)}{(BW \times AT)}$$

where:
ING_r = Ingestion intake, adjusted for absorption (mg/kg-day)
CW = Chemical concentration in water (mg/L)
CR = Contact rate (L/h)
ABS_s = Bioavailability/gastrointestinal (GI) absorption factor (%)
ET = Exposure time (h/event)
EF = Exposure frequency (events/year)
ED = Exposure duration (years)
BW = Body weight (kg)
AT = Averaging time (period over which exposure is averaged – days)

Box 13.4 Equation for estimating ingestion exposure to contaminated food products

$$ING_f = \frac{(C_f \times FIR \times CF \times FI \times ABS_s \times EF \times ED)}{(BW \times AT)}$$

where:
ING_f = Ingestion intake, adjusted for absorption (mg/kg-day)
C_f = Chemical concentration in food (mg/kg or mg/L)
FIR = Average food ingestion rate (mg or L/meal)
CF = Conversion factor (10^{-6} kg/mg for solids and 1.00 for fluids)
FI = Fraction ingested from contaminated source (unitless)
ABS_s = Bioavailability/gastrointestinal (GI) absorption factor (%)
EF = Exposure frequency (meals/year)
ED = Exposure duration (years)
BW = Body weight (kg)
AT = Averaging time (period over which exposure is averaged – days)

may be estimated using the algorithm shown in Box 13.4B (USEPA, 1987a, 1988, 1989a).

- *Ingestion of animal products.* Exposure resulting from the ingestion of animal products, ING_a, is a function of the type of meat ingested (including animal milk products and eggs), the gastrointestinal absorption factor, and the fraction of animal products ingested that are affected by pollutants. The calculation is done for each animal product type according to the relationship shown in Box 13.4C (CAPCOA, 1990; USEPA, 1989a).
- *Ingestion of mother's milk.* Exposure through the ingestion of a mother's milk, ING_m, is a function of the average chemical concentration in the

Box 13.4A Equation for estimating ingestion exposure to contaminated plant products

$$ING_p = \frac{(CP_z \times PIR_z \times FI_z \times ABS_s \times EF \times ED)}{(BW \times AT)}$$

where:
ING_p = Exposure intake from ingestion of plant products, adjusted for absorption (mg/kg-day)
CP_z = Chemical concentration in plant type Z (mg/kg)
PIR_z = Average consumption rate for plant type Z (kg/day)
FI_z = Fraction of plant type Z ingested from contaminated source (unitless)
ABS_s = Bioavailability/gastrointestinal (GI) absorption factor (%)
EF = Exposure frequency (days/year)
ED = Exposure duration (year)
BW = Body weight (kg)
AT = Averaging time (period over which exposure is averaged – days)

Box 13.4B Equation for estimating ingestion exposure to contaminated seafood

$$ING_{sf} = \frac{(CW \times FIR \times CF \times BCF \times FI \times ABS_s \times EF \times ED)}{(BW \times AT)}$$

where:
ING_{sf} = Total exposure, adjusted for absorption (mg/kg-day)
CW = Chemical concentration in surface water (mg/L)
FIR = Average fish ingestion rate (g/day)
CF = Conversion factor (= 10^{-3} kg/g)
BCF = Chemical-specific bioconcentration factor (L/kg)
FI = Fraction ingested from contaminated source (unitless)
ABS_s = Bioavailability/gastrointestinal (GI) absorption factor (%)
EF = Exposure frequency (days/year)
ED = Exposure duration (year)
BW = Body weight (kg)
AT = Averaging time (period over which exposure is averaged – days)

mother's milk, the amount of mother's milk ingested, and the gastrointestinal absorption factor – estimated according to the relationship shown in Box 13.4D (CAPCOA, 1990; USEPA, 1989a).

- *Receptor exposure through pica and incidental ingestion of soil/sediment.* Exposures resulting from the incidental ingestion of contaminants sorbed onto soils is determined by multiplying the concentration of the contaminant in the medium of concern by the amount of soil ingested per day and the degree of absorption; the applicable relationship is shown in Box 13.5 (CAPCOA, 1990; USEPA, 1988, 1989a, 1989b). In general, it normally is assumed that all

Box 13.4C Equation for estimating ingestion exposure to contaminated animal products

$$ING_a = \frac{(CAP_z \times APIR_z \times FI_z \times ABS_s \times EF \times ED)}{(BW \times AT)}$$

where:
ING_a = Exposure intake through ingestion of plant products, adjusted for absorption (mg/kg-day)
CAP_z = Chemical concentration in food type Z (mg/kg)
$APIR_z$ = Average consumption rate for food type Z (kg/day)
FI_z = Fraction of product type Z ingested from contaminated source (unitless)
ABS_s = Bioavailability/gastrointestinal (GI) absorption factor (%)
EF = Exposure frequency (days/years)
ED = Exposure duration (years)
BW = Body weight (kg)
AT = Averaging time (period over which exposure is averaged – days)

Box 13.4D Equation for estimating ingestion exposure to contaminated mother's milk from breast-feeding

$$ING_m = \frac{(CMM \times IBM \times ABS_s \times EF \times ED)}{(BW \times AT)}$$

where:
ING_m = Exposure intake through ingestion of mother's milk, adjusted for absorption (mg/kg-day)
CMM = Chemical concentration in mother's milk – which is a function of mother's exposure through all routes and the contaminant body half-life (mg/kg)
IBM = Daily average ingestion rate for breast milk (kg/day)
ABS_s = Bioavailability/gastrointestinal (GI) absorption factor (%)
EF = Exposure frequency (days/year)
ED = Exposure duration (years)
BW = Body weight (kg)
AT = Averaging time (period over which exposure is averaged – days)

---

ingested soil during receptor exposures comes from a contaminated source, so that the *FI* term becomes unity.

13.2.3 Potential Receptor Dermal Exposures

The major types of dermal exposures that could affect environmental management decisions consist of dermal contacts with contaminants adsorbed onto soils and dermal absorption from contaminated waters (see Figure 13.2). In general, dermal intake is determined by the chemical concentration in the medium of concern, the

> **Box 13.5** Equation for estimating pica and incidental ingestion exposure to contaminated soils/sediments
>
> $$ING_s = \frac{(C_s \times SIR \times CF \times FI \times ABS_s \times EF \times ED)}{(BW \times AT)}$$
>
> where:
> ING_s = Ingestion intake, adjusted for absorption (mg/kg-day)
> C_s = Chemical concentration in soil (mg/kg)
> SIR = Average soil ingestion rate (mg soil/day)
> CF = Conversion factor (10^{-6} kg/mg)
> FI = Fraction ingested from contaminated source (unitless)
> ABS_s = Bioavailability/gastrointestinal (GI) absorption factor (%)
> EF = Exposure frequency (days/year)
> ED = Exposure duration (years)
> BW = Body weight (kg)
> AT = Averaging time (period over which exposure is averaged – days)

body surface area in contact with the medium, the duration of the contact, the flux of the medium across the skin surface, and the absorbed fraction – conservatively estimated as follows:

$$\text{Dermal exposure to soil (mg/kg-day)} = \frac{(SS \times SA \times CS \times UF \times CF)}{BW}$$

$$\text{Dermal exposure to water (mg/kg-day)} = \frac{(WS \times SA \times CW \times UF)}{BW}$$

where SS is surface dust on skin (mg/cm²/day); CS is chemical concentration in soil (mg/kg); CF is conversion factor (= 1.00E-06 kg/mg); WS is water contacting skin (L/cm²/day); CW is chemical concentration in water (mg/L); SA is exposed skin surface area (cm²); UF is uptake factor; and BW is body weight (kg).

Potential receptor dermal exposures through dermal contacts with contaminated soils/sediments and from the dermal absorption of chemicals present in contaminated waters are annotated below.

- *Receptor exposure through soils contact/dermal absorption.* The dermal exposures to chemicals in soils and sediments from a contaminated site may be estimated by applying the equation shown in Box 13.6 (CAPCOA, 1990; DTSC, 1994; USEPA, 1988, 1989a, 1989b).
- *Receptor exposure through dermal contact with waters and seeps.* Dermal exposures to chemicals in water may occur during domestic use (such as bathing and washing), or through recreational activities (such as swimming or fishing). The dermal intakes of chemicals in ground or surface water and/or from seeps from a contaminated site may be estimated by using the equation shown in Box 13.7 (USEPA, 1988, 1989a, 1989b).

> **Box 13.6** Equation for estimating dermal exposures through contacts with contaminated soils
>
> $$DEX_s = \frac{(C_s \times CF \times SA \times AF \times ABS_s \times SM \times EF \times ED)}{(BW \times AT)}$$
>
> where:
> DEX_s = Absorbed dose (mg/kg-day)
> C_s = Chemical concentration in soil (mg/kg)
> CF = Conversion factor (10^{-6} kg/mg)
> SA = Skin surface area available for contact, i.e., surface area of exposed skin (cm²/event)
> AF = Soil to skin adherence factor, i.e., soil loading on skin (mg/cm²)
> ABS_s = Skin absorption factor for chemicals in soil (%)
> SM = Factor for soil matrix effects (%)
> EF = Exposure frequency (events/year)
> ED = Exposure duration (years)
> BW = Body weight (kg)
> AT = Averaging time (period over which exposure is averaged – days)

> **Box 13.7** Equation for estimating dermal exposures through contacts with contaminated waters
>
> $$DEX_w = \frac{(C_w \times CF \times SA \times PC \times ABS_s \times ET \times EF \times ED)}{(BW \times AT)}$$
>
> where:
> DEX_w = Absorbed dose from dermal contact with chemicals in water (mg/kg-day)
> C_w = Chemical concentration in water (mg/L)
> CF = Volumetric conversion factor for water (1 L/1000 cm³)
> SA = Skin surface area available for contact, i.e., surface area of exposed skin (cm²)
> PC = Chemical-specific dermal permeability constant (cm/h)
> ABS_s = Skin absorption factor for chemicals in water (%)
> ET = Exposure time (h/day)
> EF = Exposure frequency (days/year)
> ED = Exposure duration (years)
> BW = Body weight (kg)
> AT = Averaging time (period over which exposure is averaged – days)

13.2.4 Computation of 'Intake Factors' for Exposure Assessment: Illustration of the Calculation of Chemical Intakes and Doses

Several exposure parameters are generally required in order to model the various exposure scenarios typically associated with environmental contamination problems. Usually, default values may be obtained from literature for some of the requisite parameters used in the estimation of chemical intakes and doses. Table 13.1 – which

Table 13.1 An example listing of case-specific exposure parameters

Parameter	Children aged up to 6	Children aged 6–12	Adult	Reference sources
Physical characteristics				
Average body weight (kg)	16	29	70	(a,b,c)
Average total skin surface area (cm^2)	6980	10 470	18 150	(a,b,e,h)
Average lifetime (yrs)			70	(a,b,c,e)
Average lifetime exposure period (yrs)	5	6	58	(b,e)
Activity characteristics				
Inhalation rate (m^3/h)	0.25	0.46	0.83	(b,e)
Retention rate of inhaled air (%)	100	100	100	(e)
Frequency of fugitive dust inhalation (365 days/yr)				
– off-site residents, schools and passers-by	365	365	365	(b,e)
– off-site workers	–	–	260	(b,e)
Duration of fugitive dust inhalation (outside) (h/day)				
– off-site residents, schools and passers-by	12	12	12	(b,e)
– off-site workers	–	–	8	(b,e)
Amount of soil ingested incidentally (mg/day)	200	100	50	(a,b,c,e,h,i)
Frequency of soil contact (days/yr)				
– off-site residents, schools and passers-by	330	330	330	(b,e)
– off-site workers	–	–	260	(b,e)
Duration of soil contact (h/day)				
– off-site residents, schools and passers-by	12	8	8	(b,e)
– off-site workers	–	–	8	(b,e)
Skin area contacted by soil (%)	20	20	10	(b,e,h)
Material characteristics				
Soil to skin adherence factor (mg/cm^2)	0.75	0.75	0.75	(a,b,e,f,g)
Soil matrix attenuation factor (%)	15	15	15	(d)

Note: The exposure factors represented here are for potential maximum exposures (for conservative estimates), and could be modified as appropriate to reflect the most reasonable exposure patterns anticipated. For instance, soil exposure will be reduced by snow cover and rainy days, thus reducing potential exposures for children playing in contaminated areas.

(a) USEPA (1989d); (b) USEPA (1989a); (c) USEPA (1988); (d) Hawley (1985); (e) Estimate based on site-specific conditions; (f) Lepow et al. (1974); (g) Lepow et al. (1975); (h) Sedman (1989); (i) Calabrese et al. (1989).

is by no means complete – gives typical parameters that represent a generic set of values commonly used in some applications. More detailed information on such parameters can be obtained from several sources (e.g., Calabrese et al., 1989; CAPCOA, 1990; Lepow et al., 1974, 1975; OSA, 1992; USEPA, 1987b, 1988, 1989a, 1989b, 1991).

A spreadsheet for automatically calculating exposure 'intake factors' – based on the algorithms presented above – may be developed to facilitate the computational efforts involved (Table 13.2). Some example evaluations for potential receptor groups purported to be exposed through inhalation, soil ingestion (i.e., incidental or pica behavior), and dermal contact are discussed in the following sections for demonstration only. The same set of units are maintained throughout, as given in the preceding sections.

13.2.4.1 Inhalation exposures

The daily inhalation intake of contaminated fugitive dust for various population groups is presented below for both carcinogenic and noncarcinogenic effects. The assumed parameters used in the computational demonstration are given in Table 13.1, and the electronic spreadsheet automation process is shown in Table 13.2.

- *Estimation of lifetime average daily dose (LADD) for carcinogenic effects.* For the fugitive dust inhalation pathway, the *LADD* (also, the carcinogenic chronic daily intake [*CDI*]) is estimated for the different population groups (pre-selected as representative of the critical receptors in the risk assessment) – and the results are shown below.

    - The carcinogenic *CDI* for children aged up to 6 years is calculated to be:

    $$CInh_{(1-6)}$$
    $$= \frac{(CA \times IR \times RR \times ABS_s \times ET \times EF \times ED)}{(BW \times AT)}$$
    $$= \frac{([CA] \times 0.25 \times 1 \times ABS_s \times 12 \times 365 \times 5)}{(16 \times [70 \times 365])}$$
    $$= 1.34 \times 10^{-02} \times ABS_s \times [CA]$$

    - The carcinogenic *CDI* for children aged 6 to 12 years is calculated to be:

    $$CInh_{(6-12)}$$
    $$= \frac{(CA \times IR \times RR \times ABS_s \times ET \times EF \times ED)}{(BW \times AT)}$$
    $$= \frac{([CA] \times 0.46 \times 1 \times ABS_s \times 12 \times 365 \times 6)}{(29 \times [70 \times 365])}$$
    $$= 1.63 \times 10^{-02} \times ABS_s \times [CA]$$

Table 13.2 Example spreadsheet for calculating case-specific 'intake factors' for an exposure assessment

PATHWAY===>	Fugitive Dust Inhalation Pathway							
Group	IR	RR	ET	EF	ED	BW	AT	INH Factor
C(1-6)@NCarc	0.25	1	12	365	5	16	1825	1.88E-01
C(1-6)@Carc	0.25	1	12	365	5	16	25550	1.34E-02
C(6-12)@NCarc	0.46	1	12	365	6	29	2190	1.90E-01
C(6-12)@Carc	0.46	1	12	365	6	29	25550	1.63E-02
ResAdult@NCarc	0.83	1	12	365	58	70	21170	1.42E-01
ResAdult@Carc	0.83	1	12	365	58	70	25550	1.18E-01
JobAdult@NCarc	0.83	1	8	260	58	70	21170	6.76E-02
JobAdult@Carc	0.83	1	8	260	58	70	25550	5.60E-02

PATHWAY===>	Soil Ingestion Pathway							
Group	IR	CF	FI	EF	ED	BW	AT	ING Factor
C(1-6)@NCarc	200	1.00E-06	1	330	5	16	1825	1.13E-05
C(1-6)@Carc	200	1.00E-06	1	330	5	16	25550	8.07E-07
C(6-12)@NCarc	100	1.00E-06	1	330	6	29	2190	3.12E-06
C(6-12)@Carc	100	1.00E-06	1	330	6	29	25550	2.67E-07
ResAdult@NCarc	50	1.00E-06	1	330	58	70	21170	6.46E-07
ResAdult@Carc	50	1.00E-06	1	330	58	70	25550	5.35E-07
JobAdult@NCarc	50	1.00E-06	1	260	58	70	21170	5.09E-07
JobAdult@Carc	50	1.00E-06	1	260	58	70	25550	4.22E-07

PATHWAY===>	Soil Dermal Contact Pathway							
Group	SA	CF	AF	SM	EF	ED	BW	AT	DEX Factor
C(1-6)@NCarc	1396	1.00E-06	0.75	0.15	330	5	16	1825	8.87E-06
C(1-6)@Carc	1396	1.00E-06	0.75	0.15	330	5	16	25550	6.34E-07
C(6-12)@NCarc	2094	1.00E-06	0.75	0.15	330	6	29	2190	7.34E-06
C(6-12)@Carc	2094	1.00E-06	0.75	0.15	330	6	29	25550	6.30E-07
ResAdult@NCarc	1815	1.00E-06	0.75	0.15	330	58	70	21170	2.64E-06
ResAdult@Carc	1815	1.00E-06	0.75	0.15	330	58	70	25550	2.19E-06
JobAdult@NCarc	1815	1.00E-06	0.75	0.15	260	58	70	21170	2.08E-06
JobAdult@Carc	1815	1.00E-06	0.75	0.15	260	58	70	25550	1.72E-06

Notes:
Notations and units are as defined in the text.
INH Factor = Inhalation factor for calculation of doses and intakes.
ING Factor = Soil ingestion factor for calculation of doses and intakes.
DEX Factor = Dermal exposure/skin adsorption factor for calculation of doses and intakes.
C(1-6)@NCarc = Noncarcinogenic effects for children aged 1–6 years.
C(1-6)@Carc = Carcinogenic effects for children aged 1–6 years.
C(6-12)@NCarc = Noncarcinogenic effects for children aged 6–12 years.
C(6-12)@Carc = Carcinogenic effects for children aged 6–12 years.
ResAdult@NCarc = Noncarcinogenic effects for resident adults.
ResAdult@Carc = Carcinogenic effects for resident adults.
JobAdult@NCarc = Noncarcinogenic effects for adult workers.
JobAdult@Carc = Carcinogenic effects for adult workers.

- The carcinogenic *CDI* for adult residents is calculated to be:

$$CInh_{(adultR)}$$
$$= \frac{(CA \times IR \times RR \times ABS_s \times ET \times EF \times ED)}{(BW \times AT)}$$

$$= \frac{([CA] \times 0.83 \times 1 \times ABS_s \times 12 \times 365 \times 58)}{(70 \times [70 \times 365])}$$

$$= 1.18 \times 10^{-01} \times ABS_s \times [CA]$$

- The carcinogenic *CDI* for adult workers is calculated to be:

$$CInh_{(adultW)}$$
$$= \frac{(CA \times IR \times RR \times ABS_s \times ET \times EF \times ED)}{(BW \times AT)}$$

$$= \frac{([CA] \times 0.83 \times 1 \times ABS_s \times 8 \times 260 \times 58)}{(70 \times [70 \times 365])}$$

$$= 5.60 \times 10^{-02} \times ABS_s \times [CA]$$

- *Estimation of average daily dose (ADD) for noncarcinogenic effects.* For the fugitive dust inhalation pathway, the *ADD* (also, the noncarcinogenic *CDI*) is estimated for the different population groups (pre-selected as representative of the critical receptors in the risk assessment) – and the results are shown below.

  - The noncarcinogenic *CDI* for children aged up to 6 years is calculated to be:

$$NCInh_{(1-6)}$$
$$= \frac{(CA \times IR \times RR \times ABS_s \times ET \times EF \times ED)}{(BW \times AT)}$$

$$= \frac{([CA] \times 0.25 \times 1 \times ABS_s \times 12 \times 365 \times 5)}{(16 \times [5 \times 365])}$$

$$= 1.88 \times 10^{-01} \times ABS_s \times [CA]$$

  - The noncarcinogenic *CDI* for children aged 6 to 12 years is calculated to be:

$$NCInh_{(6-12)}$$
$$= \frac{(CA \times IR \times RR \times ABS_s \times ET \times EF \times ED)}{(BW \times AT)}$$

$$= \frac{([CA] \times 0.46 \times 1 \times ABS_s \times 12 \times 365 \times 6)}{(29 \times [6 \times 365])}$$

$$= 1.90 \times 10^{-1} \times ABS_s \times [CA]$$

- The noncarcinogenic CDI for adult residents is calculated to be:

$$NCInh_{(adultR)}$$

$$= \frac{(CA \times IR \times RR \times ABS_s \times ET \times EF \times ED)}{(BW \times AT)}$$

$$= \frac{([CA] \times 0.83 \times 1 \times ABS_s \times 12 \times 365 \times 58)}{(70 \times [58 \times 365])}$$

$$= 1.42\text{E-}01 \times ABS_s \times [CA]$$

- The noncarcinogenic CDI for adult workers is calculated to be:

$$NCInh_{(adultW)}$$

$$= \frac{(CA \times IR \times RR \times ABS_s \times ET \times EF \times ED)}{(BW \times AT)}$$

$$= \frac{([CA] \times 0.83 \times 1 \times ABS_s \times 8 \times 260 \times 58)}{(70 \times [58 \times 365])}$$

$$= 6.76 \times 10^{-2} \times ABS_s \times [CA]$$

13.2.4.2 Ingestion exposures

The daily ingestion intake of contaminated soils for various population groups are calculated for both carcinogenic and noncarcinogenic effects. The assumed parameters used in the computational demonstration are given in Table 13.1, and the electronic spreadsheet automation process is shown in Table 13.2.

- *Estimation of lifetime average daily dose (LADD) for carcinogenic effects.* For the soil ingestion pathway, the *LADD* (also, the carcinogenic *CDI*) is estimated for the different population groups (pre-selected as representative of the critical receptors in the risk assessment) – and the results are shown below.

  - The carcinogenic *CDI* for children aged up to 6 years is calculated to be:

$$CIng_{(1-6)}$$

$$= \frac{(CS \times IR \times CF \times FI \times ABS_s \times EF \times ED)}{(BW \times AT)}$$

$$= \frac{([CS] \times 200 \times 1.00\text{E-}06 \times 1 \times ABS_s \times 330 \times 5)}{(16 \times [70 \times 365])}$$

$$= 8.07\text{E-}07 \times ABS_s \times [CS]$$

- The carcinogenic *CDI* for children aged 6 to 12 years is calculated to be:

$$CIng_{(6-12)}$$

$$= \frac{(CS \times IR \times CF \times FI \times ABS_s \times EF \times ED)}{(BW \times AT)}$$

$$= \frac{([CS] \times 100 \times 1.00\text{E-}06 \times 1 \times ABS_s \times 330 \times 6)}{(29 \times [70 \times 365])}$$

$$= 2.67\text{E-}07 \times ABS_s \times [CS]$$

- The carcinogenic *CDI* for adult residents is calculated to be:

$$CIng_{(\text{adultR})}$$

$$= \frac{(CS \times IR \times CF \times FI \times ABS_s \times EF \times ED)}{(BW \times AT)}$$

$$= \frac{([CS] \times 50 \times 1.00\text{E-}0.6 \times 1 \times ABS_s \times 330 \times 58)}{(70 \times [70 \times 365])}$$

$$= 5.35\text{E-}07 \times ABS_s \times [CS]$$

- The carcinogenic *CDI* for adult workers is calculated to be:

$$CIng_{(\text{adultW})}$$

$$= \frac{(CS \times IR \times CF \times FI \times ABS_s \times EF \times ED)}{(BW \times AT)}$$

$$= \frac{([CS] \times 50 \times 1.00\text{E-}06 \times 1 \times ABS_s \times 260 \times 58)}{(70 \times [70 \times 365])}$$

$$= 4.22\text{E-}07 \times ABS_s \times [CS]$$

- *Estimation of average daily dose (ADD) for noncarcinogenic effects.* For the soil ingestion pathway, the *ADD* (also, the noncarcinogenic *CDI*) is estimated for the different population groups (pre-selected as representative of the critical receptors in the risk assessment) – and the results are shown below.

  – The noncarcinogenic *CDI* for children aged up to 6 years is calculated to be:

  $NCIng_{(1-6)}$

  $$= \frac{(CS \times IR \times CF \times FI \times ABS_s \times EF \times ED)}{(BW \times AT)}$$

  $$= \frac{([CS] \times 200 \times 1.00\text{E-}06 \times 1 \times ABS_s \times 330 \times 5)}{(16 \times [5 \times 365])}$$

  $$= 1.13\text{E-}05 \times ABS_s \times [CS]$$

  – The noncarcinogenic *CDI* for children aged 6 to 12 years is calculated to be:

  $NCIng_{(6-12)}$

  $$= \frac{(CS \times IR \times CF \times FI \times ABS_s \times EF \times ED)}{(BW \times AT)}$$

  $$= \frac{([CS] \times 100 \times 1.00\text{E-}06 \times 1 \times ABS_s \times 330 \times 6)}{(29 \times [6 \times 365])}$$

  $$= 3.12\text{E-}06 \times ABS_s \times [CS]$$

  – The noncarcinogenic *CDI* for adult residents is calculated to be:

  $NCIng_{(adultR)}$

  $$= \frac{(CS \times IR \times CF \times FI \times ABS_s \times EF \times ED)}{(BW \times AT)}$$

$$= \frac{([CS] \times 50 \times 1.00\text{E-}06 \times 1 \times ABS_s \times 330 \times 58)}{(70 \times [58 \times 365])}$$

$$= 6.46\text{E-}07 \times ABS_s \times [CS]$$

- The noncarcinogenic CDI for adult workers is calculated to be:

$$NCIng_{(\text{adultW})}$$

$$= \frac{(CS \times IR \times CF \times FI \times ABS_s \times EF \times ED)}{(BW \times AT)}$$

$$= \frac{([CS] \times 50 \times 1.00\text{E-}06 \times 1 \times ABS_s \times 260 \times 58)}{(70 \times [58 \times 365])}$$

$$= 5.09\text{E-}07 \times ABS_s \times [CS]$$

13.2.4.3 Dermal exposures

The daily dermal intake of contaminated soils for various population groups are calculated for both carcinogenic and noncarcinogenic effects. The assumed parameters used in the computational demonstration are given in Table 13.1, and the electronic spreadsheet automation process is shown in Table 13.2.

- *Estimation of lifetime average daily dose (LADD) for carcinogenic effects.* For the soil dermal contact pathway, the *LADD* (also, the carcinogenic *CDI*) is estimated for the different population groups (pre-selected as representative of the critical receptors in the risk assessment) – and the results are shown below.

  - The carcinogenic CDI for children aged up to 6 years is calculated to be:

$$CDEX_{(1-6)}$$

$$= \frac{(CS \times CF \times SA \times AF \times ABS_s \times SM \times EF \times ED)}{(BW \times AT)}$$

$$= \frac{([CS] \times 1.00\text{E-}06 \times 1396 \times 0.75 \times ABS_s \times 0.15 \times 330 \times 5)}{(16 \times [70 \times 365])}$$

$$= 6.34\text{E-}07 \times ABS_s \times [CS]$$

- The carcinogenic *CDI* for children aged 6 to 12 years is calculated to be:

$CDEX_{(6-12)}$

$$= \frac{(CS \times CF \times SA \times AF \times ABS_s \times SM \times EF \times ED)}{(BW \times AT)}$$

$$= \frac{([CS] \times 1.00\text{E-}06 \times 2094 \times 0.75 \times ABS_s \times 0.15 \times 330 \times 6)}{(29 \times [70 \times 365])}$$

$$= 6.30\text{E-}07 \times ABS_s \times [CS]$$

- The carcinogenic *CDI* for adult residents is calculated to be:

$CDEX_{(adultR)}$

$$= \frac{(CS \times CF \times SA \times AF \times ABS_s \times SM \times EF \times ED)}{(BW \times AT)}$$

$$= \frac{([CS] \times 1.00\text{E-}06 \times 1815 \times 0.75 \times ABS_s \times 0.15 \times 330 \times 58)}{(70 \times [70 \times 365])}$$

$$= 2.19\text{E-}06 \times ABS_s \times [CS]$$

- The carcinogenic *CDI* for adult workers is calculated to be:

$CDEX_{(adultW)}$

$$= \frac{(CS \times CF \times SA \times AF \times ABS_s \times SM \times EF \times ED)}{(BW \times AT)}$$

$$= \frac{([CS] \times 1.00\text{E-}06 \times 1815 \times 0.75 \times ABS_s \times 0.15 \times 260 \times 58)}{(70 \times [70 \times 365])}$$

$$= 1.72\text{E-}06 \times ABS_s \times [CS]$$

- *Estimation of average daily dose (ADD) for noncarcinogenic effects.* For the soil dermal contact pathway, the *ADD* (also, the noncarcinogenic *CDI*) is estimated for the different population groups (pre-selected as representative of the critical receptors in the risk assessment) – and the results are shown below.
  - The noncarcinogenic *CDI* for children aged up to 6 years is calculated as follows:

$NCDEX_{(1-6)}$

$$= \frac{(CS \times CF \times SA \times AF \times ABS_s \times SM \times EF \times ED)}{(BW \times AT)}$$

$$= \frac{([CS] \times 1.00\text{E-}06 \times 1396 \times 0.75 \times ABS_s \times 0.15 \times 330 \times 5)}{(16 \times [5 \times 365])}$$

$$= 8.87\text{E-}06 \times ABS_s \times [CS]$$

- The noncarcinogenic *CDI* for children aged 6 to 12 years is calculated to be:

$NCDEX_{(6-12)}$

$$= \frac{(CS \times CF \times SA \times AF \times ABS_s \times SM \times EF \times ED)}{(BW \times AT)}$$

$$= \frac{([CS] \times 1.00\text{E-}06 \times 2094 \times 0.75 \times ABS_s \times 0.15 \times 330 \times 6)}{(29 \times [6 \times 365])}$$

$$= 7.34\text{E-}06 \times ABS_s \times [CS]$$

- The noncarcinogenic *CDI* for adult residents is calculated to be:

$NCDEX_{(adultR)}$

$$= \frac{(CS \times CF \times SA \times AF \times ABS_s \times SM \times EF \times ED)}{(BW \times AT)}$$

$$= \frac{([CS] \times 1.00\text{E-}06 \times 1815 \times 0.75 \times ABS_s \times 0.15 \times 330 \times 58)}{(70 \times [58 \times 365])}$$

$$= 2.64\text{E-}06 \times ABS_s \times [CS]$$

- The noncarcinogenic *CDI* for adult workers is calculated to be:

$NCDEX_{(adultW)}$

$$= \frac{(CS \times CF \times SA \times AF \times ABS_s \times SM \times EF \times ED)}{(BW \times AT)}$$

$$= \frac{([CS] \times 1.00\text{E-}06 \times 1815 \times 0.75 \times ABS_s \times 0.15 \times 260 \times 58)}{(70 \times [58 \times 365])}$$

$$= 2.08\text{E-}06 \times ABS_s \times [CS]$$

13.3 DETERMINATION OF TOXICOLOGICAL PARAMETERS USED IN HUMAN HEALTH RISK ASSESSMENTS

In the processes involved in the assessment of human health risks due to environmental contaminants, it often becomes necessary to compare receptor chemical intakes with doses shown to cause adverse effects in humans or experimental animals. The dose at which no effects are observed in human populations or experimental animals is referred to as the 'no-observed-effect level' (*NOEL*). Where data identifying a *NOEL* are lacking, a 'lowest-observed-effect level' (*LOEL*) may be used as the basis for determining safe threshold doses. For acute effects, short-term exposures/doses shown to produce no adverse effects are involved; this is called the 'no-observed-adverse-effect level' (*NOAEL*). A *NOAEL* is an experimentally determined dose at which there has been no statistically or biologically significant indication of the toxic effect of concern. In cases where a *NOAEL* has not been demonstrated experimentally, the term 'lowest-observed-adverse-effect level' (*LOAEL*) is used. In general, for chemicals possessing carcinogenic potentials, the *LADD* is compared with the *NOEL* identified in long-term bioassay experimental tests; for chemicals with acute effects, the *MDD* is compared with the *NOEL* observed in short-term animal studies.

An elaboration on the derivation of the relevant toxicological parameters commonly used in human health risk assessments follows below, with further in-depth discussions to be found in the literature elsewhere (e.g., Dourson and Stara, 1983; USEPA, 1985b, 1986a, 1989a, 1989c, 1989d).

13.3.1 Toxicity Parameters for Noncarcinogenic Effects

Traditionally, risk decisions on systemic toxicity are made using the concept of 'acceptable daily intake' (*ADI*), or by using the so-called reference dose (*RfD*). The *ADI* is the amount of a chemical (in mg/kg body weight/day) to which a receptor can be exposed to on a daily basis over an extended period of time – usually a lifetime – without suffering a deleterious effect. The *RfD* is defined as the maximum amount of a chemical (in mg/kg body weight/day) that the human body can absorb without experiencing chronic health effects. For exposure of humans to the noncarcinogenic effects of environmental chemicals, the *ADI* or *RfD* is used as a measure of exposure considered to be without adverse effects. Although often used interchangeably, *RfDs* are based on a more rigorously defined methodology and is therefore preferred over *ADIs*.

The *RfD* provides an estimate of the continuous daily exposure of a noncarcinogenic substance for the general human population (including sensitive subgroups) which appears to be without an appreciable risk of deleterious effects. *RfDs* are established as thresholds of exposure to toxic substances below which there should be no adverse health impact. These thresholds are established on a substance-specific basis for oral and inhalation exposures, taking into account evidence from both human epidemiologic and laboratory toxicologic studies.

The reference concentration (*RfC*) of a chemical – like the *RfD* – represents an estimate of the exposure that can occur on a daily basis over a prolonged period, with a reasonable anticipation that no adverse effect will occur from that exposure. In contrast to *RfDs*, however, *RfCs* are expressed in units of concentration in an environmental medium (e.g., mg/m^3 or $\mu g/L$). *RfCs* generally presuppose continuous exposure, with an average inhalation rate and body weight; it may therefore be inappropriate to use them in 'nonstandard' exposure scenarios. In any case, both the *RfD* and *RfC* represent estimates of the exposure that can occur on a daily basis over a prolonged period, with a reasonable expectation that no adverse effect will occur from that exposure.

In general, because of differing assessments regarding the appropriate measurement of exposure, different measurements of noncancer toxic hazard are normally employed for different routes of exposure. For example, oral or dermal exposures are generally compared to *RfDs* – expressed in units of dose (i.e., mg of chemical per kg of body weight per day), whereas inhalation exposures are compared to *RfCs* – expressed in mg of chemical per m^3 of air or μg of chemical per liter of water. In fact, the toxicity parameters are dependent on the route of exposure; however, oral *RfDs* will normally be used for both ingestion and dermal exposures. Also, subchronic *RfD* is typically used to refer to cases involving only a portion of the lifetime, whereas chronic *RfD* refers to lifetime exposures.

In assessing the chronic and subchronic effects of noncarcinogens and also noncarcinogenic effects associated with carcinogens, the experimental dose value (e.g., *NOEL*) is usually divided by a safety (or uncertainty) factor to yield the *RfD*, as illustrated below.

13.3.1.1 Derivation of *RfDs* and *RfCs*

The *RfD* is a 'benchmark' dose operationally derived from the *NOAEL* by consistent application of general 'order-of-magnitude' uncertainty factors (*UFs*) (also called 'safety factors') that reflect various types of data sets used to estimate *RfDs*. In addition, a modifying factor (*MF*) is sometimes used that is based on professional judgment of the entire database of the specific chemical.

More generally stated, *RfDs* (and *ADIs*) are calculated by dividing a *NOEL* (i.e., the highest level at which a chemical causes no observable changes in the species under investigation), a *NOAEL* (i.e., the highest level at which a chemical causes no observable adverse effect in the species being tested), or a *LOAEL* (i.e., that dose rate of chemical at which there are statistically or biologically significant increases in frequency or severity of adverse effects between the exposed and appropriate control groups) which are derived from human or animal toxicity studies, by one or more uncertainty and modifying factors.

*RfD*s are typically calculated using a single exposure level and uncertainty factors that account for specific deficiencies in the toxicological database. Both the exposure level and uncertainty factors are selected and evaluated in the context of the available chemical-specific literature. After all toxicological, epidemiologic, and supporting data have been reviewed and evaluated, a key study is selected that reflects optimal data on the critical effect. Dose-response data points for all reported effects are examined as a component of this review. USEPA (1989b) discusses specific issues of particular significance in this endeavor – including the types of response levels (ranked in order of increasing severity of toxic effects as *NOEL*, *NOAEL*, *LOAEL*, and *FEL* [the Frank effect level, defined as overt or gross adverse effects]) considered in deriving *RfD*s for systemic toxicants.

The *RfD* (or *ADI*) can indeed be determined from the *NOAEL* (or *LOAEL*) for the critical toxic effect by consistent application of uncertainty factors (*UF*s) and a modifying factor (*MF*), in accordance with the following relationship:

$$\text{Human dose (e.g., } ADI \text{ or } RfD) = \frac{\text{experimental dose (e.g., } NOAEL)}{(UF \times MF)}$$

or, specifically:

$$RfD = \frac{NOAEL}{(UF \times MF)}$$

or, more generally:

$$RfD = \frac{NOAEL \text{ or } LOAEL}{\left(\prod_{i=1}^{n} UF_i \times MF\right)}$$

Some hypothetical example situations involving the determination of *RfD*s based on information on *NOAEL* and then also on *LOAEL* are given below.

- *Determination of the RfD for a hypothetical example using the NOAEL.* Consider the case of a study made on 250 animals (e.g., rats) that is of subchronic duration, yielding a *NOAEL* dosage of 5 mg/kg-day. Then,

$$UF = 10H \times 10A \times 10S = 1000$$

  In addition, there is a subjective adjustment (*MF*) based on the high number of animals (250) per dose group:

$$MF = 0.75$$

  These factors then give $UF \times MF = 750$, so that

$$RfD = \frac{NOAEL}{(UF \times MF)} = \frac{5}{750} = 0.007 \text{ (mg/kg-day)}$$

- *Determination of the RfD for a hypothetical example using the LOAEL.* If the *NOAEL* is not available, and if 25 mg/kg-day had been the lowest dose tested that showed adverse effects, then

$$UF = 10H \times 10A \times 10S \times 10L = 10\,000$$

Again using the subjective adjustment of *MF* = 0.75, one obtains

$$RfD = \frac{LOAEL}{(UF \times MF)} = \frac{25}{7500} = 0.003 \text{ (mg/kg-day)}$$

Overall, the uncertainty factor used in deriving the *RfD* reflects scientific judgment regarding the various types of data used to estimate *RfD* values. It is used to offset the uncertainties associated with extrapolation of data, etc. Generally, the *UF* consists of multiples of 10 (although values less than 10 could also be used), each factor representing a specific area of uncertainty inherent in the extrapolations from the available data. For example, a factor of 10 may be introduced to account for the possible differences in responsiveness between humans and animals in prolonged exposure studies. A second factor of 10 may be used to account for variation in susceptibility among individuals in the human population. The resultant *UF* of 100 has been judged to be appropriate for many chemicals. For other chemicals, with databases that are less complete (e.g., those for which only the results of subchronic studies are available), an additional factor of 10 (leading to a *UF* of 1000) might be judged to be more appropriate. For certain other chemicals, based on well-characterized responses in sensitive humans (as in regard to the effect of fluoride on human teeth), an *UF* as small as 1 might be selected (Dourson and Stara, 1983). Box 13.8 provides the general guidelines for the process of selecting uncertainty and modifying factors for the derivation of *RfD*s (Dourson and Stara, 1983; USEPA, 1986b, 1989a, 1989b)

In general, the choice of the *UF* and *MF* values reflect the uncertainty associated with the estimation of an *RfD* from different human or animal toxicity databases. For instance, if sufficient data from chronic duration exposure studies are available on the threshold region of a chemical's critical toxic effect in a known sensitive human population, then the *UF* used to estimate the *RfD* may be set at unity (1). That is, these data are judged to be sufficiently predictive of a population subthreshold dose, so that additional *UF*s are not needed (USEPA, 1989b).

The derivation of an *RfC* is a parallel process, based on a 'no-observed-adverse-effect concentration' (*NOAEC*) or 'lowest-observed-adverse-effect concentration' (*LOAEC*). Alternatively, an *RfC* may be derived from an *RfD*, taking into account the exposure conditions of the study used to derive the *RfD*.

When no toxicological information exists for a chemical, concepts of structure–activity relationships may have to be employed to derive acceptable intake levels by

Box 13.8 General guidelines for selecting uncertainty and modifying factors in the derivation of RfDs

Standard uncertainty factors (UFs):

- Use a 10-fold factor when extrapolating from valid experimental results in studies using prolonged exposure to average healthy humans. This factor is intended to account for the variation in sensitivity among the members of the human population, due to heterogeneity in human populations, and is referenced as '10H'. Thus, if *NOAEL* is based on human data, a safety factor of 10 is usually applied to the *NOAEL* dose to account for variations in sensitivities between individual humans
- Use an additional 10-fold factor when extrapolating from valid results of long-term studies on experimental animals when results of studies of human exposure are not available or are inadequate. This factor is intended to account for the uncertainty involved in extrapolating from animal data to humans and is referenced as '10A'. Thus, if *NOAEL* is based on animal data, the *NOAEL* dose is divided by an additional safety factor of 10, to account for differences between animals and humans
- Use an additional 10-fold factor when extrapolating from less than chronic results on experimental animals when there are no useful long-term human data. This factor is intended to account for the uncertainty involved in extrapolating from less than chronic (i.e., subchronic or acute) *NOAELs* to chronic *NOAELs* and is referenced as '10S'
- Use an additional 10-fold factor when deriving an *RfD* from a LOAEL, instead of a *NOAEL*. This factor is intended to account for the uncertainty involved in extrapolating from *LOAELs* to *NOAELs* and is referenced as '10L'
- Use an additional up to 10-fold factor when extrapolating from valid results in experimental animals when the data are 'incomplete'. This factor is intended to account for the inability of any single animal study to adequately address all possible adverse outcomes in humans, and is referenced as '10D'

Modifying factor (MF):

- Use professional judgment to determine the *MF*, which is an additional uncertainty factor that is greater than zero and less than or equal to 10. The magnitude of the *MF* depends upon the qualitative professional assessment of scientific uncertainties of the study and database not explicitly treated above – e.g., the completeness of the overall database and the number of species tested. The default value for the *MF* is 1

influence and analogy to closely related or similar compounds. In such cases, some reasonable degree of conservatism is suggested in any judgment call to be made.

13.3.1.2   Inter-conversions of noncarcinogenic toxicity parameters

RfD values for inhalation exposure are usually reported both as a concentration in air (mg/m^3) and as a corresponding inhaled dose (mg/kg-day). When determining the toxicity value for inhalation pathways, the inhalation reference concentration (*RfC* [mg/m^3]) should be used when available. The *RfC* can also be converted to equivalent *RfD* values (in units of dose [mg/kg-day]) by multiplying the *RfC* by a ventilation rate of 20 m³/day (for adults) and dividing it by an average adult body weight of 70 kg. That is,

$$RfD_i \text{ [mg/kg-day]} = \frac{RfC\text{[mg/m}^3\text{]} \times 20 \text{ m}^3/\text{day}}{70 \text{ kg}} = 0.286 \ RfC$$

Conversely, RfD values associated with oral exposures, and reported in mg/kg-day, can be converted to a corresponding concentration in drinking water – called the 'drinking water equivalent level' ($DWEL$), as follows:

$$DWEL \text{ [mg/L in water]} = \frac{\text{oral } RfD \text{ (mg/kg-day)} \times \text{body weight (kg)}}{\text{ingestion rate (L/day)}}$$

$$= \frac{RfD_o \text{ (mg/kg-day)} \times 70 \text{ kg}}{2 \text{ (L/day)}} = 35 RfD_o$$

assuming a 2 L/day water consumption by a 70-kg adult.

Also, in the risk characterization process, a comparison is generally made between the RfD and the estimated exposure dose (EED); the EED should include all sources and routes of exposure involved. If the EED is less than the RfD, the need for regulatory concern may be small. An alternative measure also considered useful to risk managers is the 'margin of exposure' (MOE), which is the magnitude by which the $NOAEL$ of the critical toxic effect exceeds the EED, where both are expressed in the same units.

Suppose the EED for humans exposed to a chemical substance (with a RfD of 0.005 mg/kg-day) under a proposed use pattern is 0.02 mg/kg-day (i.e., the EED is greater than the RfD), then:

$$NOAEL = RfD \times (UF \times MF) = 0.005 \times 1000 = 5 \text{ mg/kg-day}$$

and

$$MOE = \frac{NOAEL}{EED} = \frac{5(\text{mg/kg-day})}{0.02(\text{mg/kg-day})} = 250$$

Because the EED exceeds the RfD (and the MOE is less than the $UF \times MF$ of 1000), the risk manager will need to look carefully at the data set, the assumptions for both the RfD and the exposure estimates, and the comments of the risk assessors. In addition, the risk manager will need to weigh the benefits associated with the case, and other nonrisk factors, in reaching a decision on the so-called 'regulatory dose' (RgD), defined by:

$$RgD = \frac{NOAEL}{MOE}$$

The *MOE* may indeed be used as a surrogate for risk; as the *MOE* becomes larger, the risk becomes smaller.

13.3.2 Toxicity Parameters for Carcinogenic Effects

Two estimates for expressing carcinogenic hazards based on the dose–response function find common application: the slope factor (*SF*) expresses the slope of the dose–response function in dose-related units (viz. $[\text{mg/kg-day}]^{-1}$), while the unit risk factor (*URF*) expresses the slope in concentration-based units (viz. $[\mu\text{g/m}^3]^{-1}$). Typically, *SF*s are used when evaluating risks from oral or dermal exposures, while *URF*s are used in evaluating risks from inhalation exposures.

The *SF*, also called cancer potency factor or potency slope, is a measure of the carcinogenic toxicity or potency of a chemical. It is the cancer risk (proportion affected) per unit of dose (i.e., risk per mg/kg-day). The *SF* is the plausible upper-bound estimate of the probability of a response per unit intake of a chemical over a lifetime. It is used in risk assessments to estimate an upper-bound lifetime probability of an individual developing cancer as a result of exposure to a particular level of potential carcinogen.

In evaluating risks from chemicals found in certain environmental settings, dose–response measures may be expressed as risk per concentration unit – yielding the *URF* (also called unit cancer risk, *UCR* or unit risk, *UR*) values. These measures may include the unit risk factor for air (i.e., inhalation *URF*) and the unit risk for drinking water (i.e., oral *URF*). The continuous lifetime exposure concentration units for air and drinking water are usually expressed in micrograms per cubic meter ($\mu\text{g/m}^3$) and micrograms per liter ($\mu\text{g/L}$), respectively.

13.3.2.1 Derivation of *SF*s and *URF*s

Assuming a no-threshold situation for carcinogenic effects, an estimate of the excess cancer per unit dose, called the unit cancer risk, or the cancer slope factor is used to develop risk decisions for environmental contamination problems. Under the nonthreshold assumption, exposure to any level of a carcinogen is considered to have a finite risk of inducing cancer.

The determination of carcinogenic toxicity parameters often involves the use of a variety of mathematical extrapolation models (see Chapter 11). Scientific investigators have developed numerous models to extrapolate and estimate low-dose carcinogenic risks to humans from high-dose carcinogenic effects usually observed in experimental animal studies. Such models yield an estimate of the upper limit in lifetime risk per unit of dose (or the unit cancer risk). Major regulatory agencies (such as the USEPA) generally use the so-called linearized multistage model to estimate the cancer potency.

The *linearized multistage model* uses animal tumor incidence data to compute maximum likelihood estimates (*MLE*) and upper 95% confidence limits (UCL_{95}) of risk associated with a particular dose. The true risk is very unlikely to be greater than the *UCL*, may be lower than the *UCL*, and could be even as low as zero. The linearized multistage model yields upper-bound estimates of risk which are a linear function of dose at low doses and are frequently used as a basis for regulation.

The multistage model may be expressed mathematically as follows:

$$P(d) = 1 - \exp[-(q_0 + q_1 d + q_2 d^2 + \ldots + q_k d^k)]$$

where $P(d)$ is the lifetime probability of developing a tumor at a given dose, d, of carcinogen; q_0 is a constant that accounts for the background incidence of cancer (i.e., occurring in the absence of the carcinogen under consideration); and $q_1, q_2, \ldots q_k$ are coefficients that allow the data to be expressed to various powers of the dose of carcinogen, in order to obtain the best fit of the model to the data. To determine the extra risk above the background rate at dose, d, the above equation takes the form:

$$P_e(d) = 1 - \exp[-(q_1 + q_2 d^2 + \ldots + q_k d^k)]$$

At low doses, the extra risk is approximated by:

$$P_e(d) = q_1 d$$

The linearized multistage model, known to make several conservative assumptions, results in highly conservative risk estimates, yielding over-estimates of actual *URFs* for carcinogens; in fact, the actual risks may be substantially lower than that predicted by the upper bounds of this model (Paustenbach, 1988). Even so, such an approach is generally preferred since it allows analysts to err on the side of safety.

When no toxicological information exists for a chemical, structural similarity factors, etc. can be used to estimate cancer potency units for chemicals lacking such values, but that are suspected carcinogens. This is achieved, for instance, by estimating the geometric mean of a number of similar compounds whose *URFs* are known, and using this as a surrogate value for the chemical with unknown *URF*.

13.3.2.2 Inter-conversion of carcinogenic toxicity parameters

The *URF* estimates the upper-bound probability of a 'typical' or 'average' person contracting cancer when continuously exposed to one microgram per cubic meter (1 $\mu g/m^3$) of the chemical over an average (70-year) lifetime. Potency estimates are also given in terms of the potency slope factor (*SF*), which is the probability of contracting cancer due to exposure to a given lifetime dose, in units of mg/kg-day.

The *SF* can be converted to *URF* (also, unit risk, *UR* or unit cancer risk, *UCR*), by adopting several assumptions. The most critical factor is that the endpoint of concern must be a systematic tumor, so that potential target organs experience the same blood concentration of the active carcinogen regardless of the method of administration. This implies an assumption of equivalent absorption by the various routes of administration. The basis for these conversions is the assumption that, at low doses, the dose–response curve is linear, so that:

$$P(d) = SF \times [DOSE]$$

where $P(d)$ is the response (probability) as a function of dose; SF is the cancer potency slope factor ($[\text{mg/kg-day}]^{-1}$); and $[DOSE]$ is the amount of chemical intake (mg/kg-day). The inter-conversions between URF and SF are given below.

- *Inter-conversions of the inhalation potency factor.* Risks associated with unit chemical concentration in air may be estimated as follows:

air unit risk = risk per $\mu g/m^3$ (air)

$$= \text{slope factor (risk per mg/kg-day)} \times \frac{1}{\text{body weight (kg)}}$$
$$\times \text{inhalation rate (m}^3\text{/day)} \times 10^{-3} \text{ (mg/}\mu\text{g)}$$

Thus, the inhalation potency can be converted to an inhalation URF by applying the following conversion factor:

$$[(\text{kg-day})/\text{mg}] \times [1/70 \text{ kg}] \times [20 \text{ m}^3/\text{day}] \times [1 \text{ mg}/1000 \ \mu\text{g}] = 2.86 \times 10^{-4}$$

Thus, the lifetime excess cancer risk from inhaling 1 $\mu g/m^3$ concentration for a full lifetime is:

$$URF_i \ (\mu g/m^3)^{-1} = (2.86 \times 10^{-4}) \times SF_i$$

Conversely, the SF_i can be derived from the URF_i as follows:

$$SF_i = (3.5 \times 10^3) \times URF_i$$

The assumptions used involve a 70-kg body weight, and an average inhalation rate of 20 m^3/day.
- *Inter-conversions of the oral potency factor.* Risks associated with unit chemical concentration in water may be estimated as follows:

water unit risk = risk per $\mu g/L$ (water)

$$= \text{slope factor (risk per mg/kg-day)} \times \frac{1}{\text{body weight (kg)}}$$
$$\times \text{ingestion rate (L/day)} \times 10^{-3} \text{ (mg/}\mu\text{g)}$$

Thus, the ingestion potency can be converted to an oral URF value by applying the following conversion factor:

$$[(\text{kg-day})/\text{mg}] \times [1/70 \text{kg}] \times [2 \text{ L/day}] \times [1 \text{ mg}/1000 \ \mu\text{g}] = 2.86 \times 10^{-5}$$

Thus, the lifetime excess cancer risk from ingesting 1 μg/L concentration for a full lifetime is:

$$URF_o \, (\mu g/L)^{-1} = (2.86 \times 10^{-5}) \times SF_o$$

Or, alternatively the potency, SF_o, can be derived from the unit risk as follows:

$$SF_o = (3.5 \times 10^4) \times URF_o$$

The assumptions used involve a 70-kg body weight, and an average water ingestion rate of 2 L/day.

13.3.3 The Use of Surrogate Toxicity Parameters

In general, toxicity parameters used in risk assessments are dependent on the route of exposure. However, oral RfDs and SFs will normally be used for both ingestion and dermal exposures to some chemicals that affect receptors through a systemic action; this will be inappropriate if the chemical affects the receptor contacts through direct local action at the point of application. Thus, in several (but certainly not all) situations, it is appropriate to use oral SFs and RfDs as surrogate values to estimate systemic toxicity as a result of dermal absorption of a chemical (DTSC, 1994; USEPA, 1989b, 1992). It is noteworthy, however, that use of the oral SF or oral RfD directly does not correct for differences in absorption and metabolism between the oral and dermal routes.

In addition to the uncertainties caused by route differences, further uncertainty is introduced by the fact that the oral dose–response relationships are based on potential (i.e., administered) dose, whereas dermal dose estimates are absorbed doses. Ideally, these differences in route and dose type should be resolved via pharmacokinetic modeling. Alternatively, if estimates of the gastrointestinal absorption fraction are available for the compound of interest in the appropriate vehicle, then the oral dose–response factor, unadjusted for absorption, can be converted to an absorbed dose basis as follows:

$$RfD_{absorbed} = RfD_{administered} \times ABS_{GI}$$

$$SF_{absorbed} = \frac{SF_{administered}}{ABS_{GI}}$$

Typically, absorption fractions corresponding roughly to 10% and 1% are applied to organic and inorganic chemicals, respectively.

So far, direct toxic effects on the skin have not been accounted for. This means, it may be inappropriate to use the oral slope factor to evaluate the risks associated with dermal exposure to carcinogens such as benzo(a)pyrene, which cause skin cancer through a direct action at the point of application. Thus, the use of an oral

SF or oral *RfD* for the dermal route may result in an over- or under-estimation of the risk or hazard, depending on the chemical involved. Consequently, the use of the oral toxicity value as a surrogate for a dermal value will tend to increase the uncertainty in the estimation of risks and hazards; however, this is not generally expected to significantly under-estimate the risk or hazard relative to the other routes of exposure that are evaluated in the risk assessment (DTSC, 1994; USEPA, 1992).

Furthermore, in the evaluation of the inhalation pathways, when an inhalation *SF* or *RfD* is not available for a compound, the oral *SF* or *RfD* may be used in its place for screening analyses. Similarly, inhalation *SF*s and *RfD*s may be used as surrogates for both ingestion and dermal exposures for those chemicals lacking oral toxicity values.

In other situations, toxicity values to be used in characterizing risks are available only for certain chemicals within a chemical class. In such cases, rather than simply eliminating those chemicals without toxicity values from a quantitative evaluation – as has been the case in some past studies – it usually is prudent to group data for such class of chemicals (e.g., according to structure–activity relationships or other similarities) for consideration in the risk assessment; such grouping should not be based solely on toxicity class or carcinogenic classifications. Significant uncertainties will likely result by using this type of approach. Hence, if and when this type of grouping is carried out, it should be acknowledged and documented in the risk assessment summary – indicating the fact that the action may have produced over- or under-estimates of the true risk.

The introduction of additional uncertainties in an approach that relies on surrogate toxicity parameters cannot be over-emphasized, and such uncertainties should be well documented as part of the risk evaluation process.

13.3.3.1 Route-to-route extrapolation of toxicological parameters

Risk characterizations generally should consider every likely exposure route in the evaluation process. However, toxicity data may not always be available for each route of concern, in which case the use of surrogate values – that may include extrapolation from data for another exposure route – may be required. Extrapolations may be possible for some cases where there is reliable information on the degree of absorption of materials by both routes of exposure – assuming the substance is not locally more active by one route. In any case, this type of extrapolation can provide useful approximations to employ – at least for preliminary risk assessments.

For systemic effects away from the site of entry, an inhalation toxicity parameter, TP_{inh} (mg/m^3), may be converted to an oral value, TP_{oral} (mg/kg-day), or vice versa, using the following relationship (van Leeuwen and Hermens, 1995):

$$TP_{inh} \times IR \times t \times BAF_{inh} = TP_{oral} \times BAF_{oral} \times BW$$

where *IR* is the inhalation rate (m^3/h); *t* is the time (h); BAF_r is the bioavailability for route *r*, for which default values should be used if no data exists (e.g., use 1 for oral exposure, 0.75 for inhalation exposure, and 0 [in the case of very low or very high lipophilicity or high molecular weight] or 1 [in the case of intermediate lipophilicity and low molecular weight] for dermal exposure); and BW is the body weight (kg).

A dermal toxicity parameter for systemic effects, TP_{derm} (mg/kg-day) can be derived from the TP_{oral} (mg/kg-day) or the TP_{inh} (mg/m^3) values as follows (van Leeuwen and Hermens, 1995):

$$TP_{derm} = TP_{oral} \times \frac{BAF_{oral}}{BAF_{derm}}$$

$$TP_{derm} = \frac{TP_{inh} \times IR \times t}{BW} \times \frac{BAF_{inh}}{BAF_{derm}}$$

It is noteworthy that, route-to-route extrapolation introduces additional uncertainty into the overall risk assessment process; such uncertainty can be reduced by using physiologically-based pharmacokinetic (PB-PK) models. If sufficient pharmacokinetic data are available, PB-PK models are particularly useful for predicting disposition differences due to exposure route differences (van Leeuwen and Hermens, 1995).

In principle, it is generally not possible to extrapolate between exposure routes for some substances that produce localized effects dependent upon the route of exposure. For example, a toxicity value based on localized lung tumors that result only from inhalation exposure to a substance would not be appropriate for estimating risks associated with dermal exposure to the substance. Thus, it may be appropriate only to extrapolate dermal toxicity values from values derived for oral exposure. In fact, it is not recommended that oral toxicity reference values be extrapolated casually from inhalation toxicity values, although such extrapolations may be performed on a case-by-case basis (USEPA, 1989b).

In general, inhalation values should *not* be extrapolated from oral values – if at all avoidable. Even so, situations arise when it becomes necessary to rely on such approximations to make effective environmental management decisions. As an illustrative example, the processes involved in such extrapolation exercises are given below for carcinogenic and noncarcinogenic effects of chemical constituents.

- *Noncarcinogenic effects.* Reference concentrations (*RfC*s) for inhalation exposures may be extrapolated from oral reference doses (*RfD*s) for adults by using the following relationship:

$$\text{Extrapolated } RfC \text{ (mg/m}^3\text{)} = RfD_{oral} \text{ (mg/kg-day)} \times \frac{70 \text{ (kg)}}{20 \text{ (m}^3/\text{day)}}$$

It should be noted that, for this simplistic approximation, dosimetric adjustments have not been made to account for respiratory tract deposition efficiency and distribution; physical, biological, and chemical factors; and other aspects of

exposure (e.g., discontinuous exposures) that affect uptake and clearance (USEPA, 1996). Consequently, this simple extrapolation method relies on the implicit assumption that the route of administration is irrelevant to the dose delivered to a target organ – an assumption not supported by the principles of dosimetry or pharmacokinetics.
- *Carcinogenic effects.* For carcinogens, unit risk factors (*URF*s) for inhalation exposures may be extrapolated from oral carcinogenic slope factors (*SF*s) for adults by using the following relationship:

$$\text{Extrapolated } URF\ [(\mu g/m^3)^{-1}] = SF_{\text{oral}}\ [(\text{mg/kg-day})^{-1}]$$
$$\times \frac{20\ (m^3/\text{day})}{70\ (\text{kg})} \times 10^{-3}\ (\text{mg}/\mu g)$$

Using the extrapolated *URF*, risk-specific air concentrations can be calculated as a lifetime average exposure concentration, as follows:

$$\text{Extrapolated air concentration } (\mu g/m^3) = \frac{\text{target risk (e.g. } 10^{-6})}{URF\ [(\mu g/m^3)^{-1}]}$$

13.3.3.2 Toxicity equivalence factors

A toxicity equivalence factor (*TEF*) procedure is one used to derive quantitative dose–response estimates for agents that are members of a category or class of agents. TEFs are based on shared characteristics that can be used to order the class members by carcinogenic potency when cancer bioassay data are inadequate for this purpose. The ordering is by reference to the characteristics and potency of a well-studied member or members of the class. Other class members are indexed to the reference agent(s) by one or more shared characteristics to generate their *TEF*s. Examples of shared characteristics that may be used include: receptor-binding characteristics, results of assays of biological activity related to carcinogenicity, or structure–activity relationships.

To date, adequate data to support the use of *TEF*s has been found in only one class of compounds – dioxins (USEPA, 1989c). The *TEF*s are usually indexed at increments of a factor of 10. Very good data, however, may permit a smaller increment to be used.

13.4 RISK CHARACTERIZATION

Typically, the health risks to potentially exposed populations resulting from exposure to environmental contaminants are characterized through a calculation of noncarcinogenic hazard quotients and indices and/or carcinogenic risks (CAPCOA, 1990; CDHS, 1986; USEPA, 1986a, 1989a). These parameters can then be com-

> **Box 13.9   Major sources of uncertainty in human health risk assessments**
>
> Uncertainty in health effects/toxicity data
>
> - Uncertainty in extrapolating from high dose to low dose
> - Uncertainty in extrapolating data from experimental animals to humans
> - Uncertainty due to differences between individuals
>
> Uncertainty in measuring or calculating exposure point concentrations
>
> - Uncertainty in transposing environmental concentrations into exposure point concentrations
> - Uncertainty in assumptions used to model exposure point concentrations
>
> Uncertainty in calculating exposure dose
>
> - Uncertainty in source terms (i.e., environmental sampling and monitoring data)
> - Uncertainty in estimating exposure dose using environmental models

pared with benchmark criteria/standards in order to arrive at risk decisions about an environmental contamination problem. In any event, it should be recognized that several sources of uncertainties will generally be associated with human health risk assessments, such as the common ones identified in Box 13.9. It is therefore important to recognize the inevitability of such issues – as this knowledge will facilitate the design of an effectual environmental management program.

In general, several important considerations affect the processes involved in completing the risk characterization task. Some particularly important issues are discussed in the proceeding sections, followed by a presentation of the relevant algorithms for estimating human health risks that follow from exposure to environmental contaminants.

13.4.1   Absorption Adjustments

Absorption adjustments may be necessary to ensure that the exposure estimate and the toxicity value being compared during the risk characterization are both expressed as absorbed doses or both expressed as administered doses (i.e., intakes). Adjustments may also be necessary for different vehicles of exposure (e.g., water, food, or soil) – although, in most cases, the unadjusted toxicity value will provide a reasonable or conservative estimate of risk. Furthermore, adjustments may be needed for different absorption efficiencies, depending on the medium of exposure. In particular, correction for fractional absorption is generally appropriate when interaction with environmental media or other contaminants may alter absorption from that expected for the pure compound; and/or when assessment of exposure was via a different route of contact to that which was utilized in the experimental studies used to establish the toxicity parameters (i.e., *SF*s and *RfD*s).

In general, only limited toxicity reference values exist for dermal exposure; consequently, oral values are frequently used to assess risks from dermal exposures

(USEPA, 1989d). On the other hand, most *RfD*s and some carcinogenic *SF*s usually are expressed as the amount of substance administered per unit time and unit body weight, whereas exposure estimates for the dermal route of exposure are eventually expressed as absorbed doses. Thus, for dermal exposure to contaminants in water or in soil, it may become necessary to adjust an oral toxicity value from an administered to an absorbed dose – generally carried out as shown below (USEPA, 1989d).

- *Adjustment of an administered dose to an absorbed dose RfD.* For adjustment of administered dose *RfD* (RfD_{adm}) of a chemical with oral absorption efficiency, *ABS*, in the species on which the *RfD* is based, to an absorbed dose *RfD* (RfD_{abs}) – simply multiply the unadjusted *RfD* by the absorption efficiency percent, i.e., $RfD_{abs} = RfD_{adm} \times ABS$, which can be compared with the amount estimated to be absorbed dermally.
- *Adjustment of an administered dose to an absorbed dose SF.* For adjustment of administered dose *SF* (SF_{adm}) of a chemical with oral absorption efficiency, *ABS*, in the species on which the *SF* is based, to an absorbed dose *SF* (SF_{abs}) – simply divide the unadjusted *SF* by the absorption efficiency percent, i.e., $SF_{abs} = SF_{adm}/ABS$, which can be used to estimate the cancer risk associated with the estimated absorbed dose for the dermal route of exposure.
- *Adjustment of an exposure estimate to an absorbed dose.* If the toxicity value is expressed as an absorbed rather than an administered dose, then it may become necessary to convert the exposure estimate from an intake into an absorbed dose for comparison. For adjustment of unadjusted exposure estimate or intake (CDI_{adm}) of a chemical with absorption efficiency, *ABS*, for the contaminant in the media of concern, to an adjusted exposure or absorbed dose (CDI_{abs}) – simply multiply the unadjusted *CDI* by the absorption efficiency percent, i.e., $CDI_{abs} = CDI_{adm} \times ABS$, which can be used in comparisons with the *RfD* or *SF* that is based on an absorbed, not administered dose.

It is noteworthy that, for evaluations of the dermal exposure pathway, if the oral toxicity value is already expressed as an absorbed dose, it is not necessary to adjust the toxicity value. Also, exposure estimates should not be adjusted for absorption efficiency if the toxicity values are based on administered dose.

Absorption factors should not be used to modify exposure estimates in those cases where absorption is inherently factored into the toxicity/risk parameters used for the risk characterization. Thus, 'correction' for fractional absorption is appropriate only for those values derived from experimental studies based on absorbed dose. That is, absorbed dose should be used in risk characterization only if the applicable toxicity parameter (e.g., *SF* or *RfD*) has been adjusted for absorption; otherwise, intake (unadjusted for absorption) are used for the calculation of risk levels. Absorption efficiency adjustment procedures are discussed elsewhere in the literature (e.g., USEPA, 1989d, 1992). In the absence of reliable information, 100% absorption is normally used for most chemicals; for metals, approximately a 10% absorption may be considered as a reasonable upper-bound for other than the inhalation exposure route.

13.4.2 Aggregate Effects of Chemical Mixtures

While some potential environmental hazards involve significant exposure to only a single compound, most instances of environmental contamination involve concurrent or sequential exposures to a mixture of compounds that may induce similar or dissimilar effects over exposure periods ranging from short-term to lifetime (USEPA, 1984a, 1986b).

In fact, the constituents associated with most environmental contamination problems tend to be heterogeneous and variable mixtures that may contain several distinct compounds, distributed over wide spatial regions and across several environmental compartments. The toxicology of complex mixtures is not well understood, complicating the problem in regard to the potential for these compounds to cause various health and environmental effects. Nonetheless, there is the need to assess the cumulative health risks for the chemical mixtures despite potential large uncertainties that may exist. The risk assessment process must address the multiple endpoints or effects and also the uncertainties in the dose-response functions for each effect.

In general, potential receptors are typically exposed not to isolated pollutants, but rather to a complex, dilute mixture of many substances. Considering how many chemicals exist in the environment, there is a virtually infinite number of combinations that could constitute potential synergisms and antagonisms. In the absence of any concrete evidence of what the interactive effects might be, an additive method that simply sums individual chemical effects on a target organ is employed in the evaluation of chemical mixtures.

The common method of approach in the assessment of chemical mixtures assumes additivity of effects for carcinogens when evaluating multiple carcinogens, even though alternative procedures that are more realistic and/or less conservative have been proposed for certain situations by some investigators (e.g., Bogen, 1994; Chen et al., 1990; Gaylor and Chen, 1996; Kodell and Chen, 1994; Slob, 1994). In any case, prior to the summation of aggregate risks, estimated cancer risks should preferably be segregated by weight-of-evidence category for the environmental contaminants, the goal being to provide a clear understanding of the risk contribution of each category of carcinogen.

For multiple pollutant exposures to noncarcinogens and noncarcinogenic effects of carcinogens, constituents should be grouped by the same mode of toxicological action (i.e., those which induce the same physiologic endpoint – such as liver or kidney toxicity). Toxicological endpoints that will normally be considered with respect to chronic toxicity include cardiovascular systems (CVS); central nervous system (CNS); gastrointestinal system (GI); immune system; reproductive system (including teratogenic and developmental effects); kidney; liver; and the respiratory system. Listings of environmental chemicals with noncarcinogenic toxic effects on specific target organ/system can be found in such databases as IRIS (Integrated Risk Information System), as well as in the literature elsewhere (e.g., Cohrssen and Covello, 1989; USEPA, 1996). Cumulative noncarcinogenic risk is evaluated through the use of a hazard index that is generated for each health or physiologic 'endpoint'. In fact, in a strict sense, constituents should not be grouped together unless they have the same toxicological endpoint. Thus, it becomes necessary to segregate chemicals by organ-specific toxicity, since strict additivity without consideration for target-organ toxicities could over-estimate potential hazards (USEPA, 1986b, 1989d). Consequently, the hazard index is preferably calculated only after putting chemicals into groups with same physiologic endpoints.

Finally, in combining multi-chemical risk estimates across exposure pathways, it should be noted that, if two pathways do not affect the same individual or subpopulation, then neither pathway's individual risk estimate or hazard index affects the other – and therefore risks should not be combined. Thus, one should not automatically sum risks from all exposure pathways evaluated for an environmental contamination problem – unless it has been determined/established that such aggregation is appropriate.

13.4.3 Estimation of Carcinogenic Risks to Human Health

For potential carcinogens, risk is defined by the incremental probability of an individual developing cancer over a lifetime as a result of exposure to a carcinogen. The risk of contracting cancer can be estimated by combining information about the carcinogenic potency of a chemical and exposure to the substance. Specifically, carcinogenic risks are estimated by multiplying the route-specific cancer slope factor (which is the upper 95% confidence limit of the probability of a carcinogenic response per unit intake over a lifetime of exposure) by the estimated intakes; this yields the excess or incremental individual lifetime cancer risk.

In general, risks associated with the inhalation and noninhalation pathways may be estimated in accordance with the following generic relationships:

Risk for inhalation pathways = ground-level concentration (GLC) [$\mu g/m^3$]
$\times$ unit risk [$(\mu g/m^3)^{-1}$]

Risk for noninhalation pathways = dose [mg/kg-day]
$\times$ potency slope [(mg/kd-day)$^{-1}$]

The carcinogenic effects of the constituents associated with potential environmental contamination problems are typically calculated using the linear low-dose and one-hit models, represented by the following relationships (USEPA, 1989d):

Linear low-dose model $CR = CDI \times SF$

One-hit model, $CR = 1 - \exp(-CDI \times SF)$

where CR is the probability of an individual developing cancer (dimensionless); CDI is the chronic daily intake for long-term exposure (i.e., averaged over a 70-year lifetime) (mg/kg-day); and SF is the slope factor ([mg/kg-day]$^{-1}$).

> Box 13.10 General equations for calculating carcinogenic risks to human health
>
> - For the linear low-dose model used at low levels of carcinogenic risks:
>
> $$\text{Total cancer risk, } TCR_{\text{lo-risk}} = \sum_{j=1}^{p} \sum_{i=1}^{n} (CDI_{ij} \times SF_{ij})$$
>
> - For the one-hit model used at high levels of carcinogenic risks:
>
> $$\text{Total cancer risk, } TCR_{\text{hi-risk}} = \sum_{j=1}^{p} \sum_{i=1}^{n} [1 - \exp(-CDI_{ij} \times SF_{ij})]$$
>
> where:
> TCR = Probability of an individual developing cancer (dimensionless)
> CDI_{ij} = Chronic daily intake for the ith contaminant and jth exposure route/pathway (mg/kg-day)
> SF_{ij} = Slope factor for the ith contaminant and jth exposure route ($[\text{mg/kg-day}]^{-1}$)
> n = Total number of carcinogens
> p = Total number of pathways or exposure routes

The linear low-dose model is based on the linearized multistage model – which assumes that there are multiple stages for cancer; the one-hit model assumes that there is a single stage for cancer, and that one molecular or radiation interaction induces malignant change – making it very conservative.

In reality, and for all practical purposes, the linear low-dose cancer risk model is valid only at low risk levels (i.e., estimated risks <0.01). For situations where chemical intakes may be high (i.e., potential risks >0.01), the one-hit model represents the more appropriate algorithm to use.

As noted above in Section 13.4.2, the method of approach for assessing the cumulative health risks from chemical mixtures generally assumes additivity of effects for carcinogens when evaluating multiple carcinogens. Thus, for multiple carcinogenic chemicals and multiple exposure routes/pathways, the aggregate cancer risk for all exposure pathways and all contaminants associated with a potential environmental contamination problem can be estimated using the algorithms shown in Box 13.10. The combination of risks across exposure pathways is based on the assumption that the same receptors would consistently experience the reasonable maximum exposure via the multiple pathways. Hence, if specific pathways do not affect the same individual or receptor group, risks should not be combined under those circumstances.

As a rule-of-thumb, incremental risks of between 10^{-4} and 10^{-7} are generally perceived as being reasonable and adequate for the protection of human health and the environment, with 10^{-6} often used as the 'point-of-departure'. In reality, however, populations may be exposed to the same constituents from sources unrelated to a specific project location. Consequently, it is preferred that the estimated carcinogenic risk is well below the 10^{-6} benchmark level, to allow for a reasonable margin of protectiveness for populations potentially at risk. Indeed, if any calculated cancer risk exceeds the 10^{-6} benchmark, then the health-based criterion for the chemical mixture has been exceeded and the need for interim corrective measures must be considered.

13.4.3.1 Population excess cancer burden

The two important parameters or measures often used for describing carcinogenic effects are the individual cancer risk and the estimated number of cancer cases – the cancer burden. The individual cancer risk from simultaneous exposure to several carcinogens is assumed to be the sum of the individual cancer risks from each individual chemical. The risk experienced by the individual receiving the greatest exposure is referred to as the 'maximum individual risk'.

To assess the population cancer burden associated with an environmental contamination problem, the number of cancer cases due to an emission source or the presence of a potential environmental contamination problem within a given community can be estimated by multiplying the individual risk experienced by a group of people by the number of people in that group. Thus, if 10 million people (as an example) experience an estimated cancer risk of 10^{-6} over their lifetimes, it would be estimated that 10 (i.e., 10 million $\times$ 10^{-6}) additional cancer cases could occur. The number of cancer incidents in each receptor area can be added to estimate the number of cancer incidents over an entire region. Hence, the excess cancer burden, B_{gi}, is given by:

$$B_{gi} = R_{gi} \times P_g$$

where B_{gi} is the population excess cancer burden for the ith chemical for the exposed group, G; R_{gi} is the excess lifetime cancer risk for the ith chemical for the exposed population group, G; and P_g is the number of persons in the exposed population group, G.

Assuming cancer burden from each carcinogen is additive, then the total population group excess cancer burden is:

$$B_g = \sum_{i=1}^{N} B_{gi} = \sum_{i=1}^{N} (R_{gi} \times P_g)$$

and

$$\text{Total population burden, } B = \sum_{g=1}^{G} B_g = \sum_{g=1}^{G} \left\{ \sum_{i=1}^{N} B_{gi} \right\} = \sum_{g=1}^{G} \left\{ \sum_{i=1}^{N} (R_{gi} \times P_g) \right\}$$

Insofar as possible, cancer risk estimates are expressed in terms of both individual and population risk. For the population risk, the individual upper-bound estimate of excess lifetime cancer risk for an average exposure scenario is simply multiplied by the size of the potentially exposed population. Generally speaking, the risk assessment is case-specific and the calculated risks should be combined for pollutants originating from a given locale or group of locations in the case-study affecting same receptor groups.

> Box 13.11 General equation for calculating noncarcinogenic risks to human health
>
> $$\text{Total hazard index} = \sum_{j=1}^{p} \sum_{i=1}^{n} \frac{E_{ij}}{RfD_{ij}}$$
>
> $$= \sum_{j=1}^{p} \sum_{i=1}^{n} [HQ]_{ij}$$
>
> where:
> E_{ij} = Exposure level (or intake) for the ith contaminant and jth exposure route/pathway (mg/kg-day)
> RfD_{ij} = Acceptable intake level (or reference dose) for ith contaminant and jth exposure route (mg/kg-day)
> $[HQ]_{ij}$ = Hazard quotient for ith contaminant and jth pathway
> n = Total number of chemicals showing noncarcinogenic effects
> p = Total number of pathways or exposure routes

13.4.4 Estimation of Noncarcinogenic Hazards to Human Health

The potential noncancer health effects of contaminants associated with an environmental contamination problem are usually expressed by the hazard quotient (HQ) and/or the hazard index (HI). The HQ is defined by the ratio of the estimated chemical exposure level to the route-specific reference dose, represented as follows (USEPA, 1989d):

$$\text{Hazard quotient, } HQ = \frac{E}{RfD}$$

where E is the chemical exposure level or intake (mg/kg-day) and RfD is the reference dose (mg/kg-day).

As noted previously in Section 13.4.2, for multiple pollutant exposures to non-carcinogens and noncarcinogenic effects of carcinogens, constituents are normally grouped by the same mode of toxicological action. Cumulative risk is evaluated through the use of a hazard index that is generated for each health or toxicological 'endpoint'. Chemicals with the same endpoint are generally included in a hazard index calculation. For multiple noncarcinogenic effects of several chemical compounds and multiple exposure routes/pathways, the aggregate noncancer risk for all exposure pathways and all contaminants associated with a potential environmental contamination problem can be estimated using the algorithm shown in Box 13.11.

The combination of hazard quotients across exposure pathways is based on the assumption that the same receptors would consistently experience the reasonable maximum exposure via the multiple pathways. Thus, if specific pathways do not affect the same individual or receptor group, hazard quotients should not be combined under those circumstances. Furthermore, in the strictest sense, constituents should not be grouped together unless the toxicological endpoint is known to be the same, otherwise the process will likely over-estimate and over-state potential health effects.

In accordance with general guidelines on the interpretation of hazard indices, for any given chemical, there may be potential for adverse health effects if the hazard index exceeds unity (1). It is noteworthy that, since the *RfD* incorporates a large margin of safety, it is possible that no toxic effects may occur even if this benchmark level is exceeded. However, as a rule-of-thumb in the interpretation of the results from *HI* calculations, a reference value of less than or equal to unity (i.e., $HI \leq 1$) should be taken as the acceptable benchmark. For HI values greater than unity (i.e., $HI > 1$), the higher the value, the greater is the likelihood of adverse noncarcinogenic health impacts. In fact, since populations may be exposed to the same constituents from sources unrelated to a specific locale, it is preferred that the estimated noncarcinogenic hazard index be well below the benchmark level of unity, to allow for additional margin of protectiveness for populations potentially at risk. Indeed, if any calculated hazard index exceeds unity, then the health-based criterion for the chemical mixture has been exceeded and the need for interim corrective measures must be considered.

13.4.4.1 Chronic versus subchronic noncarcinogenic effects

Human receptor exposures to environmental contaminants can occur over long-term periods (i.e., chronic exposures), or over short-term periods (i.e., subchronic exposures). Chronic exposures for humans usually range in duration from about seven years to a lifetime; subchronic human exposures typically range in duration from about two weeks to seven years (USEPA, 1989a). Of course, shorter-term exposures of less than two weeks could also be anticipated.

The chronic noncancer hazard index is represented by the following modification to the general equation presented above:

$$\text{Total chronic hazard index} = \sum_{j=1}^{p} \sum_{i=1}^{n} \frac{CDI_{ij}}{RfD_{ij}}$$

where CDI_{ij} is the chronic daily intake for the ith contaminant and jth exposure route/pathway, and RfD_{ij} is the chronic reference dose for the ith contaminant and jth exposure route.

The subchronic noncancer hazard index is represented by the following modification to the general equation presented above:

$$\text{Total subchronic hazard index} = \sum_{j=1}^{p} \sum_{i=1}^{n} \frac{SDI_{ij}}{RfD_{sij}}$$

where SDI_{ij} is the subchronic daily intake for the ith contaminant and jth exposure route/pathway, and RfD_{sij} is the subchronic reference dose for the ith contaminant and jth exposure route.

Appropriate chronic and subchronic toxicity parameters and intakes are used in the estimation of noncarcinogenic effects associated with the different exposure durations.

13.4.5 Risk Computations: Illustration of the Processes for Calculating Carcinogenic Risks and Noncarcinogenic Hazards

Illustrative example evaluations for potential receptor groups purported to be exposed through inhalation, soil ingestion (i.e., incidental or pica behavior), and dermal contact are discussed in the following sections – to demonstrate the computational mechanics for estimating chemical risks. The same set of units are maintained throughout as given above in related prior discussions.

13.4.5.1 Carcinogenic risk calculations

In accordance with the relationships presented earlier on in this chapter, the potential carcinogenic risks associated with environmental contaminants present in various matrices may be calculated for all relevant exposure routes – as shown in the demonstration examples given below.

- *Carcinogenic effects for contaminants in water.* The carcinogenic risk associated with a potential receptor exposure to chemical constituents in water can be estimated using the following annotated relationship:

$Risk_{water}$

$= [CDI_o \times SF_o] + [CDI_i \times SF_i]$

$= [(CDI_{ing} + CDI_{der}) \times SF_o] + [CDI_i \times SF_i]$

$= \{[(INGf \times C_w) + (DEXf \times C_w)] \times SF_o\} + \{[(INHf \times C_w) \times SF_i]\}$

More generally, the carcinogenic risk may be calculated from 'first principles' as follows:

$Risk_{water}$

$= \left\{ SF_o \times C_w \times \dfrac{(WIR_{adult} \times FI \times ABS_{gi} \times EF \times ED_{adult})}{(BW_{adult} \times AT \times 365 \text{ day/yr})} \right\}$

$+ \left\{ SF_o \times C_w \times \dfrac{(WIR_{child} \times FI \times ABS_{gi} \times EF \times ED_{child})}{(BW_{child} \times AT \times 365 \text{ day/yr})} \right\}$

$+ \left\{ SF_o \times C_w \times \dfrac{(SA_{adult} \times K_p \times CF \times FI \times ABS_{gi} \times EF \times ED_{adult} \times ET_{adult})}{(BW_{adult} \times AT \times 365 \text{ day/yr})} \right\}$

$$+ \left\{ SF_o \times C_w \times \frac{(SA_{child} \times K_p \times CF \times FI \times ABS_{gi} \times EF \times ED_{child} \times ET_{child})}{(BW_{child} \times AT \times 365 \text{ day/yr})} \right\}$$

$$+ \left\{ SF_i \times C_w \times \frac{(IR_{adult} \times FI \times ABS_{gi} \times EF \times ED_{adult})}{(BW_{adult} \times AT \times 365 \text{ day/yr})} \right\}$$

$$+ \left\{ SF_i \times C_w \times \frac{(IR_{child} \times FI \times ABS_{gi} \times EF \times ED_{child})}{(BW_{child} \times AT \times 365 \text{ day/yr})} \right\}$$

As an example, substitution of the exposure assumptions presented in Table 13.3 into the above equation, yields the following reduced equation:

$$\text{Risk}_{water} = (SF_o \times C_w \times 0.0149) + (SF_o \times C_w \times 0.0325 \times K_p) + (SF_i \times C_w \times 0.0149)$$

Consequently, by substituting the chemical-specific parameters in the reduced risk equation, potential carcinogenic risks associated with the particular constituent are derived.

As a simple illustrative example calculation of human health carcinogenic risk, consider a situation where PCBs from abandoned electrical transformers have leaked into a groundwater reservoir that serves as a community water supply source. Environmental sampling and analysis conducted in a routine testing of the public water supply system showed an average PCB concentration of 2 μg/L. The question then is: what is the individual lifetime cancer risk for a person who uses this water for drinking water purposes? Assuming the only exposure route of concern is from water ingestion (a reasonable assumption for this situation), and using a cancer oral SF of 7.7 (obtained from Table D.1 in Appendix D), then the cancer risk attributable to this exposure scenario is calculated to be:

$\text{Risk} = SF_o \times CDI_o$

$= SF_o \times C_w \times 0.0149$

$= 7.7 \times (2 \ \mu\text{g/L} \times 10^{-3} \ \text{mg}/\mu\text{g}) \times 0.0149$

$= 2.3 \times 10^{-4}$

Similar evaluations can indeed be carried out for the various media and exposure routes.

- *Carcinogenic effects for contaminants in soils.* The carcinogenic risk associated with a potential receptor exposure to chemical constituents in soils can be estimated using the following annotated relationship:

Table 13.3 Definitions and exposure assumptions for example risk computations associated with exposure to environmental contaminants in water and soil

Parameter	Parameter definition and exposure assumption
SF_o	Oral cancer potency slope (obtained from the literature or Appendix D) ($[mg/kg-day]^{-1}$)
SF_i	Inhalation cancer potency slope (obtained from the literature or Appendix D) ($[mg/kg-day]^{-1}$)
C_w	Chemical concentration in water (obtained from the sampling and/or modeling) (mg/L)
C_s	Chemical concentration in soil (obtained from the sampling and/or modeling) (mg/kg)
C_a	Chemical concentration in air (obtained from the sampling and/or modeling) (mg/m^3)
K_p	Chemical-specific dermal permeability coefficient from water (obtained from the literature, e.g., DTSC, 1994) (cm^2/h)
AF	Soil to skin adherence factor (1 mg/cm^2)
SA	Skin surface area available for water contact (adult = 23 000 cm^2; child = 7200 cm^2); Skin surface area available for soil contact (adult = 5800 cm^2; child = 2000 cm^2)
WIR	Average water intake rate – where intake from inhalation of volatile constituents may be assumed as equivalent to the amount of ingested water (adult = 2L/day; child = 1L/day)
SIR	Average soil ingestion rate (adult = 100 mg/day; child = 200 mg/day)
IR	Inhalation rate (adult = 20 m^3/day; child = 10 m^3/day)
CF	Conversion factor for water (1L/1000 cm^3); Conversion factor for soil (10^{-6} kg/mg)
FI	Fraction ingested from contaminated source (1)
ABS_{gi}	Bioavailability/gastrointestinal (GI) absorption factor (100%)
ABS_s	Chemical-specific skin absorption fraction of chemical from soil (%)
EF	Exposure frequency for water (350 days/year); exposure frequency for soil (soil ingestion = 350 days/year; dermal contact – adult = 100 days/year, child = 350 days/year)
ED	Exposure duration (adult = 24 years; child = 6 years)
ET	Exposure time during showering/bathing (adult = 0.25 h/day; child = 0.14 h/day)
BW	Body weight (adult = 70 kg; child = 15 kg)
AT	Averaging time (period over which exposure is averaged = 70 years or [70 × 365] days)

$Risk_{soil}$

$$= [CDI_o \times SF_o] + [CDI_i \times SF_i]$$

$$= [(CDI_{ing} + CDI_{der}) \times SF_o] + [CDI_i \times SF_i]$$

$$= \{[(INGf \times C_s) + (DEXf \times C_s)] \times SF_o\} + \{[(INHf \times C_a) \times SF_i]\}$$

More generally, the carcinogenic risk may be calculated from 'first principles' as follows:

$\text{Risk}_{\text{soil}}$

$$= \left\{ SF_o \times C_s \times \frac{(SIR_{\text{adult}} \times CF \times FI \times ABS_{\text{gi}} \times EF \times ED_{\text{adult}})}{(BW_{\text{adult}} \times AT \times 365 \text{ day/yr})} \right\}$$

$$+ \left\{ \frac{SF_o \times C_s \times (SIR_{\text{child}} \times CF \times FI \times ABS_{\text{gi}} \times EF \times ED_{\text{child}})}{(BW_{\text{child}} \times AT \times 365 \text{ day/yr})} \right\}$$

$$+ \left\{ SF_o \times C_s \times \frac{(SA_{\text{adult}} \times AF \times CF \times FI \times ABS_{\text{gi}} \times ABS_s \times EF \times ED_{\text{adult}})}{(BW_{\text{adult}} \times AT \times 365 \text{ day/yr})} \right\}$$

$$+ \left\{ \frac{SF_o \times C_s \times (SA_{\text{child}} \times AF \times CF \times FI \times ABS_{\text{gi}} \times ABS_s \times EF \times ED_{\text{child}})}{(BW_{\text{child}} \times AT \times 365 \text{ day/yr})} \right\}$$

$$+ \left\{ \frac{SF_i \times C_a \times (IR_{\text{adult}} \times FI \times ABS_{\text{gi}} \times ED \times ED_{\text{adult}})}{(BW_{\text{adult}} \times AT \times 365 \text{ day/yr})} \right\}$$

$$+ \left\{ \frac{SF_i \times C_a \times (IR_{\text{child}} \times FI \times ABS_{\text{gi}} \times ED \times ED_{\text{child}})}{(BW_{\text{child}} \times AT \times 365 \text{ day/yr})} \right\}$$

As an example, substitution of the exposure assumptions presented in Table 13.3 into the above equation, yields the following reduced equation:

$$\text{Risk}_{\text{soil}} = (SF_o \times C_s \times [1.57 \times 10^{-6}]) + (SF_o \times C_s \times [1.88 \times 10^{-5}] \times ABS_s) + (SF_i \times C_a \times 0.149)$$

Consequently, by substituting the chemical-specific parameters in the reduced risk equation, potential carcinogenic risks associated with the particular constituent are derived.

13.4.5.2 Noncarcinogenic hazard calculations

In accordance with the relationships presented earlier on in this chapter, the potential noncancer risks associated with environmental contaminants present in various matrices may be calculated for all relevant exposure routes – as shown in the demonstration examples given below for childhood exposure from infancy through age six.

- *Noncarcinogenic effects for contaminants in water*. The noncarcinogenic risk associated with a potential receptor exposure to chemical constituents in water can be estimated using the following annotated relationship:

Hazard$_{water}$

$$= \left[CDI_o \times \frac{1}{RfD_o}\right] + \left[CDI_i \times \frac{1}{RfD_i}\right]$$

$$= \left[(CDI_{ing} + CDI_{der}) \times \frac{1}{RfD_o}\right] + \left[CDI_i \times \frac{1}{RfD_i}\right]$$

$$= \left\{\left[(INGf \times C_w) + (DEXf \times C_w)\right] \times \frac{1}{RfD_o}\right\} + \left\{\left[(INHf_i \times C_w) \times \frac{1}{RfD_i}\right]\right\}$$

More generally, the noncarcinogenic risk may be calculated from 'first principles' as follows:

Hazard$_{water}$

$$= \left\{\frac{1}{RfD_o} \times C_w \times \frac{(WIR_{child} \times FI \times ABS_{gi} \times EF \times ED_{child})}{(BW_{child} \times AT \times 365 \text{ day/yr})}\right\}$$

$$+ \left\{\frac{1}{RfD_o} \times C_w \times \frac{(SA_{child} \times K_p \times CF \times FI \times ABS_{gi} \times EF \times ED_{child} \times ET_{child})}{(BW_{child} \times AT \times 365 \text{ day/yr})}\right\}$$

$$+ \left\{\frac{1}{RfD_i} \times C_w \times \frac{(IR_{child} \times FI \times ABS_{gi} \times EF \times ED_{child})}{(BW_{child} \times AT \times 365 \text{ day/yr})}\right\}$$

As an example, substitution of the exposure assumptions presented in Table 13.4 into the above equation, yields the following reduced equation:

$$\text{Hazard}_{water} = \left(\frac{1}{RfD_o} \times C_w \times 0.0639\right) + \left(\frac{1}{RfD_o} \times C_w \times 0.0644 \times K_p\right)$$

$$+ \left(\frac{1}{RfD_i} \times C_w \times 0.0639\right)$$

Consequently, by substituting the chemical-specific parameters in the reduced risk equation, potential noncarcinogenic risks associated with the particular constituent are derived.

As a simple illustrative example calculation of human health noncarcinogenic risk, consider a situation where an aluminum container is used for the storage of water meant for household consumption. Laboratory testing of the water revealed that some aluminum consistently gets dissolved in this drinking water – averaging concentrations of approximately 10 mg/L. The question then is:

Table 13.4 Definitions and exposure assumptions for example hazard computations associated with exposure to environmental contaminants in water and soil

Parameter	Parameter definition and exposure assumption
RfD_o	Oral reference dose (obtained from the literature or Appendix D) ([mg/kg-day])
RfD_i	Inhalation reference dose (obtained from the literature or Appendix D) ([mg/kg-day])
C_w	Chemical concentration in water (obtained from the sampling and/or modeling) (mg/L)
C_s	Chemical concentration in soil (obtained from the sampling and/or modeling) (mg/kg)
C_a	Chemical concentration in air (obtained from the sampling and/or modeling) (mg/m^3)
K_p	Chemical-specific dermal permeability coefficient from water (obtained from the literature, e.g., DTSC, 1994) (cm^2/h)
AF	Soil to skin adherence factor (1 mg/cm^2)
SA	Skin surface area available for water contact (child = 7200 cm^2); Skin surface area exposed/available for soil contact (child = 2000 cm^2)
WIR	Average water intake rate – where intake from inhalation of volatile constituents may be assumed as equivalent to the amount of ingested water (child = 1L/day)
SIR	Average soil ingestion rate (child = 200 mg/day)
IR	Inhalation rate (child = 10 m^3/day)
CF	Conversion factor for water (1L/1000 cm^3); Conversion factor for soil (10^{-6} kg/mg)
FI	Fraction ingested from contaminated source (1)
ABS_{gi}	Bioavailability/gastrointestinal (GI) absorption factor (100%)
ABS_s	Chemical-specific skin absorption fraction of chemical from soil (%)
EF	Exposure frequency (350 days/year)
ED	Exposure duration (child = 6 years)
ET	Exposure time during showering/bathing (child = 0.14 h/day)
BW	Body weight (child = 15 kg)
AT	Averaging time (period over which exposure is averaged = 6 years or [6 × 365] days)

what is the individual noncancer risk for a person who uses this water for drinking water purposes? Assuming the only exposure route of concern is from water ingestion (a reasonable assumption for this situation), and using a noncancer toxicity index (i.e., an RfD) of 1.0 (obtained from Table D.1 in Appendix D), then the noncancer risk attributable to this exposure scenario is calculated to be:

$$\text{Hazard index} = (1/RfD_o) \times CDI_o$$

$$= (1/RfD_o) \times C_w \times 0.0639$$

$$= 1.0 \times 10 \text{ mg/L} \times 0.0639$$

$$= 0.6$$

Similar evaluations can indeed be carried out for the various media and exposure routes.

- *Noncarcinogenic effects for contaminants in soils.* The noncarcinogenic risk associated with a potential receptor exposure to chemical constituents in soils can be estimated using the following annotated relationship:

$$\text{Hazard}_{\text{soil}}$$

$$= \left[CDI_o \times \frac{1}{RfD_o}\right] + \left[CDI_i \times \frac{1}{RfD_i}\right]$$

$$= \left[(CDI_{\text{ing}} + CDI_{\text{der}}) \times \frac{1}{RfD_o}\right] + \left[CDI_i \times \frac{1}{RfD_i}\right]$$

$$= \left\{\left[(INGf \times C_s) + (DEXf \times C_s)\right] \times \frac{1}{RfD_o}\right\} + \left\{\left[(INHf_i \times C_a \times \frac{1}{RfD_i}\right]\right\}$$

More generally, the noncarcinogenic risk may be calculated from 'first principles' as follows:

$$\text{Hazard}_{\text{soil}}$$

$$= \left\{\frac{1}{RfD_o} \times C_s \times \frac{(SIR_{\text{child}} \times CF \times FI \times ABS_{gi} \times EF \times ED_{\text{child}})}{(BW_{\text{child}} \times AT \times 365 \text{ day/yr})}\right\}$$

$$+ \left\{\frac{1}{RfD_o} \times C_s \times \frac{(SA_{\text{child}} \times AF \times CF \times FI \times ABS_{gi} \times ABS_s \times EF \times ED_{\text{child}})}{(BW_{\text{child}} \times AT \times 365 \text{ day/yr})}\right\}$$

$$+ \left\{\frac{1}{RfD_i} \times C_a \times \frac{(IR_{\text{child}} \times FI \times ABS_{gi} \times EF \times ED_{\text{child}})}{(BW_{\text{child}} \times AT \times 365 \text{ day/yr})}\right\}$$

As an example, substitution of the exposure assumptions presented in Table 13.4 into the above equation, yields the following reduced equation:

$$\text{Hazard}_{\text{soil}} = \left(\frac{1}{RfD_o} \times C_s \times [1.28 \times 10^{-5}]\right)$$

$$+ \left(\frac{1}{RfD_o} \times C_s \times [1.28 \times 10^{-4}] \times ABS_s\right)$$

$$+ \left(\frac{1}{RfD_i} \times C_a \times 0.0639 \right)$$

Consequently, by substituting the chemical-specific parameters in the reduced hazard equation, potential noncarcinogenic risks associated with the particular constituent are derived.

13.5 HUMAN HEALTH RISK ASSESSMENT IN PRACTICE

Quantitative human health risk assessment is often an integral part of most environmental management programs that are designed for environmental contamination problems. The basic tasks involved in a comprehensive human health risk assessment typically will consist of the following components (Asante-Duah, 1996):

- Data evaluation
  - Assess the quality of available data.
  - Identify, quantify, and categorize environmental contaminants.
  - Screen and select chemicals of potential concern.
  - Carry out statistical analysis of relevant environmental data.

- Exposure assessment
  - Compile information on the physical setting of the problem location.
  - Identify source areas, significant migration pathways, and potentially impacted or receiving media.
  - Determine the important environmental fate and transport processes for the chemicals of potential concern, including cross-media transfers.
  - Identify populations potentially at risk.
  - Determine likely and significant receptor exposure pathways.
  - Develop representative conceptual model(s) for the problem situation.
  - Develop exposure scenarios (to include both the current and potential future land uses for the locale).
  - Estimate/model exposure point concentrations for the chemicals of potential concern found in the significant environmental media.
  - Compute potential receptor intakes and resultant doses for the chemicals of potential concern (for all potential receptors and significant pathways of concern).

- Toxicity assessment
  - Compile toxicological profiles (to include the intrinsic toxicological properties of the chemicals of potential concern, such as their acute, subchronic, chronic, carcinogenic, and reproductive effects).
  - Determine appropriate toxicity indices (such as the acceptable daily intakes or reference doses, cancer slope or potency factors, etc.).

- Risk characterization
    - Estimate carcinogenic risks from carcinogens.
    - Estimate noncarcinogenic hazard quotients and indices for systemic toxicants.
    - Perform sensitivity analyses, evaluate uncertainties associated with the risk estimates, and summarize the risk information.

Carcinogenic and noncarcinogenic toxicity indices relevant to the estimation of human health risks – represented by the cancer slope factor and reference dose, respectively – are presented in Table D.1 (Appendix D) for selected chemical constituents. A more complete and up-to-date listing may be obtained from a variety of toxicological databases – such as IRIS (developed and maintained by the USEPA).

Some worked examples for the practical application of the human health risk assessment methodology to 'real-world' problems are presented in Chapter 18.

SUGGESTED FURTHER READING

Eschenroeder, A., R.J. Jaeger, J.J. Ospital, and C. Doyle. 1986. Health risk assessment of human exposure to soil amended with sewage sludge contaminated with polychlorinated dibenzodioxins and dibenzofurans. *Veterinary and Human Toxicology*, 28: 356–442.

Goldman, M. 1996. Cancer risk of low-level exposure. *Science*, 271(5257): 1821–1822.

Johnson, D.L., K. McDade, and D. Griffith. 1996. Seasonal variation in paediatric blood lead levels in Syracuse, NY, USA. *Environmental Geochemistry and Health*, 18: 81–88.

Kimmel, C.A. and D.W. Gaylor. 1988. Issues in qualitative and quantitative risk analysis for developmental toxicology. *Risk Analysis*, 8: 15–20.

Krewski, D., C. Brown and D. Murdoch. 1984. Determining safe levels of exposure: safety factors for mathematical models. *Fundam. Appl. Toxicol.*, 4: S383–S394.

Lee, V.M., M.F. Dahab, and I. Bogardi. 1995. Nitrate-risk assessment using fuzzy-set approach. *Journal of Environmental Engineering*, 121(3): 245–256.

OTA (Office of Technology Assessment). 1993. *Researching Health Risks*. OTA, US Congress, US Government Printing Office, Washington, DC.

Tardiff, R.G. and J.V. Rodricks (eds). 1987. *Toxic Substances and Human Risk*. Plenum Press, New York.

USEPA (US Environmental Protection Agency). 1991. *Risk Assessment Guidance for Superfund: Volume I – Human Health Evaluation Manual (Part C, Risk Evaluation of Remedial Alternatives)*. Interim (December 1991). Office of Emergency and Remedial Response, Washington, DC. PB90-155581, OSWER Directive: 9285.7-01C.

Zartarian, V.G. and J.O. Leckie. 1998. Dermal exposure: the missing link. *Environmental Science and Technology*, 32(5): 134A–137A.

REFERENCES

Asante-Duah, D.K. 1996. *Managing Contaminated Sites: Problem Diagnosis and Development of Site Restoration*. J. Wiley, Chichester, UK.

Bogen, K.T. 1994. A note on compounded conservatism. *Risk Analysis*, 14: 379–381.

Calabrese, E.J., et al. 1989. How much soil do young children ingest: an epidemiologic study. *Regulatory Toxicology and Pharmacology*, 10: 123–137.

CAPCOA (California Air Pollution Control Officers Association). 1990. *Air Toxics 'Hot Spots' Program. Risk Assessment Guidelines.* CAPCOA, Los Angeles, CA.
CDHS (California Department of Health Services). 1986. *The California Site Mitigation Decision Tree Manual.* CDHS, Toxic Substances Control Division, Sacramento, CA.
Chen, J.J., D.W. Gaylor, and R.L. Kodell. 1990. Estimation of the joint risk from multiple-compound exposure based on single-compound experiments. *Risk Analysis,* 10: 285–290.
Cohrssen, J.J. and V.T. Covello. 1989. *Risk Analysis: A Guide to Principles and Methods for Analyzing Health and Environmental Risks.* National Technical Information Service, US Dept. of Commerce, Springfield, VA.
DOE (US Department of Energy). 1987. *The Remedial Action Priority System (RAPS): Mathematical Formulations.* US Dept. of Energy, Office of Environment, Safety and Health, Washington, DC.
Dourson, M.L. and J.F. Stara. 1983. Regulatory history and experimental support of uncertainty (safety) factors. *Regulatory Toxicology and Pharmacology,* 3: 224–238.
DTSC (Department of Toxic Substances Control). 1994. *Preliminary Endangerment Assessment Guidance Manual (A Guidance Manual for Evaluating Hazardous Substance Release Sites).* California Environmental Protection Agency, DTSC, Sacramento, CA.
Gaylor, D.W. and J.J. Chen. 1996. A simple upper limit for the sum of the risks of the components in a mixture. *Risk Analysis,* 16(3): 395–398.
Hawley, J.K. 1985. Assessment of health risk from exposure to contaminated soil. *Risk Analysis,* 5(4): 289–302.
Hoddinott, K.B. (ed.). 1992. *Superfund Risk Assessment in Soil Contamination Studies.* American Society for Testing and Materials, ASTM Publication STP 1158, Philadelphia, PA.
Huckle, K.R. 1991. *Risk Assessment – Regulatory Need or Nightmare?* Shell Publications, Shell Centre, London, UK.
Kodell, R.L. and J.J. Chen. 1994. Reducing conservatism in risk estimation for mixtures of carcinogens. *Risk Analysis,* 14: 327–332.
Lepow, M.L., M. Bruckman, L. Robino, S. Markowitz, M. Gillette, and J. Kapish, 1974. Role of airborne lead in increased body burden of lead in Hartford children. *Environmental Health Perspectives,* 6: 99–101.
Lepow, M.L., L. Bruckman, M. Gillette, S. Markowitz, R. Robino, and J. Kapish, 1975. Investigations into sources of lead in the environment of urban children. *Environmental Research,* 10: 415–426.
McKone, T.E. 1987. Human exposure to volatile organic compounds in household tap water: the indoor inhalation pathway. *Environmental Science and Technology,* 21: 1194.
McKone, T.E. 1989. Household exposure models. *Toxicology Letters,* 49: 321–339.
McKone, T.E. and J.I. Daniels. 1991. Estimating human exposure through multiple pathways from air, water, and soil. *Regulatory Toxicology and Pharmacology,* 13: 36–61.
NRC (National Research Council). 1983. *Risk Assessment in the Federal Government: Managing the Process.* NAS Press, Washington, DC.
NRC. 1991a. *Frontiers in Assessing Human Exposure to Environmental Toxicants.* NAS Press, Washington, DC.
NRC. 1991b. *Human Exposure Assessment for Airborne Pollutants: Advances and Opportunities.* NAS Press, Washington, DC.
OSA (Office of Scientific Affairs). 1992. *Supplemental Guidance for Human Health Multimedia Risk Assessments of Hazardous Waste Sites and Permitted Facilities.* Cal EPA, DTSC, Sacramento, CA.
Patton, D.E. 1993. The ABCs of risk assessment. *EPA Journal,* 19: 10–15.
Paustenbach, D.J. (ed.). 1988. *The Risk Assessment of Environmental Hazards: A Textbook of Case Studies.* J. Wiley, New York.
Ricci, P.F. (ed.). 1985. *Principles of Health Risk Assessment.* Prentice-Hall, Englewood Cliffs, NJ.
Ricci, P.F. and M.D. Rowe (eds). 1985. *Health and Environmental Risk Assessment.* Pergamon Press, New York.

Sedman, R. 1989. The development of applied action levels for soil contact: a scenario for the exposure of humans to soil in a residential setting. *Environmental Health Perspectives*, 79: 291–313.

Slob, W. 1994. Uncertainty analysis in multiplicative models. *Risk Analysis*, 14: 571–576.

USEPA (US Environmental Protection Agency). 1984a. *Approaches to Risk Assessment for Multiple Chemical Exposures*. Environmental Criteria and Assessment Office, Cincinnati, OH. EPA/600/9-84/008.

USEPA. 1984b. Proposed guidelines for carcinogen, mutagenicity, and developmental toxicant risk assessment. *Federal Register*, 49: 46294–46331.

USEPA. 1985a. *Principles of Risk Assessment: A Nontechnical Review*. Office of Policy Analysis, Washington, DC.

USEPA. 1985b. *Toxicology Handbook*. Office of Waste Programs Enforcement, Washington, DC.

USEPA. 1986a. Guidelines for carcinogen risk assessment. *Federal Register*, 51(185): 33992–34003, CFR 2984, September 24, 1986.

USEPA. 1986b. Guidelines for the health risk assessment of chemical mixtures. *Federal Register*, 51(185): 34014–34025, CFR 2984, September 24, 1986.

USEPA. 1986c. *Methods for Assessing Environmental Pathways of Food Contamination: Methods for Assessing Exposure to Chemical Substances*, Vol. 8. Exposure Evaluation Division, Office of Toxic Substances. EPA/560/5-85/008.

USEPA. 1986d. *Superfund Public Health Evaluation Manual*. Office of Emergency and Remedial Response, Washington, DC. EPA/540/1-86/060.

USEPA. 1987a. *Handbook for Conducting Endangerment Assessments*. USEPA, Research Triangle Park, NC.

USEPA. 1987b. *RCRA Facility Investigation (RFI) Guidance*. Washington, DC. EPA/530/SW-87/001.

USEPA. 1988. *Superfund Exposure Assessment Manual*. Office of Remedial Response, Washington, DC. EPA/540/1-88/001, OSWER Directive 9285.5-1.

USEPA. 1989a. *Exposure Factors Handbook*. Office of Health and Environmental Assessment, Washington, DC. EPA/600/8-89/043.

USEPA. 1989b. *Interim Methods for Development of Inhalation Reference Doses*. Office of Health and Environmental Assessment, Washington, DC. EPA/600/8-88/066F.

USEPA. 1989c. *Interim Procedures for Estimating Risks Associated with Exposures to Mixtures of Chlorinated Dibenzo-p-dioxins and -Dibenzofurans (CDDs and CDFs)*. Risk Assessment Forum, Washington, DC. EPA/625/3-89/016.

USEPA. 1989d. *Risk Assessment Guidance for Superfund. Volume I – Human Health Evaluation Manual (Part A)*. Office of Emergency and Remedial Response, Washington, DC. EPA/540/1-89/002.

USEPA. 1991. *Risk Assessment Guidance for Superfund, Volume I: Human Health Evaluation Manual. Supplemental Guidance. 'Standard Default Exposure Factors'*. (Interim Final). March, 1991. Office of Emergency and Remedial Response, Washington, DC. OSWER Directive 9285.6-03.

USEPA. 1992. *Dermal Exposure Assessment: Principles and Applications*. Office of Research and Development, Washington, DC. EPA/600/8-91/011B.

USEPA. 1996. *Soil Screening Guidance: Technical Background Document*. Office of Solid Waste and Emergency Response, Washington, DC. EPA/540/R-95/128.

Van Leeuwen, C.J. and J.L.M. Hermens (eds). 1995. *Risk Assessment of Chemicals: An Introduction*. Kluwer Academic Publishers, Dordrecht, The Netherlands.

Chapter Fourteen

Ecological Risk Assessments

An ecological risk assessment (ERA) is defined as a process that evaluates the likelihood that adverse ecological effects may occur, or are occurring as a result of exposure to one or more environmental stressors – usually imposed by human activities (Bartell et al., 1992; Linthurst et al., 1995; NRC, 1983; Richardson, 1995; Solomon, 1996; USEPA, 1986, 1988b, 1992, 1994, 1995). ERAs typically evaluate the ecological effects resulting from such human activities as the release of chemicals into the environment, habitat destruction – including wetland destruction, etc.

A typical ERA involves the qualitative and/or quantitative appraisal of the actual or potential effects of environmental contamination problems on plants and animals other than humans and domesticated species (USEPA, 1989b, 1990). It is primarily concerned with the adverse effects of risk agents on populations of particular animal, plant, or microbial species and on the structure and function of ecosystems (Cohrssen and Covello, 1989).

In general, ecological risk assessment techniques can be employed to better develop responsible environmental management programs. The scope of applications for some common applications and uses of ecological risk assessment are identified in Chapter 20. The present chapter elaborates the major components of the ecological risk assessment methodology, as may be applied to the evaluation of an environmental contamination problem.

14.1 THE ECOLOGICAL RISK ASSESSMENT METHODOLOGY

Figure 14.1 illustrates the fundamental elements of an ERA – in similar terms to the human health risk assessment presented earlier on in Chapter 13 – whose individual components (Box 14.1) are discussed in subsequent sections below. The basic components of an ERA program will generally consist of several tasks (Box 14.2), normally conducted in phases to ensure program cost-effectiveness. Where it is deemed necessary, a more detailed assessment that comprises a biological diversity analysis and population studies may become part of the overall ERA process.

In general, the uniqueness of ecological systems makes it difficult to comprehensively prescribe and standardize procedures for ecological risk assessments (Kolluru et al., 1996). This uniqueness does indeed argue in support of flexibility in implementing any standard sets of protocols for ecological assessment programs. Although several general considerations familiar to many risk assessment procedures may apply similarly to ERAs, it must be recognized that, in assessing

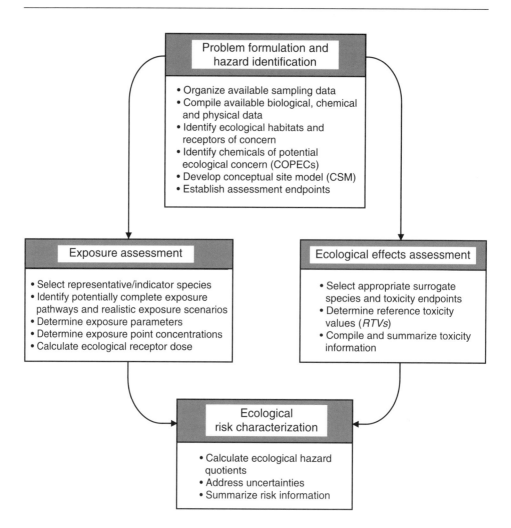

Figure 14.1 The ecological risk assessment process

ecosystem risks, the process generally becomes more complex. For the most part, this added complexity may be attributed to the fact that toxicity effects assessment (which is a very important component of the overall risk assessment process) usually must be determined for several different families of organisms; in addition, the risk assessment must also consider the role of the various organisms in the ecosystem structure and function.

Overall, the process used to evaluate environmental or ecological risks parallels that used in the evaluation of human health risks (as discussed in the preceding chapter). In both cases, potential risks are determined by the integration of information on chemical exposures with toxicological data for the contaminants of potential concern. Unlike endangerment assessment for human populations, however, ERAs often lack significant amounts of critical and credible data

Ecological risk assessments

Box 14.1 Key fundamental elements of an ecological risk assessment

Evaluation of site characteristics:

- Nature and extent of contaminated area
- Sensitive environments

Contaminant evaluation:

- Identification and characterization
- Biological and environmental concentrations
- Toxicity of contaminants
- Potential benchmark/reference criteria

Potential for exposure:

- Actual or potential sources of contaminant release
- Media to which contaminants can be or are being released
- Organisms that can come into contact with the contaminants
- Environmental conditions under which transport and/or exposure may be taking place

Selection of assessment and measurement endpoints:

- Ecological endpoints
- Evaluation of potentially affected habitats
- Evaluation of potentially affected populations

Ecological risk characterization:

- Qualitative risk characterization
- Quantitative risk characterization

Box 14.2 General basic tasks in an ecological risk assessment program

- Compilation of relevant site data for the study area
- Determination of background (ambient) concentrations of the site contaminants in the study area
- Identification of the contaminants of potential ecological concern
- Identification and location of habitats and environments at the study area and its vicinity
- Selection of indicator species or habitats
- Ecotoxicity assessment and/or bioassay of selected indicator species
- Development of an appropriate conceptual site model (CSM) and/or foodchain diagram
- Establishment of appropriate assessment endpoints for all chemicals of potential ecological concern
- Characterization of exposure based on environmental fate and transport of the chemicals of potential ecological concern, as predicted using the CSM and ecological foodchain considerations
- Development of ecological risk characterization parameters for the indicator species/habitats

necessary for a comprehensive quantitative evaluation. Nonetheless, the pertinent data requirements should be identified and categorized insofar as practicable, so that reliable risk estimates can be generated.

14.1.1 Ecological Hazard Evaluation

The design of an ERA program for an environmental contamination problem typically involves a process to clearly define the common elements of populations, communities, and ecosystems – which then forms a basis for the development of a logical framework that can be used to characterize risks. Invariably, the development of an ERA requires the identification of one or several ecological assessment endpoints. These endpoints define the environmental resources which are to be protected, and which, if found to be impacted, determine the need for corrective actions. The selection of appropriate site-specific assessment endpoints is therefore crucial to the development of a cost-effective ecological characterization and/or environmental management program that is protective of potential ecological receptors.

Ultimately, a contaminant entering the environment will cause adverse effects if, and only if, it exists in a form and concentration sufficient to cause harm; it comes in contact with organisms or environmental media with which it can interact; and the ensuing interaction is detrimental to life functions.

14.1.1.1 Identification of the nature of ecosystems

The different types of ecosystems have unique combinations of physical, chemical, and biological characteristics, and thus may respond to contamination in their own unique ways. The physical and chemical structure of an ecosystem may determine how contaminants affect its resident species, and the biological interactions may determine where and how the contaminants move in the environment and which species are exposed to particular concentrations. It is therefore imperative to clearly identify the types of ecosystem(s) associated with the particular environmental contamination problem.

Ecosystems may be classified into two broad categories: terrestrial and aquatic ecosystems, with wetlands serving as a zone of transition between terrestrial and aquatic environments. The following specific ecosystems will normally be investigated in an ERA:

- Terrestrial ecosystems (to be categorized according to the vegetation types that dominate the plant community and terrestrial animals).
- Wetlands (which are areas in which topography and hydrology create a zone of transition between terrestrial and aquatic environments).
- Freshwater ecosystems (in which environment, the dynamics of water temperature, and movement of water can significantly affect the availability and toxicity of contaminants).
- Marine ecosystems (which are of primary importance because of their vast size and critical ecological functions).

- Estuaries (which support a multitude of diverse communities, are more productive than their marine or freshwater sources, and are important breeding grounds for numerous fish, shellfish, and bird species).

The types of ecosystems vary with climatic, topographical, geological, chemical, and biotic factors. Each ecosystem type has unique combinations of physical, chemical, and biological characteristics, and thus may respond to environmental contamination and stress in its own unique way.

The ecosystem types pertaining to a case-specific study should be defined and integrated into the overall ERA. In fact, a wide variety of other possible measures of community structure can be employed in ERA programs.

14.1.1.2 Evaluation of ecological habitats and community structure

Different evaluation strategies are generally employed in ERAs, depending on the level of refinement required to define the conditions within an ecological community. For instance, an evaluation of the condition of aquatic communities may proceed from two directions – as discussed below.

- *Examination of lower trophic levels.* The first direction could consist of examining the structure of the lower trophic levels as an indication of the overall health of the aquatic ecosystem. This approach emphasizes the base of the aquatic foodchain, and may involve studies of plankton (microscopic flora and fauna), periphyton (including bacteria, yeast, molds, algae, and protozoa), macrophyton (aquatic plants), and benthic macroinvertebrates (e.g., insects, annelid worms, mollusks, flatworms, roundworms, and crustaceans). Benthic macroinvertebrates are commonly used in studies of aquatic communities. These organisms usually occupy a position near the base of the foodchain. Just as importantly, however, their range within the aquatic environment is restricted, so that their community structure may be referenced to a particular stream reach or portion of lake substrate. By contrast, fish are generally mobile within the aquatic environment, and evidence of stress or contaminant load may not be amenable to interpretation with reference to specific releases. The presence or absence of particular benthic macroinvertebrate species, sometimes referred to as 'indicator species', may provide evidence of a response to environmental stress. A 'species diversity index' provides a quantitative measure of the degree of stress within the aquatic community; this is an example of the common basis for interpreting the results of studies pertaining to aquatic biological communities. Measures of species diversity are most useful for comparison of streams with similar hydrologic characteristics or for the analysis of trends over time within a single stream (USEPA, 1989b).
- *'Focused' examination of select species.* The second approach to evaluating the condition of an aquatic community could focus on a particular group or species, possibly because of its commercial or recreational importance or because a substantial historic database already exists. This is done through selective sampling of specific organisms, most commonly fish, and evaluation of standard 'condition factors' (e.g., length, weight, girth). In many cases, receiving water

bodies are recreational fisheries, monitored by some government regulatory agencies or similar entity. In such cases, it is common to find some historical record of the condition of the fish population, and it may be possible to correlate contaminant release records with alterations in the status of the fish population.

Additional detail regarding the application of other measures of community structure can be found in the literature of ecological assessments and related subjects (e.g., Barnthouse and Suter, 1986; Carlsen, 1996; NRC, 1989; USEPA, 1973).

In general, the different levels of an ecological community are studied to determine whether they exhibit any evidence of stress. If the community appears to have been disturbed, the goal will be to characterize the source(s) of the stress and, specifically, to focus on the degree to which the release of environmental contaminants has caused the disturbance or possibly exacerbated an existing problem.

14.1.1.3 Selecting assessment endpoints

Assessment endpoints are explicit expressions of actual environmental value that is to be protected (Suter, 1993; USEPA, 1992). The principal criteria used in the selection of assessment endpoints include: their ecological relevance, their susceptibility to the stressor, and whether they represent management goals (to include a representation of societal values). Ecologically relevant endpoints reflect important characteristics of the system and are functionally related to other endpoints (USEPA, 1992). These are endpoints that help sustain the natural structure and function of an ecosystem.

Ecological resources are considered susceptible when they are sensitive to a human-induced stressor to which they are exposed. Delayed effects and multiple stressor exposures add complexity to evaluations of susceptibility. Conceptual models need to reflect these factors. If a species is unlikely to be exposed to the stressor of concern, it is inappropriate as an assessment endpoint.

Although assessment endpoints must be defined in terms of measurable attributes, selection is not dependent on the ability to measure those attributes directly or on whether methods, models, and data are currently available. If the response of an assessment endpoint cannot be directly measured, it may be predicted from measures of responses by surrogate or similar entities (Suter, 1993; USEPA, 1992). Measures that will be used to evaluate assessment endpoint response to exposures for the risk assessment are often identified during conceptual model development and specified in the analysis plan.

14.1.1.4 Selection of indicator species

It generally is not feasible to evaluate every species that may be present at a locale that is affected by an environmental contamination problem. Consequently, selected target or indicator species will normally be chosen in an ERA study. Then, by using reasonably conservative assumptions in the overall assessment, it is rationalized that adequate protection of selected indicator species will provide

Ecological risk assessments 305

Box 14.3 Guiding criteria for the selection of ERA target species

- Species that are threatened, endangered, rare, or of special concern
- Species that are valuable for several purposes of interest to human populations (i.e., of economic and societal values)
- Species critical to the structure and function of the particular ecosystem which they inhabit
- Species that serve as indicators of important changes in the ecosystem
- Relevance of species at the site and its vicinity

---

protection for all other significant environmental species as well. That said, it is noteworthy, that not every organism may be suitable for use as indicator species in the evaluation of contaminant impacts on ecological systems. Thus, several general considerations and specific criteria should be used to guide the selection of target species in an ERA (Box 14.3) (USEPA, 1989a, 1989b, 1990).

In general, it is important to consider the effects of environmental contaminants on both an endangered population as well as on the habitats critical to its survival. Consequently, the presence of threatened or endangered species, and/or habitats critical to their survival should be documented, and the location of such species determined. Similarly, sensitive sport or commercial species and habitats, essential for their reproduction and survival, should be identified. Information on these may be obtained from appropriate national, federal, provincial, state, regional, local, and/or private institutions and other organizations.

14.1.1.5 Screening for chemicals of potential ecological concern

A very important early step in the assessment of ecological risks associated with environmental contamination problems is the screening of chemicals detected at the impacted locale, in order to identify the specific constituents that do indeed represent a potential risk. Part of this screening process usually will involve a comparison of measured contaminant concentrations to 'benchmark' values (e.g., a national ambient water quality criteria [AWQC] for protection of aquatic life, a sediment quality criteria [SQC], etc.); the 'benchmark' values generally represent constituent concentrations that are regarded to be nontoxic. In fact, the most appropriate screening strategy is to use multiple benchmark values along with background threshold concentrations, knowledge of contaminant composition and its nature, and physico-chemical properties to identify the chemicals of potential ecological concern (COPECs) (Suter, 1996).

Several alternative approaches for calculating ecotoxicological screening benchmarks that will facilitate the COPEC selection process are presented in the literature elsewhere (e.g., Ankley et al., 1996; Suter, 1996; USEPA, 1988a; van Leeuwen, 1990). Naturally, the relative utility of any given benchmark depends on the reliability of the source information on which such a value is based or from which it is derived. Indeed, the choice of method for calculating benchmarks can significantly influence their sensitivity and utility – and thus the need for careful evaluation of the alternative methods of choice in the derivation of such benchmark/reference values.

> **Box 14.4   Key factors influencing the ecological effects of environmental contaminants**
>
> Nature of contamination
>
> - Chemical category
> - Physical and chemical properties
> - Frequency of release
> - Toxicity
>
> Physical/chemical characteristics of the environment
>
> - Temperature, pH, salinity, hardness, etc.
> - Soil composition, etc.
>
> Biological factors
>
> - Susceptibility of species
> - Characteristics governing population abundance and distribution
> - Temporal variability in communities
> - Movement of chemicals in foodchains

14.1.1.6   Identification of the nature and ecological effects of environmental contaminants

Although a contaminant may cause illness and/or death to individual organisms, its effects on the structure and function of ecological assemblages or interlinkages may be measured in terms quite different from those used to describe individual effects. Consequently, an ecological hazard evaluation should include a wider spectrum of ecological effects on individual organisms as well as the ecological interlinkages. Furthermore, the biological, chemical, and environmental factors perceived to influence the ecological effects of contaminants should be identified and succinctly described. In general, a variety of environmental variables (Box 14.4) can significantly influence the nature and extent of the effects of a contaminant on ecological receptors, and may therefore determine the degree of impacts that contaminants exert on ecological systems.

Environmental effects generally include changes in aquatic and terrestrial natural resources brought about by exposure to chemical substances. The introduction of contaminants into an ecosystem can indeed cause direct and/or indirect harm to organisms. Typical major consequences from ecosystem exposures to contaminants may include the specific effects identified below (Calmano and Forstner, 1996; NRC, 1981; Pickering and Henderson, 1966; USEPA, 1989b).

- *Reduction in population size.* Contaminants can cause reductions in populations of organisms through numerous mechanisms affecting species births, mortalities, and migratory tendencies.
- *Changes in community structure.* Contaminants introduced into ecosystems may create opportunities for unanticipated and unpredictable changes in community with respect to species composition and relative abundance. Because most

environmental contaminants of concern exhibit toxic effects, they often reduce the number and kinds of species that can survive in the habitat. This may then result in a community dominated by large numbers of a few species that are tolerant of the contaminant, or a community in which no species predominate but most of the component populations contain fewer organisms. In fact, a contaminant need not be directly toxic to affect community structure.
- *Changes in ecosystem structure and function.* As contaminants modify the species composition and relative abundance of populations in a community, the often complex patterns of matter and energy flow within the ecosystem may also change. If certain key species are reduced or eliminated, this may interrupt the flow of energy and nutrients to other species not directly experiencing a toxic effect.

Knowledge of such environmental effects is important in analyzing potential risks from chemical releases, migration pathways, and potential receptor exposures. Based on the nature of exposures to both human and nonhuman receptors, responsible environmental management programs can be developed to address the culprit environmental contamination problem.

14.1.2 Exposure Assessment: The Characterization of Ecological Exposures

The objectives of the exposure assessment are to define contaminant behaviors; identify potential ecological receptors; determine exposure routes by which contaminants may reach ecological receptors; and estimate the degree of contact and/or intakes of the COPECs by the potential receptors.

The exposure characterization process describes the contact or co-occurrence of stressors with ecological receptors. The characterization is based on measures of exposure and of ecosystem and receptor characteristics. These measures are used to analyze sources, the distribution of the stressor in the environment, and the extent and pattern of contact or co-occurrence.

In general, exposure is analyzed by describing the sources and releases, the distribution of the stressor or contaminant in the environment, and the extent and pattern of contact or co-occurrence.

14.1.2.1 Development of conceptual models

Conceptual site models (CSMs) for ERAs are developed based on information about stressors, potential exposures, and predicted effects on an ecological entity (the assessment endpoint). The complexity of the CSM depends on the complexity of the problem, number of stressors, number of assessment endpoints, nature of effects, and characteristics of the ecosystem. The general design process/protocol for developing CSMs that was discussed previously in Chapter 10 will normally be employed to complete the appropriate CSM diagram.

In a representative situation, a chemical may be released into the environment, and is then subject to physical dispersal into the air, water, soils, and/or sediments.

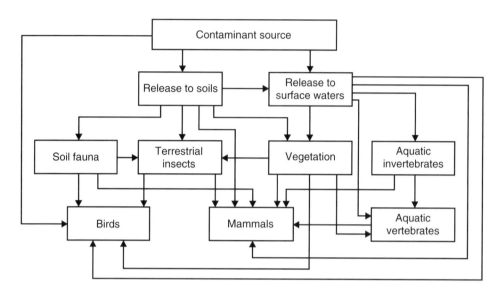

Figure 14.2 A simplified schematic of a foodchain diagram showing contaminant migration pathways within an ecosystem

The chemical may then be transported spatially and into the biota and perhaps be chemically or otherwise modified or transformed and degraded by abiotic processes (such as photolysis, hydrolysis, etc.) and/or by microorganisms present in the environment. The resulting transformation products may have different environmental behavior patterns and toxicological properties from the parent chemical. Nonetheless, it is the nature of exposure scenarios that determine the potential for any adverse impacts.

The exposure analysis process evaluates the interaction of identified environmental stressors with the ecological component. In general, the nature of exposure scenarios determines the potential for any adverse impacts. Hence, a foodchain (also called foodweb) – that gives a simplified and generic conceptual representation of typical interlinkages resulting from the consumption, uptake, and absorption processes associated with an ecological community – is normally constructed for the target species, to facilitate the development of realistic exposure scenarios (Figure 14.2). Subsequently, the relevant exposure routes are selected based on the behavior patterns and/or ecological niches of the target species and communities (Figure 14.3). This means that the nature of the target organisms (e.g., birds, fish, etc.) must be identified, together with the nature of exposure (such as acute, chronic or intermittent), as part of the ERA process.

14.1.2.2 Calculation of chemical intakes by potential ecological receptors

The amount of a target species exposure to environmental contamination is based on the maximum plausible exposure concentrations of the chemicals in the affected environmental matrices. The total daily exposure (in mg/kg-day) of target species

Ecological risk assessments 309

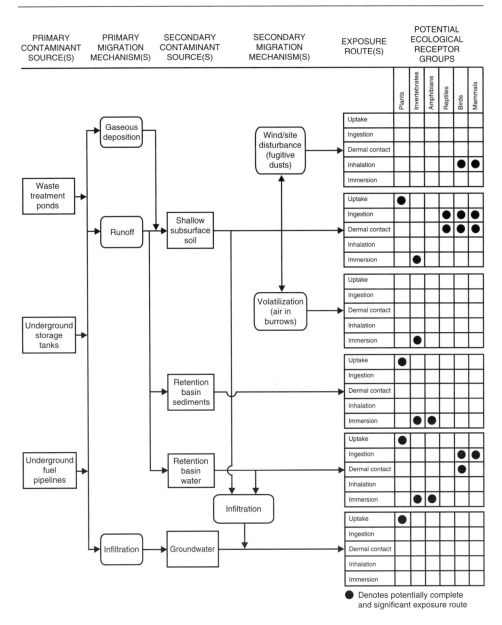

Figure 14.3 A simplified conceptual model diagram for an environmental contamination problem associated with an ecological system

can be calculated by summing the amounts of constituents ingested and absorbed from all sources (e.g., soil, vegetation, surface water, fish tissue, and other target species), and also that absorbed through inhalation and dermal contacts.

Analytical procedures used to estimate the receptor exposures to chemicals in various contaminated media (such as a wildlife or a game species' daily chemical

---

> **Box 14.5** Estimation of wildlife or game exposure to environmental chemicals
>
> $$E = C \times F \times \left[\sum_i D_{di} D_{ci} D_{ui}\right] \times BAF \times [1/BW]$$
>
> where:
> E = Exposure or average daily dose (μg/kg-day)
> C = Chemical concentration in media (e.g., soil) averaged over appropriate exposure period (70-year if to be subsequently consumed by humans, etc.)
> F = Food consumption rate (g/day)
> D_{di} = Component i of diet
> D_{ci} = Consumption factor of i
> D_{ui} = Bioaccumulation or uptake coefficient for i
> BAF = Bioavailability factor
> BW = Body weight (kg)

exposure and the resulting body burden) are similar to those discussed under human health risk assessment in Chapter 13. For example, wildlife or game daily chemical exposure may be estimated by applying the algorithm shown in Box 14.5. As an illustration of the practical application of this algorithm, it is recognized that a deer's daily exposure to a chemical (e.g., dioxin) in an ecological set-up (together with the resulting body burden) will usually need to be estimated before a human consumer's (e.g., a hunter's) oral exposure from eating deer meat can be estimated. In this case, the deer's average daily uptake of dioxins will be given by:

$$E = C \times F \times [G_d\ G_c\ G_u + I_d\ I_c\ I_b] \times BAF \times (1/BW)$$

where G_d is the grass component of diet; G_c is the grass consumption factor; G_u is the soil chemical uptake coefficient for grass; I_d is the rodent component of diet; I_c is the rodent consumption factor; and I_b is the rodent bioavailability factor. Further details on the relevant models and algorithms can be found elsewhere in the literature of endangerment assessment (e.g., CAPCOA, 1990; Paustenbach, 1988; Sutter, 1993; USEPA, 1993).

14.1.3 Ecotoxicity Assessment: The Characterization of Ecological Effects

The ecotoxicological assessment phase of an ERA comprises of a determination of the ecological response to potential environmental stressors. Characterization of ecological effects describes the effects that are elicited by a stressor, links these effects with the assessment endpoints, and evaluates how the effects change with varying stressor levels. The conclusions of the ecological effects characterization are summarized in a stressor–response profile (similar to the dose–response curve in a human health risk assessment) – and also as noted previously in Chapter 11.

During ecological response analysis, the relationship between the stressor and the magnitude of ecological effects involved is quantified, and cause-and-effect relationships are evaluated. In addition, extrapolations will generally be made from measurement endpoints to assessment endpoints – resulting in the generation of a stressor–response profile which becomes an input to the risk characterization (Zehnder, 1995).

Similar to the human health endangerment assessment (discussed in Chapter 13), the scientific literature is reviewed to obtain ecotoxicity information for the chemicals of potential ecological concern that are associated with the environmental contamination problem. Additional data may have to be developed via field sampling and analysis and/or bioassays. Subsequently, critical toxicity values for the contaminants of concern are derived for the target ecological receptor species and ecological communities of concern, to be used in characterizing risks associated with the locale.

In general, evaluating the ecotoxicity of a particular substance requires careful specification of the endpoints of concern – which entails describing the organism tested or observed; the nature of the effects; the concentration or dose needed to produce the effect; the duration of exposure needed to produce the effect; and the environmental conditions under which the effects were observed (Calow, 1993; MacCarthy and Mackay, 1993; USEPA, 1989b).

14.1.4 Ecological Risk Characterization

During risk characterization, the likelihood of adverse effects occurring as a result of exposure to an environmental stressor is evaluated. Ecological risk characterization consists of steps similar to that discussed under the human endangerment assessment (Chapter 13). That is, the doses determined for the ecological receptors and community during the exposure assessment are integrated with the appropriate toxicity values and information derived in the ecotoxicity assessment, in order to arrive at plausible ecological risk estimates. This entails both temporal and spatial components, requiring an evaluation of the probability or likelihood of an adverse effect occurring; the degree of permanence and/or reversibility of each effect; the magnitude of each effect; and receptor populations or habitats that will be affected.

Typically, a quantitative ecological risk characterization is accomplished by using the ecological risk quotient (ErQ) method – similar to the hazard quotient employed in a human health risk characterization. In the ErQ approach, the exposure point concentration or estimated daily dose is compared to a benchmark critical toxicity parameter (e.g., a national ambient water quality criterion), as follows:

$$\text{Ecological risk quotient } (ErQ) = \frac{\begin{bmatrix} \text{exposure point concentration} \\ or \text{ estimated daily dose} \end{bmatrix}}{\begin{bmatrix} \text{benchmark critical ecotoxicity} \\ \text{parameter; } or \text{ a surrogate} \end{bmatrix}}$$

The denominator represents the concentration that produces an assessment endpoint (e.g., toxic effects) in target species. Where data for specific species or

> Box 14.6 Major sources of uncertainty and variability in ERAs
>
> - Uncertainty due to natural complexity of ecosystems – i.e., uncertainty associated with response of ecosystems (or their components) to anthropogenic stress – which may involve numerous factors
> - Uncertainty due to stochasticity of ecosystem measures
> - Source term uncertainty – i.e., associated with estimates of the rate and the spatial and temporal pattern of release of a chemical from a source or set of sources
> - Uncertainty in extrapolation models – i.e., extrapolating affects data (including, extrapolation of toxicity data from one species to another or from low dose to high dose for same species, etc.)
> - Uncertainty concerning magnitude of effects
> - Uncertainty associated with different test endpoints
> - Environmental variability in time and space
> - Variations in sensitivity among individuals and life stages, and between species
> - Stochastic birth and death processes

endpoints are unavailable, other toxicity data (e.g., LC_{50} or *LOEL*) may be used to derive a surrogate parameter. When *NOEL*-type values are used as surrogate, however, the *ErQ* will have a significantly more conservative meaning as an indicator of risk.

The *ErQ* estimates the risk of an environmental contaminant to an indicator species, independent of the interactions among species or between different chemicals of potential ecological concern. In general, if *ErQ* ≤1, then an acceptable risk is indicated. Conversely, an *ErQ* >1 calls for action or further refined investigations, due to the possibility of unacceptable levels of risk to potential ecological receptors.

14.1.4.1 Uncertainty analysis

Three common qualitatively distinct sources of uncertainty to evaluate in all risk assessments are the inherent variability, parameter uncertainty, and model errors. Invariably, all ecological risk estimates rely on numerous assumptions and consideration of the many uncertainties that are inherent in the ERA process (Box 14.6). In particular, it is noteworthy that virtually all risk assessments have data gaps that must be addressed, but it is not always possible to obtain more information – especially due to lack of time and/or monetary resources, or a practical means to acquire more data. Under such circumstances, extrapolations (e.g., between taxa, between responses, from laboratory to field data, between geographical areas, between spatial scales, and/or between temporal scales) may be the only way to bridge gaps in available data, in order to link measures of effect with assessment endpoints. Consequently, it is important to address the level of confidence or the degree of uncertainty associated with the estimated risk attributable to an environmental contamination problem.

Uncertainties are typically associated with both toxicity information (such as ecotoxicity values and site-specific dose–response assessments) and exposure

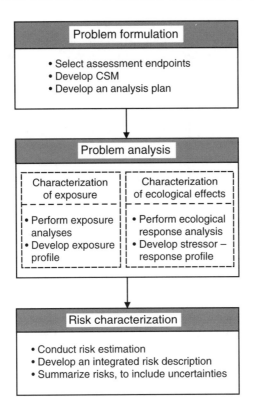

Figure 14.4 A generalized conceptual framework for ecological risk assessment

assessment information. Consequently, factors that may significantly increase the uncertainty of the ERA should be identified and addressed in at least a qualitative and, where possible, quantitative manner.

14.2 A GENERAL FRAMEWORK FOR ECOLOGICAL RISK ASSESSMENTS

An ERA consists of a process for organizing and analyzing data, information, assumptions, and uncertainties in order to evaluate the likelihood of adverse ecological effects from an environmental contamination problem. A general framework for the conduct of an ERA will typically consist of the three primary phases – *problem formulation*, *problem analysis*, and *risk characterization* – depicted in Figure 14.4 (USEPA, 1992).

The ERA process is indeed iterative by nature. For instance, it may take more than one pass through problem formulation to complete planning for the risk assessment, or information gathered in the analysis phase may suggest further problem formulation activities such as modification of the previously selected endpoints. To maximize efficient use of the often limited resources, therefore, ERAs are frequently designed in sequential tiers that proceed from simple, relatively inexpensive evaluations to more costly and complex assessments. Ultimately, the value of the risk assessment depends on whether it is used to make quality environmental management decisions.

14.2.1 The Problem Formulation Phase

Problem formulation is a formal process for generating and evaluating preliminary hypotheses about why ecological effects have occurred, or may occur, from ensuing human activities. It involves delimiting goals and assessment endpoints, preparing the conceptual model, and developing an analysis plan.

Successful completion of problem formulation depends on the quality of several investigatory elements – especially relating to assessment endpoints that adequately reflect management goals and the ecosystem they represent, and conceptual models that describe key relationships between a stressor and assessment endpoint or among several stressors and assessment endpoints (USEPA, 1992).

Assessment endpoints are critical to problem formulation because they link the risk assessment to management concerns, and they are central to conceptual model development. A conceptual model as employed in problem formulation is a written description and visual representation of predicted responses by ecological entities to stressors to which they are exposed; the model includes ecosystem processes that influence these responses.

The design of an analysis plan is the final stage of the problem formulation. It includes the most important pathways and relationships identified during problem formulation that will be pursued in the problem analysis phase of the ERA process. Several criteria – including the availability of information; the strength of information about relationships between stressors and effects; the assessment endpoints and their relationship to ecosystem function; the relative importance or influence and mode of action of stressors; and the completeness of known exposure pathways – become an important basis for the selection of critical relationships in the CSM that are to be pursued in the analysis. The analysis plan does indeed provide the basis for making selections of data sets that will be used for the ERA.

Essential to the development of the problem formulation elements are the effective integration and evaluation of available information on sources of stressors and stressor characteristics, exposure characteristics, ecosystem potentially at risk, and ecological effects.

14.2.2 The Problem Analysis Phase

The problem analysis phase of the ERA consists of hazard identification plus dose–response and exposure assessments – which entails evaluating exposure to stressors and the relationship between stressor levels and ecological effects. This involves the technical evaluation of data to facilitate the development of conclusions about ecological exposure and relationships between the stressor and ecological effects.

During this phase, measures of exposure (e.g., source attributes, stressor levels in the environment), measures of effects (e.g., results of laboratory or field studies), and measures of ecosystem and receptor attributes (e.g., life history characteristics) are used to evaluate issues that were identified in the problem formulation phase.

The analysis phase is composed of two principal activities – the characterization of exposure, and the characterization of ecological effects – which ultimately result in the development of exposure and stressor–response profiles. The CSM and analysis plan developed during problem formulation provide the framework for the analysis phase.

The end result of the problem analysis phase typically will consist of summary profiles that describe exposure and the stressor–response relationship. When combined, these profiles provide the basis for reaching conclusions about risk during the risk characterization phase.

14.2.3 The Risk Characterization Phase

The purpose of the risk characterization phase of the ERA is to evaluate the likelihood that adverse effects have occurred or will occur as a result of exposure to a stressor. Its key elements consist of estimating risk through integration of exposure and stressor–response profiles, describing risk by discussing lines of evidence, and determining ecological adversity.

A risk estimation is used to determine the likelihood of adverse effects to assessment endpoints by integrating exposure and effects data and evaluating any associated uncertainties. The process uses exposure and stressor–response profiles derived *a priori* in the problem analysis phase of the ERA.

14.3 ECOLOGICAL RISK ASSESSMENT IN PRACTICE

A contaminant entering the environment may cause adverse effects only if the contaminant exists in a form and concentration sufficient to cause harm; the contaminant comes in contact with organisms or environmental media with which it can interact; and the interaction that takes place is detrimental to life functions. The ecological assessment seeks to determine the nature, magnitude, and transience or permanence of observed or expected effects of contaminants introduced into an ecosystem. This type of assessment is usually directed at investigating the loss of habitat, reduction in population size, changes in community structure, and changes in ecosystem structure and function. In fact, knowledge of environmental effects is important in analyzing both nonhuman and potential human risks associated with chemical releases into the environment. In practice, ERAs may be said to involve the application of the science of ecotoxicology to public policy (Suter, 1993).

Ideally, the ERA would estimate the potential for occurrence of adverse effects that are manifested as changes in the diversity, health, and behavior of the constellation of organisms that share a given environment over time. Ecological areas included in an ERA should therefore not be limited by property boundaries of a study area, if affected environments or habitats are located beyond the property boundaries.

A qualitative and/or quantitative ERA is often an integral part of most environmental management programs that are designed to address environmental contamination problems (Figure 14.5). Similar to the human health endangerment assessment presented earlier in Chapter 13, the basic components and tasks involved in a comprehensive ERA may consist of the following tasks:

- Data evaluation

  - Assess the quality of available data.
  - Identify, quantify, and categorize environmental contaminants.
  - Screen and select chemicals of potential ecological concern.
  - Carry out statistical analysis of relevant environmental data.

- Exposure assessment

  - Compile information on the physical setting of the locale.
  - Identify source areas, significant migration pathways, and potentially impacted or receiving media.
  - Determine the important environmental fate and transport processes for the chemicals of potential concern, including cross-media transfers.
  - Identify populations potentially at risk.
  - Determine likely and significant receptor exposure pathways.
  - Develop representative conceptual site model(s).
  - Develop exposure scenarios (to include both the current and potential future land uses for the site).
  - Estimate/model exposure point concentrations for the chemicals of potential concern found in the significant environmental media.
  - Compute potential receptor intakes and resultant doses for the chemicals of potential ecological concern (for all potential receptors and significant pathways of concern).

- Ecotoxicity assessment

  - Compile toxicological profiles (to include the intrinsic toxicological properties of the chemicals of potential concern, such as their acute, subchronic, chronic, carcinogenic, and reproductive effects).
  - Determine appropriate toxicity indices (such as the AWQC, SQC, etc.).

- Risk characterization

  - Qualify risks or hazards, as appropriate.
  - Quantify risks or hazards, by estimating *ErQ*s or similar risk parameters.
  - Perform sensitivity analyses, evaluate uncertainties associated with the risk estimates, and summarize the risk information.

Unlike endangerment assessment for human populations, however, ERAs often lack the significant amount of critical and credible data necessary for a compre-

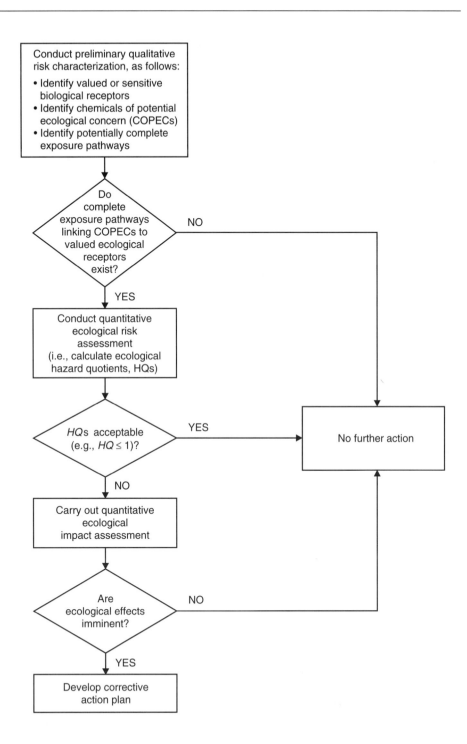

Figure 14.5 A procedural framework for ecological risk assessment in practice

> **Box 14.7 Selection criteria for various major aspects of an ERA**
>
> | ERA issue/item | Selection criteria |
> |---|---|
> | Valued biological receptors | • Regulatory importance
• Ecological importance
• Commercial importance
• Recreational importance |
> | Chemicals of potential ecological concern (COPECs) – considered during hazard identification | • Accessibility
• Concentration
• Bioaccumulation/bioconcentration potential
• Potency/toxicity
• Bioavailability
• Persistence |
> | Potentially complete exposure pathways/ exposure scenario | • Location of COPECs
• Migration potential of COPECs
• Behavior and ecology of valued biological receptors |
> | Establishing [exposure] assessment endpoints – which form the basis for collecting data, for determination of contaminant effects, and for evaluation of migration needs from releases or mitigation activities (i.e., the explicit expressions of values that are to be protected) | • Societal relevance
• Biological relevance
• Measurability or predictability
• Susceptibility to chemicals |
> | Establishing [exposure assessment] measurement endpoints – which link the existing conditions on-site to the goals expressed by the assessment endpoints (i.e., are more easily collected data that act as a surrogate for the assessment endpoints) | • Relationship to assessment endpoint
• Ability to be measured
• Availability of existing data
• Relationship to known contaminants and pathways
• Degree of natural variability
• Temporal and spatial relational scale |

hensive quantitative evaluation. Nonetheless, the pertinent data requirements should be identified and categorized insofar as practicable. Several particularly important criteria to adopt in various significant aspects of an ERA process are enumerated in Box 14.7.

14.3.1 The General Purpose

The objectives of an ERA, conducted to address environmental contamination problems, consist of identifying and estimating the potential ecological impacts, with the specific focus being to determine the following:

- biological and ecological characteristics of the study area;
- types, forms, amounts, distributions, and concentrations of the contaminants of concern;
- migration pathways to, and exposure of ecological receptors to pollutants;
- habitats potentially affected and populations potentially exposed to contaminants; and
- actual and/or potential ecological effects/impacts and overall nature of risks.

ERAs can be divided into two general types based on habitats: terrestrial (land), and aquatic (freshwater and marine), with avifauna belonging in either or both groups. Typical ecological effects include changes in the aquatic and terrestrial natural resources brought about by exposure to environmental contaminants. Although the assessment focuses on the impacts of contaminants on the terrestrial and aquatic flora and fauna that inhabit an impacted locale and vicinity, ecological assessments may also identify new or unexpected exposure pathways which could potentially affect human populations through the foodchain or through changes in the ecosystem.

14.3.2 General Considerations in Ecological Investigations

Ecological risk analysis can be performed at several hierarchical levels, and risks can be expressed as a qualitative or quantitative estimate, depending on the available data (Zehnder, 1995). The ERA allows for the identification of habitats and organisms that may be affected by the chemicals of potential concern that are associated with an environmental contamination problem.

Typically, the following elements are given in-depth consideration during the ecological investigation associated with environmental contamination problems:

- Definition and role of the ERA within the context of the environmental contamination problem assessment and characterization.
- Establishment of a concept of acceptable ecological risk.
- Evaluation and selection of appropriate ecological endpoints at the population, community, and ecosystem levels.
- Evaluation of exposure and biomarkers of exposure.
- Validation of strategy adopted for the ERA (including the basis for its acceptability and appropriateness for the case-specific problem).
- Design of field sampling programs, data analysis and evaluation plans, and ecological monitoring programs.
- Determination of acute and chronic risks and secondary hazards.
- Application of ERA results to environmental restoration plans (e.g., for derivation of site-specific remediation objectives).

In general, the environmental transport media of greatest interest in ERAs usually relate to surface water and soil – because these are the media that are most frequently contacted by the organisms of interest. Whereas surface water is of primary interest to aquatic ecosystems, terrestrial ecosystems involve both soil and surface water. The reason that both media are of concern in terrestrial assessments is because many terrestrial receptors contact surface water bodies for such reasons as drinking, development through some of the life stages (e.g., tadpole stage of frogs and toads, and larval stage of dragonflies), and living in or near the water (e.g., beaver, muskrat, and some snakes). Consequently, areas of contaminated soil and territories near contaminated surface water bodies as well as near contaminated soils will normally require ecological risk assessments.

The collection and review of the existing information on terrestrial and aquatic ecosystems, wetlands and floodplains, threatened and endangered species, soils, and other topics relevant to the study should form a prime basis for identifying any data gaps. For instance, survey information on soil types, vegetation cover, and residential migratory wildlife may be required for terrestrial habitats, whereas comparative information needed for freshwater and marine habitats will most likely include survey data on abundance, distribution, and kinds of populations of plants and animals living in the water column and in or on the bottom. The biological and ecological information collected should include a general identification of the flora and fauna associated within and around the site, with particular emphasis placed on sensitive environments, especially endangered species and their habitats and those species consumed by humans or found in human foodchains. Furthermore, the biological, chemical, and environmental factors perceived to influence the ecological effects of contaminants should be completely identified and adequately described.

14.4 A COMPARATIVE LOOK AT THE PARALLEL NATURE OF ECOLOGICAL AND HUMAN HEALTH ENDANGERMENT ASSESSMENTS

Traditionally, most endangerment assessments have focused almost exclusively on risks to human health, often ignoring potential ecological effects. This bias has resulted in part from anthropocentrism and in part from the common but mistaken belief that protection of human health automatically protects nonhuman organisms (Suter, 1993). In fact, human health risks in most situations are more substantial than ecological risks; considering that the mitigative actions taken to alleviate risks to human health are often sufficient to mitigate potential ecological risks at the same time, extensive ecological investigations are usually not required for most environmental management programs. However, it should be recognized that in some situations nonhuman organisms, populations, or ecosystems may be more sensitive than human receptors. Consequently, ecological risk assessment programs have become an important part of the management of environmental contamination problems, to facilitate the attainment of an adequate level of protectiveness for both human and ecological populations potentially at risk. Their purpose is to contribute to the protection and management of the environment through scientifically justifiable and credible evaluation of the ecological effects of human

> Box 14.8 Typical general elements of the parallel activities involved in human health and ecological risk assessments
>
> *Human health risk assessment*
>
> - Identify chemicals of potential concern (COCs) to human health
> - Quantify release, migration, and fate of contaminants that could affect human receptors
> - Determine human populations potentially-at-risk (PARs)
> - Identify exposure routes for PARs
> - Conduct health effects studies
> - Characterize human health risks, recognizing and addressing associated uncertainties
>
> *Ecological risk assessment*
>
> - Identify COPECs for ecological receptors
> - Quantify release, migration, and fate of contaminants that could affect ecological systems and/or receptors
> - Determine potentially affected habitats and potentially exposed ecological populations
> - Identify exposure routes for potentially exposed ecological populations
> - Conduct ecological effects studies and tests
> - Characterize ecological risks, recognizing and addressing associated uncertainties

activities. Ultimately, ecological risk analysis becomes a very important instrument to integrate ecological issues in the formulation of environmental management decisions.

The process used to evaluate environmental or ecological risks is analogous, and parallels that used in the evaluation of human health risks that was presented in Chapter 13 (see Box 14.8 for comparison of analogous tasks/activities). In both cases, potential risks are determined by the integration of information on chemical exposures with toxicological data for the contaminants of potential concern.

In fact, numerous and diverse risk assessment methodologies have been developed for the protection of human health, most of which can be useful to ERAs as well. In adapting the relevant tools from human health risk assessment to ERA problems, however, it must be recognized that nonhuman organisms, populations, or ecosystems may be more sensitive than humans under similar types of circumstances; such differences may be attributable to a variety of reasons, including the following (Suter, 1993):

- Some types of exposure scenarios are peculiar only to nonhuman receptors (e.g., root uptake and drinking from waste sumps).
- Certain chemicals are likely to be more toxic to some nonhuman species than to humans – considering the heterogeneity in the former.
- There are mechanisms of action at the ecosystem level (such as eutrophication by nutrient chemicals) that have no human analogues.
- Nonhuman organisms may be exposed more intensely to chemicals, even when the routes of exposure are similar. After all, in general, humans normally are not immersed in a particular ambient environment.

- Most birds and mammals have higher metabolic rates than humans, thus resulting in the former group receiving a larger dose per unit body weight.
- Nonhuman organisms are highly coupled to their environments, so that even when they are resistant to a chemical, they may still experience secondary effects – such as loss of food or physical habitat.

In cases of both human health and ecological endangerment assessments, the process paradigm offers a rigorous form of evaluation to qualify and/or quantify potential effects of environmental contaminants on a variety of populations potentially at risk.

SUGGESTED FURTHER READING

Bergman, H.L and E.J. Dorward-King (eds). 1997. *Reassessment of Metals Criteria for Aquatic Life Protection.* SETAC Technical Publication Series, SETAC Press, Pensacola, FL.

Blancato, J.N., R.N. Brown, C.C. Dary, and M.A. Saleh (eds). 1996. *Biomarkers for Agrochemicals and Toxic Substances: Applications and Risk Assessment.* ACS Symposium Series No. 643, American Chemical Society, Washington, DC.

CCME (Council of Canadian Ministers of the Environment). 1996. *A Framework for Ecological Risk Assessment: General Guidance.* Report CCME-EPC-111. CCME, Winnipeg, Manitoba, Canada.

Grothe, D.R., K.L. Dickson, and D.K. Reed-Judkins (eds). 1996. *Whole Effluent Toxicity Testing: An Evaluation of Methods and Prediction of Receiving System Impacts.* SETAC Press, Pensacola, FL.

Hoffman, D.J., B.A. Rattner, and R.J. Hall. 1990. Wildlife toxicology. *Environmental Science and Technology*, 24: 276.

Hudson, R., R. Tucker, and M. Haegeli 1984. *Handbook of Toxicity of Pesticides to Wildlife*, 2nd edition. USFWS Resources Publication No. 153, Washington, DC.

Mayer, F.L. and M.R. Ellersick. 1986. *Manual of Acute Toxicity: Interpretation and Database for 410 Chemicals and 66 Species of Freshwater Animals.* US Dept. of Interior, Fish and Wildlife Service, Resource Publ. 160, Washington, DC.

Menzie-Cura and Associates. 1996. *An Assessment of the Risk Assessment Paradigm for Ecological Risk Assessment.* Report Prepared for the Presidential/Congressional Commission on Risk Assessment and Risk Management, Washington, DC.

Mitsch, W.J. and S.E. Jorgesen (eds). 1989. *Ecological Engineering, An Introduction into Ecotechnology.* J. Wiley, New York.

Schroeder, R.L. 1985. *Habitat Suitability Index Models: Northern Bobwhite.* US Dept. of Interior, Fish and Wildlife Service, Washington, DC.

Schuurmann, G. and B. Markert. 1966. *Ecotoxicology.* J. Wiley, New York.

REFERENCES

Ankley, G.T., D.M. Di Toro, D.J. Hansen and W.J. Berry. 1996. Technical basis and proposal for deriving sediment quality criteria for metals. *Environmental Toxicology and Chemistry*, 15(2): 2056–2066.

Barnthouse, L.W. and G.W. Suter II (eds). 1986. *User's Manual for Ecological Risk Assessment.* ORNL-6251, Oak Ridge National Lab., Oak Ridge, TN.

Bartell, S.M., R.H. Gardner, and R.V. O'Neill. 1992. *Ecological Risk Estimation.* Lewis Publishers, Chelsea, MI.

Calow, P. (ed.). 1993. *Handbook of Ecotoxicology*. Blackwell Scientific Publications, London, UK.
Calmano, W. and U. Forstner (eds). 1996. *Sediments and Toxic Substances: Environmental Effects and Ecotoxicity*. Springer-Verlag, Berlin, Germany.
CAPCOA (California Air Pollution Control Officers Association). 1990. *Air Toxics 'Hot Spots' Program. Risk Assessment Guidelines*. CAPCOA, Los Angeles, CA.
Carlsen, T.M. 1996. Ecological risk to fossorial vertebrates from volatile organic compounds in soil. *Risk Analysis*, 16(2): 211–219.
Cohrssen, J.J. and V.T. Covello. 1989. *Risk Analysis: A Guide to Principles and Methods for Analyzing Health and Environmental Risks*. National Technical Information Service, US Dept. of Commerce, Springfield, VA.
Kolluru, R.V., S.M. Bartell, R.M. Pitblado, and R.S. Stricoff (eds). 1996. *Risk Assessment and Management Handbook (for Environmental, Health, and Safety Professionals)*. McGraw-Hill, New York.
Linthurst, R.A., P. Bourdeau, and R.G. Tardiff (eds). 1995. *Methods to Assess the Effects of Chemicals on Ecosystems*. SCOPE 53/IPCS Joint Activity 23/SGOMSEC 10, J. Wiley, Chichester, UK.
MacCarthy, L.S. and D. Mackay. 1993. Enhancing ecotoxicological modeling and assessment. *Environ. Sci. Technol.*, 27(9): 1719–1728.
NRC (National Research Council). 1981. *Testing for Effects of Chemicals on Ecosystems*. NAS Press, Washington, DC.
NRC. 1983. *Risk Assessment in the Federal Government: Managing the Process*. NAS Press, Washington, DC.
NRC. 1989. *Biological Markers of Air-Pollution Stress and Damage in Forests*. NAS Press, Washington, DC.
Paustenbach, D.J. (ed.). 1988. *The Risk Assessment of Environmental Hazards: A Textbook of Case Studies*. J. Wiley, New York.
Pickering, Q.H. and C. Henderson. 1966. The acute toxicity of some heavy metals to different species of warm water fish. *Air and Water Pollution International Journal*, 10(6/7): 453–463.
Richardson, M. (ed.). 1995. *Environmental Toxicology Assessment*. Taylor & Francis, London, UK.
Solomon, K.R. 1996. Overview of recent developments in ecotoxicological risk assessment. *Risk Analysis*, 16(5): 627–633.
Suter II, G.W. 1993. *Ecological Risk Assessment*. Lewis Publishers, Boca Raton, FL.
Suter II, G.W. 1996. Toxicological benchmarks for screening contaminants of potential concern for effects on freshwater biota. *Environmental Toxicology and Chemistry*, 15(7): 1232–1241.
USEPA (US Environmental Protection Agency). 1973. *Biological Field and Laboratory Methods for Measuring the Quality of Surface Waters and Effluents*. National Environmental Research Center, Cincinnati, OH. EPA/670/4-73/001.
USEPA. 1986. *Ecological Risk Assessment. Hazard Evaluation Division Standard Evaluation Procedure*. Washington, DC.
USEPA. 1988a. *Estimating Toxicity of Industrial Chemicals to Aquatic Organisms Using Structure Activity Relationships*. Office of Toxic Substances. EPA/560/6-88/001.
USEPA. 1988b. *Review of Ecological Risk Assessment Methods*. Office of Policy, Planning and Evaluation, Washington, DC.
USEPA. 1989a. *Ecological Assessments of Hazardous Waste Sites: A Field and Laboratory Reference Document*. Office of Research and Development – Corvallis Environmental Research Laboratory, OR. EPA/600/3-89/013.
USEPA. 1989b. *Risk Assessment Guidance for Superfund. Volume II – Environmental Evaluation Manual*. Office of Emergency and Remedial Response, Washington, DC. EPA/540/1-89/001.
USEPA. 1990. *State of the Practice of Ecological Risk Assessment Document*. Office of Pesticides and Toxic Substances, USEPA draft report. Washington, DC.

USEPA. 1992. *Framework for Ecological Risk Assessment*. February, 1992, Risk Assessment Forum, Washington, DC. EPA/630/R-92/001.
USEPA. 1993. *Wildlife Exposure Factors Handbook*. Office of Research and Development, Washington, DC. EPA/600/R-93/187a & 187b.
USEPA. 1994. *Ecological Risk Assessment Issue Papers*. Risk Assessment Forum, Washington, DC. EPA/630/R-94/009.
USEPA. 1995. *Ecological Risk: A Primer for Risk Managers*. Washington, DC. EPA/734/R-95/001.
van Leeuwen, K. 1990. Ecotoxicological effects assessment in the Netherlands. *Environmental Management*, 14: 779–792.
Zehnder, A.J.B. (ed.). 1995. *Soil and Groundwater Pollution: Fundamentals, Risk Assessment and Legislation*. Kluwer Academic Publishers, Dordrecht, The Netherlands.

Chapter Fifteen

Probabilistic Risk Assessments

Probabilistic risk assessment (PRA) is a method generally used to quantify the frequency of occurrence, the degree of system response, and the magnitude of consequences associated with accident events or system failures for hazardous facilities, industrial facilities, and other technological systems. Although its most extensive application has been in the nuclear industry, PRA has also been successfully used to evaluate other issues relating to the study of chemical hazard problems – such as those arising out of releases from hazardous waste treatment, storage, and disposal facilities (TSDFs) and hazardous materials transport accidents. For instance, probabilistic techniques may be applied to the assessment of the structural integrity of hazardous waste facilities and containments, or for the evaluation of accident scenarios in the transportation of hazardous materials. Such applications would be based on quantitative measures of the probability that a facility or containment failure will occur, or that an accident will occur during transportation and/or equipment operation.

This chapter introduces the fundamental principles and practical concepts of PRA methodologies that may find significant usefulness in the investigation of certain types of environmental hazard situations.

15.1 THE PROBABILISTIC RISK ASSESSMENT METHODOLOGY

A major objective of the PRA process is to assign a probability (or frequency) to possible consequences of system failure. Thus, with this method, risk is defined in terms of frequency and magnitude of consequences (Glickman and Gough, 1990; Henley and Kumamoto, 1981), viz.:

$$\text{Risk}\left[\frac{\text{consequence}}{\text{time}}\right] = \text{frequency}\left[\frac{\text{events}}{\text{unit time}}\right] \times \text{magnitude}\left[\frac{\text{consequence}}{\text{event}}\right]$$

or equivalently, the failure probability is defined in terms of hazard–outcome–consequence probabilities as follows:

$$\text{Partial failure probability, } Pr\{\text{Fail}\} = Pr\{\text{Hazard}\} \times Pr\{\text{Outcome}\} \times Pr\{\text{Consequence}\}$$

where $Pr\{\text{Hazard}\}$ is the probability of an accident or failure event; $Pr\{\text{Outcome}\}$ is the conditional probability of system response and outcome due to the failure event; and *Pr{Consequence}* is the conditional probability of adverse effects from accident sequence(s).

Subsequently, the product of frequency and magnitude, or the failure probability are summed over all incident sequences or failure pathways to yield total failure probability represented as follows:

$$\text{Total failure probability} = \sum_i \text{risk} \left[\frac{\text{consequence}}{\text{time}} \right]$$

$$= \sum_i \left\{ \text{frequency} \left[\frac{\text{events}}{\text{unit time}} \right] \times \text{magnitude} \left[\frac{\text{consequence}}{\text{event}} \right] \right\}$$

or

$$\text{Total failure probability} = \sum_{i=1}^{n} Pr\{\text{Fail}\}_i$$

$$= \sum_{i=1}^{n} [Pr\{\text{Hazard}\} \times Pr\{\text{Outcome}\} \times Pr\{\text{Consequence}\}]$$

The concept of probability of failure required in the risk evaluation is usually defined by using the likelihood of structural breach and/or an accident event. Estimation of the applicable probability values are achieved by the use of reliability theory and/or expert judgments, or by use of stochastic simulations and historical information. A Bayesian approach may always be employed to update estimates on the basis of additional information acquired in time.

In general, decision/logic trees typically provide an approach of analyzing a sequence of uncertain events or conditions at hazardous facilities, that could potentially lead to adverse consequences (Figure 15.1). Two types of decision/logic trees are commonly used to identify sets of events leading to system failure: event trees – that use deductive logic, and fault trees – which use inductive logic. In fact, event and fault trees can prove to be very useful techniques for identification and quantification of accidents and failures of technological systems – including failure scenarios at hazardous facilities.

A PRA may use fault-tree or event-tree analysis, or combinations thereof. In many situations, fault-tree analysis is used to supplement event-tree modeling by utilizing the former to establish the appropriate probabilities of the event-tree branches. Event-tree concepts can also aid in developing exposure and accident scenarios. The technique can indeed help designers anticipate risk in order to correct problems at the design stage rather than through retrofit technologies.

15.1.1 Event-Tree Modeling and Analysis

The event tree is a diagram that illustrates the chronological ordering of event scenarios in a problem that calls for a decision analysis protocol. Each event

Figure 15.1 Conceptual representation of the PRA process

pathway is shown by a branch of the event tree. The event-tree model can be used to display the paths of the events or actions that relate to the safety or potential for failure of hazardous facilities, and also the anticipated consequences for the various pathways. It provides a diagrammatic representation of event sequences that begin with a so-called initiating event or hazard and terminate in one or more undesirable consequences. Events identified as part of a failure scenario can be displayed in a tree structure that represents a sequence of events in progression, displaying branching points where several possibilities can be anticipated that can lead to an event at the top. This technique basically is an algorithm in which it is possible to assign probabilities to each of the events. Then, by simply multiplying or adding probabilities, the overall chance of failure can be calculated for a given period of time. In this regard, event trees generally provide requisite tools that can be used to analyze conditions associated with hazardous situations or facilities that could potentially result in adverse consequences.

Event-tree analysis (ETA) is a methodology that models risk as a chain of interconnected events. It can be used for hazard identification and for estimation of probability for a sequence of events leading to hazardous situations. The event-tree approach will indeed allow the evaluation of a range of possible scenarios, rather than the overly conservative approach of making an assessment for the 'worst-case' scenario only. The approach allows for a systematic consideration of all potential loading conditions that may be brought to bear on a system, the potential exposure scenarios following system breach, and the consequences of all potential exposures to any population/receptors at risk. Figure 15.2 illustrates the logic used to construct an event tree for a typical exposure scenario, such as that involving an accident situation resulting in contaminant release to various environmental compartments.

The event-tree concepts can be used to simplify the exposure assessment in particular, by transforming and representing complex situations with simplistic but adequate and manageable scenarios. This structure consequently provides a systematic approach for building a technically defensible information base for decisions on potential hazards, by providing a mechanism for tackling environmental problems in a logical and comprehensive manner.

15.1.1.1 Conceptual elements of an event-tree analysis

An event tree uses deductive logic, starting with an initiating event, and then using forward logic to enumerate all possible sequences of subsequent events that will help determine other possible outcomes and consequences. In probabilistic evaluations, the event-tree structure requires that each event level be defined by its probability, which is conditional on preceding events in the tree structure.

328 Risk assessment in environmental management

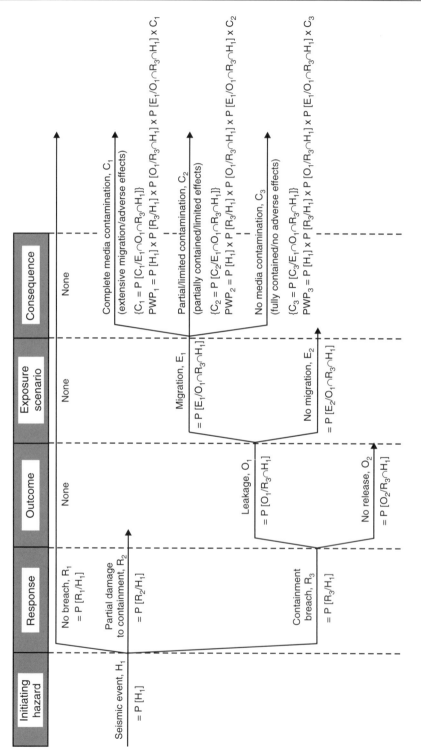

Figure 15.2 Illustration of the event-tree analysis process

Overall, the event-tree concept offers a more efficient way to perform a probabilistic risk analysis, as and when necessary. For instance, PRA techniques utilizing the event-tree model can be used in the safety evaluation of various environmental management systems. The processes involved help in the development of a systematic framework to aid risk management decisions. Several important conceptual elements may be employed in such application of an ETA to environmental management systems; common examples relate to the pathway probability and risk cost/impact assessment concepts, as discussed below.

- *The pathway probability concept.* Where appropriate, in a probabilistic risk analysis, the probability of a consequence due to the occurrence of a hazardous situation is defined by a so-called pathway probability (*PWP*), which is the product of an initiating probability value and the conditional probabilities of subsequent events. The *PWP* is given by the following relationship:

$$PWP = Pr\{H \cap R \cap O \cap E \cap C\}$$

$$= Pr\{H\} \times Pr(R/H) \times Pr\{O/H \cap R\} \times Pr\{E/H \cap R \cap O\} \times Pr\{C/H \cap R \cap O \cap E\}$$

where $Pr\{H\}$ is the probability of a specific hazard (H) of an initiating event occurring; $Pr\{R/H\}$ is the conditional probability of system response (R), given H; $Pr\{O/H \cap R\}$ is the conditional probability of an outcome (O), given H and R; $Pr\{E/H \cap R \cap O\}$ is the conditional probability of exposure (E), given H, R and O; and $Pr\{C/H \cap R \cap O \cap E\}$ is the conditional probability of a specific consequence (C), given H, R, O, and E.

The initiating event is that cause or sequence of actions that creates an effect or response. Such an event will generally lead to follow-up conditions or consequential impacts. The fundamental assumption here is that the preceding event is the cause of subsequent events – i.e., the current event is dependent on the previous. In the case of mutual independence, this relationship becomes:

$$P\{H \cap R \cap O \cap E \cap C\} = P\{H\} \times P\{R\} \times P\{O\} \times P\{E\} \times P\{C\}$$

To determine the desired probability requires that one knows the conditional probabilities along a specific risk path – or in the case of mutually independent events, the *a priori* probability of occurrence for each event (i.e., for mutually independent events, only the probability of occurrence for each event would be required for a similar evaluation).

- *Risk costs and impacts assessment.* Where applicable, the cost associated with the probability of failure – the so-called risk cost (*RC*) – of an environmental management system may be estimated based on anticipated economic or environmental consequences and the *PWP*. This parameter may be computed according to the following relationships:

Partial RC, $C_i = \sum_j (PWP_i \times EC_{ij})$, for ith pathway

Total RC, $C = \sum_{i=1}^{N} C_i$, for all existing N pathways

$$= \sum_{i=1}^{N} \left\{ \sum_{j=1}^{IZ} (PWP_i \times EC_{ij}) \right\}, \text{ for all existing } IZ \text{ impact zones}$$

where EC_{ij} is the economic and/or environmental damages/costs in impact zone j associated with the ith pathway.

The likely impacts on populations potentially at risk (which may include potential life loss) that is attributable to a hazardous situation depends on the exposed population, the likelihood of exposure, and the PWP – computed in accordance with the following relationships:

Life impacted/event = $PAR_{ij} \times$ exposure probability (EP_{ij})

Life impacted, $LI_i = \sum_j (PAR_{ij} \times EP_{ij} \times PWP_i)$, for ith pathway

Total LI, $LI = \sum_i LI_i$, for all existing N pathways

$$= \sum_{i=1}^{N} \left\{ \sum_{j=1}^{IZ} (PAR_{ij} \times EP_{ij} \times PWP_i) \right\}, \text{ for all existing } IZ \text{ impact zones}$$

where PAR_{ij} represents the number of people (i.e., population-at-risk, PAR) in impact zone j associated with the ith pathway, and EP_{ij} is the associated exposure probability. In this case, the average individual risk is given by:

$$\text{Individual risk} = \frac{\{\text{total } LI\}}{\left\{ \sum_j PAR_j \right\}}$$

Results from the event-tree model may, in general, be put into a spreadsheet format for better comprehension. Such a formulation also allows for easy comparison of alternative remedial action programs identified to help address the hazardous situation. The effect of each remedial alternative in reducing the risks associated with an environmental management system can then be evaluated and compared.

15.1.2 Fault-Tree Modeling and Analysis

A fault tree represents the combination and sequences of events which could cause specific system failure. It is traced back from a particular system failure event

Probabilistic risk assessments 331

Symbols	Definitions
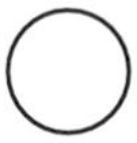 ○	*Basic event – Circle.* This indicates a basic initiating fault event requiring no further development; signifies that the appropriate limit of resolution has been reached.
 ◇	*Undeveloped event – Diamond.* This describes a specific fault event that is not developed further, either because the event results in insignificant consequence or because there is no relevant information on this.
 ⌂	*External event, or house.* This represents events that are not in themselves faults; it is used to signify an event that is normally expected to occur (e.g., a phase change in a dynamic system). This event acts as a switch by being set to 0 or 1 to reflect boundary conditions. It is a special type of terminal event that can be 'turned on' (= 1) or 'turned off' (= 0) by the analyst. It is used primarily to study the failure behaviour of systems under different scenarios.
 ▭	*Intermediate event – Box.* This is a fault event that occurs because of one or more antecedent causes acting through logic gates.
 (OR symbol)	*OR gate.* This is used to indicate that the output event occurs if at least one of the input events occur. There should be at least two input events to an OR gate.
 (AND symbol)	*AND gate.* This is used to indicate that the output event occurs if and only if all of the input events occur. All the input events need not occur simultaneously; they may occur at different times. There should be at least two input events to an AND gate.
 (Priority AND symbol)	*PRIORITY AND gate.* This is a special type of AND gate in which the output event occurs only if all input events occur in a specified ordered sequence.
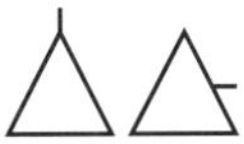 △ △	*Transfer symbols – Triangles.* Used as a matter of convenience to avoid extensive duplication in the fault tree; a line from the apex of the triangle denotes a transfer in, and a line from the side of the triangle denotes a transfer out.

Figure 15.3   Selected common fault-tree symbols

(called the top event) and spreads down through lower level events until it reaches the basic failure events. A fault tree is generally tailored to its top event. Hence, the fault tree includes only the fault events and logical interrelationships that contribute to the top event. Furthermore, the postulated fault events that appear on the fault tree may not be exhaustive. They may include only the events considered and/or determined by the analyst to be significant.

The fault tree itself is a graphic model of the various parallel and sequential combinations of faults that will result in the top event. Figure 15.3 shows commonly used fault-tree symbols found in the PRA literature. A circle, diamond, or 'house', represents a primary event (i.e., an event that is not developed further and has no inputs). The two basic types of fault-tree logic gates are the 'OR gate' and the 'AND gate'; used in combination with the 'NOT operator' (commonly shown as a dot above the gate), these gates can be used to define other specialized fault-tree gates.

Fault-tree analysis (FTA) is a technique used to analyze complex systems, usually to predict the expected probability of failure of a system. It seeks to relate the occurrence of an undesirable 'top event' to one or more antecedent 'basic events'. The top event may be, and usually is, related to the basic events via certain intermediate events. A fault-tree diagram exhibits the causal chain linking the basic events to the intermediate events and the latter to the top event. In this chain, the logical connection between events is indicated by the so-called 'logic gates'.

FTA may be used for hazard identification, although it is primarily used in risk assessment as a tool to provide an estimate of failure probabilities. The strength of FTA as a qualitative tool is its ability to identify the combinations of the basic causes that can lead to a failure situation. It is noteworthy that often the FTA process tends to rely on cause-and-effect relationships obtained from the application of other hazard evaluation techniques.

A schematic representation of a fault tree is given in Figure 15.4; this illustrates a typical fault-tree structure, but does not necessarily depict the level of complexity in a typical fault tree. A fault tree has a branching structure defined by logic gates located at branch intersections. The logic gates define the causal relationship between lower level events and higher level events. The basic concepts of fault-tree construction and analysis are well documented elsewhere in literature (e.g., CMA, 1985; Dhillon and Singh, 1981; Lees, 1980; NRC, 1983).

Fault trees can be used to estimate the probability of occurrence of the top event, given estimates of the probabilities of occurrence of the basic events. They may be used to map all relevant possibilities and to determine the probability of the final outcome. To accomplish this latter goal, the probabilities of all component stages, as well as their logical connections, must be completely specified. By assigning probability values to the basic events, the calculation of the probability of the top event can be achieved by performing algebraic manipulations based on some basic probability principles.

15.1.2.1 Illustrative elements of a fault-tree analysis

The fault-tree approach is a inductive process, whereby the top event is postulated and the possible pathways for that event to occur are systematically derived. The fault tree is essentially qualitative, but is very often quantified due to its binary logic and adaptability to Boolean expressions.

In a FTA, an undesired state of a system is specified and the system is then analyzed in the context of its environment and operation to determine all the feasible and credible ways in which the undesirable event could possibly occur. The FTA always starts with the definition of the undesirable event whose probability is to be determined. The tree is developed to lower levels until the lowest events – called primary faults – are reached. Once all the primary event probabilities are assigned, the computation of the probability of the top event becomes an issue of

Probabilistic risk assessments 333

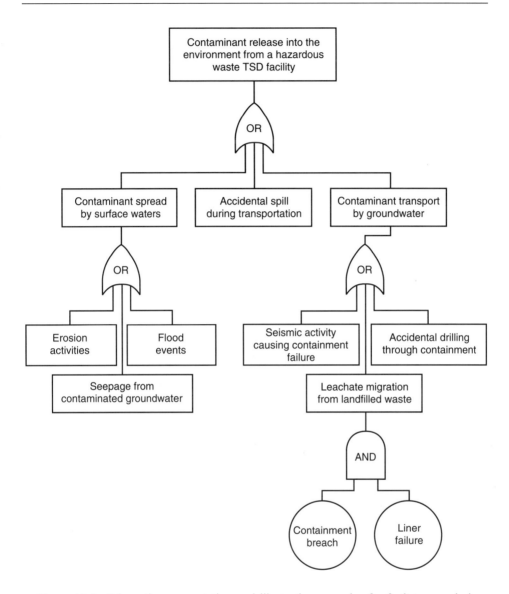

Figure 15.4 Schematic representation and illustrative example of a fault-tree analysis process

Boolean algebra manipulations, for which there are several computer codes for completion.

Consider a simple example of a fault-tree model that involves a waste containment system. Suppose that the top event is 'contaminant release', resulting from the following basic events: containment breach and liner failure, *or* air emissions (from surface cracks or fissures). The fault-tree diagram for this system is illustrated in Figure 15.5.

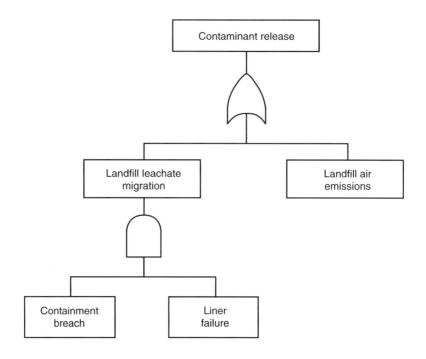

Figure 15.5 Illustrative fault-tree analysis

One of the purposes of FTA is the calculation of the probability of the top event. Let A, B, and C represent the basic events of waste containment breach, liner failure, and air emissions (from surface cracks or fissures), respectively. If T represents the top event (i.e., contaminant release), then the following relationship holds true:

$$T = AB + C$$

This relationship indicates that T occurs if both A and B occur or if C occurs. Assuming that A, B, and C are independent, then application of the probability addition theorem yields:

$$P(T) = P(AB) + P(C) - P(ABC)$$

Since A, B, and C are independent, then:

$$P(AB) = P(A) \times P(B) \text{ and } P(ABC) = P(A) \times P(B) \times P(C)$$

Therefore,

$$P(T) = P(A)P(B) + P(C) - P(A)P(B)P(C)$$

Thus, if the individual probabilities are known, or can be estimated, then the probability for the top event occurring can be obtained.

15.1.3 Other PRA Techniques and Tools

Event trees and fault trees are not the only decision analytical tools available for performing PRAs. Although event-tree and fault-tree analyses are the most powerful methods in PRA, other relatively simpler as well as more complex methods may also be employed, as and when appropriate. In fact, several so-called system analysis methods exist that can be used in addition to, or in support of, the event- and fault-tree approaches. Some of the more popular techniques are enumerated below.

- *Cause–consequence analysis (CCA)* is a functional combination of fault-tree and event-tree analyses. A CCA usually becomes very useful as a communication tool because the cause–consequence diagram generally displays the relationships between the failure consequences (that are modeled by the ETA) and their basic causative events (modeled by the FTA).
- *FMEA (failure modes and effects analysis)* is a qualitative process that identifies failure modes for the components of concern and traces their effects on other components, subsystems, and systems. This approach provides an orderly examination of the hazardous conditions in a system, and is simple to apply. It may include an assessment of criticality and probability of occurrence of each potential component failure mode; if the criticality of the effects is also considered in the analysis, then it is usually referred to as the failure modes, effects, and criticality analysis (FMECA). It is an inductive analysis that systematically details, on a component-by-component basis, all possible failure modes and identifies their resulting effects on the system.
- *RBDs (reliability block diagrams)* are models generated by an inductive process whereby a given system, divided into blocks representing distinct elements, is represented according to system–success pathways and scenarios. The model generally is used to represent active elements in a system in a manner that allows an exhaustive search for, and the identification of, all pathways for success.
- *Hazard analysis (HAZAN)*, or hazard quantification, is limited to the identification of hazards, and considerations of strategies to employ to avoid the hazards. It involves the estimation of the expected frequencies or probabilities of events with adverse or potentially adverse consequences.
- *Hazard and operability study (HAZOP)* is a systematic, inductive technique for identifying hazards and operability problems through an entire system. It uses guide-words to identify deviations leading to hazardous situations. After the serious hazards have been identified via a HAZOP study (or some other qualitative approach for that matter), a quantitative examination would usually be performed. HAZOP highlights specific deviations for which mitigative measures need to be developed.

The above listing is by no means exhaustive, since variations or even completely different ones may be encountered in the literature and elsewhere (e.g., Ang and Tang, 1984; Apostolakis, 1974; Cheremisinoff and Graffia, 1996; GAO, 1985; Harr, 1987; Hauptmanns and Werner, 1991; Kolluru et al., 1996; McCormick, 1981; NRC, 1983; NUREG, 1983; Sidall, 1983; Theodore et al., 1989; USBR, 1986; USEPA, 1987; Waller and Covello, 1984; World Bank, 1985; Zogg, 1987). Whichever method of analysis is to be used, however, the process should be so designed to optimally fulfil the requirements of the specific task.

15.2 PROBABILISTIC RISK ASSESSMENT IN PRACTICE

Two basic kinds of probabilistic risk analysis can be differentiated – inductive and deductive methods. The various methods all have their strengths and weaknesses, depending on how and where they are used. In any event, the PRA method generally aids in estimating the probabilities of events that have the potential to cause adverse consequences. The method may find several general and specific applications in environmental management programs, including the more common example situations identified below.

- *Hazardous materials TSDF design and failure investigations.* Safety aspects of the design of a hazardous materials TSDF can be evaluated by the use of PRA methods. The PRA process serves as an analytical technique for integrating diverse aspects of facility design and operation, to enable an evaluation of potential risks associated with a given hazardous facility. In this case, the risk evaluation will address the consequences associated with the probability of failure. Incremental risks due to failure, as a result of modifications in design criteria, can also be assessed. For instance, the probability of failure for using only one liner versus the failure probability in using multiple liners in a hazardous waste facility design may be evaluated and compared using PRA techniques. Also, potential risk reduction by inclusion of an early monitoring system, etc. in the design of a hazardous waste facility may be evaluated by the use of techniques and concepts derived by the PRA process. Similarly, risks associated with the failure of hazardous materials containments and other facilities may be evaluated by use of PRA concepts.

  In general, the adequacy of facility design and operation is assessed by identifying those sequences of potential events that dominate risk, and also by establishing which sectors and features of the facility contribute most to the occurrence of accident scenarios. Thus, a PRA may help reveal the features of a facility and system that may merit greater attention, and thus provide a better focused plan in safety improvement programs. The information base developed in a PRA will identify dominant accident scenarios and facility aspects contributing significantly to risk. Such information could be used in developing emergency-response plans. Also, the information developed during the assessment could help in making management decisions on the allocation of

limited resources for safety improvements – by directing attention to the features and scenarios dominating the facility risk.
- *Analysis of hazardous materials transportation risks.* Transportation of hazardous material (e.g., excavated contaminated soils, etc.) could, in the event of a spill, result in environmental and public health exposures and risks. Transport risk is therefore of major concern, particularly with regards to long-hauling of hazardous materials. Transport problems can indeed be a key issue for hazardous materials movements, not only from the point of view of economics and risk assessment, but also from the social and psychological perspectives as well. This is due in part to historical records of spills of virgin chemicals, oils, etc. that have occurred during transportation – and indeed other transportation accidents (see, e.g., Davidson, 1990; Saccomanno et al., 1989).

Potential spill incidents may result from any accidental spill – due to mishandling, loading/unloading mishaps, vehicle accident, etc. The number of expected spill incidents, N_{spill}, may be estimated according to the following relationship:

$$N_{spill} \{\text{by mode of transport}\} = [\text{spill incidents/vehicle mile}]$$
$$\times [\text{route miles/trip}]$$
$$\times [\text{number of vehicle trips}]$$

or,

$$N_{spill} \{\text{by mode of transport}\} = [\text{spill incidents/vehicle ton-mile}]$$
$$\times [\text{route miles/trip}]$$
$$\times [\text{number of vehicle trips}]$$
$$\times [\text{tons hauled/trip}]$$

The PRA approach may indeed find important uses in assessing risks from transportation of hazardous materials (Nicolet-Monnier and Gheorghe, 1996; Theodore et al., 1989). The risks of transporting hazardous materials may be defined in terms of the accident probability; the spill or release probabilities in an accident situation; the hazard classes for different damage scenarios (including the hazard areas for impacted zones); and the expected consequences on populations and environment within the accident/impacted corridor. The accident-induced releases of hazardous materials can be analyzed using fault- and/or event-tree methods of approach. Different levels of risks may be associated with different shipments and would depend on the material properties, the spill/release scenarios, and the overall transportation environment.

In the transport of hazardous materials, an accident during transportation will not necessarily cause a release. Therefore, transportation risks would be estimated as the product of the probability of an accident and the conditional probability of release from a given accident. Transportation risks can generally be analysed for a system by examining several variables – including the road network, loading/unloading accidents, and traffic density.
- *Quasi-PRA applications.* In several situations, risk is estimated without due consideration being given to the probability of causative events. Thus, the

assumption is that failure has already occurred or has an absolute chance of occurring. This scenario whereby risks are projected and based on the certainty assumption that breach or failure has already occurred may not be realistic in all situations. It will therefore be more pragmatic to estimate actual risks, given an estimated probability of the causative event or the incidence of failure. For instance, 'true' human health or environmental risks associated with a likely future scenario (rather than a prevailing situation) may be estimated as follows:

Actual risk = [probability of failure incidence or causative event]
× [estimated health or environmental risk value]

This 'modified' risk estimate will represent the true risks imposed by a hazardous situation, and may be applicable under many different circumstances.

Overall, the evaluation of risks associated with environmental management systems (such as hazardous waste facilities) is generally made from a broad spectrum of perspectives – including those for the public, regulatory agencies, owner/designer/operator (which may be public or private), and the insurer (who is the potential liability bearer) – and a risk analysis must address, insofar as possible, all these different perspectives concurrently. PRA techniques generally will help to achieve this important objective in a logical way. Typically, PRA techniques facilitate the development of general risk management and risk prevention programs.

15.2.1 A PRA Demonstration Problem

To illustrate the potential application of PRA methodologies in an environmental management situation, a very simplified evaluation scheme is presented concerning transportation risks in an accident corridor. Consider, for instance, the transport of hazardous wastes over a distance of 500 km in a specially equipped tanker/truck. For this study location, about 200 shipments of such wastes are undertaken in a year. Accident frequencies have been evaluated – recognizing the container may be damaged in the event of an accident along the transportation corridor. In such circumstances, the severity of public exposure to chemicals depends on several factors – such as meteorological conditions, the likelihood of containment breach, extent of release due to spillage, and the dispersion pattern of released chemicals. Based on the transportation risk data/information for this hypothetical case shown in Table 15.1, it is estimated that:

Average annual number of trucks damaged = [truck damage frequency]
× [no. of shipments]
× [travel distance]

$$= (2.0 \times 10^{-7}) \times (200) \times (500) = 2.0 \times 10^{-2}$$

Table 15.1 Hazardous waste transportation risk and health impacts data

Parameter	Risk value	Population impacted	Exposure probability (EP)*
Average number of trucks damaged	2.0×10^{-7}/truck km	–	–
Release likelihood from accident	0.4	–	–
Release severity index:			
(1) Small (limited)	0.5	0	0.50
(2) Medium (average)	0.3	500	0.07
(3) Large (extensive)	0.1	1000	0.02

*EP refers to the likelihood of exposures to lethal concentrations of chemicals released into environment.

For the different release severity indices (ranging from small to large), the following additional computations may be completed:

- Limited severity scenario
  - Average annual frequency of release of limited severity from accident
    = [no. of damaged trucks] × [release probability] × [release severity index]
    = $(2.0 \times 10^{-2}) \times (0.4) \times (0.5) = 4 \times 10^{-3}$
  - Average annual frequency of exposure (for limited severity scenario)
    = [annual frequency of release] × [exposure probability]
    = $(4 \times 10^{-3}) \times (0.5) = 2 \times 10^{-3}$
- Medium severity scenario
  - Average annual frequency of release of medium severity from accident
    = $(2.0 \times 10^{-2}) \times (0.4) \times (0.3) = 2.4 \times 10^{-3}$
  - Average annual frequency of exposure (for medium severity scenario)
    = $(2.4 \times 10^{-3}) \times (0.07) = 1.68 \times 10^{-4}$
- Extensive severity scenario
  - Average annual frequency of release of extensive severity from accident
    = $(2.0 \times 10^{-2}) \times (0.4) \times (0.1) = 8 \times 10^{-4}$
  - Average annual frequency of exposure (for extensive severity scenario)
    = $(8 \times 10^{-4}) \times (0.02) = 1.60 \times 10^{-5}$

In general, potential health impacts of any toxic releases depends on the severity of releases, meteorological conditions, as well as the population density along the transportation corridor. Table 15.2 summarizes these results, and Figure 15.6 is a risk curve showing a plot of the annual frequency of exposure versus the number of people affected; these results can be used in management decisions with respect to mode of transportation and traffic network design for the shipments of hazardous materials. Additionally, the average individual lifetime exposure probability can be estimated based on the average human life expectancy (say, 70 years), and these results can be compared to other activities that populations in the community are exposed to; subsequently, appropriate risk management and risk prevention programs can be developed to mitigate potential problems.

Table 15.2 Health impact data for F–N relationship

Population size	Average annual exposure frequency
0	2.00×10^{-3}
500	1.68×10^{-4}
1000	1.60×10^{-5}

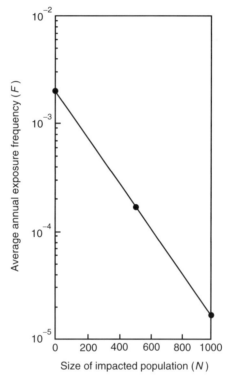

Figure 15.6 PRA application problem: F–N/Risk curve for a hazardous waste transportation hypothetical example

15.3 UTILITY OF PROBABILISTIC RISK ASSESSMENTS

PRA techniques provide a structured and systematic framework for evaluating the safety of environmental management systems – such as potentially hazardous materials TSDFs. The methodologies involved provide an effective way to build the comprehensive and defensible information base necessary for dealing with potential hazards from such facilities.

In fact, the potential for failures of hazardous waste facilities, and the inherent uncertainties associated with such facilities, all pose some degree of hazard; failures are generally considered to be the result of facility breach, followed by the release and migration of potentially harmful chemical contaminants into or through the environment. Failures may range from design flaws and deficiencies or faults, to operational and traffic accidents, to natural and human-made disasters.

PRA generally offers techniques used to identify possible hazards and potential consequences that follow from a system failure. Such analyses can then be used to improve overall system design and/or operation in order to minimize potential risks.

SUGGESTED FURTHER READING

Bohnenblust, H. and S. Pretre. 1990. Appraisal of individual radiation risk in the context of probabilistic exposures. *Risk Analysis*, 10(2): 247–253.
Ess, T.H. 1981. Risk estimation. In: *Proceedings of National Conference on Risk and Decision Analysis for Hazardous Waste Disposal*, Aug. 24–27, Baltimore, Maryland, pp. 155–163.
Frankiel, E.G. 1984. *Systems Reliability and Risk Analysis*. Martinus Nijhoff Publishers, Boston/The Hague.
Grossel, S.S. and D.A. Crowl (eds). 1995. *Handbook of Highly Toxic Materials Handling and Management*. Marcel Dekker, New York.
Louvar, J.F. and B.D. Louvar. 1998. *Health and Environmental Risk Analysis: Fundamentals with Applications*. Prentice-Hall, Upper Saddle River, NJ.
Sundararajan, C. 1991. *Guide to Reliability Engineering (Data, Analysis, Applications, Implementation, and Management)*. Van Nostrand Reinhold, New York.

REFERENCES

Ang, A.H-S. and W.H. Tang. 1984. *Probability Concepts in Engineering Planning and Design. Vol. II: Decision, Risk, and Reliability*. J. Wiley, New York.
Apostolakis, G. 1974. *Mathematical Methods of Probabilistic Safety Analysis*. University of California Report No. UCLA-ENG-7464, Los Angeles, CA.
Cheremisinoff, N.P. and M. Graffia. 1996. *Safety Management Practices for Hazardous Materials*. Marcel Dekker, New York.
CMA (Chemical Manufacturers Association). 1985. *Risk Analysis in the Chemical Industry*. Government Institutes, Rockville, MD.
Davidson, A. 1990. *In the Wake of the Exxon Valdez: The Devastating Impact of the Alaska Oil Spill*. Douglas & McIntyre, Vancouver, Canada.
Dhillon, B.S. and C. Singh. 1981. *Engineering Reliability*. J. Wiley, New York.
GAO (US General Accounting Office). 1985. *Probabilistic Risk Assessment: An Emerging Aid to Nuclear Power Plant Safety Regulation*. GAO/RCED-85-11, June 19, 1985.
Glickman, T.S. and M. Gough (eds). 1990. *Readings in Risk*. Resources for the Future, Washington, DC.
Harr, M.E. 1987. *Reliability-Based Design in Civil Engineering*. McGraw-Hill, New York.
Hauptmanns, U. and W. Werner. 1991. *Engineering Risks (Evaluation and Valuation)*. Springer-Verlag, Berlin, Germany.
Henley, E.J. and H. Kumamoto. 1981. *Reliability Engineering and Risk Assessment*. Prentice-Hall, Englewood Cliffs, NJ.
Kolluru, R.V., S.M. Bartell, R.M. Pitblado, and R.S. Stricoff (eds). 1996. *Risk Assessment and Management Handbook (for Environmental, Health, and Safety Professionals)*. McGraw-Hill, New York.

Lees, F.B. 1980. *Loss Prevention in the Process Industries.* Volume 1. Butterworths, Boston, MA.

McCormick, N.J. 1981. *Reliability and Risk Analysis: Methods and Nuclear Power Applications.* Academic Press, New York.

Nicolet-Monnier, M. and A.V. Gheorghe. 1996. *Quantitative Risk Assessment of Hazardous Materials Transport Systems.* Kluwer Academic Publishers, Dordrecht, The Netherlands.

NRC (National Research Council). 1983. *Risk Assessment in the Federal Government: Managing the Process.* NAS Press, Washington, DC.

NUREG (Nuclear Regulatory Commission). 1983. *PRA Procedures Guide – A Guide to the Performance of Probabilistic Risk Assessment for Nuclear Power Plants.* Report NUREG/CR-2300. Office of Nuclear Regulatory Research, Washington, DC.

Saccomanno, F.F., J.H. Shortreed, M. Van Averde and J. Higgs. 1989. Comparison of risk measures for the transport of dangerous commodities by truck and rail. Presented at the 68th Annual Meeting of the Transportation Research Board, Washington, DC, January, 1989.

Sidall J.N. 1983. *Probabilistic Engineering Design: Principles and Applications.* Marcel Dekker, New York.

Theodore, L., J.P. Reynolds, and F.B. Taylor. 1989. *Accident and Emergency Management.* Wiley–Interscience, New York.

USBR (US Bureau of Reclamation). 1986. *Guidelines to Decision Analysis.* ACER Technical Memo No. 7, Denver, CO.

USEPA (US Environmental Protection Agency). 1987. *Technical Guidance for Hazard Analysis.* Washington, DC, December 1987.

Waller, R.A. and V.T. Covello (eds). 1984. *Low-Probability/High-Consequence Risk Analysis: Issues, Methods, and Case Studies.* Plenum Press, New York.

World Bank. 1985. *Manual of Industrial Hazard Assessment Techniques.* Office of Environment and Scientific Affairs, Washington, DC, October 1985.

Zogg, H.A. 1987. *'Zurich' Hazard Analysis.* Zurich Insurance Group, Risk Engineering, Zurich, Switzerland.

PART V

DETERMINATION OF ACCEPTABLE RISK-BASED LIMITS FOR ENVIRONMENTAL CHEMICALS

This part of the book presents methodologies for determining risk-based chemical constituent levels that may be required for use in environmental manangement programs. It comprises the following specific chapters:

- Chapter 16, *General Protocols for Establishing Acceptable Chemical Concentrations and Environmental Quality Criteria*, presents generic procedures that may be employed or adapted to help establish what may constitute reasonably acceptable or tolerable chemical levels in the environment, and to help answer the ever-daunting 'how clean is clean'/'how safe is safe' questions. It also offers a brief annotation of the application of risk assessment techniques to determine reasonably 'safe' or 'acceptable' concentrations of chemical constituents that may appear in a variety of consumer products.
- Chapter 17, *Development of Risk-Based Remediation Goals*, elaborates the development of site-specific cleanup criteria and remediation goals for contaminated site problems.

Chapter Sixteen

General Protocols for Establishing Acceptable Chemical Concentrations and Environmental Quality Criteria

An important and yet controversial issue that comes up in attempts to establish 'safe' or 'acceptable' chemical concentrations for environmental management problems relates to the notion of an 'action level' (AL). The AL is considered as the concentration of a chemical in a particular medium that, when exceeded, presents significant risk of adverse impact to potential receptors (CDHS, 1986). In fact, in a number of situations, the AL tends to drive the environmental management decision made about an environmental contamination problem. However, the ALs may not always result in 'acceptable' or 'tolerable' risk levels due to the nature of the critical exposure scenarios, receptors, and other conditions that are specific to the particular hazard situation. Under such circumstances, it becomes necessary to develop more stringent and health-protective levels that will meet the 'acceptable' risk level criteria.

This chapter elaborates a number of analytical relationships that can be adopted or used to estimate environmental quality criteria and/or benchmark restoration goals necessary for environmental management decisions. It also offers a brief specific discussion of how risk assessment may facilitate a determination of what constitutes a reasonably 'safe' or 'acceptable' concentration of chemicals appearing in a variety of consumer products.

16.1 REQUIREMENTS AND CRITERIA FOR ESTABLISHING RISK-BASED TARGET LEVELS FOR ENVIRONMENTAL CHEMICALS

Risk-based target levels may generally be established for various environmental matrices by manipulating the risk and exposure models previously presented in Parts III and IV, and further elaborated in the literature elsewhere (e.g., ASTM, 1994, 1995; USEPA, 1991, 1996a, 1996b). This involves a back-calculation process to yield a media concentration that is based on health-protective exposure parameters (i.e., results in a noncancer hazard index ≤ 1 and/or a carcinogenic risk $\leq 10^{-6}$, for example). The target levels are typically established for both carcinogenic and noncarcinogenic effects of the environmental contaminants, with the more stringent value usually being selected as an environmental quality criterion (Figure 16.1); invariably, the carcinogenic limit tends to be more stringent in most situations where

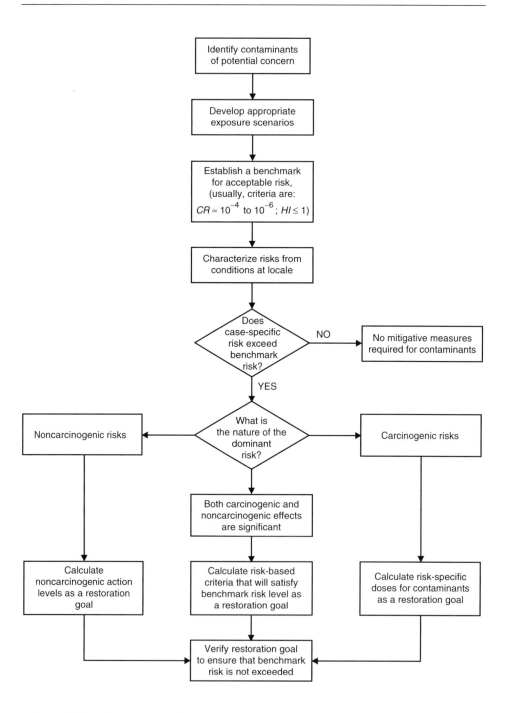

Figure 16.1 General protocol for developing risk-based target levels of environmental quality criteria for environmental chemicals

both values exist. In addition, the following criteria, assuming dose additivity, must be met by the environmental quality goal:

$$\sum_{j=1}^{p} \sum_{i=1}^{n} \frac{CMAX_{ij}}{RBCL_{ij}} < 1$$

where $CMAX_{ij}$ is the prevailing maximum concentration of contaminant i in environmental matrix j, and $RBCL_{ij}$ is the risk-based chemical level for contaminant i in medium j. Furthermore, the following general guidelines may be used to facilitate the process of establishing media-specific $RBCL$s:

- In developing environmental quality criteria, it usually is necessary to establish a target level of risk for the contaminants of concern; environmental standards are generally established within the risk range of 10^{-7} to 10^{-4} (with a lifetime excess cancer risk of 10^{-6} normally used as a point-of-departure) and a hazard index of 1.
- It is recommended that the cumulative risk posed by multiple contaminants does not exceed a 10^{-4} cancer risk and/or a hazard index of unity.
- If sensitive populations (including sensitive ecosystems and habitats, or threatened or endangered species) are to be protected, then more stringent standards may be required.
- If nearby populations are exposed to hazardous constituents from other sources, lower target levels may generally be required than would ordinarily be necessary.
- If exposures to certain hazardous constituents occur through multiple pathways, lower target levels should generally be prescribed.

When these conditions are satisfied, then the $RBCL$ represents a maximum acceptable contaminant level that will likely be protective of public health and the environment. In general, exceeding the $RBCL$ will usually call for the development and implementation of a corrective action and/or risk management plan.

16.2 DERIVATION OF HEALTH-BASED ACTION LEVELS FOR ENVIRONMENTAL CHEMICALS

Health-based criteria for carcinogens, commonly referred to as 'virtually safe doses' (VSDs), may be used as the action level of a carcinogenic chemical constituent. A VSD is the daily dose of a carcinogenic chemical that, over a lifetime, will result in an incidence of cancer equal to a specified risk level. This corresponds to environmental concentrations that, under case-specific intake assumptions, yield a specified cumulative lifetime cancer risk (e.g., of 10^{-6} for Class A and B carcinogens, 10^{-5} for Class C carcinogens, etc.; or 10^{-7} for situations when sensitive populations such as nursing homes or schools are potentially exposed, 10^{-4} when the potential for exposures is rather remote, etc.). The VSD is calculated based on the appropriate risk level, and this becomes the action level for the specific chemical.

> Box 16.1 General equation for calculating media action levels for carcinogenic chemical constituents
>
> $$AL_c = \frac{(R \times BW \times LT \times CF)}{(SF \times I \times A \times ED)}$$
>
> where:
> AL_c = Action level (equal to the *VSD*) in medium of concern (e.g., mg/kg in soil; mg/L in water)
> R = Specified benchmark risk level, usually set at 10^{-6} (dimensionless)
> BW = Body weight (kg)
> LT = Assumed lifetime (years)
> CF = Conversion factor (equals 10^6 for ingestion exposure from solid materials; 1.00 for ingestion of fluids)
> SF = Cancer slope factor ($[mg/kg\text{-}day]^{-1}$)
> I = Intake assumption (mg/day for solid material ingestion rate; L/day for fluid ingestion)
> A = Absorption factor (dimensionless)
> ED = Exposure duration (years)

Health-based criteria for noncarcinogenic effects of chemical constituents, commonly referred to as allowable daily intakes (*ADI*s), are generally estimated from the threshold exposure limit below which no adverse effects are anticipated.

For substances that are both carcinogenic and systemically toxic, the lower of the *VSD* or *ADI* criteria should be used for the relevant environmental management decision.

16.2.1 Action Levels for Carcinogenic Chemicals

The governing equation for calculating action levels for carcinogenic constituents is shown in Box 16.1. This model – derived from algorithms and concepts presented earlier on in Chapter 13 – assumes that there is only one chemical constituent involved in the problem situation. In other situations where several chemicals may be of concern, it is assumed (for simplification purposes) that each carcinogen has a different mode of biological action and target organs. Each of the carcinogens is, therefore, assigned 100% of the 'acceptable' excess carcinogenic risk (typically equal to 1×10^{-6}) in calculating the health-protective concentration levels. That is, excess carcinogenic risk is not allocated among the carcinogens because each individual carcinogen may have different mode of biological action and target organs.

16.2.1.1 Example calculations

Consider a hypothetical situation whereby some residential population may be consuming water contaminated with methylene chloride. Then, the allowable human exposure due to ingestion of 2 liters of the water containing methylene chloride (with oral $SF = 7.5 \times 10^{-3}$ [mg/kg-day]$^{-1}$) by a 70-kg weight adult over a 70-year lifetime is given by:

$$VSD = \frac{(R \times BW \times LT \times CF)}{(SF \times I \times A \times ED)}$$

$$VSD_{\text{methchl}} = \frac{[10^{-6} \times 70 \times 70 \times 1]}{[0.0075 \times 2 \times 1 \times 70]} = 0.005 \text{ mg/L} = 5 \text{ } \mu\text{g/L}$$

That is, the *VSD* or allowable exposure level for methylene chloride based on an acceptable excess lifetime cancer risk level of 10^{-6} is estimated to be 5 μg/L.

Next, in another situation, consider the case of a contaminated site impacting a multipurpose surface water body due to overland flow. This surface water body is used both as a culinary water supply source and for recreational purposes. Assuming – in addition to the water intake – an average daily consumption of aquatic organisms, *DIA*, of 6.5 g/day, and a *BCF* of 0.91 L/kg for methylene chloride, then the human exposure levels for ingestion of both water and fish is determined from the following modified equation:

$$VSD_{\text{methchl}} = \frac{[R \times BW \times LT \times CF]}{[SF \times (I + (DIA \times BCF)) \times A \times ED]}$$

$$= \frac{[10^{-6} \times 70 \times 70 \times 1]}{[0.0075 \times (2 + (0.0065 \times 0.91)) \times 1 \times 70]} \approx 0.005 \text{ mg/L} = 5 \text{ } \mu\text{g/L}$$

Thus, the allowable exposure levels for drinking water and eating aquatic organisms contaminated with methylene chloride is also approximately 5 μg/L in this particular case.

16.2.2 Action Levels for Noncarcinogenic Chemicals/Systemic Toxicants

The governing equation for calculating action levels for noncarcinogenic effects of chemical constituents is shown in Box 16.2. This model – derived from algorithms and concepts presented earlier on in Chapter 13 – assumes that there is only one chemical constituent involved. In situations where several chemicals may be of concern, it is assumed (for simplification purposes) that each chemical has a different organ-specific noncarcinogenic effect. Otherwise, the right-hand side may be multiplied by a percentage factor to account for the contribution to the hazard index by each noncarcinogenic chemical subgroup.

16.2.2.1 Example calculations

Consider a hypothetical situation whereby some residential population may be consuming water contaminated with ethylbenzene. Then, the allowable human exposure concentration due to ingestion of 2 liters of water containing ethylbenzene (with an *RfD* of 0.1 mg/kg-day) by a 70-kg weight adult is given by:

> Box 16.2 General equation for calculating media action levels for noncarcinogenic effects of chemical contaminants
>
> $$AL_{nc} = \frac{(RfD \times BW \times CF)}{(I \times A)}$$
>
> where:
> AL_{nc} = Action level in medium of concern (e.g., mg/kg in soil; mg/L in water)
> RfD = Reference dose (mg/kg-day)
> BW = Body weight (kg)
> CF = Conversion factor (equals 10^6 for ingestion exposure from solid materials; 1.00 for fluid ingestion)
> I = Intake assumption (mg/day for solid material ingestion rate; L/day for fluid ingestion)
> A = Absorption factor (dimensionless)

$$ADI = \frac{[RfD \times BW]}{[DW \times A]}$$

$$ADI_{ebz} = \frac{[0.1 \times 70]}{[2 \times 1]} = 3500 \ \mu g/L$$

Next, in another situation, consider the case of a contaminated site impacting a multipurpose surface water body due to overland flow. This surface water body is used both as a culinary water supply source and for recreational purposes. Assuming – in addition to the water intake – an average daily consumption of aquatic organisms, DIA, of 6.5 g/day, and a BCF of 37.5 L/kg for ethylbenzene, then the human exposure levels for ingestion of both water and fish is determined from the following modified equation:

$$ADI_{ebz} \ [mg/L] = \frac{[RfD \times BW]}{[2 + (0.0065 \times BCF)] \times 1}$$

$$= \frac{[0.1 \times 70]}{[2 + (0.0065 \times 37.5)]} = 3120 \ \mu g/L$$

Thus, the allowable exposure concentration (represented by the water AL) for drinking water and eating aquatic organisms contaminated with ethylbenzene is approximately 3120 $\mu g/L$.

16.3 HEALTH-PROTECTIVE RISK-BASED CHEMICAL CONCENTRATIONS

After defining the critical pathways and exposure scenarios associated with an environmental contamination problem, it often is possible to calculate the various media concentrations that do not pose significant risks to an exposed population. This level represents the risk-based concentrations (*RBCs*).

To determine the *RBC* for a chemical compound, algebraic manipulations of the hazard index or carcinogenic risk equations and the exposure estimation equations discussed in Chapter 13 can be used to arrive at the appropriate analytical relationships. This step-wise computational process involves a back-calculation process that results in an acceptable media concentration, and that is based on health-protective exposure parameters (i.e., it generally yields a noncancer hazard index ≤ 1 and/or a carcinogenic risk $\leq 10^{-6}$). For chemicals with carcinogenic effects, a target risk of 1×10^{-6} is generally used in the back-calculation; a target hazard index of 1.0 is generally used for noncarcinogenic effects.

16.3.1 *RBCs* for Carcinogenic Constituents

As discussed in Chapter 13 the cancer risk (*CR*) for the significant human exposure routes (comprised of inhalation, ingestion, and dermal exposures) may be represented as follows:

$$CR = \sum \left\{ \sum_{i=1}^{p} CDI_p \times SF_p \right\}$$

$$= [CDI_i \times SF_i]_{\text{inhalation}} + [CDI_o \times SF_o]_{\text{ingestion}} + [CDI_d \times SF_o]_{\text{dermal contact}}$$

$$= C_m \{[INHf \times SF_i] + [INGf \times SF_o] + [DEXf \times SF_o]\}$$

where the *CDIs* represent the chronic daily intakes, adjusted for absorption (mg/kg-day); *INHf*, *INGf*, and *DEXf* represent the inhalation, ingestion, and dermal contact 'intake factors', respectively (see Chapter 13); C_m is the contaminant concentration in the environmental matrix of concern; and the *SFs* are the route-specific cancer slope factors; the subscripts i, o and d refer to the inhalation, oral ingestion, and dermal contact exposures, respectively. This model can be reformulated to calculate the carcinogenic *RBC* (*RBC*$_c$) for the environmental media of interest. This is derived by back-calculating from the chemical intake equations presented in Chapter 13 for inhalation, ingestion, and dermal contact exposures. Hence,

$$RBC_c = C_m = \frac{CR}{\{[INHf \times SF_i] + [INGf \times SF_o] + [DEXf + SF_o]\}}$$

For illustrative purposes, assume that there is only one chemical constituent present in soils at a hypothetical contaminated site, and that only exposures via the dermal and ingestion routes contribute to or dominate the total target carcinogenic risk (of $CR = 10^{-6}$). Then,

$$\sum CDI = \frac{CR}{SF_o} = RSD$$

or,

$$(CDI_{ing} + CDI_{der}) = \frac{CR}{SF_o}$$

i.e.,

$$\frac{(RBC_c \times SIR \times CF \times FI \times ABS_{si} \times EF \times ED)}{(BW \times AT \times 365)}$$

$$+ \frac{(RBC_c \times CF \times SA \times AF \times ABS_{sd} \times SM \times EF \times ED)}{(BW \times AT \times 365)}$$

$$= \frac{CR}{SF_o}$$

Consequently,

$$RBC_c = \frac{(BW \times AT \times 365) \times (RSD)}{(CF \times EF \times ED)\{(SIR \times FI \times ABS_{si}) + (SA \times AF \times ABS_{sd} \times SM)\}}$$

where RSD represents the risk-specific dose, defined by the ratio of the target risk to the slope factor.

16.3.2 *RBCs for Noncarcinogenic Effects of Chemical Constituents*

As discussed in Chapter 13, the hazard index (HI) for the significant human exposure routes (comprised of inhalation, ingestion, and dermal exposures) is given by:

$$HI = \sum \left\{ \sum_{i=1}^{p} \frac{CDI_p}{RfD_p} \right\}$$

$$= \left[\frac{CDI_i}{RfD_i} \right]_{inhalation} + \left[\frac{CDI_o}{RfD_o} \right]_{ingestion} + \left[\frac{CDI_i}{RfD_o} \right]_{dermal\ contact}$$

$$= C_m \left\{ \left[\frac{INHf}{RfD_i} \right] + \left[\frac{INGf}{RfD_o} \right] + \left[\frac{DEXf}{RfD_o} \right] \right\}$$

where the CDIs represent the chronic daily intakes, adjusted for absorption (mg/kg-day); $INHf$, $INGf$, and $DEXf$ represent the inhalation, ingestion, and dermal contact 'intake factors', respectively (see Chapter 13); C_m is the contaminant concentration in the environmental matrix of concern; and the *RfD*s are the route-specific reference doses; the subscripts i, o, and d refer to the inhalation, oral ingestion and dermal contact exposures, respectively. This model can be reformulated to calculate the noncarcinogenic *RBC* (RBC_{nc}) for the environmental media of interest. This is derived by back-calculating from the chemical intake equations presented in Chapter 13 for inhalation, ingestion, and dermal contact exposures. Hence,

$$RBC_{nc} = C_m = \frac{1}{\left\{\left[\dfrac{INHf}{RfD_i}\right] + \left[\dfrac{INGf}{RfD_o}\right] + \left[\dfrac{DEXf}{RfD_o}\right]\right\}}$$

For illustrative purposes, assume that there is only one chemical constituent present in soils at a hypothetical contaminated site, and that only exposures via the dermal and ingestion routes contribute to, or dominate the total target hazard index (of *HI* =1). Then,

$$\sum CDI = RfD$$

or,

$$(CDI_{ing} + CDI_{der}) = RfD_o$$

i.e.,

$$\frac{(RBC_{nc} \times SIR \times CF \times FI \times ABS_{si} \times EF \times ED)}{(BW \times AT \times 365)}$$

$$+ \frac{(RBC_{nc} \times CF \times SA \times AF \times ABS_{sd} \times SM \times EF \times ED)}{(BW \times AT \times 365)}$$

$$= RfD_o$$

Consequently,

$$RBC_{nc} = \frac{(BW \times AT \times 365) \times (RfD_o)}{(CF \times EF \times ED)\{(SIR \times FI \times ABS_{si}) + (SA \times AF \times ABS_{sd} \times SM)\}}$$

assuming a benchmark hazard index of unity.

16.4 A 'PREFERABLE' HEALTH-PROTECTIVE CHEMICAL LEVEL

Often, the *RBC* that has been established based on an acceptable risk level or hazard index is for a single contaminant in one environmental matrix. Therefore the risk and hazard associated with multiple contaminants in a multimedia setting are not fully accounted for during the back-modeling process used to establish the *RBC*s. In contrast, the evaluation of risks associated with a given environmental contamination problem usually involves a set of equations designed to estimate hazard and risk for several contaminants, and for a multiplicity of exposure pathways. Under this latter type of scenario, the computed 'acceptable' risks could indeed exceed the health-protective limits; consequently, it becomes necessary to establish a modified *RBC* for the requisite environmental management decision. To obtain the modified *RBC*, the 'acceptable' contaminant level is estimated in the same way as previously elaborated (in Sections 16.2 and 16.3), but with the cumulative effects of multiple contaminants being taken into account through a process of apportioning target risks and hazards among all the chemicals of potential concern.

16.4.1 The Modified *RBC* for Carcinogenic Chemicals

The modified *RBC* for carcinogenic constituents may be derived by application of a 'risk disaggregation factor', that allows for the apportionment of risk amongst all chemicals of concern. The *RBC* may be estimated by proportionately aggregating – or rather disaggregating – the target cancer risk between the chemicals of potential concern. The assumption used for apportioning the excess carcinogenic risk may be that all carcinogens have the same mode of biological actions and target organs; otherwise, excess carcinogenic risk is not apportioned among carcinogens, but rather each assumes the same value in the computational efforts. A more comprehensive approach to 'partitioning' or combining risks would involve more complicated mathematical manipulations, such as by the use of linear programming algorithms.

The acceptable risk level may be apportioned between the chemical constituents contributing to the overall target risk by assuming that each constituent contributes equally or proportionately to the total acceptable risk. The 'risk fraction' obtained for each constituent can then be used to derive the modified *RBC* by working from the relationships established previously for the computation of *RBC*s (Section 16.3); by using the approach to estimating media *RBC*s, this is carried out according to the following approximate relationship:

$$RBC_{c\text{-mod}} = \frac{[\%] \times CR}{\{[INHf \times SF_i] + [INGf \times SF_o] + [DEXf \times SF_o]\}}$$

All the terms are the same as defined previously in Section 16.3, and [%] represents the proportionate contribution from a specific chemical constituent to the overall target risk level. One may also choose to use weighting factors (based, for instance, on carcinogenic classes such that class A carcinogens are given twice as much weight as class B, etc., or chemicals posing carcinogenic risk via all exposure routes are given more weight than those presenting similar risks via specific routes only) in apportioning the chemical contributions to the target risk levels.

Overall, such an approach will ensure that the sum of risks from all the chemicals involved over all exposure pathways is less than or equal to the set target risk (i.e., $\leq 10^{-6}$, as an example).

16.4.2 The Modified *RBC* for Noncarcinogenic Constituents

The modified *RBC* for noncarcinogenic constituents may be derived by application of a 'hazard disaggregation factor', that allows for the apportionment of hazard index amongst all chemicals of concern. The *RBC* may be estimated by proportionately aggregating – or rather disaggregating – the noncancer hazard index among the chemicals of potential concern.

The acceptable hazard level may be apportioned between the chemical constituents contributing to the overall hazard index, assuming each constituent contributes equally or proportionately to the total acceptable hazard index. The 'hazard fraction' obtained for each constituent can then be used to derive the modified *RBC* by working from the relationships established previously for the computation of *RBC*s (Section 16.3); by using the approach to estimating media *RBC*s, this is carried out according to the following approximate relationship for noncarcinogenic effects of chemicals having the same toxicological endpoints:

$$RBC_{\text{nc-mod}} = \frac{[\%] \times 1}{\left\{ \left[\frac{INHf}{RfD_i}\right] + \left[\frac{INGf}{RfD_o}\right] + \left[\frac{DEXf}{RfD_o}\right] \right\}}$$

All the terms are the same as defined previously in Section 16.3, and [%] represents the proportionate contribution from a specific chemical constituent to the overall target hazard index for noncarcinogenic effects of chemicals with same physiologic endpoint.

Overall, such an approach will ensure that the sum of hazard quotients across all exposure pathways for all chemicals (with the same physiologic endpoints) is less than or equal to the hazard index criterion of 1.0.

16.5 HEALTH-PROTECTIVE CHEMICAL CONCENTRATIONS IN CONSUMER PRODUCTS

A number of peoples around the world are exposed to a barrage of chemical compounds on a daily basis – through their use of a variety of consumer products. Generally, risk assessments – which allow the consumer exposures to be estimated by measurements and/or models – assist in the management of potential health problems expected or anticipated from the use of such consumer products. The exposure assessment component of the process involved tends to be particularly complicated – though not unsurmountable – by the huge diversity in usage and composition of consumer products, resulting in the intermittent exposures to varying amounts and types of products that contain varying concentrations of chemical compounds (van Veen, 1996; Vermeire et al., 1993).

16.5.1 Formulation of Potential Consumer Exposures

Consumer exposures may occur via the inhalatory, dermal, and/or oral routes. In general, careful consideration of the types and extent of potential consumer exposures, combined with hazard assessment and exposure–response information, is necessary for the conduct of a credible human health risk assessment in relation to consumer products usage. The hazard assessment for a consumer product or component thereof is associated with any human health effects, and the exposure–response assessments involve an examination of the relationship between the degree of exposure to a product or component and the magnitude of any specific adverse effect(s). Additionally, an exposure assessment – which is critical to determining potential risks – requires realistic data to assess the extent of possible skin, inhalation, and ingestion exposures to products and components (Corn, 1993).

The likely types of consumer exposures to a variety of consumer products may indeed fall into several categories, including the following particularly significant ones (Al-Saleh and Coate, 1995; Corn, 1993; OECD, 1993):

- *Skin exposures* – from normal usage of product – expressed by the following form of generic relationship,

$$\text{Dermal exposure} = \frac{\{[CONC] \times [PERM] \times [AREA] \times [EXPOSE]\}}{[BW]}$$

where $CONC$ is the concentration of material; $PERM$ is the skin permeability constant; $AREA$ is the area of exposed skin; $EXPOSE$ is the exposure duration; and BW is the average body weight.

- *Oral exposures* – from normal usage of product – expressed by the following form of generic relationship,

$$\text{Oral exposure} = \frac{\{[CONC] \times [CONSUME] \times [ABSORB] \times [EXPOSE]\}}{[BW]}$$

where $CONC$ is the concentration of material; $CONSUME$ is the consumption amount/rate of material; $ABSORB$ is the percent (%) absorption; $EXPOSE$ is the exposure duration; and BW is the average body weight.

- *Inhalation exposures to volatiles* – from normal usage of product – expressed by the following form of generic relationship,

$$\text{Inhalation exposure to volatiles} = \frac{\{[VAPOR] \times [INHALE] \times [RETAIN] \times [EXPOSE]\}}{[BW]}$$

where *VAPOR* is the vapor phase concentration of material; *INHALE* is the inhalation rate; *RETAIN* is the lung retention rate; *EXPOSE* is the exposure duration; and *BW* is the average body weight.
- *Inhalation exposures to particulates* – from normal usage of product – expressed by the following form of generic relationship,

$$\text{Inhalation exposure to particulates} = \frac{\left\{ \begin{array}{c} [PARTICLE] \times [RESPIRABLE] \\ \times [INHALE] \times [ABSORB] \times [EXPOSE] \end{array} \right\}}{[BW]}$$

where *PARTICLE* is the total aerosol or particulate concentration of material; *RESPIRABLE* is the % of respirable material; *INHALE* is the inhalation rate; *ABSORB* is the % absorbed; *EXPOSE* is the exposure duration; and *BW* is the average body weight.

In addition, accidental exposures may occur via the same routes from dermal contact, oral ingestion, and/or inhalation exposures.

16.5.1.1 Dietary exposures to chemicals

Human dietary exposure to chemicals in food (and indeed similar consumables) depends both on food consumption patterns and the residue levels of a particular chemical on/in food, generally expressed by the following conceptual relationship (Driver et al., 1996):

$$\text{Dietary exposure} = f(\text{Consumption, Chemical concentration})$$

Typically, multiplying the average consumption of a particular food product by the average chemical concentration on/in that food yields the average level of ingestion of that chemical from the food product. In reality, however, estimation of dietary exposure to chemicals – such as pesticides or food additives – becomes a complex endeavor, especially because of the following likely reasons (Driver et al., 1996):

- occurrence of a particular chemical in more than one food item;
- variation in chemical concentrations in food products;
- person-to-person variations in the consumption of various food products;
- variation in dietary profiles across age, gender, ethnic groups, and geographic regions; and
- possible reductions in chemical concentrations due to transformation during transport, storage, and food preparation.

Consequently, both food consumption and chemical concentration data are best represented or characterized by dynamic distributions that reflect a wide range of values, rather than by a single value. The distribution of dietary exposures and risks is determined by using both the distribution of food consumption levels and the distribution of chemical concentrations in food (see, e.g., Brown et al., 1988; Driver et al., 1996; NRC, 1993; Rodricks and Taylor, 1983; USEPA, 1986). In fact, the inherent variability and uncertainty in food consumption and chemical concentration data result in a high degree of variability in dietary exposure and risk of a given chemical. For instance, the dietary habits of a home gardener may result in an increase or decrease in exposure – especially resulting from their consumption rate and contaminated fractions. In this regard, individual consumers may ingest significantly different quantities of produce and, depending on their fruit/vegetable preferences, may rely on crops that are efficient accumulators of chemicals or otherwise.

16.5.2 Assessing the Chemical Safety of Consumer Products

Consumer product safety is a function of exposure and toxicity, determined primarily based on the exposure patterns/rates and the toxicity of the chemical components of concern – represented by the following conceptual expressions:

$$\text{Risk} = f(\text{Exposure}, \text{Toxicity})$$

or,

$$\text{Safety} \, \alpha \, \frac{1}{\text{Risk}} = \frac{1}{f(\text{Exposure}, \text{Toxicity})}$$

For a particular consumer product to be classified as reasonably safe, the chemical-specific exposure dose should generally be less than the chemical's 'acceptable' daily intake – defined as the daily intake level for a chemical which represents no anticipated significant risk to the consumer or exposed individual.

16.5.2.1 Determination of tolerable chemical concentrations

Chemicals in consumer products (including those occurring in dietary materials or foods) may be classified into two broad categories – carcinogenic and noncarcinogenic materials. The methods for deriving the 'acceptable' daily intakes and/or tolerable concentrations for such chemicals are generally based on procedures/protocols presented earlier on in Section 16.3, and briefly annotated below.

- *'Acceptable' daily intake and 'tolerable' concentration for carcinogens.* The 'acceptable' daily intake for carcinogenic materials appearing in a consumer product may be estimated by the following approximate relationships:

$$ADI_{\text{carcinogen}} = \frac{[TR \times AT \times 365 \text{ days/years}]}{[ED \times EF \times SF]}$$

Thence, the 'tolerable' chemical concentration for carcinogens ($TC_{carcinogen}$) (mg/kg or mg/L) in the consumer product will be defined by,

$$TC_{carcinogen} = \frac{[ADI_{carcinogen} \times BW]}{[FR \times CR \times ABS]} \times CF$$

where $ADI_{carcinogen}$ is the tolerable daily intake for the carcinogenic materials (mg/kg-day); TR is the generally acceptable risk level (usually set at 10^{-6}); AT is the averaging time (years); ED is the exposure duration (years); EF is the exposure frequency (days/year); SF is the cancer potency or slope factor ([mg/kg-day]$^{-1}$); BW is the average body weight (kg); FR is the fraction of consumed material that is assumed to be contaminated; CR is the consumption rate (kg/day or L/day); ABS is the % absorption rate; and CF is a conversion factor to help maintain the dimensional tractability of the algorithm.

- *'Acceptable' daily intake and 'tolerable' concentration for noncarcinogens.* The 'acceptable' daily intake for noncarcinogenic materials appearing in a consumer product may be estimated by the following approximate relationship:

$$ADI_{noncarcinogen} = \frac{[HQ \times AT \times 365 \text{ days/year} \times RfD]}{[ED \times EF]}$$

Thence, the 'tolerable' chemical concentration for noncarcinogens ($TC_{noncarcinogen}$) (mg/kg or mg/L) in the consumer product will be defined by,

$$TC_{noncarcinogen} = \frac{[ADI_{noncarcinogen} \times BW]}{[FR \times CR \times ABS]} \times CF$$

where $ADI_{noncarcinogen}$ is the tolerable daily intake for the noncarcinogenic materials (mg/kg-day); HQ is the generally acceptable hazard level (usually set at 1); AT is the averaging time (years); ED is the exposure duration (years); EF is the exposure frequency (days/year); RfD is the noncancer reference dose or acceptable daily intake (mg/kg-day); BW is the average body weight (kg); FR is the fraction of consumed material that is assumed to be contaminated; CR is the consumption rate (kg/day or L/day); ABS is the % absorption rate; and CF is a conversion factor to help maintain the dimensional tractability of the algorithm.

16.6 MISCELLANEOUS METHODS FOR ESTABLISHING ENVIRONMENTAL QUALITY GOALS

Several possibilities exist to use various analytical tools in the development of alternative and media-specific contaminant concentration limits and environmental restoration goals. Some select general procedures commonly employed in establishing environmental quality goals are briefly annotated below.

- *Determination of risk-specific concentrations in air.* By employing the risk assessment concepts and methodologies discussed in Chapter 13, risk-specific concentrations of chemicals in air may be estimated from the unit risk in air as follows:

$$\text{Air concentration } [\mu g/m^3] = \frac{[\text{specified risk level}] \times [\text{body weight}]}{SF_i \times [\text{inhalation rate}] \times 10^{-3}}$$

$$= \frac{[\text{specified risk level}]}{\text{unit risk factor, } URF_i} = \frac{1 \times 10^{-6}}{URF_i}$$

The assumptions generally used for such computations involve a specified risk level of 10^{-6}, a 70-kg body weight, and an average inhalation rate of 20 m³/day.

- *Determination of risk-specific concentrations in water.* By employing the risk assessment concepts and methodologies discussed in Chapter 13, risk-specific concentrations of chemicals in drinking water can be estimated from the oral slope factor; the water concentration corrected for an upper-bound increased lifetime risk of R is given by:

$$\text{Water concentration } [mg/L] = \frac{[\text{specified risk level}] \times [\text{body weight}]}{SF_o \times [\text{ingestion rate}]}$$

or,

$$= \frac{\text{specified risk level}}{\text{unit risk factor, } URF_o}$$

$$= \frac{1 \times 10^{-6} \times 70 \text{ kg}}{SF_o \, (mg/kg/day)^{-1} \times 2 \text{ L/day}} = \frac{3.5 \times 10^{-5}}{SF_o}$$

The assumptions generally used for such computations involve a specified risk level of 10^{-6}, a 70-kg body weight, and an average water ingestion rate of 2 L/day.

- *Use of contaminant partitioning coefficients.* Environmental quality criteria may be derived from correlational analyses of the partitioning of chemicals between environmental compartments. For instance, based on the assumption that the distribution of contaminants among various compartments in sediment is controlled by continuous equilibrium exchanges, chemical-specific partition coefficients can be used to predict contaminant concentrations in sediment, biota, and/or water. That is, sediment–water equilibrium partitioning concepts can be used to predict contaminant concentrations in sediment and/or water. Similarly, sediment–biota partitioning coefficients can be determined and used to predict distribution of the contaminant between sediment and benthic organism and/or interstitial water and benthic organism (assuming bioaccumulation factors are constant and independent of organism or sediment). Other equilibrium relationships can indeed be employed for the estimation of various environmental quality criteria.

- *Use of attenuation-dilution factors.* In the past, environmental restoration goals have been established that either account for attenuation alone or dilution only (e.g., Brown, 1986; Dawson and Sanning, 1982; Santos and Sullivan, 1988; USEPA, 1987). To take account of both contaminant attenuation and dilution effects, a relationship that integrates total attenuation and net dilution concurrently seems appropriate, if this is to become the basis for a realistic environmental restoration program. For example, in the application of attenuation–dilution factors as the basis for developing site cleanup limits for a potentially contaminated site, the following relationship that accounts for both total attenuation and net dilution concurrently may be adopted for such purposes:

$$SCL = \{Std\} \times \prod_m \{AF\} \times \prod_m \{DF\}$$

where SCL is source/soil cleanup level (mg/kg); Std is receiving media criteria or regulatory standard to be attained (e.g., drinking water standard); $\Pi_m \{AF\}$ is cumulative attenuation factor, defining the loss of contaminant during transport (i.e., product of the intermedia distribution constants); and $\Pi_m \{DF\}$ is cumulative dilution factor during transport (i.e., product of the ratios of receiving media to source concentrations). For a single intermedia transfer, the soil cleanup level may be estimated by a simplified relationship; if, for instance, leachate migrates from soil into groundwater that is a source of a drinking water supply, the SCL may be estimated as follows:

$$SCL = DWS \times AF \times DF$$

where DWS is drinking water standard; AF is attenuation of contaminant in soil (typical values in the range 1 to 1000); and DF is dilution of contaminant by groundwater (typical values in the range 1 to 100).

In general, alternate cleanup goals based simply on attenuation or dilution mechanisms may not be acceptable unless they are accompanied by an in-depth evaluation and discussion of all pertinent processes involved.

- *Mass balance analyses.* Simple mass balance analytical relationships can be applied between various environmental compartments in order to derive environmental restoration goals. For intermedia transfer of contaminants, current contaminant loadings may be coupled with allowable loadings (as represented by available media standards), and a back-modeling procedure can be used to obtain the required environmental quality criteria, according to the following simple relationship:

$$C_{max} = \frac{C_{std}}{C_r} \times C_s$$

where C_{max} is the maximum acceptable source concentration; C_{std} is the receiving media environmental quality criteria for target receptors (i.e., regulatory standard); C_r is the receiving media concentration; and C_s is the prevailing source concentration.

For example, consider a situation where water quality standards exist that should be met for a creek adjoining a contaminated site. By performing back-calculations, based on contaminant concentrations in the creek as a result of the current constituents loading from the site, a conservative estimate can be made for C_{max}. The use of such a concentration limit in a corrective action process ensures that the creek is not adversely impacted, based on the exposure scenario defined for the site. Consequently, if the site is cleaned up to such levels, then the surface water quality is not expected to be impacted.

Broadly speaking, these approaches represent reasonably conservative ways of setting environmental restoration goals. The use of such methods of approach will usually ensure that risks are not under-estimated, which should result in the adequate protection of public health.

16.7 INCORPORATING DEGRADATION RATES INTO THE ESTIMATION OF ENVIRONMENTAL QUALITY CRITERIA

The effects of contaminant decay are generally not incorporated into the estimation of environmental restoration goals. On the other hand, since exposure scenarios used in calculating the *RBC* or similar criteria consider the fact that exposures could be occurring over long time periods (up to a lifetime of 70 years), it is prudent, in a detailed analysis, to consider the fact that degradation or other transformation of the chemical at the source could occur. In such cases, the degradation properties of the contaminants of concern should be closely evaluated. Subsequently, an adjusted *RBC* (or its equivalent) can be estimated that is based on the original *RBC* (or equivalent), a degradation rate coefficient, and a specified exposure duration. The newly calculated limit would represent the true environmental quality goal, given by:

$$RBC_a = \frac{RBC}{\text{degradation factor }(DGF)}$$

where RBC_a is the adjusted *RBC* or its equivalent (i.e., an environmental quality goal based on the *RBC* and a degradation rate coefficient).

Assuming first-order kinetics, an approximation of the degradation effects can be obtained by the following term:

$$DGF = \frac{(1 - e^{-kt})}{kt}$$

where k is a chemical-specific degradation rate constant (days^{-1}), and t is the time period over which exposure occurs (days). Consequently,

$$RBC_a = RBC \times \frac{kt}{(1 - e^{-kt})}$$

For a first-order decaying substance, k is estimated from the following relationship:

$$T_{1/2} \text{ [days]} = \frac{0.693}{k} \quad \text{or} \quad k \text{[days}^{-1}\text{]} = \frac{0.693}{T_{1/2}}$$

where $T_{1/2}$ is the half-life, which is the time after which the mass of a given substance will be one-half its initial value. The relationship for RBC_a assumes that a first-order degradation/decay is occurring during the complete exposure period; decay/degradation is initiated at time, $t = 0$ years; and the RBC is the average allowable concentration over the exposure period. In fact, if significant degradation is likely to occur, the RBC_a calculations become much more complicated. In that case, predicated source contaminant levels must be calculated at frequent intervals and summed over the exposure period.

16.8 UTILITY OF ENVIRONMENTAL BENCHMARKS

Often, pre-established environmental quality criteria (EQC) are used to define environmental target goals if they are determined to represent an 'acceptable' benchmark level for the case-specific situation. However, such EQCs may not always be available or they may not be adequate if the presence of multiple contaminants, multiple pathways, or other extraneous factors result in an 'unacceptable' aggregate risk for the particular situation. Under such circumstances, the 'acceptable' level may be represented by a risk-based benchmark – derived for the various pathways from elaborately defined exposure scenarios. In fact, the use of risk assessment principles to establish case-specific benchmarks for environmental contamination problems represents an even better and more sophisticated approach to designing cost-effective environmental management programs – in comparison with the use of generic benchmarks.

Typically, the risk-based benchmarks are developed by 'back-modeling' from a target risk level in order to obtain an acceptable risk-based concentration, or a maximum acceptable concentration. Invariably, the type of exposure scenarios envisioned and the exposure assumptions used may predicate the concomitant benchmark level. Ultimately, the use of such an approach aids in the development and/or selection of appropriate environmental management strategies capable of achieving a set of performance goals – such that public health and/or the environment are not jeopardized.

In general, the benchmarks may be used to determine the degree of contamination; evaluate the need for further investigation of problem situation;

provide guidance on the need for corrective actions; establish environmental restoration goals and strategies; and verify the adequacy of possible remedial actions.

SUGGESTED FURTHER READING

Beck, L.W., A.W. Maki, N.R. Artman and E.R. Wilson. 1981. Outline and criteria for evaluating the safety of new chemicals. *Regulatory Toxicology and Pharmacology*, 1: 19–58.
CCME (Council of Canadian Ministers of the Environment). 1995. *A Protocol for the Derivation of Environmental and Human Health Soil Quality Guidelines*. Draft. CCME, Winnipeg, Manitoba, Canada.
Ess, T.H. 1981. Risk acceptability. In: *Proceedings of National Conference on Risk and Decision Analysis for Hazardous Waste Disposal*, Aug. 24–27, Baltimore, Maryland, pp. 164–174.
Munro, I.C. and D.R. Krewski. (1981). Risk assessment and regulatory decision making. *Food and Cosmetic Toxicology*, 19: 549–560.
Page, G.W. and M. Greenberg. 1982. Maximum contaminant levels for toxic substances in water: a statistical approach. *Water Resources Bulletin*, 18.6 (December): 955–962.
Splitstone, D.E. 1991. How clean is clean . . . statistically? *Pollution Engineering*, 23: 90–96.

REFERENCES

Al-Saleh, I.A. and L. Coate. 1995. Lead exposure in Saudi Arabia from the use of traditional cosmetics and medical remedies. *Environmental Geochemistry and Health*, 17: 29–31.
ASTM (American Society for Testing and Materials). 1994. *Risk-Based Corrective Action Guidance*. ASTM, Philadelphia, PA.
ASTM. 1995. *Standard Guide for Risk-Based Corrective Action Applied at Petroleum-Release Sites*. ASTM (E1739-95), Philadelphia, PA.
Brown, H.S. 1986. A critical review of current approaches to determining 'How clean is clean' at hazardous waste sites. *Hazardous Wastes and Hazardous Materials*, 3(3): 233–260.
Brown, H.S., R. Guble and S. Tatelbaum. 1988. Methodology for assessing hazards of contaminants to seafood. *Regulatory Toxicology and Pharmacology*, 8: 76–100.
CDHS (California Department of Health Services). 1986. *The California Site Mitigation Decision Tree Manual*. CDHS, Toxic Substances Control Division, Sacramento, CA.
Corn, M. (ed.). 1993. *Handbook of Hazardous Materials*. Academic Press, San Diego, CA.
Dawson, G.W. and D. Sanning. 1982. Exposure-response analysis for setting site restoration criteria. In: *Proceedings National Conference on Management of Uncontrolled Hazardous Waste Sites*, USEPA, Washington, DC, pp. 386–389.
Driver, J.H., M.E. Ginevan, and G.K. Whitmyre. 1996. Estimation of dietary exposure to chemicals: a case study illustrating methods of distributional analyses for food consumption data. *Risk Analysis*, 16(6): 763–771.
NRC (National Research Council). 1993. *Pesticides in the Diets of Infants and Children*. NAS Press, Washington, DC.
OECD (Organization for Economic Cooperation and Development). 1993. *Occupational and Consumer Exposure Assessment*. OECD Environment Monograph 69, OECD, Paris, France.
Rodricks, J. and M.R. Taylor. (1983). Application of risk assessment to food safety decision making. *Regulatory Toxicology and Pharmacology*, 3: 275–307.

Santos, S.L. and J. Sullivan. 1988. The use of risk assessment for establishing corrective action levels at RCRA sites. *Hazardous Wastes and Hazardous Materials*, 3: 275–307.

USEPA (US Environmental Protection Agency). 1986. *Methods for Assessing Exposure to Chemical Substances, Volume 8: Methods for Assessing Environmental Pathways of Food Contamination*. Exposure Evaluation Division, Office of Toxic Substances. September, 1986, EPA/560/5-85/008.

USEPA. 1987. *RCRA Facility Investigation (RFI) Guidance*. Washington, DC. EPA/530/SW-87/001.

USEPA. 1991. *Risk Assessment Guidance for Superfund: Volume I – Human Health Evaluation Manual (Part B, Development of Risk-based Preliminary Remediation Goals)*. Interim (October 1991). Office of Emergency and Remedial Response, Washington, DC. PB92-963333, OSWER Directive: 9285.7-01B.

USEPA. 1996a. *Soil Screening Guidance: Technical Background Document*. Office of Emergency and Remedial Response, Washington, DC. EPA/540/R-95/128.

USEPA. 1996b. *Soil Screening Guidance: User's Guide*. Office of Emergency and Remedial Response, Washington, DC. EPA/540/R-96/018.

van Veen, M.P. 1996. A general model for exposure and uptake from consumer products. *Risk Analysis*, 16(3): 331–338.

Vermeire, T.G., P. van der Poel, R. van de Laar, and H. Roelfzema. 1993. Estimation of consumer exposure to chemicals: application of simple models. *Science of the Total Environment*, 135: 155–176.

Chapter Seventeen

Development of Risk-based Remediation Goals

There are several arguments that can be presented in favor of using risk-based scientific approaches in environmental decisions. But one of its most common applications involves the use of risk-based methods to help close and/or remedy contaminated sites.

An important consideration in the development of site restoration programs for contaminated sites is the level of cleanup to be achieved during possible remedial activities. The cleanup level is a site-specific criterion that a remedial action would have to satisfy in order to keep exposures of potential receptors to environmental contaminants within an 'acceptable' level. The acceptable level corresponds to a contaminant concentration in specific environmental media that, when exceeded, may result in significant risks of adverse impact to potential receptors. This target level generally tends to drive the cleanup process for a contaminated land, and therefore represents the maximum acceptable contaminant level for site restoration decisions.

For most contaminated land problems, soils and groundwater apparently represent the most significant media of concern. The computational procedures for developing risk-based restoration goals for these environmental media are therefore elaborated in this chapter. The same principles can indeed be extended in the formulation and development of target levels for a variety of other environmental matrices.

17.1 FACTORS AFFECTING THE DEVELOPMENT OF RISK-BASED SITE RESTORATION GOALS

Due to the possibility for different cleanup levels to be imposed on similar sites potentially contaminated to the same degree, it is important that a systematic approach is used in the development of case-specific cleanup criteria for contaminated sites. It is also important that the determination of the extent of cleanup required at a contaminated site is based on an assessment of the potential risks to both human health and the environment. In general, the use of risk-based methods should *not* ignore the importance of resource conservation as part of an overall environmental management program.

In fact, the use of risk-based cleanup levels is likely to result in timely, cost-effective, and adequate site restoration programs. As a rule-of-thumb, remedies

> Box 17.1 Important factors affecting the process of establishing contaminated site cleanup criteria
>
> - Nature and level of risks involved
> - Regulatory requirements and/or guidelines
> - Migration and exposure pathways (from contaminant sources to receptors)
> - Individual site characteristics affecting exposure
> - Current and future beneficial uses of the affected land and subsurface resources
> - Variability in exposure scenarios
> - Probability of occurrence of exposure to the populations potentially at risk
> - Possibilities of receptor exposures to elevated levels of other contamination not related to site activities
> - Sensitivity and vulnerability of the populations potentially at risk
> - Potential effects of site contamination on human and ecological receptors
> - Reliability of scientific and technical information/data relating to exposure assessment, toxicity data, risk models, and potential remediation strategies

whose cumulative effects fall within the risk range of approximately 10^{-4} to 10^{-7} for carcinogens, or meet an acceptable hazard level of unity for noncarcinogenic effects, are generally considered protective of human health. Where necessary, however, the potential ecological impacts should also be determined before a final site restoration decision is made. Indeed, media cleanup goals should generally be established at contaminant levels protective of both human health and the environment. Often, however, cleanup levels established for the protection of human health will also be protective of the environment at the same time. But there may be instances where adverse environmental effects may occur at or below contaminant levels that adequately protect human health. Consequently, sensitive ecosystems (e.g., wetlands) as well as threatened and endangered species or habitats that may be affected by releases of hazardous contaminants or constituents should, insofar as possible, be evaluated separately as part of the process used to establish media cleanup criteria needed for site restoration initiatives.

In general, several exposure- and technical-related factors will tend to affect the processes involved in the development of remediation objectives and cleanup goals (see, e.g., ASTM, 1994, 1995; Bowers et al., 1996; CCME, 1991, 1996; CRWQCB, 1989; Fitchko, 1989; Liptak and Lombardo, 1996; LUFT, 1989; Odermatt and Menatti, 1996; Pierzynski et al., 1994; USEPA, 1989, 1991, 1996a, 1996b; WPCF, 1990) – including the key determinants identified in Box 17.1. In particular, the type of exposure scenarios envisioned for a contaminated site and its vicinity usually will significantly affect whatever is considered to be an acceptable cleanup level. Thus, entirely different cleanup levels may be needed for similar pieces of equally contaminated sites, based on the differences in the exposure scenarios. That is, the same amount of contamination at similar sites does not necessarily call for the same level of cleanup. In general, however, the cleanup must attain contaminant levels that are protective of all receptors for both current and future land uses. After defining the critical pathways and exposure scenarios associated with a project site, it often is possible to calculate the various media concentrations at or below which potential receptor exposures will pose no significant risks to the exposed populations. Thus, the risk-based approach should be protective while providing a sound scientific basis for the efficient use of the usually limited cleanup funds.

Ultimately, the scale and urgency of response actions at contaminated sites depends on the degree to which contaminant levels exceed their respective benchmarks or risk-based criteria. Where site remediation is not feasible, the environmental quality criteria or risk-based criteria can be used to guide land-use restrictions or other forms of risk management actions that are protective of human health and the environment.

17.2 RISK-BASED SOIL CLEANUP LEVELS

Soils present at potentially contaminated sites can be a major long-term reservoir for chemical contaminants, with the capacity to release contamination into several other environmental media. In fact, the importance of soil remediation levels for contaminated sites cannot be over-emphasized, considering the significance of contaminated soils in the corrective action investigation process. Yet, because of the inherent variability of soils (both spatially and temporally), a generic list of 'acceptable soil contaminant levels' that would be protective of human health and the environment in all possible situations is generally not available for universal use at potentially contaminated sites.

Consider, for illustrative purposes, a potentially contaminated site that is being considered for remediation so that it may be re-developed for either residential or industrial purposes. Contaminant levels in residential soils in which children might play (which allows for pica behavior in toddlers and other infants) must necessarily be lower than the same contaminant levels in soils present at a site designated for large industrial complexes (which effectively prevent direct exposures to contaminated soils). Also, the release potential of several chemical constituents will usually be different in sandy soils and clayey soils, thus affecting the possible exposure scenarios and therefore the acceptable soil contaminant levels that are designated for the different types of soils. Consequently, it is generally preferable to establish and use site-specific cleanup criteria for most contaminated site problems, especially where soil exposures are critical to the site restoration decisions.

To determine the risk-based cleanup levels for a chemical compound present in soils at a contaminated site, algebraic manipulations of the hazard index or carcinogenic risk equation and the exposure estimation equations discussed in Parts III and IV can be used to arrive at the appropriate analytical relationships that define the risk-based soil cleanup criteria necessary for a site restoration program. The step-wise computational process involves performing back-calculations from the risk and exposure models in order to arrive at an acceptable soil concentration (ASC) that is based on health-protective exposure parameters. For chemicals with carcinogenic effects, a target cancer risk of 10^{-6} is typically used in the back-calculation process; a target hazard index of 1.0 is generally used for noncarcinogenic effects.

> Box 17.2 General equation for calculating risk-based soil cleanup level for a carcinogenic chemical constituent
>
> $$ASC_c = \frac{TCR}{\left(\frac{EF \times ED \times CF}{BW \times AT \times 365}\right) \times \{[SF_i \times IR \times RR \times ABS_a \times AEF \times CF_a] + [(SF_o \times SIR \times FI \times ABS_{si}) + (SF_o \times SA \times AF \times ABS_{sd} \times SM)]\}}$$
>
> $$= \frac{(TCR) \times (BW \times AT \times 365)}{(EF \times ED \times CF) \times \{[SF_i \times IR \times RR \times ABS_a \times AEF \times CF_a] + SF_o[(SIR \times FI \times ABS_{si}) + (SA \times AF \times ABS_{sd} \times SM)]\}}$$
>
> where:
> ASC_c = Acceptable soil concentration (i.e., acceptable risk-based cleanup level) of carcinogenic contaminant in soil (mg/kg)
> TCR = Target cancer risk, usually set at 10^{-6} (dimensionless)
> SF_i = Inhalation slope factor ($[\text{mg/kg-day}]^{-1}$)
> SF_o = Oral slope factor ($[\text{mg/kg-day}]^{-1}$)
> IR = Inhalation rate (m³/day)
> RR = Retention rate of inhaled air (%)
> ABS_a = Percent chemical absorbed into bloodstream (%)
> AEF = Air emissions factor, i.e., PM_{10} particulate emissions or volatilization (kg/m³)
> CF_a = Conversion factor for air emission term (10^6)
> SIR = Soil ingestion rate (mg/day)
> CF = Conversion factor (10^{-6} kg/mg)
> FI = Fraction ingested from contaminated source (dimensionless)
> ABS_{si} = Bioavailability absorption factor for ingestion exposure (%)
> ABS_{sd} = Bioavailability absorption factor for dermal exposures (%)
> SA = Skin surface area available for contact, i.e., surface area of exposed skin (cm²/event)
> AF = Soil to skin adherence factor, i.e., soil loading on skin (mg/cm²)
> SM = Factor for soil matrix effects (%)
> EF = Exposure frequency (days/year)
> ED = Exposure duration (years)
> BW = Body weight (kg)
> AT = Averaging time (i.e., period over which exposure is averaged) (years)

In general, for substances that are both carcinogenic and systemically toxic, media *ASCs* are independently calculated for the carcinogenic and noncarcinogenic effects; the lower of the two criteria is then selected in the corrective action decision. Invariably, the carcinogenic *ASC* tends to be more stringent in most situations where both values exist.

17.2.1 Soil Cleanup Level for Carcinogenic Contaminants

Box 17.2 contains the general equation for calculating the risk-based site restoration criteria for a single carcinogenic chemical present in soils at a contaminated site. This has been derived by back-calculating from the risk and chemical exposure equations associated with the inhalation of soil emissions, ingestion of soils, and dermal contact with soils. It is noteworthy that, where appropriate and necessary, this general equation may also be re-formulated to incorporate the receptor age-adjustment exposure factors developed and presented earlier on in Chapter 10.

17.2.1.1 An illustrative example

In a simplified example of the application of the above equation (for calculating a media-specific *ASC* for a carcinogenic chemical), consider a hypothetical site located within a residential setting where children may be exposed to site contamination during recreational activities. It has been determined that soils at this playground for young children in the neighborhood are contaminated with methylene chloride. It is expected that children aged 1 to 6 years could be ingesting up to 200 mg of contaminated soils per day during outdoor activities at the impacted playground.

The *ASC* associated with the *ingestion only exposure* of 200 mg of soil (contaminated with methylene chloride which has an oral *SF* of 7.5×10^{-3} [mg/kg-day]$^{-1}$) on a daily basis by a 16-kg child over a 5-year exposure period is conservatively estimated to be:

$$ASC_{mc} = \frac{[10^{-6} \times 16 \times 70 \times 365]}{[0.0075 \times 200 \times 1 \times 1 \times 365 \times 5 \times 10^{-6}]} = 149 \text{ mg/kg}$$

That is, the allowable exposure concentration (represented by the *ASC*) for methylene chloride in soils within this residential setting, assuming a benchmark excess lifetime cancer risk level of 10^{-6}, is estimated to be approximately 149 mg/kg. Thus, if environmental sampling and analysis indicates contamination levels in excess of 149 mg/kg at this residential playground, then immediate corrective action (such as restricting access to the playground as an interim measure) should be implemented. It is noteworthy that other potentially significant exposure routes (e.g., dermal contact and inhalation), as well as other sources of exposure (e.g., via drinking water and food) have not been accounted for in this illustrative example – all of which may further require the need to lower the calculated *ASC* for any site restoration decisions.

In general, regulatory guidance would probably require reducing the contaminant concentration, ASC_{mc}, to about 20% of the calculated value in view of the fact that there could be other sources of exposure (e.g., air, food, etc.). This thinking should generally be factored into the overall environmental management program, and particularly in the risk management decisions about contaminated site management problems.

17.2.2 Soil Cleanup Criteria for the Noncarcinogenic Effects of Site Contaminants

Box 17.3 contains the general equation for calculating the risk-based site restoration criteria for the noncarcinogenic effects of a single chemical constituent found in soils at a contaminated site. This has been derived by back-calculating from the hazard and chemical exposure equations associated with the inhalation of soil

> Box 17.3 General equation for calculating risk-based soil cleanup level for the noncarcinogenic effects of a chemical constituent
>
> $$ASC_{nc} = \frac{THQ}{\left(\frac{EF \times ED \times 10^{-6}}{BW \times AT \times 365}\right) \times \left\{ \left[\frac{IR \times RR \times ABS_a}{RfD_i} \times AEF \times CF_a\right] + \left[\left(\frac{SIR}{RfD_o} \times FI \times ABS_{si}\right)\right] + \left[\frac{SA \times AF \times ABS_{sd} \times SM}{RfD_o}\right]\right\}}$$
>
> $$= \frac{(THQ) \times (BW \times AT \times 365)}{(EF \times ED \times CF) \times \left\{ \left[\frac{IR \times RR \times ABS_a}{RfD_i} \times AEF \times CF_a\right] + \frac{1}{RfD_o}[(SIR \times FI \times ABS_{si}) + (SA \times AF \times ABS_{sd} \times SM)]\right\}}$$
>
> where:
> ASC_{nc} = Acceptable soil concentration (i.e., acceptable risk-based cleanup level) of noncarcinogenic contaminant in soil (mg/kg)
> THQ = Target hazard quotient (usually equal to 1) (dimensionless)
> RfD_i = Inhalation reference dose (mg/kg-day)
> RfD_o = Oral reference dose (mg/kg-day)
> IR = Inhalation rate (m³/day)
> RR = Retention rate of inhaled air (%)
> ABS_a = Percent chemical absorbed into bloodstream (%)
> AEF = Air emission factor, i.e., PM_{10} particulate emissions or volatilization (kg/m³)
> CF_a = Conversion factor for air emission term (10^6)
> SIR = Soil ingestion rate (mg/day)
> CF = Conversion factor (10^{-6} kg/mg)
> FI = Fraction ingested from contaminated source (dimensionless)
> ABS_{si} = Bioavailability absorption factor for ingestion exposure (%)
> ABS_{sd} = Bioavailability absorption factor for dermal exposures (%)
> SA = Skin surface area available for contact, i.e., surface area of exposed skin (cm²/event)
> AF = Soil to skin adherence factor, i.e., soil loading on skin (mg/cm²)
> SM = Factor for soil matrix effects (%)
> EF = Exposure frequency (days/year)
> ED = Exposure duration (years)
> BW = Body weight (kg)
> AT = Averaging time (i.e., period over which exposure is averaged, equals ED for noncarcinogens) (years)

emissions, ingestion of soils, and dermal contact with soils. It is noteworthy that, where appropriate and necessary, this general equation may also be re-formulated to incorporate the receptor age-adjustment exposure factors developed and presented earlier on in Chapter 10.

17.2.2.1 An illustrative example

In a simplified example of the application of the above equation (for calculating a media-specific ASC for the noncarcinogenic effects of a chemical constituent), consider a hypothetical site located within a residential setting where children may be exposed to site contamination during recreational activities. It has been determined that soils at this playground for young children in the neighborhood are contaminated with ethylbenzene. It is expected that children aged 1 to 6 years could be ingesting up to 200 mg of contaminated soils per day during outdoor activities at the impacted playground.

The *ASC* associated with the *ingestion only exposure* of 200 mg of soil (contaminated with ethylbenzene which has an oral *RfD* of 0.1 mg/kg-day) on a daily basis by a 16-kg child over a 5-year exposure period is conservatively estimated to be:

$$ASC_{ebz} = \frac{0.1 \times [1 \times 16 \times 5 \times 365]}{[200 \times 1 \times 1 \times 365 \times 5 \times 10^{-6}]} \approx 8000 \text{ mg/kg}$$

That is, the allowable exposure concentration (represented by the *ASC*) for ethylbenzene in soils within this residential setting is estimated to be approximately 8000 mg/kg. Thus, if environmental sampling and analysis indicates contamination levels in excess of 8000 mg/kg at this residential playground, then immediate corrective action (such as restricting access to the playground as an interim measure) should be implemented. It is noteworthy that other potentially significant exposure routes (e.g., dermal contact and inhalation) as well as other sources of exposure (e.g., via drinking water and food) have not been accounted for in this illustrative example – all of which may further require the need to lower the calculated *ASC* for any site restoration decisions.

In general, regulatory guidance would probably require reducing the contaminant concentration, ASC_{ebz}, to about 20% of the calculated value in view of the fact that there could be other sources of exposure (e.g., air, food, etc.). This thinking should generally be factored into the overall environmental management program, and particularly in the risk management decisions about contaminated site management problems.

17.3 RISK-BASED WATER CLEANUP LEVELS

To determine the risk-based cleanup levels for a chemical compound present in water at a contaminated site, algebraic manipulations of the hazard index or carcinogenic risk equation and the exposure estimation equations discussed in Parts III and IV can be used to arrive at the appropriate analytical relationships that define the risk-based water cleanup criteria necessary for a site restoration program. The step-wise computational process involves performing back-calculations from the risk and exposure models in order to arrive at an acceptable water concentration (*AWC*) that is based on health-protective exposure parameters. For chemicals with carcinogenic effects, a target risk of 1×10^{-6} is typically used in the back-calculation process; a target hazard index of 1.0 is generally used for noncarcinogenic effects.

In general, for substances that are both carcinogenic and systemically toxic, media *AWC*s are independently calculated for the carcinogenic and noncarcinogenic effects; the lower of the two criteria is then selected in the corrective action

> **Box 17.4** General equation for calculating risk-based water cleanup level for a carcinogenic chemical constituent
>
> $$AWC_c = \frac{TCR}{\left(\frac{EF \times ED}{BW \times AT \times 365}\right) \times \{[SF_i \times IR_w \times RR \times ABS_a \times CF_a] + [(SF_o \times WIR \times FI \times ABS_{si}) + (SF_o \times SA \times K_p \times ET \times ABS_{sd} \times CF)]\}}$$
>
> $$= \frac{TCR \times (BW \times AT \times 365)}{(EF \times ED) \times \{[SF_i \times IR_w \times RR \times ABS_a \times CF_a] + SF_o[(WIR \times FI \times ABS_{si}) + (SA \times K_p \times ET \times ABS_{sd} \times CF)]\}}$$
>
> where:
> AWC_c = Acceptable water concentration (i.e., acceptable risk-based cleanup level) of carcinogenic contaminant in water (mg/L)
> TCR = Target cancer risk, usually set at 10^{-6} (dimensionless)
> SF_i = Inhalation slope factor ($[\text{mg/kg-day}]^{-1}$)
> SF_o = Oral slope factor ($[\text{mg/kg-day}]^{-1}$)
> IR_w = Intake from the inhalation of volatiles (sometimes equivalent to the amount of ingested water) (m³/day)
> RR = Retention rate of inhaled air (%)
> ABS_a = Percent chemical absorbed into bloodstream (%)
> CF_a = Conversion factor for volatiles inhalation term (1000 L/1 m³ = 10³ L/m³)
> WIR = Water ingestion rate (L/day)
> CF = Conversion factor (1 L/1000 cm³ = 10^{-3} L/cm³)
> FI = Fraction ingested from contaminated source (dimensionless)
> ABS_{si} = Bioavailability absorption factor for ingestion exposure (%)
> ABS_{sd} = Bioavailability absorption factor for dermal exposures (%)
> SA = Skin surface area available for contact, i.e., surface area of exposed skin (cm²/event)
> K_p = Chemical-specific dermal permeability coefficient from water (cm²/h)
> ET = Exposure time during water contacts (e.g., during showering/bathing activity) (h/day)
> EF = Exposure frequency (days/year)
> ED = Exposure duration (years)
> BW = Body weight (kg)
> AT = Averaging time (i.e., period over which exposure is averaged) (years)

decision. Invariably, the carcinogenic *AWC* tends to be more stringent in most situations where both values exist.

17.3.1 Water Cleanup Level for Carcinogenic Contaminants

Box 17.4 contains the relevant equation to use in the development of risk-based site restoration criteria for a single carcinogenic constituent present in water at a contaminated site. This has been derived by back-calculating from the risk and chemical exposure equations associated with the inhalation of contaminants in water (for volatile constituents only), ingestion of water, and dermal contact with water. It is noteworthy that, where appropriate and necessary, this general equation may also be re-formulated to incorporate the receptor age-adjustment exposure factors developed and presented earlier on in Chapter 10.

17.3.1.1 An illustrative example

In a simplified example of the application of the above equation (for calculating a media-specific AWC for a carcinogenic chemical), consider the case of a contaminated site that is impacting an underlying water supply aquifer due to contaminant migration into groundwater. This groundwater resource is used for culinary water supply purposes.

The AWC associated with the *ingestion only exposure* of 2 liters of water (contaminated with methylene chloride which has an oral SF of 7.5×10^{-3} [mg/kg-day]$^{-1}$) on a daily basis by a 70-kg adult over a 70-year lifetime is given by the following approximation:

$$AWC_{mc} = \frac{[10^{-6} \times 70 \times 70 \times 365]}{[0.0075 \times 2 \times 1 \times 365 \times 70]} = 0.005 \text{ mg/L} = 5 \text{ μg/L}$$

That is, the allowable exposure concentration (represented by the AWC) for methylene chloride, assuming a benchmark excess lifetime cancer risk level of 10^{-6}, is estimated to be 5 μg/L. Obviously, the inclusion of other pertinent exposure routes (such as inhalation of vapors and also dermal contacts during showering/bathing and washing activities) will likely result in the need for a lower AWC in any site restoration decision.

In general, most regulatory guidance would probably require reducing AWC_{mc} to about 20% of the value calculated here, in view of the fact that there may be other sources of exposure (e.g., air, food, etc) not yet accounted for. In fact, such considerations should generally be factored into the overall assessment, and in particular into the environmental decision-making and risk management processes.

17.3.2 Water Cleanup Level for the Noncarcinogenic Effects of Site Contaminants

Box 17.5 contains the relevant equation to use in the development of risk-based site restoration criteria for a single noncarcinogenic constituent present in water at a contaminated site. This has been derived by back-calculating from the hazard and chemical exposure equations associated with the inhalation of contaminants in water (for volatile constituents only), ingestion of water, and dermal contact with water. It is noteworthy that, where appropriate and necessary, this general equation may also be re-formulated to incorporate the receptor age-adjustment exposure factors developed and presented earlier on in Chapter 10.

17.3.2.1 An illustrative example

In a simplified example of the application of the above equation (for calculating a media-specific AWC for the noncarcinogenic effects of a chemical constituent), consider the case of a contaminated site impacting a multipurpose groundwater supply source due to contaminant migration into an underlying aquifer. This groundwater resource is used for culinary water supply purposes.

> **Box 17.5 General equation for calculating risk-based water cleanup level for noncarcinogenic effects of a chemical constituent**
>
> $$ASC_{nc} = \frac{THQ}{\left(\frac{EF \times ED}{BW \times AT \times 365}\right) \times \left\{\left[\frac{IR_w \times RR \times ABS_a \times CF_a}{RfD_i}\right] + \left[\left(\frac{WIR}{RfD_o}\right) \times FI \times ABS_{si}\right] + \left[\frac{SA \times K_p \times ET \times ABS_{sd} \times CF}{RfD_o}\right]\right\}}$$
>
> $$= \frac{THQ \times (BW \times AT \times 365)}{(EF \times ED) \times \left\{\left[\frac{IR_w \times RR \times ABS_a \times CF_a}{RfD_i}\right] + \frac{1}{RfD_o}[(WIR \times FI \times ABS_{si}) + (SA \times K_p \times ET \times ABS_{sd} \times CF)]\right\}}$$
>
> where:
> AWC_{nc} = Acceptable water concentration (i.e., acceptable risk-based cleanup level) of noncarcinogenic contaminant in water (mg/L)
> THQ = Target hazard quotient (usually equal to 1)
> RfD_i = Inhalation reference dose (mg/kg-day)
> RfD_o = Oral reference dose (mg/kg-day)
> IR_w = Inhalation intake rate (m³/day)
> RR = Retention rate of inhaled air (%)
> ABS_a = Percent chemical absorbed into bloodstream (%)
> CF_a = Conversion factor for volatiles inhalation term (1000 L/1 m³ = 10^3 L/m³)
> WIR = Water intake rate (L/day)
> CF = Conversion factor (1 L/1000 cm³ = 10^{-3} L/cm³)
> FI = Fraction ingested from contaminated source (dimensionless)
> ABS_{si} = Bioavailability absorption factor for ingestion exposure (%)
> ABS_{sd} = Bioavailability absorption factor for dermal exposures (%)
> SA = Skin surface area available for contact, i.e., surface area of exposed skin (cm²/event)
> K_p = Chemical-specific dermal permeability coefficient from water (cm²/hr)
> ET = Exposure time during water contacts (e.g., during showering/bathing activity) (h/day)
> EF = Exposure frequency (days/year)
> ED = Exposure duration (years)
> BW = Body weight (kg)
> AT = Averaging time (i.e., period over which exposure is averaged) (years)

The *AWC* associated with the *ingestion only exposure* of 2 liters/day of water (contaminated with ethylbenzene which has an oral *RfD* of 0.1 mg/kg-day) on a daily basis by a 70-kg adult is approximated by:

$$AWC_{ebz} = \frac{0.1 \times [1 \times 70 \times 70 \times 365]}{[2 \times 1 \times 1 \times 365 \times 70]} = 3500 \ \mu g/L$$

Thus, the allowable exposure concentration (represented by the *AWC*) for ethylbenzene is estimated to be 3500 µg/L. Of course, additional exposures via inhalation and dermal contacts during showering/bathing and washing activities may also have to be incorporated to yield an even lower *AWC*, in order to arrive at a more responsible site restoration decision.

In general, most regulatory guidance would probably require reducing AWC_{ebz} to about 20% of the value calculated here, in view of the fact that there may be other sources of exposure (e.g., air, food, etc) not yet accounted for. In fact, such considerations should generally be factored into the overall assessment, and in particular into the environmental decision-making and risk management processes.

17.4 THE CLEANUP DECISION IN SITE RESTORATION PROGRAMS

Cleanup of contaminated sites is an important environmental issue in a number of regions of the world. In general, all cleanup actions typically target the remediation of previous contamination, using health-based cleanup goals. Consequently, similar cleanup criteria may be used in all site restoration programs, regardless of the nature of corrective action program being considered.

To arrive at a responsible decision on the acceptable cleanup criteria to adopt for a contaminated site problem, the corrective action program should, among other things, carefully evaluate the following:

- *Level of risk* indicated by the contaminants of concern.
- *Background level* of the contaminants of concern at upgradient, upstream, and/ or upwind locations relative to the source(s) of contamination or release(s).
- *Natural attenuation effects* of the contaminants of concern (via processes such as evaporation, photolysis, dilution, biodegradation, etc.).
- *Asymptotic level* of the contaminants of concern in the impacted media – which represents the attainment of contaminant levels below which continued remediation produces negligible reductions in contaminant levels.
- *Best available technologies (BATs)* that can be proven to offer feasible and cost-effective remediation methods and processes in the site restoration program.

Other principal considerations relate to the overall cost of cleanup, the time required to complete site remediation, and the possibility of a cleanup activity creating potential liability problems.

The use of site-specific cleanup criteria will normally result in significant cost-savings, because this allows a site management team to employ cost-effective corrective action strategies to achieve significant risk reduction for the particular situation. Indeed, the cleanup criteria could become the driving force behind remediation costs – and it has been demonstrated that the development and use of risk-based cleanup guidelines results in timely and cost-effective remediation (e.g., Liptak and Lombardo, 1996). It is therefore prudent to allocate adequate resources to develop site-specific cleanup criteria, that should facilitate the selection and design of cost-effective remedial action alternatives.

Often, if site-specific cleanup criteria are to be developed for a site, then a substantial wealth of information must also be collected on site soil and groundwater characteristics. Typical soil characteristics required to determine site-specific cleanup criteria include porosity, particle size, moisture content, organic carbon content, partition coefficients, soil pH, depth of contamination; general aquifer characteristics required to determine site-specific cleanup criteria include effective porosity, hydraulic conductivity, bulk density, longitudinal and transverse dispersivities, aquifer saturated thickness, hydraulic gradient, depth to water table, average groundwater velocity (Lesage and Jackson, 1992). These parameters are also useful in a general sense for interpreting the results of a site investigation, which are used to characterize the site and subsequently to develop the corrective action objectives.

Overall, there is good justification for using risk-based chemical limits in the site restoration decision process – especially for the following particularly important reasons (ASTM, 1994, 1995; Lesage and Jackson, 1992; Liptak and Lombardo, 1996; Pratt, 1993):

- The cleanup levels defined are based on site-specific data and conditions, which will likely have significant impacts on cleanup costs; this is because the risk-based chemical limits are specific to the proposed land use or are set to ensure that a range of end-uses can be achieved without unacceptable risk to future land uses.
- The levels defined are sensitive to human health and environmental effects, without necessarily being sensitive to regulatory changes.
- The methodology allows for the derivation of an *asymptotic level* of the contaminants of concern in the impacted media, which represents the cleanup level corresponding to a point of diminishing returns (i.e., the point when monitoring indicates that little additional progress can be made in reducing the contaminant levels); this represents the attainment of contaminant levels below which continued remediation produces negligible reductions in contamination.

In general, the use of risk-based cleanup levels is likely to result in timely, cost-effective, and adequate site restoration programs. As a rule-of-thumb, remedies whose cumulative effects fall within the risk range of approximately 10^{-4} to 10^{-7} for carcinogens, or meet acceptable levels for noncarcinogenic effects, are generally considered protective of human health. Where necessary, however, the potential ecological impacts should also be determined before a final corrective action decision is made.

SUGGESTED FURTHER READING

Alberta Environment and Alberta Labor. 1989. *Subsurface Remediation Guidelines for Underground Storage Tanks*. Alberta MUST (Management of Underground Storage Tanks) Project, A Joint Project of the Depts of Environment and Labor, Edmonton, Alberta, Canada. Draft (August, 1989).

DoE (Department of the Environment). 1997. *Guideline Values for Contamination in Soils*. CLR Report No. 10, GV Series. DoE, London, UK.

Droppo, J.G., J.W. Buck, D.L. Strenge, and B.L. Hoopes. 1993. Risk computation for environmental restoration activities. *Journal of Hazardous Materials*, 35: 341–352.

ICRCL (Interdepartmental Committee on the Redevelopment of Contaminated Land). 1987. *Guidance on the Assessment and Redevelopment of Contaminated Land. ICRCL 59/83*, 2nd edition. Department of the Environment, Central Directorate on Environmental Protection, London, UK.

USEPA (US Environmental Protection Agency). 1991. *The Role of Baseline Risk Assessment in Superfund Remedy Selection Decisions.* Office of Solid Waste and Emergency Response, Washington, DC. OSWER Directive No. 9355.0-30.

USEPA. 1994. *Revised Interim Soil Lead Guidance for CERCLA Sites and RCRA Corrective Action Facilities.* Office of Solid Waste and Emergency Response, Washington, DC. OSWER Directive No. 9355.4-12.

REFERENCES

ASTM (American Society for Testing and Materials). 1994. *Risk-Based Corrective Action Guidance.* ASTM, Philadelphia, PA.

ASTM. 1995. *Standard Guide for Risk-Based Corrective Action Applied at Petroleum-Release Sites.* ASTM (E1739-95), Philadelphia, PA.

Bowers, T.S., N.S. Shifrin, and B.L. Murphy. 1996. Statistical approach to meeting soil cleanup goals. *Environmental Science and Technology*, 30(5): 1437–1444.

CCME (Canadian Council of Ministers of the Environment). 1991. *Interim Canadian Environmental Quality Criteria for Contaminated Sites.* Report CCME EPC-CS34, The National Contaminated Sites Remediation Program, Winnipeg, Manitoba, Canada.

CCME. 1996. *Guidance Manual for Developing Site-Specific Soil Quality Remediation Objectives for Contaminated Sites in Canada.* CCME, Winnipeg, Manitoba, Canada.

CRWQCB (California Regional Water Quality Control Board). 1989. *The Designated Level Methodology for Waste Classification and Cleanup Level Determination.* Staff Report, Central Coast Region, California Regional Water Quality Control Board (June 1989).

Fitchko, J. 1989. *Criteria for Contaminated Soil/Sediment Cleanup.* Pudvan Publishing, Northbrook, IL.

Lesage, S. and R.E. Jackson (eds). 1992. *Groundwater Contamination and Analysis at Hazardous Waste Sites.* Marcel Dekker, New York.

Liptak, J.F. and G. Lombardo. 1996. The development of chemical-specific, risk-based soil cleanup guidelines results in timely and cost-effective remediation. *Journal of Soil Contamination*, 5(1): 83–94.

LUFT. 1989. *Leaking Underground Fuel Tank Field Manual: Guidelines for Site Assessment, Cleanup, and Underground Storage Tank Closure.* State of California, Leaking Underground Fuel Tank Task Force, October, 1989.

Odermatt, J.R. and J.A. Menatti. 1996. Methodology for using contaminated soil leachability testing to determine soil cleanup levels at contaminated petroleum underground storage tank (UST) sites. *Journal of Soil Contamination*, 5(2): 157–169.

Pierzynski, G.M., J.T. Sims, and G.F. Vance. 1994. *Soils and Environmental Quality.* Lewis Publishers/CRC Press, Boca Raton, FL.

Pratt, M. (ed.). 1993. *Remedial Processes for Contaminated Land.* Institution of Chemical Engineers, Warwickshire, UK.

USEPA (US Environmental Protection Agency). 1989. *Methods for Evaluating the Attainment of Cleanup Standards. Volume I: Soils and Solid Media.* Office of Policy, Planning and Evaluation, Washington, DC. EPA/230/2-89/042.

USEPA. 1991. *Risk Assessment Guidance for Superfund: Volume I – Human Health Evaluation Manual (Part B, Development of Risk-based Preliminary Remediation Goals).* Interim (October 1991). Office of Emergency and Remedial Response, Washington, DC. PB92-963333, OSWER Directive: 9285.7-01B.

USEPA. 1996a. *Soil Screening Guidance: Technical Background Document.* Office of Emergency and Remedial Response, Washington, DC. EPA/540/R-95/128.

USEPA. 1996b. *Soil Screening Guidance: User's Guide.* Office of Emergency and Remedial Response, Washington, DC. EPA/540/R-96/018.

WPCF (Water Pollution Control Federation). 1990. *Hazardous Waste Site Remediation Management.* A Special Publication of the WPCF, Technical Practice Committee, Alexandria, VA.

PART VI

THE ROLE OF RISK ASSESSMENT IN ENVIRONMENTAL MANAGEMENT DECISIONS

This part of the book is devoted to a discussion of several aspects of the practical applications of risk assessment results to environmental manangement decisions. It comprises the following specific chapters:

- Chapter 18, *Illustrative Examples of Risk Assessment Practice*, presents a select variety of practical application scenarios for the use of risk assessment in environmental management programs.
- Chapter 19, *Design of Risk Management Programs*, provides a discussion of various decision analysis, risk management, and risk communication concepts and tools – insofar as they may affect environmental management programs. It also includes some prescriptions and decision protocols that may be utilized in the management of potential environmental problems.
- Chapter 20, *Risk Assessment Applications to Environmental Management Problems*, as a closing chapter, enumerates the general scope for the application of risk assessment as pertains to managing environmental contamination problems. It then goes on to recapitulate several pertinent requirements for the utilization of risk assessment principles, concepts, and techniques to the various application scenarios. Also included here is a qualitative risk evaluation questionnaire chart – often necessary in project scoping decisions.

Chapter Eighteen

Illustrative Examples of Risk Assessment Practice

The determination of potential risks associated with environmental contamination problems invariably plays an important role in environmental characterization activities, in corrective action planning, and also in risk mitigation and risk management strategies – as demonstrated by the variety of hypothetical example problems that follow in this chapter. These illustrative examples purposely show the different types and levels of detail in which risk assessments may be formulated – such as one that is qualitative in nature through a completely quantitative evaluation.

18.1 EVALUATION OF HUMAN HEALTH RISKS ASSOCIATED WITH AIRBORNE EXPOSURES TO ASBESTOS

There are two subdivisions of asbestos: the serpentine group containing only chrysotile (which consists of bundles of curly fibrils); and the amphibole group containing several minerals (which tend to be more straight and rigid). Asbestos is neither water-soluble nor volatile, so that the form of concern is microscopic fibers (usually reported as, or measured in the environment in units of fibers per m^3 or fibers per cubic centimeter).

Processed asbestos has typically been fabricated into a wide variety of materials that have been used in consumer products (such as cigarette filters, wine filters, hair dryers, brake linings, vinyl floor tiles, and cement pipe), and also in a variety of construction materials (e.g., asbestos-cement pipe, flooring, friction products, roofing, sheeting, coating and papers, packings and gaskets, thermal insulation, electric insulation, etc.). Notwithstanding the apparent useful commercial attributes, asbestos has emerged as one of the most complex, alarming, costly, and tragic environmental health problems (Brooks et al., 1995).

A case in point, asbestos materials are frequently removed and discarded during building renovations and demolitions. To ensure safe ambient conditions under such circumstances, it often becomes necessary to conduct an asbestos sampling and analysis – with the results used to support a risk assessment. This section presents a discussion of the investigation and assessment of the human health risks associated with worker exposures to asbestos in the ventilation systems of a building.

18.1.1 Study Objective

The primary concern of the risk assessment for the ventilation systems in the case building is to determine the level of asbestos exposures that potential receptors (especially workers cleaning the ventilation systems) could experience, and whether such exposure constitutes potential significant risks.

18.1.2 Summary Results of the Environmental Sampling and Analysis

Standard air samples are usually collected on a filter paper and fibers >5 μm long are counted with a phase contrast microscope; alternative approaches include both scanning and transmission electron microscopy and x-ray diffraction. It is generally believed that fibers 5 μm or longer are of potential concern (USEPA, 1990a, 1990b).

Following an asbestos identification survey of the case structure, air samples collected from suspect areas in the building's ventilation systems were analyzed using phase contrast microscopy (PCM), and highly suspect ones further analyzed by using transmission electron microscopy (TEM). The TEM analytical results are important because they serve as a means/methods for distinguishing asbestos particles from other fibers or dust particles.

The PCM analysis produced concentration of asbestos fibers in the range of <0.002 to a maximum of 0.008 fibers/cm^3. From the TEM, chrysotile asbestos was determined to be at <0.004 structures per cm^3 (str/cm^3) in all the environmental air samples.

18.1.3 The Risk Estimation

For asbestos fibers to cause any disease in a potentially exposed population, they must gain access to the potential receptor's body. Since they do not pass through the intact skin, their main entry routes are by inhalation or ingestion of contaminated air or water (Brooks et al., 1995) – with the inhalation pathway apparently being the most critical in typical exposure scenarios. That is, for most asbestos exposures, inhalation is expected to be the only significant exposure pathway. Consequently, intake is based on estimates of the asbestos concentration in air, the rate of contact with the contaminated air, and the duration of exposure. Subsequently, the intake is integrated with the toxicity index to determine the potential risks associated with any exposures.

Individual excess cancer risk is a function of the airborne contaminant concentration, the probability of an exposure causing risk, and the exposure duration. By using the cancer risk equations presented earlier in Chapter 13, the cancer risk from asbestos exposures may be estimated in accordance with the following relationship:

Cancer risk = [airborne fiber concentration (fibers/m^3)]
  × [exposure constant (dimensionless)]
  × [inhalation unit risk ((100 PCM fibers/m^3)$^{-1}$)]

or

$$\text{Risk probability} = \text{intake} \times UR = [C_a \times INHf] \times UR$$

The following exposure assumptions are used to facilitate the intake computation for this particular problem:

- It is assumed that workers cleaning the ventilation system will complete this task within two weeks for a 5-day work week. Hence, the maximum exposure duration is, ED = 10 days – in comparison to a 70-year lifetime daily exposure.
- Assumed exposure time is 40 minutes per working hour, for an 8-hour work day.
- Inhalation rate is 20 m³/day (or 0.83 m³/h).

The exposure evaluation utilizes the information obtained from the airborne fiber samples collected and analyzed for during the prior air sampling activities; to be conservative, the maximum concentrations measured from the analytical protocols are used in the risk estimation. Thence, the fraction of an individual's lifetime for which exposure occurs – represented by the inhalation factor – is estimated to be:

$$INHf = (40/60) \times (8/24) \times (10/365) \times (1/70) = 8.7 \times 10^{-5}$$

Next, asbestos is considered carcinogenic with a unit risk of approximately 1.9×10^{-4} (100 PCM fibers/m³)$^{-1}$ (see, e.g., DTSC/Cal EPA, 1994); see also Appendix D.

Consequently, potential risk associated with the 'possible' but unlikely (represented by an evaluation based on the PCM analysis results) and the reasonable/likely (represented by an evaluation based on TEM analysis results) asbestos concentrations are determined, respectively, as follows:

- Risk represented by results of the PCM analyses is estimated by integrating the following information:

$$\text{PCM-based airborne fiber concentration (maximum)} = 0.008 \text{ fibers/cm}^3$$
$$= 8 \times 10^3 \text{ fibers/m}^3$$

$$INHf = 8.7 \times 10^{-5}$$

$$UR = 1.9 \times 10^{-4} \text{ (100 PCM fibers/m}^3)^{-1} \equiv 1.9 \times 10^{-6} \text{ per fibers/m}^3$$

Hence, cancer risk (based on PCM concentration) = 1.32×10^{-6}
- Risk represented by results of the TEM analyses is estimated by integrating the following information:

$$\text{TEM-based airborne asbestos concentration (maximum)} = 0.004 \text{ structures/cm}^3$$
$$= 4 \times 10^3 \text{ str/m}^3$$

$$INHf = 8.7 \times 10^{-5}$$

$$UR = 1.9 \times 10^{-4} \text{ (100 PCM fibers/m}^3)^{-1} \equiv 1.9 \times 10^{-6} \text{ per fibers/m}^3$$

Hence, cancer risk (based on TEM concentration) = 6.6×10^{-7}

18.1.4 A Risk Management Decision

All risk estimates indicated here are near the lower end of the generally acceptable risk range/spectrum (i.e., 10^{-4} to 10^{-6}). Thus, it may be concluded that asbestos in the case building should represent minimal potential risks of concern for workers entering the ventilation system to clean it up. Nonetheless, it is generally advisable to incorporate adequate worker protection through the use of appropriate respirators. In general, any asbestos abatement or removal program should indeed conform to strict health and safety requirements – with on-site enforcement of the specifications being carried out by a qualified health and safety officer or industrial hygienist.

18.2 A DIAGNOSTIC HUMAN HEALTH RISK ASSESSMENT FOR A CONTAMINATED SITE PROBLEM

The purpose of this section is to present a procedural illustration of the nature of human health risk evaluation required for the development of a site restoration program – often necessary as part of a decommissioning or closure plan for an abandoned industrial facility.

A hypothetical facility that has been used for a multitude of operations is used in this illustrative example. The case site, owned by PLC Limited, is located within an industrial estate in the outskirts of London. Based on current zoning plans, it is anticipated that this land parcel could be used for a variety of commercial developments in the near future. A baseline risk assessment for this inactive site that previously housed the PLC facility is necessary, in order to help make the appropriate decision regarding utilization of this land parcel.

18.2.1 Introduction and Background

A former industrial facility, located in an industrially zoned area in the outskirts of metropolitan London, operated for over three decades before being permanently closed. Site facilities include a main plant building, office buildings, storage tanks, and post-closure areas (that consist of surface impoundments for wastewater treatment operations and sludge ponds). Past operations at the plant required the storage of raw materials in above-ground tanks, the distribution of raw materials in pipelines, and the storage of chemicals, fuels, and waste materials in underground storage tanks (USTs). Historical uses of the site included machine component cleaning (in which chlorinated hydrocarbon-based solvents were used) and electroplating (for which major associated chemicals included cadmium, nickel, and chromium). Other significant activities included sandblasting of unpainted metal parts, painting, and vehicle maintenance.

Due to the sandblasting activities, incidental spillage during materials handling, and possible leakage of underground storage and distribution systems, soils and groundwater underlying the PLC plant site have been significantly impacted; this is

Illustrative examples of risk assessment practice 387

the result of releases of chemical materials that were used in the industrial processes and related activities carried out at this facility. Preliminary remedial activities have already been implemented to remove buried drums and storage tanks, and to remove soil materials from some of the most heavily contaminated areas.

Under the current decommissioning program, the PLC facility could be zoned for a variety of commercial developments in the near future. The development of a site closure or re-development plan should therefore incorporate a diagnostic risk assessment that addresses potential impacts under all realistically feasible site uses and conditions.

18.2.2 Objective and Scope

The principal goal of any comprehensive corrective action program for the PLC site would be to prevent contaminant migration from the site to potential receptors, and therefore prevent the endangerment of human health and the environment at and in the vicinity of the site. It is apparent, however, that releases at the PLC site have caused significant soil and groundwater contamination beneath this industrial facility. Several key environmental issues must therefore be addressed in the processes involved in the diagnostic assessment required of the decommissioning plan – including the following:

- Identification of the possible site-related contaminants associated with past site activities.
- Screening for the chemicals of potential concern to human health and the environment.
- Estimation of the chemical concentrations in the impacted media of significant concern.
- Determination of the populations potentially at risk from site contaminants.
- Identification of representative and site-specific exposure scenarios.
- Characterization of the potential risks associated with the site.
- Development of site-specific cleanup criteria for the impacted matrices at the site.

This illustrative evaluation focuses on a diagnostic risk assessment that can be used to facilitate risk management decisions about this site.

18.2.3 Technical Elements of the Diagnostic Risk Assessment Process

Figure 18.1 shows the major elements required of the process for developing a decommissioning plan for the PLC site. In particular, the following specific tasks are carried out, in order to accomplish the overall goal of the diagnostic risk assessment:

- Compile and characterize the list of contaminants present at the site.
- Compile the toxicological profiles of the chemicals of potential concern (COCs).

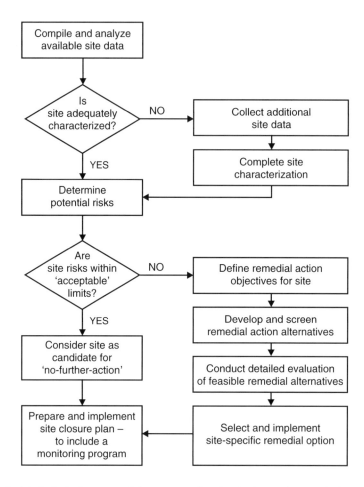

Figure 18.1 Technical elements of the process for developing a decommissioning plan for the PLC facility

- Investigate all possible contaminant migration pathways, and determine the pathways of concern.
- Identify targets in the vicinity of the site, and all other populations potentially at risk (including possible sensitive receptors).
- Develop a representative conceptual model for the site.
- Develop exposure scenarios, by integrating information on the populations potentially at risk with the likely and significant migration/exposure pathways.
- Calculate carcinogenic risks and noncarcinogenic hazard indices for the various receptor groups that have been determined to be potentially at risk.

Based on the type of exposure scenarios identified for this environmental setting, an environmental management strategy can be formulated such that the PLC site does not pose any significant risks for its intended uses.

Table 18.1 Preliminary list of possible site contaminants in soils at the PLC site

Possible site contaminant	Important synonyms, trade names, or chemical formula	Naturally occurring background level (mg/kg)	Maximum soil concentration (mg/kg)
Inorganic chemicals			
Aluminum	Al	16 400	10 400
Antimony	Sb	0.3	1.1
Arsenic	As	10	12
Barium	Ba	98	81.1
Beryllium	Be	0.7	0.33
Cadmium	Cd	1.0	2.6
Calcium	Ca	3 350	8 350
Chromium	Cr	28	57.3
Cobalt	Co	4.0	64.6
Copper	Cu	46	47.3
Iron	Fe	13 500	22 900
Lead	Pb	22	15
Magnesium	Mg	3 900	2 420
Manganese	Mn	523	170
Mercury	Hg	0.2	1.1
Molybdenum	Mo	3.0	4.8
Nickel	Ni	13	19
Potassium	K	2 610	2 200
Selenium	Se	0.2	6.1
Silver	Ag	0.2	0.55
Sodium	Na	453	588
Thallium	Tl	0.4	0.415
Vanadium	V	58	34.1
Zinc	Zn	107	432
Organic compounds			
Trichloroethene	TCE	not available	0.710
cis-1,2-Dichloroethene	*cis*-1,2-DCE	not available	0.007

18.2.4 Identification of Site Contaminants

Chemicals found in soils and groundwater at the PLC site consist of both organic and inorganic constituents, as summarized in Tables 18.1 and 18.2 (developed from the complete laboratory data package of site sampling results by adopting and using Figure 9.2 from Chapter 9 as a guide). The summary of analytical results are reported in these tables, together with the naturally occurring background threshold values, where background concentrations are available. The background levels are used as a screening indicator of possible media contamination that may be the result of past site activities.

18.2.5 Screening for Chemicals of Potential Concern

A listed site contaminant is considered to be a chemical of potential concern if it is likely to have originated from past site activities and if it could potentially result

Table 18.2 Preliminary list of possible site contaminants in groundwater at the PLC site

Possible site contaminant	Important synonyms, trade names, or chemical formula	Naturally occurring background level (μg/L)	Maximum water concentration (μg/L)
Inorganic chemicals			
Aluminum	Al	1 200	1 000
Antimony	Sb	10	5.9
Arsenic	As	7	5.0
Barium	Ba	276	133
Beryllium	Be	4.0	1.0
Calcium	Ca	197 000	54 300
Chromium	Cr	20	15
Cobalt	Co	13	6.3
Copper	Cu	58	22
Iron	Fe	3 530	3 000
Magnesium	Mg	119 000	37 800
Manganese	Mn	971	1 390
Molybdenum	Mo	6	6.5
Nickel	Ni	490	21
Potassium	K	13 300	6 550
Silver	Ag	12	4.0
Sodium	Na	420 000	124 000
Thallium	Tl	1.0	1.5
Vanadium	V	28	24
Zinc	Zn	80	70
Organic compounds			
Trichloroethene	TCE	not available	939
cis-1,2-Dichloroethene	*cis*-1,2-DCE	not available	118

in adverse effects to populations potentially at risk when such receptors are exposed to the particular compound. To obtain the chemicals of potential concern for the anticipated future use of the site, Figure 9.2 is once again used as a guide to screen the site contaminants in order to arrive at the target chemicals listed in Table 18.3. The maximum concentrations of the target chemicals in the environmental samples listed here will be used as the receptor exposure concentrations in this study.

18.2.6 Risk Characterization for Site-Specific Exposure Scenarios

Three different population groups – i.e., on-site workers, site construction workers, and off-site residential populations – are considered in developing the requisite exposure scenarios anticipated for the PLC site. Table 18.4 and Figure 18.2 provide analogous summaries of the likely migration and exposure pathways that form a basis for estimating risks associated with this site.

Calculation of potential carcinogenic risks and noncarcinogenic hazards under the existing conditions at the PLC site are performed for the three different population groups identified in the conceptual site representation (shown in Table 18.4 and Figure 18.2). The maximum concentrations of the target chemicals in the

Table 18.3 Summary of the chemicals of potential concern at the PLC site

Chemical of potential concern in soils	Important synonyms or trade names, or chemical formula	Maximum soil concentration (mg/kg)	Chemical of potential concern in groundwater	Important synonyms or trade names, or chemical formula	Maximum groundwater concentration (μg/L)
Inorganic chemicals			*Inorganic chemicals*		
Antimony	Sb	1.1	Manganese	Mn	1390
Arsenic	As	12	Molybdenum	Mo	6.5
Cadmium	Cd	2.6	Thallium	Tl	1.5
Chromium	Cr	57.3			
Cobalt	Co	64.6			
Copper	Cu	47.3			
Mercury	Hg	1.1			
Molybdenum	Mo	4.8			
Nickel	Ni	19.0			
Selenium	Se	6.1			
Silver	Ag	0.55			
Thallium	Tl	0.415			
Zinc	Zn	432			
Organic compounds			*Organic compounds*		
Trichloroethene	TCE	0.710	Trichloroethene	TCE	939
cis-1,2-Dichloroethene	*cis*-1,2-DCE	0.007	*cis*-1,2-Dichloroethene	*cis*-1,2-DCE	118

Table 18.4 Tabular analysis chart for the exposure scenarios associated with the PLC site

Contaminated exposure medium	Contaminant release source(s)	Contaminant release mechanism(s)	Potential receptor location	Receptor groups potentially at risk	Potential exposure routes	Pathway potentially complete and significant?
Air	Contaminated surface soils	Fugitive dust generation	On-site	On-site facility worker	Inhalation Incidental ingestion Dermal absorption	Yes No No
				Construction worker	Inhalation Incidental ingestion Dermal absorption	Yes No No
			Off-site	Downwind worker	Inhalation Incidental ingestion Dermal absorption	No No No
				Downwind resident	Inhalation Incidental ingestion Dermal absorption	No No No
		Volatilization	On-site	On-site facility worker	Inhalation Dermal absorption	Yes No
				Construction workers	Inhalation Dermal absorption	Yes No
			Off-site	Nearest downwind worker	Inhalation Dermal absorption	No No
				Nearest downwind resident	Inhalation Dermal absorption	No No
Soils	Contaminated soils and/or buried wastes	Direct contacting	On-site	On-site facility worker	Incidental ingestion Dermal absorption	Yes Yes
				Construction worker	Incidental ingestion Dermal absorption	Yes Yes

Illustrative examples of risk assessment practice 393

Medium	Pathway	Exposure point	Receptor	Route	Selected
		Off-site	Downwind worker	Incidental ingestion Dermal absorption	No No
			Downwind resident	Incidental ingestion Dermal absorption	No No
Surface water	Surface runoff into surface ts impoundmen	On-site	On-site facility worker	Inhalation Incidental ingestion Dermal absorption	No No No
			Construction worker	Inhalation Incidental ingestion Dermal absorption	No No No
	Erosional runoff	Off-site	Downslope resident	Inhalation Incidental ingestion Dermal absorption	No No No
			Recreational population	Inhalation Incidental ingestion Dermal absorption	No No No
Contaminated surface soils	Groundwater discharge	Off-site	Downslope resident	Inhalation Incidental ingestion Dermal absorption	No No No
Contaminated groundwater			Recreational population	Inhalation Incidental ingestion Dermal absorption	No No No
Groundwater	Infiltration/ leaching	On-site	On-site facility worker	Inhalation Incidental ingestion Dermal absorption	No No No
Contaminated soils			Construction worker	Inhalation Incidental ingestion Dermal absorption	No No No

continues overleaf

Table 18.4 (continued)

Contaminated exposure medium	Contaminant release source(s)	Contaminant release mechanism(s)	Potential receptor location	Receptor groups potentially at risk	Potential exposure routes	Pathway potentially complete and significant?
			Off-site	Downgradient resident	Inhalation Incidental ingestion Dermal absorption	Yes Yes Yes
Drainage sediments	Contaminated surface soils	Surface runoff/ episodic overland flow	On-site	On-site facility worker	Inhalation Incidental ingestion Dermal absorption	No No No
				Construction worker	Inhalation Incidental ingestion Dermal absorption	No No No
			Off-site	Nearest downgradient resident	Inhalation Incidental ingestion Dermal absorption	No No No

Illustrative examples of risk assessment practice 395

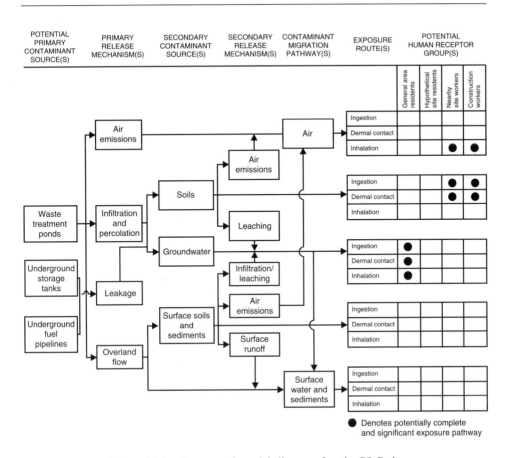

Figure 18.2 Conceptual model diagram for the PLC site

environmental samples (listed in Table 18.3) are used as the exposure point concentrations in this evaluation. Case-specific exposure parameters obtained from the literature (viz., DTSC, 1994; OSA, 1993; USEPA, 1989b, 1989c, 1991, 1992) are used in the modeling efforts involved. Toxicity values used for the risk characterization pertain to those found in recent toxicological databases (see, e.g., Appendix D) – in this case the Integrated Risk Information System. The computational processes involved are elaborated below.

- *Risk characterization associated with an on-site worker*. Table 18.5 consists of an evaluation of the potential risks associated with a nearby and/or on-site worker (following the re-development of the site for commercial activities) being exposed to the chemicals of potential concern at the PLC site – assuming the contaminated soils remain in place. It is also assumed that potential receptors may be exposed via inhalation of airborne contamination (consisting predominantly of particulate emissions from fugitive dust), through the incidental ingestion of contaminated soils, and by dermal contact with the contaminated soils at the site.

Table 18.5 Risk screening for a nearby and/or on-site worker exposure to soils at the PLC site

Chemical of potential concern	Maximum soil concentration (mg/kg)	Chemical-specific dermal absorption (ABSs)	Oral RfD (mg/kg-day)	Oral SF (1/mg/kg-day)	Inhalation RfD (mg/kg-day)	Inhalation SF (1/mg/kg-day)	Risk for air	Hazard for air	Risk for soil	Hazard for soil	Total risk (air+soil)	Total hazard (air+soil)	Risk-based soil criteria (mg/kg)	Soil criteria exceeded?
Inorganic chemicals														
Antimony (Sb)	1.1	0.01	4.00E-04				0.00E+00	2.69E-05	0.00E+00	2.91E-03			375	No
Arsenic (As)	12	0.03	3.00E-04	1.75E+00		1.20E+01	5.03E-07	3.91E-04	1.64E-05	8.77E-02			0.71	Yes
Cadmium (Cd)	2.6	0.001	5.00E-04			1.50E+01	1.36E-07	5.09E-05	0.00E+00	2.84E-03			19	No
Chromium (Cr-total)	57.3	0.01	1.00E+00				0.00E+00	5.61E-07	0.00E+00	6.06E-05			937 615	No
Cobalt (Co)	64.6	0.01	2.90E-04		2.90E-04		0.00E+00	2.18E-03	0.00E+00	2.35E-01			272	No
Copper (Cu)	47.3	0.01	3.70E-02				0.00E+00	1.25E-05	0.00E+00	1.35E-03			34 692	No
Mercury (Hg)	1.1	0.01	3.00E-04		8.60E-05		0.00E+00	1.25E-04	0.00E+00	3.87E-03			275	No
Molybdenum (Mo)	4.8	0.01	5.00E-03		5.00E-03		0.00E+00	9.39E-06	0.00E+00	1.01E-03			4 688	No
Nickel (Ni)	19.0	0.01	2.00E-02			9.10E-01	6.04E-08	9.30E-06	0.00E+00	1.00E-03			314	No
Selenium (Se)	6.1	0.01	5.00E-03				0.00E+00	1.19E-05	0.00E+00	1.29E-03			4 688	No
Silver (Ag)	0.55	0.01	5.00E-03				0.00E+00	1.08E-06	0.00E+00	1.16E-04			4 688	No
Thallium (Tl)	0.415	0.01	8.00E-05				0.00E+00	5.08E-05	0.00E+00	5.48E-03			75	No
Zinc (Zn)	432	0.01	3.00E-01				0.00E+00	1.41E-05	0.00E+00	1.52E-03			281 284	No
Organic compounds														
Trichloroethene (TCE)	0.710	0.10	6.00E-03	1.50E-02	6.00E-03	1.00E-02	2.48E-11	1.16E-06	2.34E-08	7.29E-04			30	No
cis-1,2-Dichloroethene (DCE)	0.007	0.10	1.00E-02		1.00E-02		0.00E+00	6.85E-09	0.00E+00	4.32E-06			1,620	No
							7.00E-07	0.003	1.65E-05	0.35	1.72E-05	0.35		

Notes:
1. The computational formulae and models used in this evaluation are discussed in Chapters 13 and 17.
2. Risk/hazard for air accounts for only the airborne emissions of contaminated particulates for all chemicals present at the site; strict volatilization effects are not included in this screening analysis.
3. Case-specific exposure parameters used in the calculations were obtained from the following sources – DTSC (1994), OSA (1993), and USEPA (1989b, 1989c, 1991, 1992).

Case-specific exposure parameters used in this evaluation conservatively assume that the on-site worker will be exposed at a frequency of 250 days per year over a 25-year period. Additional parameters include using a soil ingestion rate of 50 mg/day and airborne particulate emission rate of 50 $\mu g/m^3$. Other default exposure parameters indicated in the literature (e.g., DTSC, 1994; OSA, 1993; USEPA, 1989b, 1989c, 1991, 1992) were used for the calculations shown in this spreadsheet.

Based on this scenario, it is apparent that potential risks (of 1.7×10^{-5}) to an on-site worker at the PLC site could exceed a benchmark risk level of 10^{-6}. The noncarcinogenic hazard index (of 0.4), however, is within the reference index of unity. The sole 'risk driver' in this case is arsenic.

- *Risk characterization associated with a site construction worker.* Table 18.6 consists of an evaluation of the potential risks associated with a construction worker being exposed to the chemicals of potential concern at the PLC site – assuming no personal protection from the contaminated soils during site re-development activities. It is also assumed that potential receptors may be exposed via inhalation of airborne contamination (consisting predominantly of particulate emissions from fugitive dust), through the incidental ingestion of contaminated soils, and by dermal contact with the contaminated soils at the site.

  Case-specific exposure parameters used in this evaluation conservatively assume that the construction worker will be exposed at a frequency of 250 days per year over a one-year period. Additional parameters include using a soil ingestion rate of 480 mg/day and airborne particulate emission rate of 1000 $\mu g/m^3$. Other default exposure parameters indicated in the literature (viz., DTSC, 1994; OSA, 1993; USEPA, 1989b, 1989c, 1991, 1992) were used for the calculations shown in this spreadsheet.

  Based on this scenario, it is apparent that potential risks (of 2.5×10^{-6}) to a construction worker at the PLC site could marginally exceed an assumed benchmark risk level of 10^{-6}. The noncarcinogenic hazard index (of 1.6) also marginally exceeds the reference index of 1. The major 'risk drivers' are arsenic and cobalt.

- *Risk characterization for a downgradient residential population exposure to groundwater.* Table 18.7 consists of an evaluation of the potential risks associated with a hypothetical downgradient population exposure to impacted groundwater that originates from the PLC site – assuming the contaminated water is not treated before going into a public water supply system. It is also assumed that potential receptors may be exposed through the inhalation of volatiles during domestic usage of contaminated water, from the ingestion of contaminated water, and by dermal contact to contaminated waters.

  Default exposure parameters indicated in the literature (viz., DTSC, 1994; OSA, 1993; USEPA, 1989b, 1989c, 1991, 1992) were used for the calculations presented in this spreadsheet.

  Based on this scenario, it is apparent that the generally acceptable benchmark risk level of 10^{-6} and the reference hazard index of 1 may both be exceeded by several orders of magnitude, as a result of receptor exposures to raw/untreated

Table 18.6 Risk screening for a construction worker exposure to soils at the PLC site

Chemical of potential concern	Maximum soil concentration (mg/kg)	Chemical-specific dermal absorption (ABSs)	Toxicity criteria				Risk for air	Hazard for air	Risk for soil	Hazard for soil	Total risk (air+soil)	Total hazard (air+soil)	Risk-based soil criteria (mg/kg)	Soil criteria exceeded?
			Oral RfD (mg/kg-day)	Oral SF (1/mg/kg-day)	Inhalation RfD (mg/kg-day)	Inhalation SF (1/mg/kg-day)								
Inorganic chemicals														
Antimony (Sb)	1.1	0.01	4.00E-04				0.00E+00	5.38E-04	0.00E+00	1.45E-02			73	No
Arsenic (As)	12	0.03	3.00E-04	1.75E+00		1.20E+01	4.03E-07	7.83E-03	1.92E-06	2.56E-01			5	Yes
Cadmium (Cd)	2.6	0.001	5.00E-04			1.50E+01	1.09E-07	1.02E-03	0.00E+00	2.47E-02			24	No
Chromium (Cr)	57.3	0.01	1.00E+00				0.00E+00	1.12E-05	0.00E+00	3.02E-04			183 154	No
Cobalt (Co)	64.6	0.01	2.90E-04		2.90E-04		0.00E+00	4.36E-02	0.00E+00	1.17E+00			53	Yes
Copper (Cu)	47.3	0.01	3.70E-02				0.00E+00	2.50E-04	0.00E+00	6.73E-03			6 777	No
Mercury (Hg)	1.1	0.01	3.00E-04		8.60E-05		0.00E+00	2.50E-03	0.00E+00	1.93E-02			50	No
Molybdenum (Mo)	4.8	0.01	5.00E-03		5.00E-03		0.00E+00	1.88E-04	0.00E+00	5.05E-03			916	No
Nickel (Ni)	19.0	0.01	2.00E-02			9.10E-01	4.83E-08	1.86E-04	0.00E+00	5.00E-03			393	No
Selenium (Se)	6.1	0.01	5.00E-03				0.00E+00	2.39E-04	0.00E+00	6.42E-03			916	No
Silver (Ag)	0.55	0.01	5.00E-03				0.00E+00	2.15E-05	0.00E+00	5.79E-04			916	No
Thallium (Tl)	0.415	0.01	8.00E-05				0.00E+00	1.02E-03	0.00E+00	2.73E-02			15	No
Zinc (Zn)	432	0.01	3.00E-01				0.00E+00	2.82E-04	0.00E+00	7.58E-03			54 946	No
Organic compounds														
Trichloroethene (TCE)	0.710	0.10	6.00E-03	1.50E-02	6.00E-03	1.00E-02	1.98E-11	2.32E-05	1.58E-09	1.23E-03			444	No
cis-1,2-Dichloroethene (DCE)	0.007	0.10	1.00E-02		1.00E-02		0.00E+00	1.37E-07	0.00E+00	7.26E-06			946	No
							5.60E-07	0.06	1.92E-06	1.55	2.48E-06	1.6		

Notes:
1. The computational formulae and models used in this evaluation are discussed in Chapters 13 and 17.
2. Risk/hazard for air accounts for only the airborne emissions of contaminated particulates for all chemicals present at the site; strict volatilization effects are not included in this screening analysis.
3. Case-specific exposure parameters used in the calculations were obtained from the following sources – DTSC (1994), OSA (1993), and USEPA (1989b, 1989c, 1991, 1992).

Table 18.7 Risk screening for a downgradient residential population exposure to groundwater from the PLC site

Chemical of potential concern	Maximum water concentration (μg/L)	Chemical-specific K_p (cm/h)	Toxicity criteria				Risk for water	Hazard for water	Risk-based water criteria (μg/L)	Water criteria exceeded?
			Oral RfD (mg/kg-day)	Oral SF (1/mg/kg-day)	Inhalation RfD (mg/kg-day)	Inhalation SF (1/mg/kg-day)				
Inorganic chemicals										
Manganese (Mn)	1 390	1.60E-04	5.00E-03		1.40E-05		0.00E+00	1.78E+01	78	Yes
Molybdenum (Mo)	6.5	1.60E-04	5.00E-03		5.00E-03		0.00E+00	8.31E-02	78	No
Thallium (Tl)	1.5	1.60E-04	8.00E-05				0.00E+00	1.20E+00	1.3	Yes
Organic compounds										
Trichloroethene (TCE)	939	1.60E-02	6.00E-03	1.50E-02	6.00E-03	1.00E-02	3.57E-04	2.02E+01	2.6	Yes
cis-1,2-Dichloroethene (DCE)	118	1.00E-02	1.00E-02		1.00E-02		0.00E+00	1.52E+00	78	Yes
							3.57E-04	*40.7*		

Notes:
1. The computational formulae and models used in this evaluation are discussed in Chapters 13 and 17.
2. Risk/hazard for water account for both volatile chemical emissions and nonvolatile chemical contributors present at the site (DTSC, 1994).
3. Case-specific exposure parameters used in the calculations were obtained from the following sources – DTSC (1994), OSA (1993), and USEPA (1989b, 1989c, 1991, 1992).
4. K_p = chemical-specific dermal permeability coefficient for water (DTSC, 1994).

groundwater from the PLC site. The most significant contributors to site risks are from the general population exposure to manganese, thallium, TCE, and *cis*-1,2-DCE in groundwater.

For the noncarcinogenic effects considered in this evaluation, it is assumed for the sake of simplicity, that all the target chemicals have the same physiologic endpoint.

18.2.7 A Risk Management Decision

Table 18.8 summarizes the results of the diagnostic baseline risk assessment for the PLC site. In general, a cancer risk estimate greater that 10^{-6} or a noncarcinogenic hazard index greater than 1 indicate the presence of contamination that may pose a significant threat to human health. Overall, the levels of both carcinogenic and noncarcinogenic risks associated with the PLC site should *not* require time-critical removal action for contaminated soils present at this site. However, the use of untreated groundwater from aquifers underlying the site as a potable water supply source could pose significant risks to exposed individuals, or a community that uses such water for culinary purposes. Some type of corrective action or institutional control measure may therefore be necessary for this site.

The risk evaluation presented above indicates that the COCs present at the PLC site may pose some degree of risks to human receptors potentially exposed via the soil and groundwater media. Consequently, if this site is to be used for future commercial re-development projects, or if raw/untreated groundwater from this site is to be used as a potable water supply source, then a comprehensive site restoration program may be necessary to abate the imminent risks that the site poses – especially from groundwater exposures.

Whereas, limited site control measures will probably be adequate to protect construction workers at the PLC site, a more extensive corrective action program will be needed for the impacted aquifer. Thus, general response actions (comprised of an integrated soil and groundwater remediation program) should be developed for each of the potentially impacted environmental media associated with the site. Ultimately, the re-development of the site for commercial purposes may require only limited restoration activities, which may indeed be accomplished through 'incidental' site capping offered to the site following the construction of commercial buildings and pavements at the site.

18.3 A RISK-BASED STRATEGY FOR DEVELOPING A CORRECTIVE ACTION RESPONSE PLAN FOR PETROLEUM-CONTAMINATED SITES

The release of chemical substances from leaking underground storage tanks (USTs) is a common occurrence in several regions of the world. In particular, leakage of petroleum hydrocarbon products and other chemicals from USTs and pipelines is a frequent occurrence in commercial, industrial, and even domestic activities. Such leakages can result in the contamination of several environmental media, especially soils and groundwater. The subsequent impacts of such releases on the environment

Table 18.8 Summary of the risk screening for the PLC site

Receptor group	Risk parameter	Exposure routes and pathways				
		Inhalation exposure to soils	Oral exposure to soils	Total exposure to soils	Total exposure to groundwater	Overall total risk
Hypothetical downgradient resident	Cancer risk Hazard index	— —	— —	— —	3.6×10^{-4} 40.7	3.6×10^{-4} 40.7
Nearby and/or on-site worker	Cancer risk Hazard index	7.0×10^{-7} 0.003	1.6×10^{-5} 0.4	1.7×10^{-5} 0.4	— —	1.7×10^{-5} 0.4
Construction worker	Cancer risk Hazard index	5.6×10^{-7} 0.006	1.9×10^{-6} 1.6	2.5×10^{-6} 1.6	— —	2.5×10^{-6} 1.6

Note: Risks due to oral exposure is the total contribution from ingestion and dermal absorption of chemicals present in the contaminated medium.

and public health is a particularly important environmental issue. Of an even greater interest/concern is groundwater contamination by petroleum products, because many communities depend on groundwater as their primary source of drinking water.

In general, when petroleum products enter the soil, gravitational forces act to draw the fluid in a downward direction. Other forces act to retain it, either adsorbed to soil particles or trapped in soil pores. Petroleum hydrocarbon products can therefore become a major source of long-term contamination of soils and groundwater. The amount of product retained in the soil is of importance, because this could determine both the degree of contamination and the likelihood of subsequent contaminant transport into other environmental compartments.

18.3.1 Evaluation of Petroleum Product Constituents

Specific petroleum products commonly found in USTs include gasoline, diesel, heating oil, aviation fuel, waste oils, and other related petroleum hydrocarbons. Typically, motor fuel alone is a mixture of over 200 petroleum-derived chemicals plus a few synthetic products that are added to improve the fuel performance (LUFT, 1989). Several of these chemical constituents can potentially impact human and ecological receptors if released into the environment.

Petroleum fuel contaminants of major health concern include benzene, toluene, ethylbenzene, and xylene – commonly referred to by the acronym, BTEX. These BTEX compounds have the potential to move through soils to contaminate groundwaters. In fact, groundwater contamination from petroleum products, and particularly from leaking USTs, is a growing concern especially because of the potential carcinogenic and neurotoxic effects exhibited by the BTEX compounds. For instance, benzene is a known human carcinogen, whereas the others (i.e., toluene, ethylbenzene, and xylene) are noncarcinogens but are known to possess neurotoxicity effects. This means that the BTEX compounds generally constitute pre-selected target/indicator chemicals of concern for petroleum product releases. As BTEX compounds constitute the most toxic and environmentally mobile constituents, their selection as indicator chemicals assures that any corrective action cleanup criteria developed based on these will also adequately address the less toxic or the less mobile constituents.

In addition to an evaluation of BTEX compounds, analysis for total petroleum hydrocarbons (TPH) is commonly carried out for most petroleum-contaminated sites. This latter analysis detects aliphatic and aromatic constituents contained in the fuel. Detection is reported as the sum total of all hydrocarbons in the sample – rather than as individual chemical constituents. Because the lighter fractions (such as BTEX) are more mobile, these constituents can migrate or dissipate away from the main body of contamination. Less mobile hydrocarbons (such as those detected in TPH analysis) may give a more accurate indication of the actual contamination. As a consequence, soils are preferably analyzed for both the BTEX and TPH as indicators of contamination.

Further to BTEX and TPH analyses, and because of its extreme toxicity, the possible presence of organo-lead would generally be investigated where significant leaks of leaded motor fuel have occurred, or where an investigator feels that there may be potential danger of exposure to organo-lead.

It is noteworthy that diesel fuels consist primarily of aliphatic hydrocarbons (though they may also contain some limited quantities of aromatic constituents [including benzene], depending on the source and the refining process). Consequently, TPH analysis is usually the only one required for leakage and spills from diesel storages.

18.3.2 The Fate and Behavior of Petroleum Constituent Releases

Typically, in the event of contamination from a leaking UST used for petroleum products that have a density less than that of water (e.g., motor fuel), the leaking product first enters the unsaturated soil below the tank. This eventually forms a floating layer on any underlying water table if the leaking quantity is large. Also, product vapor will enter the unsaturated soils around the tank and above the water table. Water-soluble fractions of the petroleum product, such as benzene and xylene, will eventually form a plume within the groundwater. The soluble plume will spread within the groundwater by diffusion and move with the groundwater as it flows downgradient.

In the unsaturated zone, petroleum constituents tend to move downward, under the influence of gravity, and laterally due to capillary forces and the heterogeneous characteristics of soil. The movement also depends on the viscosity of the product and the rate of product release. If the amount of petroleum constituents is large enough, it will pass through the capillary zone and reach the water table. For motor fuels which are immiscible with water and also less dense than water, a layer of petroleum constituents will lie on and slightly depress the water table. Only a small amount of hydrocarbons will dissolve in the water.

Flow in the saturated zone generally transports the contamination in the direction of decreasing hydraulic potential. Petroleum constituents that reach the groundwater table will dissolve in water and be transported by groundwater. Transport of dissolved constituents in the groundwater is governed by advection and dispersion of groundwater, and by attenuation mechanisms (such as biodegradation and adsorption) of the soil media.

In general, the migration of contaminants such as petroleum constituents through the subsurface environment is governed by four major factors: the quantity or volume of release; the physical properties of the contaminants; the physical properties of the subsurface material (e.g., the adsorptive capacity of the earth materials); and the subsurface flow, such as the rates and directions of groundwater movement (Hunt et al., 1988; Yin, 1988). The quantity of the contaminant determines whether it will reach the water table. Physical properties of the leaked substance that are important to migration include solubility, specific gravity, viscosity, and surface tension; biodegradability should also be considered important to long-term contaminant migration. Physical properties of subsurface materials important to the migration of petroleum constituents include the porosity, permeability, and homogeneity. Other important physical parameters would include soil organic carbon content available for partitioning and also the available oxygen to aid aerobic biodegradation. All these parameters affect the subsurface behavior of the contaminants of concern. In addition, all processes which attenuate contaminant concentrations and/or limit the area of the contaminated zones will affect the fate of the source release.

Ultimately, an understanding of the fate and transport processes is important in determining the likelihood of medium-specific releases and exposures that result from leaking USTs.

18.3.3 Contaminant Release Analysis

In a typical situation when petroleum products have leaked from an UST, a nonaqueous phase liquid is released that can move through soil pores. Some of this liquid is left behind as disconnected fluid, called ganglia because of strong capillary forces (Hunt et al., 1988). These ganglia make up the residual saturation in the soil and are the long-term source of contaminants released to the air and groundwater. Even where no regional groundwater flow may be evident, a far field transport of selective gasoline components may indicate groundwater contamination by subsurface vapor migration away from the spill. In the unsaturated zone, denser gas is produced by volatilization of liquid gasoline, and this gas sinks toward the water table and then spreads out over the capillary fringe. Air saturated with gasoline and in contact with capillary water and groundwater generally will lose the more water-soluble components as it spreads out radially. Thus, the more volatile, less water-soluble hydrocarbons are expected to move greater distances than the more water-soluble compounds such as benzene. Indeed, available investigation data strongly suggest that groundwater contaminants detected in wells away from a petroleum fuel spill arrive mostly via the vapor phase and not by groundwater flow (Hunt et al., 1988). Thus, it is important, in some situations, to model subsurface vapor transport in order to adequately assess the potential for contamination from leaking USTs. Relevant information from this type of evaluation is also important in the choice of appropriate remediation technologies and/or cleanup strategies.

Overall, a variety of environmental fate models may be used to predict the transport and fate of contaminants. These models may range from a simple mass balance equation to multidimensional numerical solution of coupled differential equations. In any case, the basic equation governing fate and transport is based on the principle of conservation of mass for the contaminant.

Several factors affect the fate and transport of the mobile liquid, vapor, and dissolved hydrocarbon phases. Releases of petroleum constituents from leaking USTs can contaminate soils and groundwater by the migration of free product through the unsaturated zone and by dissolution of certain constituents into groundwater. It is noteworthy that, although petroleum fuel odor may be reported during soil sampling in the unsaturated zone in a site investigation/appraisal, at some site locations, no hydrocarbons may be detected. This situation suggests that the mechanisms of groundwater contamination away from the gasoline spill are vaporization of gasoline components into the soil gas, migration of denser-than-air soil gas downward to the water table, followed by radial spreading and then component partitioning into the groundwater (Hunt et al., 1988).

18.3.4 The Corrective Action and Risk Management Decision Process

It is almost a certainty that the result of most cleanup actions undertaken at petroleum-contaminated sites will leave some residual contamination. In fact, cleanup of all contaminated soil and dissolved product in groundwater is not always necessary to protect public health and the environment (LUFT, 1989). Consequently, it is important to develop cleanup goals attainable for appropriate remedial actions, based on estimates of health and environmental risks associated with the contaminants from UST releases.

In general, generic cleanup levels for contaminated soil and dissolved product are undesirable, since conditions vary from one region to another. Instead, site-specific cleanup levels are usually recommended for corrective action decisions (LUFT, 1989). Indeed, it is believed that the use of site-specific cleanup criteria could result in significant cost-savings, because such an approach can allow the use of corrective actions which are most effective in risk reduction. As part of a UST release site investigation, therefore, one should determine appropriate cleanup levels and the attainable remediation goals prior to implementing any corrective action plans.

Overall, the corrective action and risk management decision process will involve the conduct of several important investigation tasks – with the most important ones noted below.

- *Site categorization.* To facilitate a site-specific and phased approach to cost-effective corrective action decisions at petroleum-contaminated sites, a site categorization scheme may be adopted in the investigation of UST releases. The different categories will reflect the seriousness or hazardous nature of the potential release situation.

    In general, a site designation process may be used to categorize sites after observing the presence of one or more of a number of suspect conditions (such as tank closure, reported nuisance conditions, monitoring problems, and observed leakages). Existence of any of the suspect conditions provides justification to initiate a preliminary investigation that will confirm or disprove the suspected situation. Based on the preliminary site investigation and site history, the site can be assigned to one of a number of categories. For the purpose of this discussion, sites are classified into the following three categories:

    - Category 1:   LOW RISK SITES (LrS)   = No suspected soil contamination
    - Category 2:   MEDIUM RISK SITES (MrS) = Suspected or known soil contamination
    - Category 3:   HIGH RISK SITES (HrS)   = Known soil and groundwater contamination

Obviously, moving from the LrS category to the HrS sends one from a less serious to a more serious and hazardous scenario. For instance, the HrS category presents a case where both soil and groundwater contamination is confirmed. In general, if the field personnel suspect that more serious contamination has occurred than was anticipated, then a site may be re-classified from a lower risk category to an intermediary risk category, or to a higher risk category, as appropriate. Conversely, the discovery of a less serious situation may result in re-classifying a site from HrS to MrS, or to LrS category. Relevant and standard fuel leak detection and screening methods can be used to aid field personnel in the classification/categorization tasks.

- *Contamination assessment.* Typically, the LrS will require a field TPH test only. This method is recommended only when there is likelihood that a problem exists and a quick, qualitative confirmation is desired. Thus, at LrS, the field personnel may use a field TPH test to confirm the absence of soil contamination. In fact, only sites showing no evidence of possible soil contamination may use this qualitative form of analysis. The field TPH test can indeed prove helpful as a gross guidance tool.

  If a field inspection or background check indicates that a site is suspected or known to have contaminated soil, or if a site has failed the field TPH test under category 1, then *quantitative* laboratory analysis of soil samples is needed. To ensure quality results, standard procedures should be followed in the design of the sample collection and analysis procedures. For instance, soil samples should be quantitatively analyzed for BTEX compounds using an appropriate method. If any of these constituents is detected above the minimum practical quantitation limit, then the site investigation should proceed through a more detailed analysis that includes a general risk appraisal.

  If a general risk appraisal shows that groundwater is at risk, then further evaluation is required that will help determine a HrS category. The necessary procedures involve more detailed investigations and decisions above what is performed under a MrS. Among other things, groundwater gradient is determined using piezometers, which show the direction of groundwater and contaminant plume movements.

- *The site restoration program.* If levels of contamination present at the spill site and impacted media are determined to exceed 'acceptable' risk levels, then corrective actions – which may include site remediation – will generally be required. It must be recognized, however, that most cleanup actions cannot achieve a 'zero' contamination level for petroleum-contaminated sites – or for any contaminated site problem, for that matter. The development of site-specific cleanup objectives necessary for the particular situation will therefore aid in the selection of appropriate site restoration strategies.

  During the initial planning stages of a restoration program, it is recommended that soils at the contaminated site be screened to determine whether it is possible for bioremediation to meet designated cleanup specifications for the site. As a general guide, cleanup levels of TPH $\leq$ 100 ppm and BTEX $\leq$ 10 ppb have been determined to be typical attainable cleanup limits for the bioremediation of gasoline- and diesel-contaminated sites (Barnhart, 1992). Bioremediation programs usually will require bench-scale tests and/or treatability studies to enhance the development of the design criteria and parameters. After the treatability study has demonstrated that the soil system will indeed support it, a full-scale program can then be implemented.

Cleanup of petroleum fuel spill sites may generally be carried out to a 'realistic' level only. This is because, once the major chemical constituents (such as benzene) have reached approximately a 100 ppb level in groundwater, for example, continued use of most groundwater remediation technologies may not yield any substantial improvement over that achieved through natural processes. In fact, trace residual hydrocarbons left in the subsurface at such 'threshold' levels will degrade through natural processes in an equally timely manner.

Risk assessment procedures can be used to establish cleanup objectives in this regard. The site-specific risk assessment will guide the development of appropriate cleanup objectives with reference to site conditions, land uses, and exposure scenarios pertaining specifically to the case-site and its vicinity. Such an assessment will provide directions to develop effectual corrective action and risk management plans.

18.4 GENERAL SCOPE OF RISK ASSESSMENT PRACTICE

Risk assessment has several specific applications that could affect the type of decisions to be made in relation to environmental management programs. The application of the risk assessment process to environmental management problems will generally serve to document the fact that risks to human health and the environment have been evaluated and incorporated into a set of appropriate response actions. Almost invariably, every process for developing effectual environmental management programs should incorporate some concepts or principles of risk assessment. In particular, all decisions on restoration plans for potential environmental contamination problems will include, implicitly or explicitly, some elements of risk assessment.

Appropriately applied, risk assessment techniques can indeed be used to estimate the risks posed by environmental hazards under various exposure scenarios, and to further estimate the degree of risk reduction achievable by implementing various technical remedies.

SUGGESTED FURTHER READING

Bates, D.V. 1994. *Environmental Health Risks and Public Policy*. University of Washington Press, Seattle, WA.

Chrostowski, P.C., L.J. Pearsall, and C. Shaw. 1985. Risk assessment as a management tool for inactive hazardous materials disposal sites. *Environmental Management*, 9(5): 433–442.

FPC (Florida Petroleum Council). 1986. *Benzene in Florida Groundwater: An Assessment of the Significance to Human Health*. Florida Petroleum Council, Tallahassee, FL.

Hilts, S.R. 1996. A co-operative approach to risk management in an active lead/zinc smelter community. *Environmental Geochemistry and Health*, 18: 17–24.
McKone, T.E., W.E. Kastenberg, and D. Okrent. 1983. The use of landscape chemical cycles for indexing the health risks of toxic elements and radionuclides. *Risk Analysis*, 3(3): 189–205.
Suter II, G.W., B.W. Cornaby, et al. 1995. An approach for balancing health and ecological risks at hazardous waste sites. *Risk Analysis*, 15: 221–231.
Zemba, S.G., L.C. Green, E.A.C. Crouch, and R.R. Lester. 1996. Quantitative risk assessment of stack emissions from municipal waste combusters. *Journal of Hazardous Materials*, 47: 229–275.

REFERENCES

Barnhart, M. 1992. Bioremediation: do it yourself. *Soils*, Aug.–Sep.: 14–19.
Brooks, S.M., et al. 1995. *Environmental Medicine*. Mosby, Mosby-Year Book, Inc., St Louis, MO.
DTSC (Department of Toxic Substances Control). 1994. *Preliminary Endangerment Assessment Guidance Manual (A Guidance Manual for Evaluating Hazardous Substance Release Sites)*. California Environmental Protection Agency, DTSC, Sacramento, CA.
DTSC/Cal EPA. 1994. *California Cancer Potency Factors: Update, November, 1994*. California Environmental Protection Agency, Sacramento, CA.
Hunt, J.R., J.T. Geller, N. Sitar and K.S. Udell. 1988. Subsurface transport processes for gasoline components. In: *Specialty Conference Proceedings, Joint CSCE–ASCE National Conference on Environmental Engineering*, Vancouver, BC, Canada, July 13–15, 1988, pp. 536–543.
LUFT. 1989. *Leaking Underground Fuel Tank Field Manual: Guidelines for Site Assessment, Cleanup, and Underground Storage Tank Closure*. State of California, Leaking Underground Fuel Tank Task Force, October, 1989.
OSA. 1993. *Supplemental Guidance for Human Health Multimedia Risk Assessments of Hazardous Waste Sites and Permitted Facilities*. Cal EPA, DTSC. Sacramento, CA.
USEPA (US Environmental Protection Agency). 1989a. *Estimating Air Emissions from Petroleum UST Cleanups*. Office of Underground Storage Tanks, Washington, DC.
USEPA. 1989b. *Exposure Factors Handbook*. Office of Health and Environmental Assessment, Washington, DC. EPA/600/8-89/043.
USEPA. 1989c. *Risk Assessment Guidance for Superfund. Volume I–Human Health Evaluation Manual (Part A)*. Office of Emergency and Remedial Response, Washington, DC. EPA/540/1-89/002.
USEPA. 1990a. *Environmental Asbestos Assessment Manual. Superfund Method for the Determination of Asbestos in Ambient Air. Part 1: Method*. Interim Version. USEPA, Washington, DC. EPA/540/2-90/005a.
USEPA. 1990b. *Environmental Asbestos Assessment Manual. Superfund Method for the Determination of Asbestos in Ambient Air. Part 2: Technical Background Document*. Interim Version. USEPA, Washington, DC. EPA/540/2-90/005b.
USEPA. 1991. *Risk Assessment Guidance for Superfund, Volume I: Human Health Evaluation Manual. Supplemental Guidance. 'Standard Default Exposure Factors'*. (Interim Final). March, 1991. Office of Emergency and Remedial Response, Washington, DC. OSWER Directive No. 9285.6-03.
USEPA. 1992. *Dermal Exposure Assessment: Principles and Applications*. Washington, DC. EPA/600/8-91/011B.
Yin, S.C.L. 1988. Modeling groundwater transport of dissolved gasoline. In: *Specialty Conference Proceedings, Joint CSCE–ASCE National Conference on Environmental Engineering*, Vancouver, BC, Canada, July 13–15, 1988, pp. 544–551.

Chapter Nineteen

Design of Risk Management Programs

Risk management is defined as a process of weighing policy alternatives and then selecting the most appropriate regulatory action – accomplished by integrating the results of risk assessment with scientific data as well as with social, economic, and political concerns to arrive at a decision. It is a decision-making process that entails considerations of political, social, economic, and engineering information with risk-related information in order to develop, analyse, and compare regulatory options – and then to select the appropriate regulatory response to a potential health or environmental hazard situation (Cohrssen and Covello, 1989; NRC, 1994; Seip and Heiberg, 1989; van Leeuwen and Hermens, 1995). Risk management may also include the design and implementation of policies and strategies that result from this decision process.

Several important decision support tools and concepts relevant to the design of effective risk management (to include risk mitigation and risk prevention) programs for potential environmental contamination problems are presented below in this chapter.

19.1 THE GENERAL NATURE OF RISK MANAGEMENT PROGRAMS

The management of environmental contamination problems usually involves competing objectives, with the prime objective to minimize both hazards and corrective action costs under multiple constraints. Typically, once a minimum acceptable and achievable level of protection has been established via hazard assessment, alternative courses of action can be developed that weigh the magnitude of adverse consequences against the cost of corrective measures. In general, reducing hazards would require increasing costs, and cost minimization during hazard abatement will likely leave higher degrees of unmitigated hazards. Typically, a decision is made based on the alternative that accomplishes the desired objectives at the least total cost – total cost here being the sum of hazard cost and remedial cost (Figure 19.1).

Risk management uses information from hazard analyses and/or risk assessment – along with information about technical resources; social, economic, and political values; and regulatory control or response options – to determine what actions to take in order to reduce or eliminate a risk. It is comprised of actions evaluated and

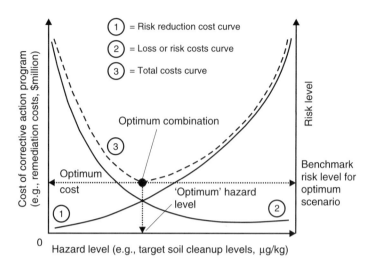

Figure 19.1 Risk reduction versus costs: a schematic of corrective action costs (e.g., cleanup or remediation costs) for varying hazard levels (e.g., chemical concentrations in environmental media or residual risk)

implemented to help in risk reduction policies, and may include concepts for prioritizing the risks, as well as an evaluation of the costs and benefits of proposed risk reduction programs.

Examples of risk management actions commonly encountered in environmental management programs include: deciding how much of a chemical a manufacturing company may discharge into a river; deciding which substances may be handled at a hazardous waste TSDF; deciding on the extent of cleanup warranted at a hazardous waste site; setting general permit levels for discharge, storage, or transport of hazardous materials; establishing levels for air contaminant emissions for air pollution control purposes; determining allowable levels of contamination in drinking water or food; industry decisions on the use of specific chemicals in manufacturing processes and related industrial activities; and hazardous waste facility design and operation by the regulated community. All these types of environmental management decisions are made based on inputs from a prior risk assessment conducted for the applicable case-specific problem. In fact, risk assessment results, serving as input to risk management, generally help in the setting of priorities for a variety of environmental management problems – further to producing more efficient and consistent risk reduction policies.

Risk management does indeed provide a context for balanced analysis and decision-making. In general, several considerations go into the risk management decisions about environmental contamination problems – including a consideration of issues such as the harmful effect of the environmental pollutants that need to be controlled; the risk–benefit–cost factor (which may include such items as the cost of pollution controls, the effects of alternative management practices, the relinquished benefits of using a pesticide or other toxic chemical, the danger of

> **Box 19.1 Key decision elements associated with the design of a risk management program**
>
> - Establish basis for contamination indicators
> - Gather and review background information for evidence of release
> - Determine area history, to help identify other possible sources of contamination
> - Identify potentially affected areas
> - Address health and safety issues associated with the case-specific situation – to include providing emergency response by mitigating release and potential hazards
> - Assess the fate and behavior characteristics of contaminants in the environment, including an identification of anticipated degradation, reaction and/or decomposition byproducts
> - Determine the critical environmental media of concern (such as air, surface water, groundwater, soils and sediments, and terrestrial and aquatic biota)
> - Delineate potential migration pathways
> - Identify and characterize populations potentially at risk (i.e., potential human and ecological receptors)
> - Determine receptor exposure pathways
> - Develop a conceptual representation or model for the problem situation
> - Evaluate potential exposure scenarios, and the possibility for human and ecosystem exposures
> - Characterize the general environmental contamination problem – to include a general identification of the contaminant types and their characteristics, a delineation of the extent of contamination for affected matrices, and a mapping of areas where contaminants may impact human health and/or environment
> - Assess the environmental and health impacts of the contaminants, if they should reach critical human and ecological receptors
> - Determine the corrective action needs, and formulate a risk management strategy – to include establishment of remedial action objectives and case-specific restoration goals; assessment of variables influencing selection of restoration goals and remedial systems; identification of remedial action alternatives; and development of general response actions
> - Design effective long-term monitoring and surveillance programs as a necessary part of an overall corrective action plan

displacing private sector initiatives, etc.); and the measure of confidence attached to several components of the risk assessment.

Risk management programs are generally designed with the goal to minimize potential negative impacts associated with environmental contamination problems. Typically, several pertinent questions relating to the nature and extent of contamination, exposure settings, migration and exposure pathways, populations potentially at risk, the nature and level of risks, environmental quality goals, regulatory policies, and availability of technically feasible remedial techniques are asked during the planning, development and implementation of corrective action programs that are directed at mitigating environmental contamination problems (Asante-Duah, 1996; BSI, 1988; Cairney, 1993; Jolley and Wang, 1993; USEPA, 1985, 1987a, 1987b, 1988, 1989, 1991; WPCF, 1988). It is very important that the environmental management program helps address all relevant issues, which will ultimately affect the type of risk management decision accepted for a environmental contamination problem (Box 19.1).

19.1.1 A System for Establishing Risk Management Needs

The selection of an appropriate and cost-effective risk management plan for an environmental contamination problem generally depends on a careful assessment of both short- and long-term risks posed by the case-specific problem. The key components of such an assessment will typically be comprised of the following general tasks: a preliminary appraisal; an environmental assessment; a risk appraisal; a risk determination; and a mitigation study. The pertinent ingredients of these tasks are presented below.

- *The preliminary appraisal.* The purpose of a preliminary appraisal, consisting of the identification of possible source(s) of contaminant releases, is to quickly assess the potential for a hazard source to adversely impact the environment and/or public health. This process typically involves: establishing a basis for contamination indicators; gathering and reviewing background information for evidence of release; determining the case history, to help identify other possible sources of contamination; identifying potentially affected areas; addressing health and safety issues associated with the case-specific situation; and addressing emergency response by mitigating release and potential hazards. The preliminary appraisal is initiated by the discovery of a potential environmental contamination problem. Conventional reconnaissance procedures may be used in this qualitative assessment that involves the collection and review of all available information (including an off-site reconnaissance to evaluate the source and nature of contamination, and the identification of any potential 'outside' polluters). Depending on the results of the preliminary survey, a problem may or may not be referred for further action. In general, the preliminary appraisal allows for a screening that will help establish a basis for more detailed environmental investigations.
- *The environmental assessment.* The objectives of an environmental assessment, involving the characterization of environmental contamination and a problem categorization (where necessary), are to identify contaminants present at a given locale and to determine case-specific characteristics that influence the migration of the contaminants. The process typically involves the following: characterizing problem locale (to include an identification of the contaminants of potential concern and a delineation of the extent of contamination for affected matrices), and determining contaminant behavior in the environment by developing a working hypothesis about contaminant fate and transport. Overall, environmental assessment activities are undertaken to define more completely the characteristics of a potential environmental contamination problem.
- *The risk appraisal.* The primary objective of the risk appraisal, consisting of a determination of the migration and exposure pathways integrated with an assessment of the environmental fate and transport of the contaminants of concern, is to determine whether potential receptors are likely at risk as a result of exposure to environmental contaminants. The process typically involves

the following: determining contaminant migration and exposure pathways; identifying populations potentially at risk; mapping areas where contaminants may impact human health and/or environment; developing realistic exposure scenarios appropriate for the specific problem area; and conducting case-specific exposure assessments. To determine whether potential receptors are truly at risk, it is necessary to identify potential migration and exposure pathways as well as contaminant exposure point concentrations in relation to acceptable threshold levels.
- *The risk determination.* The objective of the risk determination, to include an evaluation of the environmental and health impacts associated with contaminant releases as well as the development of case-specific restoration goals, is to evaluate potential risks. The calculated risk can then be compared against a benchmark risk, which then forms the basis for developing restoration goals and/or risk management plans. The process typically involves defining the magnitude or severity of health and environmental impacts.
- *The mitigation study.* The mitigation study consists of the development and implementation of a corrective action plan that may include remediation and/or monitoring programs. The process typically involves the following: establishing remediation objectives and case-specific restoration goals; identifying and evaluating variables that influence selection of restoration goals and remedial systems; developing remedial action objectives; identifying remedial action alternatives; and developing general response actions. Mitigation strategies and objectives are based on the protection of both current and potential future receptors that could become exposed to environmental contaminants. Consequently, the restoration is carried out so as to leave the affected area in such a condition as to pose no significant risks to any populations that may be potentially at risk.

Information derived from the relevant case evaluations generally helps formulate credible policies to support the risk management program. In general, the use of a systematic evaluation process will particularly facilitate rational decision-making on restoration efforts undertaken for environmental contamination problems. Risk assessment is particularly useful in determining environmental restoration goals most appropriate for an environmental contamination problem. It can also be used in the development of performance goals for various corrective action response alternatives.

By utilizing methodologies that establish restoration goals based on risk assessment principles, restoration programs can be carried out in a cost-effective and efficient manner. In the process involved, risk assessment can be used to define the level of risk, which will in turn aid in determining the level of analysis and the type of risk management actions to adopt for a given environmental management problem. The level of risk considered in such applications can be depicted by a risk–decision matrix (Figure 19.2) that will help distinguish between imminent health and/or environmental hazards and risks. In general, this can be used as an aid for policy decisions, in order to develop variations in the scope of work necessary for case-specific environmental management programs.

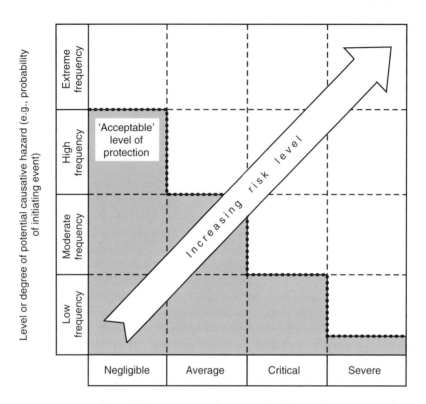

Figure 19.2 A conceptual representation defining risk profiles in a risk–decision matrix

19.1.2 Interim Corrective Action Programs

Interim corrective actions are measures used to address situations which pose imminent threats to human health or the environment, or to prevent further environmental degradation or contaminant migration pending final decisions on the necessary long-term remedial or risk management activities. For example, whenever excessive risks exist at a contaminated site, a decision should be made to implement interim corrective actions immediately, in order to protect public health and the environment. Common examples of interim corrective measures may include simply erecting a fence around a contaminated site in order to restrict/limit access to an impacted property; covering exposed contaminated soils with synthetic liner materials; applying dust suppressants to minimize emissions of contaminated fugitive dust; restricting use of contaminated groundwater, and/or providing alternative sources for culinary water supplies; temporal displacement or relocation of nearby residents away from a hazardous waste site; and the implementation of such more elaborate control measures as installing a pump-and-treat system to prevent further migration of a groundwater contaminant plume.

Typically, where it is obvious that the final remedy will require excavation and treatment or removal of contaminated 'hot-spots', such actions are better initiated as interim measures, rather than being deferred to a final remedy selection stage.

Interim corrective actions usually will not be the ultimate solution for an environmental management problem. These measures are developed primarily to minimize public exposure (to potentially acute risks/hazards) prior to developing a comprehensive corrective action program. In general, the interim measures may be relatively straightforward (such as erecting a fence or removing a small number of drums), or it may involve more elaborate control measures (such as installing a pump-and-treat system to prevent further migration of a groundwater contaminant plume).

In many situations, it is possible to identify very early in the risk management process, measures that can, and should be taken to control potential receptor exposures to contamination, or to stop further environmental degradation from occurring.

19.1.3 Conditions for 'No-Further-Action' Decisions

A major reason for conducting environmental investigation and characterization is to be able to make informed decisions regarding possible environmental restoration and risk management programs. The fundamental purpose of both an environmental restoration and a risk management program is to protect human health and the environment from the unintended consequences of environmental contamination problems.

In some situations, a decision that 'no further action' (NFA) is required for a problem may be deemed appropriate for the case-specific situation. The process involved in a NFA decision (NFAD) is intended to indicate that, based on the best available information, no further response action is necessary to ensure that the case problem does not pose significant risks to human health or the environment. Under such circumstances, a NFA closure document will usually be prepared for the locale, in accordance with applicable regulatory requirements. The NFA document usually is a stand-alone report, containing sufficient information to support the NFAD. It should therefore include case-specific evidence along with adequate technical reasoning and justification for the NFAD. Additionally, the NFA document should clearly address specific conceptual model hypotheses that have been tested to confirm that there are no likely and complete exposure pathways or scenarios associated with the problem.

Environmental quality standards specified by several regulations (such as media-specific action levels, site-specific risk-based criteria, and local or area background threshold levels) often form an important basis for a NFAD. Ultimately, however, the NFA determination rests on whether or not complete exposure pathways exist at a locale, and whether or not any of the complete pathways are significant. It is therefore logical to infer that the NFA decision criteria are indeed linked to the use of the conceptual model as a decision tool. Consequently, an evaluation of the elements or components of a conceptual model will generally serve as an additional important basis for many NFADs. This typically involves a demonstration that the source–pathway–receptor linkage cannot be completed at the particular problem situation, or that if the linkage can be completed the risk posed by the contamination present does not exceed 'acceptable' reference standards established for the area.

19.2 A FRAMEWORK FOR RISK MANAGEMENT PROGRAMS

Risk management decisions generally are complex processes that involve a variety of technical, political, and socio-economic considerations. Notwithstanding the complexity and the fuzziness of issues involved, the ultimate goal of risk management programs is to protect public health and the environment. The application of risk assessment can remove some of the ambiguity in the decision-making process; it can also aid in the selection of prudent, technically feasible, and scientifically justifiable corrective actions that will help protect public health and the environment in a cost-effective manner. However, to successfully apply the risk assessment process to a potential environmental contamination problem, the process must be tailored to the case-specific conditions and relevant regulatory constraints. Based on the results of a risk assessment, decisions can then be made relating to the types of risk management actions needed for a given environmental contamination problem. If unacceptable risk levels are identified, the risk assessment process can further be employed in the evaluation of remedial action alternatives. This will ensure that net risks to human health and the environment are truly reduced to acceptable levels via the remedial action of choice.

Corrective action response programs for environmental contamination problems may indeed vary greatly – ranging from a 'no-action' alternative to a variety of extensive and costly mitigative options. In any case, the primary objective of every risk management program is to ensure public safety and welfare by protecting human health, the environment, and public and private properties. The ability to select an appropriate and cost-effective corrective action response strategy that meets this goal will generally depend on a careful assessment of both short- and long-term risks associated with the environmental contamination problem; it also depends on the case-specific mitigative measures. In general, once quality-assured information has been compiled for a potential environmental contamination problem and a benchmark risk level is established, an acceptable risk management strategy can be determined that will be used to guide possible mitigative actions.

Figure 19.3 provides a decision framework that may be used to facilitate the decision-making process involved in environmental management programs. The process will generally incorporate a consideration of the complex interactions existing between the environmental setting, regulatory policies, and technical feasibility of remedial technologies. Specifically, the decision processes involved should help environmental analysts identify, rank/categorize, and monitor the status of potential environmental contamination problems; identify field data needs and decide on the best sampling strategy; establish appropriate remedial action goals; and choose the remedial action that is most cost-effective in managing the risks associated with the environmental contamination problem.

Design of risk management programs

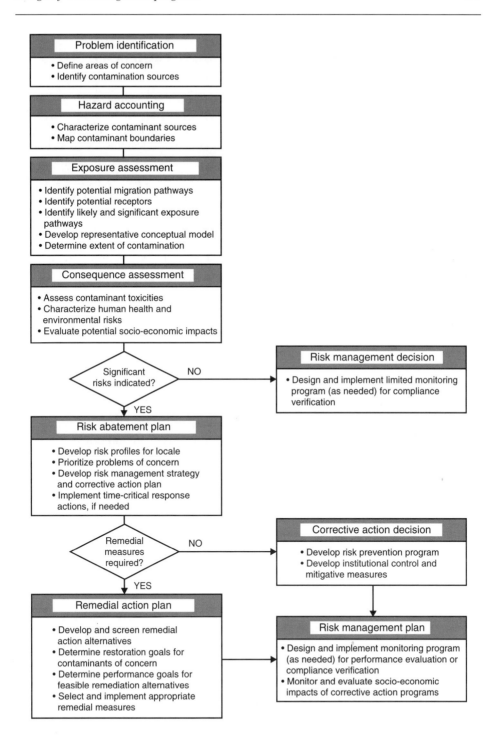

Figure 19.3 A risk management decision framework for the management of environmental contamination problems

19.3 APPLICATION OF DECISION ANALYSIS METHODS TO ENVIRONMENTAL MANAGEMENT PROGRAMS

Decision analysis is a management tool consisting of a conceptual and systematic procedure for rationally analyzing complex sets of alternative solutions to a problem, in order to improve the overall performance of the decision-making process. Decision theory provides a logical and systematic framework to structure the problem objectives and to evaluate and rank potential alternative solutions for addressing the problem. Environmental decision analyses typically involve the use of a series of techniques to design credible risk management programs – that include developing comprehensive corrective action plans, and evaluating appropriate mitigative alternatives in a technically defensible manner.

As part of an environmental management program, it is almost inevitable that the environmental analyst will often have to make choices between alternative courses of action that are based on an evaluation of risk tradeoffs and relative risks among available decision alternatives; evaluation of the cost-effectiveness of corrective action plans; or a risk–cost–benefit comparison of several management options. In fact, comparing risks, benefits, and costs amongst various risk management strategies is an important aspect of most environmental management programs. In general, a number of analytical tools – such as those annotated below – may be used to assist the processes involved (see, e.g., Bentkover et al., 1986; Haimes, 1981; Haimes et al., 1990; Keeney, 1990; Lind et al., 1991; Nathwani et al., 1990; Seip and Heiberg, 1989; USEPA, 1984).

- *Benefits assessment* – consists of an evaluation/identification of what new controls are buying in the way of benefits with respect to additional health or environmental improvements. It measures the risk-reduction effects of potential response actions.
- *Benefit–cost (or cost–benefit) analysis* – weighs the cost of control, explicitly and directly, against the monetized benefits of control, such as the avoidance of disease and the attainment of other social goods. It is a procedure for determining whether the expected benefits from a proposed action outweigh the expected costs; all costs and benefits are necessarily expressed in monetary units.

    Subjective and controversial as it might appear to express certain hazards in terms of cost, especially where public health and/or safety is concerned, it nevertheless has been used to provide an objective way of evaluating some environmental management problems. This is particularly true where risk factors are considered in the overall study. The use of a *cost–benefit analysis* – in which the risks reduced by a proposed action are translated into benefits (e.g., lives saved, improvement in quality assured life expectancy, etc.) on a monetary basis – may indeed become particularly essential to risk management (van Leeuwen and Hermens, 1995).
- *Risk–benefit analysis* – balances the economic benefits of a polluting activity against the associated risks to health and the environment. In a risk–benefit analysis, no attempt is made to measure all consequences of the activity of concern in commensurate units. A monetary estimate is obtained for the net

economic benefit of the activity (e.g., pesticide manufacture, which has useful applications but is toxic to humans at the same time). The other benefits associated with the activity (e.g., increased crop yield, etc.) are expressed in other, more realistic units (such as relating to improved quality adjusted life expectancy ($QALE$) of populations). The costs attributed to risk of adverse health effects due to use of a pesticide product are also quantified in non-monetary units. The benefits can then be balanced against the risks.

- *Cost-effectiveness analysis* – involving determination of that action which maximizes the level of risk reduction per unit cost – may be employed as an important analytical tool that looks for the least-cost path to achieve a given goal, such as the achievement of a protective standard.

  Cost-effectiveness analysis involves a comparison of the costs of alternative methods to achieve some set goal(s) of risk reduction, such as an established benchmark risk or cleanup criteria. The process compares the costs associated with different methods of achieving a specific corrective action goal. The analysis can be used to allocate limited resources among several risk abatement programs, aimed at achieving the maximum positive results per unit cost. The procedure may also be used to project and compare total costs of several corrective action plans.

  In the application of cost-effective analyses to corrective action assessments, a fixed goal is established, and then policy options are evaluated on the ability to achieve that goal in a most cost-effective manner. The goal generally consists of a specified level of 'acceptable' risk, and the remedial options are compared on the basis of the monetary costs necessary to reach the benchmark risk. Cost constraints can also be imposed so that the options are assessed on their ability to control the risk most effectively for a fixed cost. The efficacy of the corrective action alternatives in the hazard reduction process can subsequently be assessed, and the most cost-effective course of action (i.e., one with minimum cost meeting the constraint of a benchmark risk/hazard level) can then be implemented. This would then guarantee the objective of meeting the constraints at the lowest feasible cost.

- *Multi-attribute decision analysis and utility theory* – have been suggested (e.g., Keeney and Raiffa, 1976; Lifson, 1972) for the evaluation of problems involving multiple conflicting objectives, such as is the case for a number of decisions on environmental management programs. In such situations, the decision-maker is faced with the problem of having to tradeoff the performance of one objective for another. In addressing these types of problem, a mathematical structure may be developed around utility theory that presents a deductive philosophy for risk-based decisions (Keeney, 1984; Keeney and Raiffa, 1976; Lifson, 1972; Starr and Whipple, 1980).

Variations or surrogates of the above methods that may be considered as of special interest to environmental management programs are presented in the proceeding sections. In general, the purpose of the analytical tools used in risk management is to help determine the most efficient way to reduce risk. In fact, the use of structured decision support systems have proven to be efficient and cost-effective in making sound environmental decisions. Such tools can indeed play vital roles in improving the environmental decision-making process. It should be acknowledged, however, that despite the fact that decision analysis presents a systematic and flexible technique that incorporates the decision-maker's judgment, it does not necessarily provide a complete analysis of the public's perception of risk. In any case, as a common theme for all stakeholders, the greater the risk that prevails, or that is apparent or perceived, the greater the incentive to reduce it.

19.3.1 Risk–Cost–Benefit Optimization and Tradeoffs Analysis

Risk–cost–benefit analysis is a generic term for techniques encompassing risk assessment and the inclusive evaluation of risks, costs, and benefits of alternative projects or policies. In performing risk–cost–benefit analysis, one attempts to: measure risks, costs and benefits; identify uncertainties and potential tradeoffs; and then present this information coherently to decision-makers. A general form of objective function for use in a risk–cost–benefit analysis that treats the stream of benefits, costs, and risks in a net present value calculation is given by (Crouch and Wilson, 1982; Massmann and Freeze, 1987):

$$\Phi = \sum_{t=0}^{T} \frac{1}{(1+r)^t} \{B(t) - C(t) - R(t)\}$$

where Φ is objective function ($); t is time, spanning 0 to T (years); T is time horizon (years); r is discount rate; $B(t)$ is benefits in year t ($); $C(t)$ is costs in year t ($); $R(t)$ is risks in year t ($). The risk term is defined as the expected cost associated with the probability of significant impacts or failure, and is a function of the costs due to the consequences of failure in year t.

In general, tradeoff decisions made in the optimization process will be directed at improving both short- and long-term benefits of the program. In fact, any environmental management policy or program should take account of, and balance all anticipated detrimental impacts against potential positive attributes by using a tradeoffs or similar analysis. Whereas it may not be possible to quantitatively estimate all the variables involved in such an analysis, it is important to recognize the importance of all variables and to perform at least some qualitative evaluation of all parameters involved. Some of the parameters will be subjective and have high degrees of uncertainties associated with them. In any case, the tradeoffs analyses result in the following conceptual relationships:

$$\text{Total net benefits, } TNB = \sum \{\text{benefits}\} - \sum \{\text{risks + costs}\}$$

$$\text{Net benefits per person} = \frac{[\text{net benefits in commensurate units}]}{[\text{population}]}$$

Ultimately, if on the average the net benefit is positive, then the program may be acceptable and is therefore worthy of further consideration – provided risk tolerance criteria are met; if, on the average, the net benefit is negative, then the program should be rejected outright (Figure 19.4).

19.3.2 Multi-attribute Decision Analysis and Utility Theory Applications

Risk tradeoffs between increased expenditure of a remedial action and the hazard reduction achieved upon implementation of a risk management or corrective action plan may be assessed by the use of multi-attribute decision analysis and utility theory methods. Multi-attribute decision analysis and utility theory can indeed be applied in the investigation and management of a variety of environmental contamination problems, in order to determine whether one set of impacts and/or mitigation alternatives is more or less desirable than another set. With such a formulation, an explicitly logical and justifiable solution can be assessed for the usually complex decisions involved in environmental management programs.

In using expected utility maximization to evaluate alternative environmental management strategies, the preferred option will be the one that maximizes the expected utility – or equivalently, the one that minimizes the loss of expected utility. In a way, this is a nonlinear generalization of cost–benefit or risk–benefit analysis. The general processes involved in this type of analysis are annotated below.

- *Utility-attribute analysis.* Attributes measure how well a set of objectives are being achieved. Through the use of multiple attributes scaled in the form of utilities, and weighted according to their relative importance, a decision analyst can describe an expanded set of consequences associated with an environmental management program. Adopting utility as the criterion of choice among alternatives allows a multi-faceted representation of each possible consequence. Hence, in its application to environmental management problems, both hazards and costs can be converted to utility values, as measured by the relative importance that the decision-maker attaches to either attribute.

  The utility function need not be linear since the utility is not necessarily proportional to the attribute. Thus, curves of the forms shown in Figure 19.5 can be generated for the utility function. An arbitrary value (e.g., 0 or 1 or 100) of 1 can be assigned to the 'ideal' situation (i.e., a 'no hazard/no cost scenario'), and the 'doom' scenario (i.e., 'high hazard/high cost') is then assigned a corresponding relative value (e.g., -1 or 0 or 1) of 0. The shape of the curves is determined by the relative value given each attribute. The range in utilities is the same for each attribute, and attributes should, strictly speaking, be expressed as specific functions of system characteristics.

  In assigning utility value to hazard, it is commonplace to rely on various social and environmental goals which can help determine the threats posed by the hazard, rather than use the direct concept of hazard. These utility values can then be used as the basis for selection among remedial alternatives.

- *Preferences and evaluation of utility functions.* Evaluation of utility functions require skill, and when the utility function represents the preferences of a particular interest group, additional difficulties arise. Nonetheless, risk tradeoffs

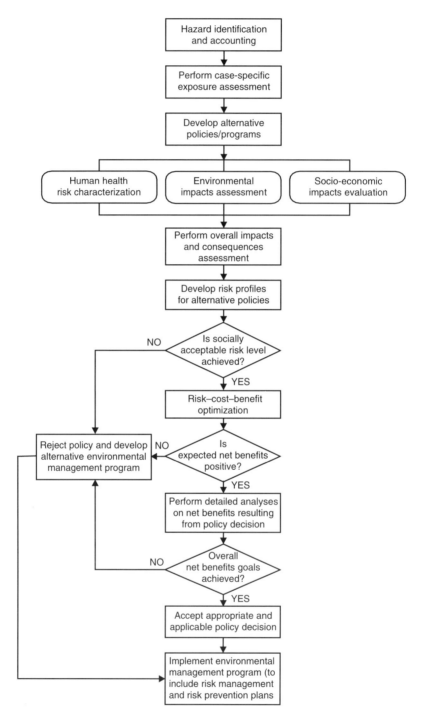

Figure 19.4 A conceptual framework for risk–cost–benefit balancing and tradeoffs analyses in an environmental management program

Design of risk management programs

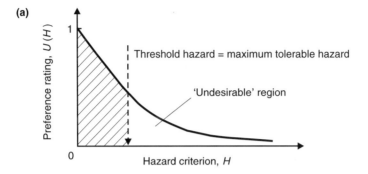

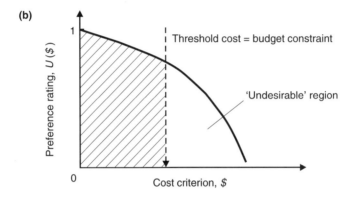

Figure 19.5   Utility functions giving the relative values of hazards and costs in similar (dimensionless) terms

may be satisfactorily determined by applying weighting factors of preferences in a utility-attribute analysis.

Preferences are directly incorporated in the utility functions by assigning an appropriate weighting factor to each utility term. The weighting factors are changed to reflect varying tradeoff values associated with alternative decisions. In general, if minimizing hazards is k times as important as minimizing costs, then weighting factors of $k/(k+1)$ and $1/(k+1)$ would be assigned to the hazard utility and the cost utility, respectively. These weighting factors would reflect, or give a measure of, the preferences for a given utility function. Past decisions can help provide empirical data that can be used for quantifying the tradeoffs and therefore the k values. The given utilities are weighted by their preferences, and are summed over all the objectives. For n alternatives, the value of the ith alternative would be determined according to:

$$V_i = \frac{k}{(k+1)} U(H_i) + \frac{1}{(k+1)} U(\$_i)$$

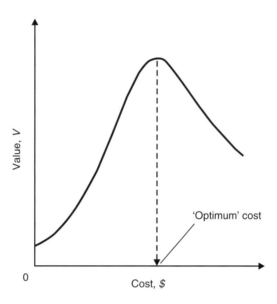

Figure 19.6 Value function for costs

where V_i is the total relative value for the ith alternative; $U(H_i)$ is the hazard utility, H, for the ith alternative; and $U(\$_i)$ is the cost utility, $\$$, associated with alternative i.

In general, the largest total relative value would ultimately be selected as the best alternative.

- *Utility optimization.* To facilitate the development of an optimum environmental management program, the total relative value can be plotted against the cost (Figure 19.6). From this plot, the optimum cost is that cost value which corresponds to the maximum total relative value. The optimum cost is equivalently obtained, mathematically, as follows:

$$\frac{\delta V}{(\delta \$)} = \frac{\delta}{(\delta \$)}\left[\frac{k}{(k+1)}U(H) + \frac{1}{(k+1)}U(\$)\right] = 0$$

or,

$$k\frac{\delta U(H)}{(\delta \$)} = -\frac{\delta U(\$)}{(\delta \$)}$$

where $\delta U(H)/(\delta \$)$ is the derivative of hazard utility relative to cost, and $\delta U(\$)/(\delta \$)$ is the derivative of cost utility relative to cost. The optimum cost is obtained by solving this equation for $\$$; this would represent the most cost-effective option for project execution.

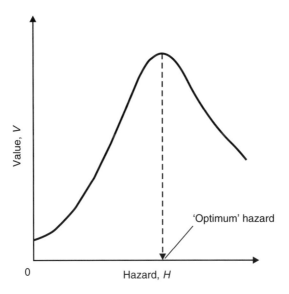

Figure 19.7 Value function for hazards

In an evaluation similar to the one presented above, a plot of total relative value against hazard provides a representation of the 'optimum hazard' (Figure 19.7). Again, this result can be evaluated in an analytical manner similar to that presented above for cost; the 'optimum hazard' is given, mathematically, by:

$$\frac{\delta V}{(\delta H)} = \frac{\delta}{(\delta H)}\left[\frac{k}{(k+1)}U(H) + \frac{1}{(k+1)}U(\$)\right] = 0$$

$$k\frac{\delta U(H)}{(\delta H)} = -\frac{\delta U(\$)}{(\delta H)}$$

where $\delta U(H)/(\delta H)$ is the derivative of hazard utility relative to hazard, and $\delta U(\$)/(\delta H)$ is the derivative of cost utility relative to hazard. Solving for H yields the 'optimum' value for the hazard.

Even though utility theory offers a rational procedure for corrective measure studies, it may transfer the burden of decision to the assessment of utility functions. Also, several subjective assumptions are used in the application of utility functions that are a subject of contemporaneous debate. The details of the paradoxes surrounding conclusions from expected utility applications are beyond the scope of this elaboration and are not discussed here.

19.4 UTILIZATION OF GIS IN RISK ASSESSMENT AND ENVIRONMENTAL MANAGEMENT PROGRAMS

An important aspect of risk assessment with growing interest relates to the coupling of environmental models with information systems – such as Geographic Information Systems (GISs) – in order to allow for effective risk mapping of a study area. GIS can indeed be used to map the location and proximity of risk to identified or selected receptors.

A GIS is a computer-based tool used to capture, manipulate, process, and display spatial or geo-referenced data (Bernhardsen, 1992; Goodchild et al., 1993, 1996). The GIS can process geo-referenced data and provide answers to such questions as the particulars of a given location, the distribution of selected phenomena and their temporal changes, the impact of a specific event, or the relationships and systematic patterns of a region (Bernhardsen, 1992). In fact, it has been suggested that, as a planning and policy tool, the GIS technology could be used to 'regionalize' the risk analysis process – moving it from its traditional focus on a micro-scale (i.e., site-specific problems) to a true macro-scale (e.g., urban or regional risk analysis, comparative risk analysis, and risk equity analysis) (D. Rejeski, in Goodchild et al., 1993).

GIS is indeed a rapidly developing technology for handling, analyzing, and modeling geographic information. It is not a source of information *per se*, but only a way to manipulate information. When the manipulation and presentation of data relates to geographic locations of interest, then our understanding of the real world is enhanced.

19.4.1 Integration of Environmental and Risk Models with GIS

Every field of environmental modeling is increasingly using spatially distributed approaches, and the use of GIS methods will likely become widespread. It is noteworthy, however, that models lacking a spatial component clearly have no use for GIS.

The specific role of environmental models integrated with GIS would largely be in their ability to communicate effectively – via the use of maps as a well-understood and accepted form of information display, and by generating a widely accepted and familiar format for the sharing of information (Goodchild et al., 1993). In general, the GIS describes the spatial environment, and the environmental modeling simulates the functioning of environmental processes (Goodchild et al., 1996). Thus, GIS can serve as a common data and analysis framework for environmental models.

It must be acknowledged that, although linkage of environmental models with GIS is frequently encountered, in the majority of prevailing cases GIS and environmental models are not really integrated – they are just used together (Goodchild et al., 1993). The GIS is often used as a pre-processor to prepare spatially distributed input data, and as a post-processor to display and possibly analyze model results further. Compared to maps, however, GIS has the inherent advantage that data storage and data presentation are separate. Consequently, data may be presented and viewed in a variety of ways.

19.4.2 The Role of GIS Applications in Environmental Management

The integration of GIS into risk assessment and environmental management programs can result in the following particularly important uses for GIS:

- as a tool in environmental modeling;
- as a tool for hazard, exposure and risk mapping; and
- as a tool for risk communication.

In fact, exposure analysis as an overlay of hazard sources and receptor locations represents an almost classical GIS application. GISs indeed have the ability to integrate spatial variables into risk assessment models, yielding maps that are powerful visual tools to communicate risk information (Goodchild et al., 1993). In principle, the conceptual mapping of risk makes it much easier to communicate hazard and risk levels to potentially affected society.

Specifically, GISs may find applications in a variety of risk assessment and environmental management programs, such as are illustrated by the following practical example scenarios (D. Rejeski, in Goodchild et al., 1993):

- *Development of thematic map layers for risk management.* In an air quality management study, by combining population density with ambient concentrations for a specific chemical ($\mu g/m^3$) and the unit risk factor for that chemical (risk/$\mu g/m^3$), a map showing the risk per unit of population may be produced that could be used to facilitate risk management decisions.
- *Data improvement for source/pathway characterization.* It is recognized that, in general, the adequate characterization of the exposure pathways affecting the fate and transport of risk-inducing agents can help improve risk estimates significantly. Thus, more spatial data could be added to a GIS model to empirically describe the transport medium and its effects on the distribution or dispersion of risk agents.
- *Decision utility in remediation planning and design.* GIS may be used to examine the spatial distribution of risks around toxic sources, such as hazardous waste sites or incinerators and smelters. The ability of GIS to aid in the calculation of volumes, for instance, would allow soil removal and transport costs to be estimated, and for an optimal remedial action decision to be made.
- *Design of risk reduction strategies.* Once risks have been mapped using GIS, it may be possible to match estimated risks to risk reduction strategies, and also to delineate spatially the regions where resources should be invested as well as the appropriate strategies to adopt for the various geographical dichotomies.
- *Investigation of risk equity assurance issues.* Considering the fact that the notion of environmental justice, based on an equitable distribution of risks, is emerging as a critical theme in environmental decision-making, GIS could become a powerful tool of choice for exploring such risk equity issues. Such application

moves beyond simply calculating risks based on somehow abstract and subjective probabilities and actually presents comparative analyses for all stakeholders and 'contesting' regions or neighborhoods.

In typical application scenarios, the GIS will generally form a central framework and integrating component that provides a variety of map types for use in an overall environmental management system. Maps or overlays may consist of simple line features (such as city boundaries) or complex topical maps serving as background for the spatially distributed environmental models; this will include model input data sets, as well as model results. A practical example of such a representation is concentration fields of pollutants from air, groundwater, or surface water models stored as grid cell files.

19.5 RISK COMMUNICATION AS A FACILITATOR OF RISK MANAGEMENT

Risk management combines socio-economic, political, legal, and engineering approaches to manage risks. Risk assessment information is used in the risk management process to help in deciding how to best protect public health and the environment. Essentially, risk assessment provides *information* on the health and environmental risks, and risk management develops and implements an *action* based on that information. This means that risk assessment can in principle be carried out objectively, whereas risk management usually involves preferences and attitudes, and should therefore be considered a subjective activity (Seip and Heiberg, 1989; NRC, 1983; USEPA, 1984). The subjectivity of the risk management task calls for the use of very effective facilitator tools/techniques – with good risk communication (i.e., the exchange of information about risk) being the logical choice. Effective risk communication is indeed important for the implementation of effective risk management programs. It is therefore important to give adequate consideration to risk communication issues when developing a risk management program to address an environmental contamination problem.

Risk communication has formally been defined as the process of conveying or transmitting information among interested parties about the following types of issues: levels of health and environmental risks; the significance or meaning of health or environmental risks; and decisions, actions, or policies aimed at managing or controlling health or environmental risks (Cohrssen and Covello, 1989). It offers a forum at which various stakeholders discuss the nature, magnitude, significance, or control of risks and related consequences with one another.

In fact, to be able to design an effectual risk management program, a variety of qualitative issues become equally important in addition to any prior risk quantification. For instance, in referring to the so-called NIMBY ('not-in-my-backyard'), NIABY ('not-in-anyone's-backyard'), NIMTO ('not-in-my-term-of-office'), LULU ('locally-unwanted-land-use'), and related syndromes (e.g., Gregory and Kunreuther, 1990) raining upon several communities, it is apparent that risk communication may dictate public perception, and therefore public acceptance of risk management strategies and overall environmental management decisions.

One goal of risk communication is to improve the agreement between the magnitude of a risk and the public's political and behavioral response to this risk – necessitating researchers to investigate a number of message characteristics and risk communication strategies (Weinstein and Sandman, 1993; Weinstein et al., 1996). The process involved provides information to a concerned public about potential health risks from exposure to toxic chemicals or similar environmental hazards. In fact, because the perception of risks often differs widely, risk communication typically requires a sensitive approach and should involve genuine dialogue (van Leeuwen and Hermens, 1995). Among several other factors, trust and credibility are believed to be key determinants in the realization of any risk communication goals. Apparently, defying a negative stereotype is key to improving perceptions of trust and credibility (Peters et al., 1997).

Only limited presentation on the risk communication topic is given in this volume, with more detailed elaboration/discussions to be found in the literature elsewhere (e.g., Cohrssen and Covello, 1989; Covello, 1992, 1993; Covello and Allen, 1988; Fisher and Johnson, 1989; Freudenburg and Pastor, 1992; Hance et al., 1990; Kasperson and Stallen, 1991; Laird, 1989; Leiss, 1989; Leiss and Chociolko, 1994; Lundgren, 1994; Morgan and Lave, 1990; NRC, 1989; Pedersen, 1989; Peters et al., 1997; Renn, 1992; Silk and Kent, 1995; Slovic, 1993; van Leeuwen and Hermens, 1995; Vaughan, 1995; Weinstein and Sandman, 1993; Weinstein et al., 1996). The limited literature available on the subject addresses several important elements/issues – including checklists for improving both the process and content of risk communication efforts.

19.5.1 Designing an Effective Risk Communication Program

A number of rules and guidelines have been suggested/proposed to facilitate effective risk communication (e.g., Cohrssen and Covello, 1989; Covello and Allen, 1988) – albeit there are no easy prescriptions. Overall, it is very important that risk communication should consider and embrace several important elements (Box 19.2), in order to minimize or even prevent suspicion/outrage from an usually cynical public. Serious consideration of the relevant elements should help move a potentially charged atmosphere to a responsible one, and one of cooperation and dialog. Ultimately, a proactive, planned program of risk communication will – at the very least – usually place the intended message in the public eye in advance of negative publicity and sensational media headlines.

A reliable tool and channel of communication should be identified to ensure effective transmittal of all relevant information. Overall, a systematic evaluation using structured decision methods, such as the use of the event-tree approach (Chapter 15), can greatly help in this direction. The event tree illustrates the cause and effect ordering of event scenarios, with each event being shown by a branch of the event tree in the context of the decision problem. The event-tree model structure

> Box 19.2 Important strategic considerations in developing an effective risk communication program
>
> - Accept and involve the public as a legitimate partner – especially all parties that have an interest or direct stake in the particular risk situation
> - Involve all stakeholders as early as possible – via taking a proactive stance, based on a coherent strategy, sound tactics, and careful planning of community relations and actions
> - Listen to your audience – recognizing communication is a two-way activity, and take note of the public's specific concerns
> - Ensure an effective two-way discourse/dialogue, to ensure adequate flow of information in both directions between the risk communication team and the interested public/parties. That is, flow of information should be from, and to all stakeholders
> - Be honest, frank, and open – since lost trust and credibility are almost impossible to regain
> - Focus should be on what the risks are, and what is already being done to keep these risks as low as reasonably possible
> - Have an even greater focus on long-term implication of risk management decisions/strategies, without necessarily discounting potential short-term consequences
> - Have an elaborate evaluation of alternative choices to proposed risk management strategies
> - Anticipate or investigate the affected party's likely perception of the prevailing or expected risks, since this could be central to any response to proposed actions or risk management strategies
> - Speak clearly and with compassion – especially minimizing excessive use of technical language and jargon
> - Avoid use of unnecessary jargon and excessive technical details, in order to allow the community to focus on the real/practical issues of interest to the population(s)-at-risk
> - Anticipate controversy and request for changes to proposed risk management plans, and then offer positive response which may form a basis for consensus between all stakeholders
> - Keep the needs and perceptions of the 'outsider' stakeholders in perspective, to ensure balanced and equitable program needs
> - Focus more on psychological needs of community, rather than economic realities/interests of your project
> - Plan carefully and evaluate the hazard/risk-generating performance – to help re-focus, if necessary
> - Coordinate and collaborate with other credible sources – such as by issuing communications jointly with other trustworthy sources like credible university scientists, area physicians, trusted local officials, and opinion leaders
> - Meet the needs of the media – recognizing that they tend to play a critical role in setting agendas and determining outcomes

can indeed aid risk communicators in improving the quality and effectiveness of their performance and presentations.

Ultimately, scientific information about health and environmental risks is generally communicated to the public through a variety of channels – ranging from warning labels on consumer products to public meetings/forums involving representatives from government, industry, the media, the populations potentially at risk, and other sectors of the general public (Cohrssen and Covello, 1989).

Important traditional techniques of risk communication usually consist of 'fact sheets', newsletters, public meetings, and similar forum types – all of which seem to work very well if rightly implemented or utilized.

But irrespective of the approach or technique adopted, it must be acknowledged that hazard perception and risk thresholds tend to be quite different in different parts of the world. In fact, there could even be variations within different sectors of society within the same region of the world. However, such variances should not affect the general design principles when one is developing a risk communication program.

19.6 MANAGING RISKS

Risk management programs are typically directed at risk reduction (i.e., taking measures to protect humans and/or the environment against previously identified risks), risk mitigation (i.e., implementing measures to remove risks), and/or risk prevention (i.e., instituting measures to completely prevent the occurrence of risks). Risk reduction, mitigation, and preventative programs generally can help in the improvement of system efficiency and reliability; they can also facilitate an increase in the level of protection to public health and safety, as well as aid in the reduction of liability.

In the arena of environmental contamination problems, it is almost imperative to systematically identify hazards throughout an entire environmental management system, assess the potential consequences due to any associated hazards, and examine corrective measures for dealing with the case-specific type of problem situation. Risk management – used in tandem with risk assessment – offers the necessary mechanism for achieving such goals.

SUGGESTED FURTHER READING

Bates, D.V. 1994. *Environmental Health Risks and Public Policy*. University of Washington Press, Seattle, WA.
Hansen, P.E. and S.E. Jorgensen (eds). 1991. *Introduction to Environmental Management*. Developments in Environmental Modelling, 18, Elsevier, Amsterdam, The Netherlands.
Finkel, A.M. and D. Golding (eds). 1994. *Worst Things First? (The Debate over Risk-based National Environmental Priorities)*. Resources for the Future, Washington, DC.
Fischoff, B. 1995. Risk perception and communication unplugged: twenty years of process. *Risk Analysis*, 15(2): 137–145.
Hilts, S.R. 1996. A co-operative approach to risk management in an active lead/zinc smelter community. *Environmental Geochemistry and Health*, 18: 17–24.
Holmes, G., B.R. Singh, and L. Theodore. 1993. *Handbook of Environmental Management and Technology*. J. Wiley, New York.
Jones, R.B. 1995. *Risk-Based Management: A Reliability-Centered Approach*. Gulf Publishing, Houston, TX.

Kunreuther, H. and M.V. Rajeev Gowda (eds). 1990. *Integrating Insurance and Risk Management for Hazardous Wastes*. Kluwer Academic Publishers, Boston, MA.
OES (California Office of Emergency Services). 1989. *Guidance for the Preparation of a Risk Management and Prevention Program*. State of California, California Office of Emergency Services, Hazardous Materials Division, Sacramento, CA.

REFERENCES

Asante-Duah, D.K. 1996. *Managing Contaminated Sites: Problem Diagnosis and Development of Site Restoration*. J. Wiley, Chichester, UK.
Bentkover, J.D., V.T. Covello and J. Mumpower. 1986. *Benefits Assessment: The State of the Art*. Riedel, Boston, MA.
Bernhardsen, T. 1992. *Geographic Information Systems*. Viak IT, Arendal, Norway.
BSI (British Standards Institution). 1988. *Draft for Development, DD175: 1988 Code of Practice for the Identification of Potentially Contaminated Land and its Investigation*. BSI, London, UK.
Cairney, T. (ed.). 1993. *Contaminated Land (Problems and Solutions)*. Blackie Academic & Professional, Glasgow/Chapman & Hall, London/Lewis Publishers, Boca Raton, FL.
Cohrssen, J.J. and V.T. Covello. 1989. *Risk Analysis: A Guide to Principles and Methods for Analyzing Health and Environmental Risks*. National Technical Information Service (NTIS), US Dept. of Commerce, Springfield, VA.
Covello, V.T. 1992. Trust and credibility in risk communication. *Health and Environmental Digest*, 6(1): 1–3.
Covello, V.T. 1993. Risk communication and occupational medicine. *Journal of Occupational Medicine*, 35(1): 18–19.
Covello, V.T. and F. Allen. 1988. *Seven Cardinal Rules of Risk Communication*. USEPA Office of Policy Analysis, Washington, DC.
Crouch, E.A.C. and R. Wilson. 1982. *Risk/Benefit Analysis*. Ballinger, Boston, MA.
Fisher, A. and F.R. Johnson. 1989. Conventional wisdom on risk communication and evidence from a field experiment. *Risk Analysis*, 9(2): 209–213.
Freudenburg, W.R. and S.R. Pastor. 1992. NIMBYs and LULUs: stalking the syndromes. *Journal of Social Issues*, 48(4): 39–61.
Goodchild, M.F., B.O. Parks and L.T. Steyaert (eds). 1993. *Environmental Modeling with GIS*. Oxford University Press, New York.
Goodchild, M.F., L.T. Steyaert, B.O. Parks, et al. (eds). 1996. *GIS and Environmental Modeling: Progress and Research Issues*. GIS World Books, Fort Collins, CO.
Gregory, R. and H. Kunreuther. 1990. Successful siting incentives. *Civil Engineering*, 60(4): 73–5.
Haimes, Y.Y. (ed.). 1981. *Risk/Benefit Analysis in Water Resources Planning and Management*. Plenum, New York.
Haimes, Y.Y., L. Duan, and V. Tulsiani. 1990. Multiobjective decision-tree analysis. *Risk Analysis*, 10(1): 111–129.
Hance, B.J., C. Chess, and P.M. Sandman. 1990. *Industry Risk Communication Manual: Improving Dialogue with Communities*. Lewis Publishers, Boca Raton, FL.
Jolley, R.L. and R.G.M. Wang (eds). 1993. *Effective and Safe Waste Management: Interfacing Sciences and Engineering with Monitoring and Risk Analysis*. Lewis Publishers, Boca Raton, FL.
Kasperson, R.E. and P.J.M. Stallen (eds). 1991. *Communicating Risks to the Public: International Perspectives*. Kluwer Academic Press, Boston, MA.
Keeney, R.L. 1984. Ethics, decision analysis, and public risk. *Risk Analysis*, 4: 117–129.
Keeney, R.L. 1990. Mortality risks induced by economic expenditures. *Risk Analysis*, 10(1): 147–159.

Keeney, R.D. and H. Raiffa. 1976. *Decisions with Multiple Objectives: Preferences and Value Tradeoffs*. J. Wiley, New York.
Laird, F.N. 1989. The decline of deference: the political context of risk communication. *Risk Analysis*, 9(2): 543–550.
Leiss, W. 1989. *Prospects and Problems in Risk Communication*. Institute for Risk Research. University of Waterloo Press, Waterloo, Ontario, Canada.
Leiss, W. and C. Chociolko. 1994. *Risk and Responsibility*. McGill-Queen's University Press, Montreal, Quebec, Canada.
Lifson, M.W. 1972. *Decision and Risk Analysis for Practicing Engineers*. Barnes and Noble, Cahners Bks, Boston, MA.
Lind, N.C., J.S. Nathwani, and E. Siddall. 1991. *Managing Risks in the Public Interest*. Institute for Risk Research, University of Waterloo, Waterloo, Ontario.
Lundgren, R. 1994. *Risk Communication: A Handbook for Communicating Environmental, Safety, and Health Risks*. Battelle Press, Columbus, OH.
Massmann, J. and R.A. Freeze. 1987. Groundwater contamination from waste management sites: the interaction between risk-based engineering design and regulatory policy 1. Methodology 2. Results. *Water Resources Research*, 23(2): 351–380.
Morgan, M.G. and L. Lave. 1990. Ethical considerations in risk communication practice and research. *Risk Analysis*, 10(3): 355–358.
Nathwani, J., N.C. Lind and E. Siddall. 1990. Risk–benefit balancing in risk management: measures of benefits and detriments. Presented at *The Annual Meeting of the Society for Risk Analysis*, 29th Oct.–1 Nov., 1989, San Francisco, CA. Institute for Risk Research Paper No. 18, Waterloo, Ontario, Canada.
NRC (National Research Council). 1983. *Risk Assessment in the Federal Government: Managing the Process*. National Academy Press, Washington, DC.
NRC. 1989. *Improving Risk Communication*. National Academy Press, Washington, DC..
NRC. 1994. *Science and Judgment in Risk Assessment*. Committee on Risk Assessment of Hazardous Air Pollutants. National Academy Press, Washington, DC.
Pedersen, J. 1989. *Public Perception of Risk Associated with the Siting of Hazardous Waste Treatment Facilities*. European Foundation for the Improvement of Living and Working Conditions, Dublin, Eire.
Peters, R.G., V.T. Covello, and D.B. McCallum. 1997. The determinants of trust and credibility in environmental risk communication: an empirical study. *Risk Analysis*, 17(1): 43–54.
Renn, O. 1992. Risk communication: towards a rational dialogue with the public. *Journal of Hazardous Materials*, 29(3): 465–519.
Seip, H.M. and A.B. Heiberg (eds). 1989. *Risk Management of Chemicals in the Environment*. NATO Committee on the Challenges of Modern Society, Vol. 12. Plenum Press, New York.
Silk, J.C. and M.B. Kent (eds). 1995. *Hazard Communication Compliance Manual*. Society for Chemical Hazard Communication/BNA Books, Washington, DC.
Slovic, P. 1993. Perceived risk, trust, and democracy. *Risk Analysis*, 13(6): 675–682.
Starr C. and C. Whipple. 1980. Risks of risk decisions. *Science*, 208: 1114.
USEPA (US Environmental Protection Agency). 1984. *Risk Assessment and Management: Framework for Decision Making*. USEPA, Washington, DC. EPA 600/9-85/002.
USEPA. 1985. *Characterization of Hazardous Waste Sites: A Methods Manual, Volume 1, Site Investigations*. Environmental Monitoring Systems Laboratory, Las Vegas, NV. EPA-600/4-84/075.
USEPA. 1987a. *Alternate Concentration Limit Guidance*. Office of Solid Waste, Waste Management Division, Washington, DC. EPA/530-SW-87/017, OSWER Directive 9481-00-6C.
USEPA. 1987b. *RCRA Facility Investigation (RFI) Guidance*. Washington, DC. EPA/530/SW-87/001.
USEPA. 1988. *Guidance for Conducting Remedial Investigations and Feasibility Studies Under*

CERCLA. Office of Emergency and Remedial Response, Washington, DC. EPA/540/G-89/004. OSWER Directive 9355.3-01.

USEPA. 1989. *Risk Assessment Guidance for Superfund. Volume I – Human Health Evaluation Manual (Part A)*. Office of Emergency and Remedial Response, Washington, DC. EPA/540/1-89/002.

USEPA. 1991. *Risk Assessment Guidance for Superfund, Volume I: Human Health Evaluation Manual. Supplemental Guidance. 'Standard Default Exposure Factors'*. (Interim Final). March, 1991. Office of Emergency and Remedial Response, Washington, DC. OSWER Directive: 9285.6-03.

van Leeuwen, C.J. and J.L.M. Hermens (eds). 1995. *Risk Assessment of Chemicals: An Introduction*. Kluwer Academic Publishers, Dordrecht, The Netherlands.

Vaughan, E. 1995. The significance of socioeconomic and ethnic diversity for the risk communication process. *Risk Analysis*, 15(2): 169–180.

Weinstein, N.D. and P.M. Sandman. 1993. Some criteria for evaluating risk messages. *Risk Analysis*, 13(1): 103–114.

Weinstein, N.D., K. Kolb, and B.D. Goldstein. 1996. Using time intervals between expected events to communicate risk magnitudes. *Risk Analysis*, 16(3): 305–308.

WPCF (Water Pollution Control Federation). 1988. *Hazardous Waste Site Remediation: Assessment and Characterization*. A Special Publication of the WPCF, Technical Practice Committee, Alexandria, VA.

Chapter Twenty

Risk Assessment Applications to Environmental Management Problems

The risk assessment process is intended to give the risk management team the best possible evaluation of all available scientific data, in order to arrive at justifiable and defensible decisions on a wide range of issues. For example, to ensure public safety in all situations, contaminant migration beyond a compliance boundary into the public exposure domain must be below some stipulated risk-based maximum exposure level – typically established through a risk assessment process.

Based on the results of a risk assessment, decisions can be made in relation to the types of risk management actions necessary to address a given environmental contamination problem or hazardous situation. In fact, risk-based decision-making will generally result in the design of better environmental management programs. This is because risk assessment can produce more efficient and consistent risk reduction policies. It can also be used as a screening tool for setting priorities.

This chapter enumerates the general scope for the application of risk assessment as pertains to managing environmental contamination problems; recapitulates several pertinent requirements for the utilization of risk assessment principles, concepts, and techniques to the various application scenarios; and presents a qualitative risk evaluation questionnaire chart that can be deemed as necessary in project scoping decisions.

20.1 GENERAL SCOPE OF THE PRACTICAL APPLICATION OF RISK ASSESSMENT TO ENVIRONMENTAL PROBLEMS

Risk assessment has several specific applications that could affect the type of decisions to be made in relation to environmental management programs. A number of practical examples of the potential application of risk assessment principles, concepts, and techniques in environmental management practice are annotated in Box 20.1 – with further general discussion of some of the more prominent application scenarios offered below. This listing is by no means complete and exhaustive, since variations or even completely different and unique problems may be resolved by use of one form of risk assessment methodology or another.

- *Environmental impact assessments.* Environmental impact assessment (EIA) is the analysis of the likely environmental consequences associated with a proposed human activity. The principal objective of an EIA is to ensure that

Box 20.1 Summary decision issues and specific problems typically addressed by the use of risk assessment concepts and techniques in environmental management programs

- Field sampling design – by helping identify data needs and/or data gaps
- Determination of potential risks associated with industrial, commercial, and residential properties, to facilitate land-use decisions and/or restrictions
- Corrective measure evaluation and selection of remedial alternatives: risks posed by alternative remedial actions can be assessed before program implementation – i.e., the risk assessment process allows performance goals of remedial alternatives to be established prior to their implementation
- Determination of cost-effective risk reduction policies through the selection of feasible remedial alternatives protective of public health and the environment
- Evaluation of remedial alternatives to determine whether the proposed remedial action itself will have any deleterious environmental and human health effects
- Prioritization of contaminated sites for remedial action: by providing consistent data for the rank-ordering of potentially contaminated sites, risk assessment helps regulatory agencies and potentially responsible parties to prioritize cleanup actions
- Development of target cleanup criteria and guidelines for contaminated sites: helps define media cleanup levels required for remediation decisions – i.e., risk assessment provides the basis for determining the levels of chemicals that can remain at a site, or in environmental matrices, without impacting public health and the environment
- Facilitation of decisions about the use of specific chemicals in manufacturing processes and industrial activities
- Preliminary screening for potential problems: evaluation and ranking of potential liabilities in environmental management practices
- Hazardous waste facility design and operation
- Design of monitoring programs (to identify chemicals present in various media, and their persistence)
- Ecological/environmental risk assessment for the identification of critical habitats and organisms exposed to environmental contaminants
- Probabilistic risk assessment for the evaluation of transportation risks associated with the movement of hazardous materials
- Implementation of general risk management and risk prevention programs for environmental management planning
- Facilitatation of property transactions by assisting developers, lenders, and buyers in the 'safe' acquisition of both residential and commercial properties
- Streamlining of site remediation activities: facilitates more cost-effective site closures for contaminated sites
- Addressing of health and safety issues associated with environmental chemicals: determining 'safe' exposure limits for toxic chemicals used or found in the workplace
- Facilitation of decisions on alternative disposal methods for potentially hazardous waste materials
- Determination of discharge requirements for industrial facilities
- Evaluation and management of potential risks due to air emissions from industrial facilities and incinerators
- Demonstration and justification for the acceptability of risks associated with 'baseline' conditions, allowing regulatory agencies to accept a 'no-action' recommendation in remedial action decisions
- Determination of potential risks anticipated from site remediation, allowing appropriate risk management and site control measures to be implemented
- Evaluation of expected performance and effectiveness of alternative remediation techniques
- Analysis of human health impacts from chemical residues found in food products (including contaminated fish and pesticide-treated produce) and a variety of consumer products

environmental considerations are incorporated into the planning for, decisions on, and implementation of development activities. Typically, this may include an identification of factors contributing the most to overall risks of exposures to the environmental hazards of concern. It may incorporate an analysis of baseline risks, and a consistent process to document potential public health and environmental threats from the relevant activity. EIAs are generally so-designed with the principal objective to help prevent – or at least minimize – a development activity's adverse impacts, while maximizing its beneficial effects.

As an example, quantitative risk assessment is often undertaken as part of the siting process for newly proposed facilities – in many cases as a regulatory requirement for EIAs, or may also be carried out for operating facilities – usually to evaluate implications from design changes and/or changes in exposure parameters. In fact, the risk assessment of stack emissions from municipal solid waste (MSW) incineration facilities seems to be one of the most important applications for this type of evaluation. This becomes necessary because MSW incinerators typically release various potentially toxic compounds – some of which escape pollution control equipment and enter the outside air; these chemicals may include metals (e.g., arsenic, cadmium, and mercury), organic compounds that may have escaped combustion or are only partially oxidized, and other pollutants such as polychlorinated dibenzo(p)-dioxins and furans (PCDDs/PCDFs) that are byproducts formed in the combustion train (Zemba et al., 1996).

As part of the key processes involved in EIAs, risk assessment can aid in the design of cost-effective field sampling, data collection, and data evaluation programs. This will then allow an adequate number of samples to be collected and analyzed – with an overall objective of determining or anticipating the presence or absence of contamination, their extent and distribution, or of verifying the attainment of any mitigative measures.

- *Air toxics risk assessment.* Often an air pathway exposure assessment (APEA) is conducted to address air contaminant releases from a variety of sources. This is a systematic approach involving the application of modeling and monitoring methods to estimate contaminant emission rates and their concentrations in air, and consists of the following key components:

    – characterization of air emission sources;
    – determination of the effects of atmospheric processes on contaminant fate and transport;
    – evaluation of populations potentially at risk (for various exposure periods); and
    – estimation of receptor intakes and doses.

The purpose of the APEA is to estimate the extent of actual or potential receptor exposures to air contaminants. This involves emission quantification, modeling of environmental transport, evaluation of environmental fate, identification of potential exposure routes and populations potentially at risk from exposures, and the estimation of both short- and long-term exposure levels.

The following information, at a minimum, needs to be collected and reviewed to support the design of an APEA program:

- Source data (to include contaminant toxicity parameters, off-site sources, etc.).
- Receptor data (including identification of sensitive receptors, local land use, etc.).
- Environmental data (such as dispersion data, meteorological information, topographic maps, soil and vegetation data, etc.).
- Previous APEA data (to include regulatory standards summary, air monitoring data, emission rate monitoring and modeling information, dispersion modeling results, etc.).

A first step in assessing air toxic impacts is to evaluate case-specific characteristics and the chemical contaminants present at the locale, and to determine whether transport of hazardous chemicals into air is of potential concern. Atmospheric dispersion and emission source modeling can be used with appropriate air sampling data as input to an APEA.

In general, once the contaminant inhalation concentrations are determined by appropriate field measurements and/or modeling practices, exposure calculations are performed. The exposure estimates are subsequently integrated with the appropriate toxicity information to generate the corresponding risk and hazard estimates.

The air pathways methods of analysis in relation to air contaminant releases find several specific applications in such situations as the estimation of VOC emission rates from landforms – represented as ponds, sludges, landfills, etc. (see, e.g., Mackay and Leinonen, 1975; Mackay and Yeun, 1983; Thibodeaux and Hwang, 1982); and particulate inhalation exposure from fugitive dust (see, e.g., CAPCOA, 1990; USEPA, 1989a).

Once the emission rates are determined, and the exposure scenarios are defined, decisions can generally be made regarding potential air impacts for specified activities. Finally, exposure and risk calculations are carried out using the appropriate algorithms.

- *Evaluation of potential risks associated with the migration of contaminant vapors into building structures.* Considering that the degree of dilution in the indoor air of a building is generally far less than the situation outdoors, contaminant vapors entering/infiltrating into a building structure may represent a significant risk to occupants of the building. In fact, the migration of subsurface contaminant vapors into buildings can become a very important source of human exposure via the inhalation route. As appropriate, therefore, a determination of the relative significance of vapor transport and inhalation as a critical exposure scenario should be given serious consideration during the processes involved in the characterization of an environmental contamination problem, and in establishing environmental quality criteria and/or remediation goals. Risk assessment methods can generally be used to make this determination – as to whether or not vapor transport and inhalation represent a significant exposure scenario worth focusing on. For example, a risk characterization scenario involving

exposure of populations to vapor emissions from cracked concrete foundations/ floors can be determined for such a situation, in order for responsible risk management and/or mitigative measures to be adopted.
- *Environmental assessment and characterization.* Environmental assessments are invariably a primary activity in the general processes involved in the management of environmental contamination problems. The objective of the environmental assessment is to determine the nature and extent of potential impacts from the release or threat of release of hazardous substances.

  An environmental investigation effort, which is a major component of the environmental assessment activity, aims at collecting representative samples from the potentially contaminated media. Depending on the adequacy of historical data and the sufficiency of details about the likely environmental contaminants, sampling programs can be designed to search for specific chemical constituents that become indicator parameters for the sample analyses. Where specific historical information is lacking, a more comprehensive sampling and analytical program will generally be required; in this case, the sampling and analysis program may be carried out in phases – moving from a more or less generic scope to one of specificity, as adequate information becomes available about the situation.

  A multimedia approach to environmental characterization is usually adopted for most environmental contamination problems, so that the significance of possible air, water, soil, and biota contamination can be established through appropriate field sampling and analysis procedures. The activities involved are expected to yield high-quality environmental data needed to support possible corrective action response decisions. To accomplish this, samples are gathered and analyzed for the contaminants of potential concern in the appropriate media of interest. Proper protocols in the field sampling and laboratory analysis procedures are used to help minimize uncertainties associated with the data collection and evaluation activities.

  Ultimately, the results from these activities will facilitate a complete analysis of the contaminants found in the environment. Risk assessment techniques and principles are typically employed to facilitate the development of effectual environmental characterization programs. Thus, the information so-obtained is used to determine current and potential future risks to human health and the environment. In addition to providing information about the nature and magnitude of potential risks associated with a potential environmental contamination problem, risk assessment also provides a basis for judging the need for mitigative actions. On this basis, corrective actions are developed and implemented, with the principal objective to protect public health and the environment.
- *Design of corrective action response and environmental restoration programs.* A variety of corrective action strategies may indeed be employed in the quest to restore contaminated sites into healthier conditions. The processes involved will generally incorporate a consideration of the complex interactions existing between the hydrogeological environment, regulatory policies, and the technical feasibility of remedial technologies. In particular, a clear understanding of the

fate and behavior of the contaminants in the environment is essential for developing successful corrective action response programs, and also to ensure that the problem is not exacerbated.

The design of corrective action response programs for contaminated site problems usually involves several formalized steps. Typically, the major activities carried out will comprise of a remedial investigation (RI) and a feasibility study (FS). The RI is conducted to gather sufficient data in order to characterize conditions at a contaminated site; this may involve extensive fieldwork and site assessment used to identify site contaminants, determine constituent concentrations and distribution across the site, characterize contaminant migration pathways and routes of exposure to local surrounding populations, and identify other site conditions that might affect site restoration options. The FS is undertaken to identify and analyze potential site restoration alternatives; typically, the FS uses a screening process to reduce the number of alternative corrective measures to a limited range of remedial options, and the short-listed options are subsequently subjected to a detailed analysis in which various tradeoffs (such as cost-effectiveness, extent of cleanup, and permanence of a cleanup action) are evaluated.

The assessment of health and environmental risks play an important role in the remedial investigation/feasibility study (RI/FS), the remedial action plan (RAP) development, and also the risk mitigation and risk management strategies in the management of contaminated site problems. In particular, to reduce costs in planning for the cleanup of contaminated sites, it is important that the decision-making process involved be well defined; risk assessment provides for such a solution.

In principle, the site restoration goal and strategy selected for a contaminated site problem may vary significantly from one site to another due to several site-specific parameters. A number of extraneous factors may also affect the selection of site restoration strategies. The selection of an appropriate corrective action plan (including the choice of acceptable cleanup goals) depends on a careful assessment of both short- and long-term risks posed by the case-specific site. As a general rule, corrective action plans should provide for the removal and/or treatment of contaminants until a level necessary to protect human health and the environment is achieved.

In general, once existing site information has been analyzed and a conceptual understanding of a site is obtained, potential remedial action objectives should be defined for all impacted media at the contaminated site. Subsequently, alternative site restoration programs can be developed to support the requisite corrective action decision. Overall, risk assessment plays a very important role in the development of remedial action objectives for contaminated sites; in the identification of feasible remedies that meet the remediation objectives; and in the selection of an optimum remedial alternative.

- *Development of contaminated site cleanup goals.* Risk assessment has become particularly useful in determining the level of cleanup most appropriate for potentially contaminated sites. By utilizing methodologies that establish cleanup criteria based on risk assessment principles, corrective action programs can be conducted in a cost-effective and efficient manner. Once realistic risk reduction levels potentially achievable by various remedial alternatives are known, the decision-maker can then use other scientific criteria (such as implementability, reliability, operability, and cost) to select a final design alternative. Subsequently, an appropriate corrective action plan can be developed and implemented for the contaminated site.

In fact, a major consideration in developing a RAP for a contaminated site is the level of cleanup to be achieved – which could become the driving force behind remediation costs. It will therefore be prudent to allocate adequate resources for the development of the appropriate cleanup criteria. The site cleanup limit concept will facilitate decisions about the effective use of limited funds to clean up a site to a level appropriate/safe for its intended use. In principle, the cleanup criteria selected for a potentially contaminated site may vary significantly from site to site due to the site-specific parameters. Similarly, mitigation measures may be case-specific for various hazardous situations and problems.

In general, preliminary remediation goals (PRGs) are usually established as cleanup objectives early in a site characterization process. The development of PRGs requires site-specific data relating to the impacted media of interest, the chemicals of potential concern, and the probable future land uses. The early determination of remediation goals facilitates the development of a range of feasible corrective action decisions, which in turn helps focus remedy selection on the most effective remedial alternative(s).

It is noteworthy that an initial list of PRGs may have to be revised as new data become available during site characterization activities. In fact, PRGs are refined into final remediation goals throughout the process leading up to the final remedy selection. Therefore, it is important to iteratively review and re-evaluate the media and chemicals of potential concern, future land uses, and exposure assumptions originally identified during project scoping.

- *Risk evaluation of remedial alternatives for contaminated sites.* A major objective of any case-specific risk assessment is to provide an estimate of the baseline risks posed by the existing conditions associated with an environmental contamination problem, and to further assist in the evaluation of restoration options. The processes involved in the risk evaluation of remedial alternatives consist of the same general steps as a baseline risk assessment, except that the baseline risk assessment typically is more refined than the risk comparison of remedial alternatives. The difference between the site risks in the absence of remedial action (i.e., the baseline risk) and the risks associated with a remedial alternative will generally help define the net benefits associated with a given remedial option. Overall, it is crucial to ensure that the projected risks posed by a remedial option do not offset any benefits associated with reducing site contamination, in order to achieve an established site restoration or risk reduction goal.

Risk assessment techniques can indeed be used to quantify the human health risks and environmental hazards created by implementing specific remedial options at contaminated sites. Furthermore, they can be used to compare the risk reductions afforded by different remedial or risk control strategies.

Several advantages may accrue from the use of risk assessment procedures to arrive at consistent and cost-effective corrective action decisions for potential environmental contamination problems. These procedures can help determine whether a particular remedial alternative will pose unacceptable risks following implementation, and to determine the specific remedial alternatives that will result in the least risk upon achieving the cleanup goals or remedial action objectives for the site. Consequently, risk assessment tools can be used to aid in the process of selecting among remedial options for contaminated sites. The evaluation of both short-term and long-term risks is an important part of the relevant analyses involved.

- *Evaluation of human health risks associated with remedial actions.* A remedial action will often increase potential short-term exposures and risks at a contaminated site above baseline conditions. However, short-term health risks should generally not be used as a selection criterion for remedial alternatives, but should be used to determine appropriate management practices during implementation of the selected remedial action.

  For instance, a remedial option at a site may involve excavation, removal and transportation of contaminated soil materials. In the absence of precautionary or mitigative measures, fugitive dust generation by heavy equipment and remedial activities may create unreasonably high short-term health hazards (that may well exceed the long-term benefits associated with the cleanup objectives). On the other hand, these and other temporary sources of contaminant releases associated with construction and implementation of a remedy may not be adequate enough grounds for rejecting the remedial alternative. Under such circumstances, appropriate management practices – such as the temporary relocation of populations potentially at risk – should be considered in order to mitigate the health risks associated with such temporary source releases.

  In general, as part of an overall corrective action response process, remedial alternatives are analyzed for both potential long- and short-term health effects associated with the implementation of each remediation option. This will help design appropriate risk control measures for the remedial activity.

- *Evaluation of ecological risk issues in corrective action decisions.* Often, especially in the past, only limited attention has been given to the ecosystems associated with contaminated sites, and also to the protection of ecological resources during site remediation activities. Instead, much of the focus has been on the protection of human health and resources directly affecting public health and safety. In recent times, however, the ecological assessment of contaminated sites has been gaining considerable attention. This is the result of prevailing knowledge or awareness of the intricate interactions between ecological receptors/systems and contaminated site cleanup processes.

  In fact, human health is frequently the major concern in corrective action assessments, but an ecological assessment may serve to expand the scope of the investigation for an environmental contamination problem – by enlarging the area under consideration, or re-defining remediation criteria, or both. A detailed ecological assessment may indeed become critical in the determination of whether or not the potential contamination at a site warrant remedial action.

To achieve adequate ecological protection and/or regulatory compliance, ecological assessments should address the overall site contamination issues, and should be coordinated with all aspects of site cleanup – including human health concerns, engineering feasibility, and economic considerations (de Serres and Bloom, 1996; Maughan, 1993; Suter et al., 1995). Also, there are several important ecological concerns associated with contaminated site cleanup programs that should be addressed early enough in site characterization programs. Furthermore, a number of legislative requirements for incorporating ecological issues in site characterization and site restoration efforts now exist in several geographical regions, and these can no longer be ignored.

It is apparent, therefore, that ecological resources must be appropriately evaluated in order to achieve the mandate of any comprehensive program designed to ensure the effective management of environmental contamination problems. Thus, remedial alternatives, in addition to being evaluated for the degree to which they protect human health, are also evaluated for their ability to protect ecological receptors and ecosystems.

Ecological data gathered before and during remedial investigations are typically used to determine the appropriate level of detail for an ecological assessment; to decide whether remedial action is necessary based on ecological considerations; to evaluate the potential ecological effects of relevant remedial options; to provide information necessary for mitigation of site threats; and to design monitoring strategies used to assess the progress and effectiveness of remediation.

It is noteworthy, in a number of situations, site remediation can destroy or otherwise affect uncontaminated ecological resources; soil removal techniques, alteration of site hydrology, and site preparation are examples of remediation activities that can result in inadvertent damage to ecological resources. Thus, the ecological impacts of a site restoration activity must be understood by the decision-makers before remediation plans are approved and implemented. In situations where adverse ecological effects are identified, corrective action alternatives with potentially less damaging impacts must be evaluated as preferred methods of choice. This means that the assessment of potential ecological impacts should be performed for each remedial alternative considered for a contaminated site.

In fact, remedial actions, by their nature, can alter or destroy aquatic and terrestrial habitats. The potential for the destruction or alteration of ecological habitats, and the consequences of ecosystem disturbances and other ecological effects, must therefore be given adequate consideration during the corrective action response process. Indeed, it is very important to integrate ecological investigation results and general concerns into the overall site cleanup process.

- *Risk-based evaluation of the beneficial re-use of contaminated materials.* Environmental quality criteria for contaminated materials typically serve as benchmarks in the assessment of the degree of contamination, and also in the determination of the possible beneficial uses of the impacted materials, and which uses are not expected to be adversarial to human health and the environment. By utilizing protocols that determine maximum acceptable or safe contaminant levels based on risk assessment methodologies, such impacted materials can be managed in a cost-effective manner. In fact, various analytical tools and models can generally be used in the development of appropriate action levels or criteria that will facilitate the implementation of effective policy decisions on the impacted materials.

As an example, risk assessment principles can be incorporated into the development of statistical relationships between the nature of contamination from highway materials and the roadway types. For instance, by using the analytical results obtained from catchbasin materials derived from roadways, hazards and/or risks associated with cleanings from the different catchbasins (associated with the different road types) can be determined. In addition, state-of-the-art risk models can be used to establish 'safe' or 'action' levels for the contaminants of concern found in the catchbasin materials. This level can then serve as a reference or trigger criteria against which future catchbasin cleanings can be compared, in order to determine if such materials are likely to present elevated risks to public health and/or the environment; the established criteria could also become an important determinant in the beneficial re-use analysis for the catchbasin materials.

- *Development of environmental management strategies.* Environmental contamination problems have reached an almost nightmarish level in most societies globally – with contaminated sites representing a significant portion of the overall problem. The effective management of environmental contamination problems has indeed become an important environmental priority that will remain a growing social concern for years to come. This is due in part to the numerous complexities and inherent uncertainties involved in the analysis of such problems.

Whatever the cause of an environmental contamination problem, the impacted media usually must be remediated. Under certain circumstances, however, restoration or cleanup may not be economically or technically feasible; in that case, risk assessment and monitoring of the situation, together with institutional control measures, may be acceptable risk management strategies *in lieu* of remediation. By using case-specific characteristics rather than default modeling parameters, risk assessments can be conducted that result in realistic determinations of risks and mitigation strategies. Such an approach also helps identify uncertainties associated with the decision-making process.

Overall, a risk assessment will generally provide the decision-maker with scientifically defensible procedures for determining whether or not a potential environmental contamination problem could represent a significant adverse health and environmental risk, which should therefore be considered a candidate for mitigative actions. In fact, the use of health and environmental risk assessments in environmental management decisions in particular and corrective action programs in general is becoming an important regulatory requirement in several places. For instance, a number of environmental regulations and laws in various jurisdictions (e.g., regulations promulgated under CERCLA, RCRA, CAA, CWA in the US) increasingly require risk-based approaches in determining cleanup goals and related decision parameters.

In fact, several issues that – directly or indirectly – affect environmental management programs may be addressed by use of risk assessment tools, as illuminated by the variety of examples given above. In such applications of risk assessment, it is important to adequately characterize the exposure and physical settings for the problem situation, in order to allow for an effective application of appropriate risk assessment methods of approach.

It is noteworthy that there are several complexities involved in real-life scenarios that are unique in characterizing environmental contamination problems. Also, the populations potentially at risk from environmental contamination problems are usually heterogeneous which can greatly influence the anticipated impacts/consequences. Critical receptors should therefore be carefully identified with respect to numbers, location (areal and temporal), sensitivities, etc., so that risks are neither under-estimated or conservatively over-estimated.

It is apparent that some form of risk assessment is inevitable if environmental management programs are to be conducted in a sensible and deliberate manner. In fact, risk assessment seems to be gaining greater grounds in making public policy decisions in the control of risks associated with environmental management systems. This situation may be attributed to the fact that the very process of performing a risk assessment does lead to a better understanding and appreciation of the nature of the risks inherent in a study, and further helps develop steps that can be taken to reduce these risks.

Overall, the application of risk assessment to environmental contamination problems helps identify critical migration and exposure pathways, receptor exposure routes, and other extraneous factors contributing most to total risks. It also facilitates the determination of cost-effective risk reduction policies. Used in the corrective action planning process, risk assessment generally serves a useful tool for evaluating the effectiveness of remedies associated with environmental contamination problems, and also for determining acceptable restoration goals. Inevitably, risk-based methods of analysis facilitate the selection of appropriate and cost-effective restoration measures and/or risk management strategies for environmental management decisions.

20.2 RISK ASSESSMENT AS A COST-EFFECTIVE TOOL IN THE FORMULATION OF ENVIRONMENTAL MANAGEMENT DECISIONS

Risk assessment is a systematic technique that can be used to make estimates of significant and likely risk factors associated with environmental contamination problems. Often, risk assessment is used as a management tool to facilitate effective decision-making on the control of environmental pollution problems. In fact, the chief purpose of risk assessment is to aid decision-making and this focus should be maintained throughout a given environmental management program.

Invariably, human health and/or environmental risk assessment should become an important element of such programs.

It is apparent that risk assessments of environmental contamination problems depend on an understanding of the fate and behavior of the contaminants of concern. Consequently, the fate and transport issues in the various environmental compartments should be carefully analyzed with the best available scientific tools and environmental models. In fact, the application of computer models – in a responsible manner – for the predictive simulation of environmental systems can be a very useful and cost-effective approach when evaluating potential risks and developing credible corrective action programs.

In general, the procedures utilized must reflect current state-of-the-art methods for conducting risk assessments. For all intents and purposes, the following are noteworthy recommendations in the exercises typically involved:

- Performing risk assessment to incorporate all likely scenarios envisaged rather than for the 'worst-case' alone allows better comparison to be made between risk assessments performed by different scientists and analysts whose views on what represents a 'worst-case' may be very subjective and therefore may vary significantly.
- Exposure scenarios and chemical transport models may contribute significant uncertainty to the risk assessment. Uncertainties, heterogeneities, and similarities should be identified and well documented throughout the risk assessment.
- Whenever possible, the synergistic, antagonistic, and potentiation (i.e., the case of a nonhazardous situation becoming hazardous due to its combination with others) effects of chemicals and other hazardous situations should be carefully evaluated for inclusion in the risk assessment decisions.
- It is prudent to assess what the 'baseline' (no-action) risks are for a potentially hazardous situation or environmental contamination problem. This will give a reflection of what the existing situation is, which can then be compared against future improved situations.
- An evaluation of the 'post-remediation' risks (i.e., residual risks remaining after the implementation of corrective actions) for a potentially hazardous situation or environmental contamination problem should generally be carried out for alternative mitigation measures. This will give a reflection of what the anticipated improved situation is, compared with the prior conditions associated with the problem situation.

The application of risk assessment to environmental contamination problems can indeed remove some of the ambiguity in the decision-making process. It can also aid in the selection of prudent, technically feasible, and scientifically justifiable corrective actions that will help protect public health and the environment in a cost-effective manner. Risk assessments do indeed provide decision-makers with scientifically defensible information for determining whether an environmental contamination problem poses a significant threat to human health or the environment. They may be conducted to assist in the development and implementation of cost-effective environmental management solutions. Ultimately, the risk assessment efforts can help minimize or eliminate potential long-term problems or liabilities that could result from hazards associated with environmental contamination problems.

20.3 A CONCLUDING NOTE

Some element of risk exists in just about every modern technological development and activity – including environmental management systems; these risks must be assessed and courses of action decided, in order to minimize any consequences attributable to such risks. Faced with the bewildering array of potential risks in the arena of environmental management, therefore, society is motivated to develop systematic tools that may help bring the unpleasant but inevitable situation under control less expensively. In recent years, focus seems to have been on the use of risk assessment techniques to provide a structured and systematic framework for the evaluation of environmental problems. The development of a structured risk assessment framework will generally facilitate systematic decision-making on the protection of public health and the environment from the effects of environmental contamination problems. Such a structure will indeed provide an effective way to build the scientifically valid and technically defensible information base necessary for tackling potential health and environmental hazards.

In the application of risk assessment methods to the wide spectrum of environmental management programs, it is important to do an initial scoping of the particular problem situation; this typically will entail a more or less qualitative review of the problem. A questionnaire chart underpinning the risk evaluation of environmental contamination problems may indeed facilitate the overall environmental management process. As an example, several important questions worthy of evaluation in a typical environmental contamination problem are noted in Box 20.2; this is especially geared towards contaminated site problems but can be modified to suit different types of application scenarios. Also, it may become necessary to employ a variety of decision support tools and databases – such as from the limited listing provided in Appendix E – that are potentially applicable to the particular environmental contamination problem that is being addressed.

In general, the benefits of risk assessment outweigh the possible disadvantages, but it must be recognized that this process will not be without tribulations. Indeed, risk assessment is by no means a panacea. Its use, however, is an attempt to widen and extend the decision-maker's knowledge-base and thus improve the decision-making capability. Overall, the method deserves the effort required for its continual refinement as an environmental management tool.

Box 20.2   Questionnaire chart to facilitate risk evaluation and corrective action assessment of an environmental contamination problem

- Do contaminant sources exist?
- Are there visible sources of contamination?
- What are the sources of the contaminants?
- Is the area potentially contaminated with hazardous/toxic chemicals? If so, what are the toxic agents involved?
- Are there confining layers or porous layers in the soil horizon – where applicable?
- Is there soil erosion, or recent cuts or fills on site – where applicable?
- What is the nature of drainage and surface flow patterns at the site and immediate vicinity – where applicable?
- What are the site characteristics, hydrological features, meteorological or climatic factors, land-use patterns, and agricultural practices affecting the transport and distribution of contaminants?
- What is the distribution of the chemicals over the problem location and vicinity?
- Are there known 'hot-spots' at the project location?
- What is an appropriate background or control region to use for corrective action investigations?
- Is there any area that poses an immediate and life-threatening exposure?
- What are the important transport processes and migration pathways that contribute to intermedia transfers and the spread of contamination and/or exposure?
- Are there current or future potential receptors that could be adversely affected by the contaminant sources? In particular, are there sensitive ecosystems or residences located downgradient, downstream, or downwind from the impacted area?
- Are there one or more pathways through which environmental contaminants might migrate from the source and reach potential receptors? What are the dominant routes of exposure at the locale?
- Have populations already been impacted, and/or are populations potentially at risk?
- What are the potential risks posed to human health and the envirmonment if no further response action is taken at the locale?
- Does the risk level exceed benchmark levels specified by environmental compliance regulations? If so, what site-specific cleanup criteria will be appropriate for the locale?
- For the indicated contaminant concentrations at the locale, which areas are considered as posing risks to the environment or surrounding populations? Which areas must therefore be remediated in order to reduce risks to an 'acceptable' level?
- Are estimated risk levels low enough, such that a 'no-action' alternative is still protective of public health and the environment?
- What contaminants and environmental media should become the target for remediation?
- How much contaminated material should be remediated to achieve an acceptable environmental restoration goal?
- Which remedial alternatives can be applied at the contaminated area in order to achieve adequate cleanup?
- Will exposure pathways be interrupted or will receptors be protected following removal or remedial actions?
- What institutional control measures and risk management strategies are required in the overall corrective action decision?

SUGGESTED FURTHER READING

DoE (Department of the Environment). 1994. *A Framework for Assessing the Impact of Contaminated Land on Groundwater and Surface Water, Volumes 1 & 2*. CLR Report No. 1, DoE, London, UK.
DoE. 1995. *A Guide to Risk Assessment and Risk Management for Environmental Protection*. HMSO, London, UK.
Ginevan, M.E. and D.E. Splitstone. 1997. Improving remediation decisions at hazardous waste sites with risk-based geostatistical analysis. *Environmental Science and Technology*, 31(2): 92A–96A.
Goldstein, B.D. 1995. The who, what, when, where, and why of risk characterization. *Policy Studies Journal*, 23(1): 70–75.
HSE (Health and Safety Executive). 1996. *The Use of Risk Assessment in Government Departments*. HSE Books, Sudbury, UK.
Katz, M. and D. Thornton. 1997. *Environmental Management Tools on the Internet: Accessing the World of Environmental Information*. St. Lucie Press, Delray Beach, FL.
NRC (National Research Council). 1994a. *Alternatives for Ground Water Cleanup*. Committee on Ground Water Cleanup Alternatives. National Academy Press, Washington, DC.
NRC. 1994b. *Ranking Hazardous Waste Sites for Remedial Action*. National Academy Press, Washington, DC.
NRC. 1995. *Improving the Environment: An Evaluation of DOE's Environmental Management Program*. National Academy Press, Washington, DC.
Patrick, D.R. (ed.). 1996. *Toxic Air Pollution Handbook*. Van Nostrand Reinhold, New York.
Power, M. and L.S. McCarty. 1998. A comparative analysis of environmental risk assessment/risk management frameworks. *Environmental Science and Technology*, 32(9): 224A–231A.
Russell, M. and M. Gruber. 1987. Risk assessment in environmental policy-making. *Science*, 236: 286–290.

REFERENCES

CAPCOA (California Air Pollution Control Officers Association). 1990. *Air Toxics 'Hot Spots' Program. Risk Assessment Guidelines*. CAPCOA, CA.
de Serres, F.J. and A.D. Bloom (eds). 1996. *Ecotoxicity and Human Health (A Biological Approach to Environmental Remediation)*. Lewis Publishers/CRC Press, Boca Raton, FL.
Mackay, D. and P.J. Leinonen. 1975. Rate of evaporation of low-solubility contaminants from water bodies. *Environmental Science and Technology*, 9: 1178–1180.
Mackay, D. and A.T.K. Yeun 1983. Mass transfer coefficient correlations for volatilization of organic solutes from water. *Environmental Science and Technology*, 17: 211–217.
Maughan, J.T. 1993. *Ecological Assessment of Hazardous Waste Sites*. Van Nostrand Reinhold, New York.
Suter II, G.W., B.W. Cornaby, et al. 1995. An approach for balancing health and ecological risks at hazardous waste sites. *Risk Analysis*, 15: 221–231.
Thibodeaux, L.J. and S.T. Hwang. 1982. Landfarming of petroleum wastes – modeling the air emission problem. *Environmental Progress*, February, 1: 42–46.
USEPA (US Environmental Protection Agency). 1989. *Estimating Air Emissions from Petroleum UST Cleanups*. Office of Underground Storage Tanks, Washington, DC (June 1989).
Zemba, S.G., L.C. Green, E.A.C. Crouch, and R.R. Lester. 1996. Quantitative risk assessment of stack emissions from municipal waste combusters. *Journal of Hazardous Materials*, 47: 229–275.

PART VII

APPENDICES

This part of the book is comprised of a set of appendices that contain: selected listing of abbreviations and acronyms; glossary of selected terms and definitions; some basic probability definitions and concepts commonly used in probabilistic risk analyses; a table of selected toxicity parameters often used in human health risk assessments for a select list of chemicals; annotation of selected environmental tools and databases potentially applicable to the management of environmental contamination problems, and of potential use in risk assessment programs; some selected units of measurements and noteworthy expressions (of potential interest to the environmental professional, analyst, or decision-maker); and a listing of suggested scientific and environmental policy journals of potential interest to risk assessment and environmental management programs.

Appendix A

Selected Abbreviations and Acronyms

ADD	average daily dose
ADI	acceptable daily intake (or allowable daily intake)
AL	action level
ANOVA	analysis-of-variance
APCD	air pollution control device
APEA	air pathway exposure analysis/assessment
APL	aqueous phase liquid
ASC	acceptable soil concentration
AWC	acceptable water concentration
AWQC	ambient water quality criteria
BATNEEC	best available technique not entailing excessive cost (mostly used by the European Union [EU])
BCF	bioconcentration (or bioaccumulation) factor
B(D)AT	best (demonstrated) available technology (mostly used in North America)
BTEX	benzene, toluene, ethylbenzene, and xylene
CAG	Carcinogen Assessment Group (of the USEPA)
CAS	Chemical Abstracts Service
CCA	cause–consequence analysis
cdf	cumulative distribution function
CDI	chronic daily intake
CI	confidence interval
CL	confidence limit
COC/COPC	chemical (or contaminant) of potential concern
COPEC	chemical (or contaminant) of potential ecological concern
(C)SF	(cancer) slope factor (also called cancer potency factor, CPF, or cancer potency slope, CPS)
CSM	conceptual site model
DDT	dichlorodiphenyl trichloroethane
DGF	degradation factor
DL	detection limit (of analyte)
DNAPL	dense nonaqueous phase liquid
DQO	data quality objective
DWEL	drinking water equivalent level
EA	endangerment assessment
EED	estimated exposure dose
EIA	environmental impact assessment
EMU	environmental management unit
EPA	Environmental Protection Agency
EPC	exposure point concentration
EQC	environmental quality criteria
ERA	environmental (or ecological) risk assessment
ErQ	ecological risk quotient
ETA	event-tree analysis
EU	European Union
FMEA	failure modes and effects analysis

FMECA	failure modes, effects and criticality analysis
FS	feasibility study
FSP	field sampling plan
FTA	fault-tree analysis
GC/MS	gas chromatography/mass spectrometry
GC/PID	gas chromatograph/photoionization detector
GIS	Geographic Information System
GLC	ground-level concentration
HAZAN	hazard analysis
HAZOP	hazard and operability study
HEAST	Health Effects Assessment Summary Tables (USEPA)
HI	hazard index
HQ	hazard quotient
HRS	Hazard Ranking System (of the USEPA)
HSP	health and safety plan
IARC	International Agency for Research on Cancer (of the UN)
IDL	instrument detection limit
IGW	investigation-generated waste (also, investigation-derived waste – IDW)
IRIS	Integrated Risk Information System (of the USEPA)
IRPTC	International Register of Potentially Toxic Chemicals (of the UN)
LADD	lifetime average daily dose
LC_{50}	mean lethal concentration
LD_{50}	mean lethal dose
LOAEC	lowest-observed-adverse-effect concentration
LOAEL	lowest-observed-adverse-effect level
LOEC	lowest-observed-effect concentration
LOEL	lowest-observed-effect level
LNAPL	light nonaqueous phase liquid
MCL	maximum contaminant level
MDD	maximum daily dose
MDL	method detection limit
MEL	maximum exposure level
MF	modifying factor
MLE	maximum likelihood estimates
MOE	margin of exposure
MSW	municipal solid waste
NAPL	nonaqueous phase liquid
ND	nondetect (for analytical results)
NFA	no further action
NFAD	no-further-action decision
NOAEC	no-observed-adverse-effect concentration
NOAEL	no-observed-adverse-effect level
NOEC	no-observed-effect concentration
NOEL	no-observed-effect level
OECD	Organization for Economic Cooperation and Development
OVA/GC	organic vapor analyzer/gas chromatograph
PA	preliminary assessment
PAH	polyaromatic hydrocarbon (or, polycyclic [or polynuclear] aromatic hydrocarbon)
PAR	population-at-risk
PB-PK	physiologically-based pharmacokinetic (model)
PCB	polychlorinated biphenyl
PCDD	polychlorinated dibenzo(p)dioxins
PCDF	polychlorinated dibenzofurans
PCE	perchloroethylene

Appendix A: Selected abbreviations and acronyms 455

PCP	pentachlorophenol
pdf	probability density funcion
PEL	permissible exposure limit
pmf	probability mass function
ppb	parts per billion (micrograms per kilogram or μg/L)
ppm	parts per million (milligrams per kilogram or mg/L)
ppt	parts per trillion (nanograms per kilogram or ng/L)
PQL	practical quantitation limit (see also, SQL)
PRA	probabilistic risk assessment
PRG	preliminary remediation goal
PRP	potentially responsible party
PWP	pathway probability
QA/QC	quality assurance/quality control
QALE	quality adjusted life expectancy
QL	quantitation limit
RAP	remedial action plan
RBC	risk-based concentration
RBCL	risk-based chemical level
RBD	reliability block diagram
RfC	reference concentration
RfD	reference dose
RfD_s	subchronic reference dose
RgD	regulatory dose
RI	remedial investigation
RI/FS	remedial investigation/feasibility study
RME	reasonable maximum exposure
RMU	risk management unit
RSCL	recommended site cleanup limit
RSD	risk-specific dose
RTV	reference toxicity value
SAP	sampling and analysis plan
SF	slope factor
SDI	subchronic daily intake
SI	site inspection
SQC	sediment quality criteria
SQL	sample quantitation limit (see also, PQL)
SWMU	solid waste management unit
TCDD	tetrachlorinated dibenzo(p)dioxins
TCE	trichloroethylene (or trichloroethene)
TEF	toxicity equivalence factor
TLV	threshold limit value
TPH	total petroleum hydrocarbon
TSD(F)	treatment, storage, and disposal (facility)
UCL	upper confidence level
UCL_{95}	95% upper confidence level
UCR	unit cancer risk (also, unit risk – UR)
UF	uncertainty factor (also, safety factor)
UR	unit risk (also, unit cancer risk – UCR)
URF	unit risk factor
UST	underground storage tank
VOC	volatile organic compound/chemical
VSD	virtually safe dose
WHO	World Health Organization

Appendix B
Glossary of Selected Terms and Definitions

Absorbed dose – The amount of a chemical substance actually entering an exposed organism via the lungs (for inhalation exposures), the gastrointestinal tract (for ingestion exposures), and/or the skin (for dermal exposures). It represents the amount penetrating the exchange boundaries of the organism after contact.

Absorption – The transport of a substance through the outer boundary of a medium. Generally used to refer to the uptake of a chemical by a cell or an organism, including the flow into the bloodstream following exposure through the skin, lungs, and/or gastrointestinal tract.

Absorption barrier – Any of the exchange barriers of the body that allow differential diffusion of various substances across a boundary; examples of absorption barriers are the skin, lung tissue, and gastrointestinal tract wall.

Absorption factor – The percent or fraction of a chemical in contact with an organism that becomes absorbed into the receptor.

Acceptable daily intake (ADI) – An estimate of the maximum amount of a chemical (in mg/kg body weight/day) to which a potential receptor can be exposed on a daily basis over an extended period of time – usually a lifetime – without suffering a deleterious effect, or without anticipating an adverse effect.

Acceptable risk – A risk level generally deemed by society to be acceptable or tolerable.

Action level (AL) – The level of a chemical in selected media of concern above which there are potential adverse health and/or environmental effects. It represents the contaminant concentration above which some corrective action (e.g., monitoring or remediation) is required by regulation.

Acute exposure – A single large exposure or dose to a chemical, generally occurring over a short period (usually 24 to 96 hours) in relation to the lifespan of the exposed organism.

Acute toxicity – The development of symptoms of poisoning or the occurrence of adverse health effects after exposure to a single dose or multiple doses of a chemical within a short period of time.

Administered dose – The mass of substance administered to an organism and in contact with an exchange boundary (e.g., gastrointestinal tract) per unit body weight per unit time (e.g., mg/kg-day).

Adsorption – The removal of contaminants from a fluid stream by concentration of the constituents onto a solid material. It is the physical process of attracting and holding molecules of other chemical substances on the surface of a solid, usually by the formation of chemical bonds. A substance is said to be adsorbed if the concentration in the boundary region of a solid (e.g., soil) particle is greater than in the interior of the contiguous phase.

Agent (also, stressor) – Any physical, chemical, or biological entity that can induce an adverse response.

Aliphatic compounds – Organic compounds in which the carbon atoms exist as either straight or branched chains; examples include pentane, hexane, octane.

Ambient – Pertaining to surrounding conditions.

Ambient medium – One of the basic categories of material surrounding or contacting an organism – e.g., outdoor air, indoor air, water, or soil – through which chemicals or pollutants can move and reach the organism.

Antagonism/antagonistic effect – The interference or inhibition of the effects of one chemical substance by the action of other chemicals. It reflects the counteracting effect of one chemical on another, thus diminishing their additive effects.

Anthropogenic – Caused or influenced by human activities or actions.

Aquifer – A geological formation, group of formations, or part of a formation which is capable of yielding significant and usable quantities of groundwater to wells and/or springs.

Arithmetic mean (also, average) – A statistical measure of central tendency for data from a normal distribution, defined for a set of n values, by the sum of values divided by n:

$$X_m = \frac{\sum_{i=1}^{n} X_i}{n}$$

Aromatic compounds – Organic compounds that contain carbon molecular ring structures; examples include benzene, toluene, ethylbenzene, xylenes. These compounds are somewhat soluble, volatile, and mobile in the subsurface environment, and are a very useful indicator of contaminant migration.

Attenuation – Any decrease in the amount or concentration of a pollutant in an environmental matrix as it moves in time and space. It is the reduction or removal of contaminant constituents by a combination of physical, chemical, and/or biological factors acting upon the contaminated media.

Average concentration – A mathematical average of contaminant concentration(s) from more than one sample, typically represented by the arithmetic mean or the geometric mean for environmental samples.

Average daily dose (ADD) – The average dose calculated for the duration of receptor exposure, and used to estimate risks for chronic noncarcinogenic effects of environmental contaminants. This is defined by:

$$ADD \text{ (mg/kg-day)} = \frac{\text{contaminant concentration} \times \text{contact rate}}{\text{body weight}}$$

Background threshold level – The normal ambient environmental concentration of a chemical constituent. It may include both naturally occurring concentrations and elevated levels resulting from nonsite-related human activities. *Anthropogenic background levels* refer to concentrations of chemicals that are present in the environment due to human-made, nonsite sources (e.g., lead depositions from automobile exhaust and 'neighboring' industry). *Naturally-occurring background levels* refer to ambient concentrations of chemicals that are present in the environment and have not been influenced by human activities (e.g., natural formations of aluminum and manganese).

Benchmark risk – A threshold level of risk, typically prescribed by regulations, above which corrective measures will almost certainly have to be implemented to mitigate the risks.

Bioaccumulation – The retention and concentration of a chemical by an organism. It is a build-up of a chemical in a living organism which occurs when the organism takes in more of the chemical than it can get rid of in the same length of time and stores the chemical in its tissue, etc.

Bioassay – Measuring the effect(s) of environmental exposures by intentional exposure of living organisms to a chemical. It represents tests used to evaluate the relative potency of a chemical by comparing its effects on a living organism with the effect of a standard preparation on the same type of organism.

Bioconcentration – The accumulation of a chemical substance in tissues of organisms (such as fish) to levels greater than levels in the surrounding media (such as water) for the organism's habitat; often used synonymously with bioaccumulation.

Bioconcentration factor (BCF) – A measure of the amount of selected chemical substances that accumulates in humans or in biota. It is the ratio of the concentration of a chemical substance in an organism, at equilibrium, to the concentration of the substance in the surrounding environmental medium.

Biodegradable – Capable of being metabolized by a biologic process or an organism.

Biodegradation – Decomposition of a substance into simpler substances by the action of microorganisms, usually in soil. It may or may not detoxify the material which is decomposed.

Biomagnification – The serial accumulation of a chemical by organisms in the food chain, with higher concentrations occurring at each successive trophic level.

Biota – All living organisms which are found within a prescribed volume or space.

Cancer – A disease characterized by malignant, uncontrolled invasive growth of body tissue cells. It refers to the development of a malignant tumor or abnormal formation of tissue.

Cancer slope factor (SF) (also, cancer potency factor [CPF] or cancer potency slope [CPS]) – Health effect information factor commonly used to evaluate health hazard potentials for carcinogens. It is a plausible upper-bound estimate of the probability of a response per unit intake of a chemical over a lifetime. That is, it is used to estimate an upper-bound probability of an individual developing cancer as a result of a lifetime of exposure to a particular level of a carcinogen.

Capillary zone – The unsaturated area between ground surface and the water table.

Carcinogen – A chemical or substance capable of producing cancer in living organisms.

Carcinogen Assessment Group (CAG) – A group within the US EPA responsible for the evaluation of carcinogen bioassay results and estimates of the carcinogenic potency of various chemicals.

Carcinogenic – Capable of causing, and tending to produce or incite cancer in living organisms.

Carcinogenicity – The ability or capacity of a chemical, physical, or biological agent to cause cancer in a living organism.

Chronic – Of long-term duration.

Chronic daily intake (CDI) – The receptor exposure, expressed in mg/kg-day, averaged over a long period of time.

Chronic exposure – The long-term, low-level exposure to chemicals, i.e., the repeated exposure to or doses of a chemical over a long period of time. It may cause latent damage that does not appear until a later period in time.

Chronic toxicity – The occurrence of symptoms, diseases, or other adverse health effects that develop and persist over time, after exposure to a single dose or multiple doses of a chemical delivered over a relatively long period of time.

Cleanup – Actions taken to abate the situation involving the release or threat of release of contaminants that could potentially affect human health and/or the environment. This typically involves a process to remove or attenuate contamination levels, in order to restore the impacted media to an 'acceptable' or usable condition.

Cleanup level – The contaminant concentration goal of a remedial action, i.e., the concentration of media contaminant level to be attained through a remedial action.

Closure – All activities involved in taking a hazardous waste facility out of service and securing it for the duration required by applicable regulations and laws. Site closures typically follow the implementation of appropriate site restoration programs, with monitoring usually becoming part of the post-closure site activities.

Compliance – Status or situation that meets with stipulated legislative or regulatory requirements.

Conceptual model diagrams – These are visual representations of the CSMs, which serve as useful tools for communicating important pathways in a clear and concise way.

Confidence interval (CI) – A statistical parameter used to specify a range and the probability that an uncertain quantity falls within this range.

Confidence limits – The upper and lower boundary values of a range of statistical probability numbers.

Confidence limits, 95 percent (95% CL) – The limits of the range of values within which a single estimation will be included 95% of the time. For large samples sizes (i.e., $n > 30$),

$$95\% \; CL = X_m \pm \frac{1.96\sigma}{\sqrt{n}}$$

where CL is the confidence level, and σ is the estimate of the standard deviation of the mean (X_m). For a limited number of samples ($n \leq 30$), a confidence limit or confidence interval may be estimated from

$$CL = X_m \pm \frac{ts}{\sqrt{n}}$$

where t is the value of the Student t-distribution (refer to standard statistical texts) for the desired confidence level and degrees of freedom, $(n-1)$.

Consequence – The impacts resulting from the response due to specified exposures, or loading or stress conditions.

Contact rate – Amount of an environmental medium (viz.: air, groundwater, surface water, soil, etc.) contacted per unit time or per event (e.g., liters of groundwater ingested or milligrams of soil ingested per day).

Containment – Refers to systems used to prevent (or significantly reduce) the further spread of contamination. Such systems may consist of pumping (and/or injection) wells, and cut-off walls designed and placed at strategic locations.

Contaminant – Any physical, chemical, biological, or radiological material that can potentially have adverse impacts on environmental media, or that can adversely impact public health and the environment. It represents any undesirable substance that is not naturally occurring and therefore not normally found in the environmental media of concern.

Contaminant migration – The movement of a contaminant from its source through other matrices/media such as air, water, or soil. A *contaminant migration pathway* is the path taken by the contaminants as they travel from the contaminated locale through various environmental media.

Contaminant plume – A body of contaminated groundwater or vapor originating from a specific source and spreading out due to influences of such factors as local groundwater conditions or soil vapor flow patterns. It represents the volume of groundwater or vapor containing the contaminants released from a pollution source.

Contaminant release – The action causing a contaminant to enter into other environmental media/matrices (e.g., air, water or soil) from its source(s) of origin.

Corrective action – Action taken to correct a problematic situation. A typical example involves the remediation of chemical contamination in soil and groundwater.

Cost-effective alternative – The *most cost-effective alternative* is the lowest cost alternative that is technologically feasible and reliable, and which effectively mitigates and minimizes environmental damage. It generally provides adequate protection of public health, welfare, and/or the environment.

Data quality objectives (DQOs) – Qualitative and quantitative statements developed by analysts to specify the quality of data that, at a minimum, is needed and expected from a particular data collection activity (or site characterization activity). This is determined based on the end use of the data to be collected.

de minimis – A legal doctrine dealing with levels associated with insignificant versus significant issues relating to human exposures to chemicals presenting very low risk. It is the level below which one need not be concerned.

Decision analysis – A process of systematic evaluation of alternative solutions to a problem where the decision is made under uncertainty. The approach comprises a conceptual and systematic procedure for analyzing complex sets of alternatives in a rational manner so as to improve the overall performance of a decision-making process.

Decision framework – A structured and systematic management tool designed to facilitate rational decision-making.

Degradation – The physical, chemical, or biological breakdown of a complex compound into simpler compounds and byproducts.

Dense nonaqueous phase liquids (DNAPLs) – Organic liquids, composed of one or more contaminants that are more dense than water, often coalescing in an immiscible layer at the bottom of a saturated geologic unit. (Typical examples include chlorinated solvents).

Dermal adsorption – The process by which materials come into contact with the skin surface, but are then retained and adhered to the permeability barrier without being taken into the body.

Dermal exposure – Exposure of an organism or receptor through skin adsorption and possible absorption.

Dermally absorbed dose – The amount of the applied material (the dose) which becomes absorbed into the body.

Detection limit – The minimum concentration or weight of analyte that can be detected by a single measurement with a known confidence level. *Instrument detection limit (IDL)* represents the lowest amount that can be distinguished from the normal 'noise' of an analytical instrument (i.e., the smallest amount of a chemical detectable by an analytical instrument under ideal conditions). *Method detection limit (MDL)* represents the lowest amount that can be distinguished from the normal 'noise' of an analytical method (i.e., the smallest amount of a chemical detectable by a prescribed or specified method of analysis).

Diffusion – The migration of molecules, atoms, or ions from one fluid to another in a direction tending to equalize concentrations.

Dispersion – The overall mass transport process resulting from both molecular diffusion (which always occurs if there is a concentration gradient in the system) and the mixing of the constituent due to turbulence and velocity gradients within the system.

Dissolved product – The water-soluble fuel components of contaminant releases.

Dose – The amount of a chemical taken in by potential receptors on exposure. It is a measure of the amount of the substance received by the receptor as a result of exposure, expressed as an amount of exposure (in mg) per unit body weight of the receptor (in kg).

Dose–response – The quantitative relationship between the dose of a chemical and an effect caused by exposure to such substance.

Dose–response curve – A graphical representation of the relationship between the degree of exposure to a chemical substance and the observed or predicted biological effects or response.

Dose–response evaluation – The process of quantitatively evaluating toxicity information and characterizing the relationship between the dose of a chemical administered or received and the incidence of adverse health effects in the exposed population.

Dump – A site used for the disposal of waste materials without environmental controls or safeguards.

Ecological assessment endpoint – An explicit expression of the environmental value that is to be protected. An assessment endpoint includes both an ecological entity and specific attributes of that entity.

Ecological entity – A general term that may refer to a species, a group of species, an ecosystem function or characteristic, or a specific habitat.

Ecological measurement endpoint – A measurable ecological characteristic that is related to the valued characteristic chosen as the assessment endpoint.

Ecological risk assessment (ERA) – A process that evaluates the likelihood that adverse ecological effects may occur or are occurring as a result of exposure to one or more stressors. The process typically involves defining and quantifying risks to nonhuman biota and determining the acceptability of the estimated risks. It consists of determining the probability and magnitude of adverse effects of environmental hazards on nonhuman biota.

Ecosystem – The interacting system of a biological community and its abiotic (i.e., nonliving) environment.

Ecotoxicity assessment – The measurement of effects of environmental toxicants on indigenous populations of organisms.

Effect (local) – The response produced from contacting a chemical, that occurs at the site of first contact.

Effect (systemic) – The response produced from contacting a chemical, that requires absorption and distribution of the chemical and tends to affect the receptor at sites away from the entry point(s).

Effective porosity – The ratio of the volume of inter-connected voids through which fluid can flow to the total volume of material.

Endangerment assessment (EA) – A case-specific risk assessment of the actual or potential danger to human health and welfare and also the environment, associated with the release of hazardous chemicals into various environmental media.

Endpoint (toxic) – A biological effect used as index of the impacts of a chemical on an organism. It is an observable or measurable biological or biochemical event used as an index of the effect of a chemical on a cell, tissue, organ, organism, etc.

Environmental fate – The 'destination' of a chemical after release or escape into the environment, and following transport through various environmental compartments. It is the movement of a chemical through the environment by transport in air, water, and soil culminating in exposures to living organisms. It represents the disposition of a material in various environmental compartments (e.g., soil, sediment, water, air, biota) as a result of transport, transformation, and degradation.

Environmental medium – A part of the environment for which reasonable boundaries can be specified. Typical environmental media addressed in chemical risk assessments may include air, surface water, groundwater, soil, sediment, fruits, vegetables, meat, dairy, and fish.

Environmental pollutant – Any entity which contaminates any ambient media, including surface water, groundwater, soil, or air.

Event-tree analysis – A procedure often used to evaluate series of events which lead to an upset or accident scenario, using deductive logic.

Exposure – The situation of receiving a dose of a substance, or coming in contact with a hazard. It represents the contact of an organism with a chemical or physical agent available at the exchange boundary (e.g., lungs, gut, skin) during a specified time period.

Exposure assessment – The qualitative or quantitative estimation, or the measurement, of the dose or amount of a chemical to which potential receptors have been exposed, or could potentially be exposed to. It comprises the determination of the magnitude, frequency, duration, route and extent of exposure (to the chemicals or hazards of potential concern).

Exposure conditions – Factors (such as location, time, etc.) that may have significant effects on an exposed population's response to a hazard situation.

Exposure duration – The length of time that a potential receptor is exposed to the contaminants of concern in a defined exposure scenario.

Exposure event – An incident of contact with a chemical or physical agent, usually defined by time (e.g., number of days or hours of contact).

Exposure frequency – The number of times (per year or per event) that a potential receptor would be exposed to hazards or contaminants in a defined exposure scenario.

Exposure parameters (or factors) – Variables used in the calculation of intake (e.g., exposure duration, inhalation rate, average body weight). These consist of standard factors that may be needed to calculate a potential receptor's exposure to toxic chemicals (in the environment).

Exposure pathway – The course a chemical or physical agent takes from a source to an

Appendix B: Glossary of selected terms and definitions 463

exposed population or organism. It describes a unique mechanism by which an individual or population is exposed to chemicals or physical agents at or originating from a contaminated source area.

Exposure point – A location of potential contact between an organism and a chemical or physical agent.

Exposure route – The avenue by which an organism contacts a chemical, such as inhalation, ingestion, and dermal contact.

Exposure scenario – A set of conditions or assumptions about sources, exposure pathways, concentrations of chemicals, and potential receptors that aids in the evaluation and quantification of exposure in a given situation.

Extrapolation – The estimation of unknown values by extending or projecting from known values/observations.

Fault-tree analysis – A procedure often used to evaluate series of events which lead to an upset or accident scenario, using inductive logic.

Feasibility study (FS) – The analysis and selection of alternative remedial or corrective actions for hazardous waste or contaminated sites. The process identifies and evaluates remedial alternatives by utilizing a variety of appropriate environmental, engineering, and economic criteria.

Field sampling plan (FSP) – A documentation that defines in detail the sampling and data gathering activities to be used in the investigation of a potentially contaminated site.

Free product – A chemical constituent that floats on groundwater, or that remains 'unadulterated' in a contaminant pool in the environment.

Fugitive dust – Atmospheric dust arising from disturbances of particulate matter exposed to the air. Fugitive dust emissions consist of the release of chemicals from contaminated surface soil into the air, attached to dust particles.

Geographic Information System (GIS) – Computer-based tool used to capture, manipulate, process, and display spatial or geo-referenced data. GIS is a rapidly developing technology for handling, analyzing, and modeling geographic information. It is not a source of information *per se*, but only a way to manipulate information. When the manipulation and presentation of data relates to geographic locations of interest, then our understanding of the real world is enhanced.

Geometric mean – A statistical measure of the central tendency for data from a positively skewed distribution (lognormal), given by:

$$X_{gm} = [(X_1)(X_2)(X_3) \ldots (X_n)]^{1/n}$$

or,

$$X_{gm} = \text{antilog} \left[\frac{\sum_{i=1}^{n} \log X_i}{n} \right]$$

Groundwater – Water beneath the ground surface. It represents underground waters, whether present in a well-defined aquifer, or present temporarily in the vadose (unsaturated soil) zone.

Hazard – The inherent adverse effect that a chemical or other object poses. It is that innate character which has the potential for creating adverse and/or undesirable consequences. It defines the chance that a particular substance will have an adverse effect on human health or the environment in a particular set of circumstances that causes an exposure to that substance.

Hazard assessment – The evaluation of system performance and associated consequences over a range of operating and/or failure conditions. It involves gathering and evaluating data on types of injury or consequences that may be produced by a hazardous situation or substance.

Hazard identification – The systematic identification of potential accidents, upset conditions, etc. It is the recognition that a hazard exists and the definition of its characteristics. The process involves determining whether exposure to an agent can cause an increase in the incidence of a particular adverse health effect in receptors of interest.

Hazard index (HI) – The sum of several hazard quotients for multiple substances and/or multiple exposure pathways.

Hazard quotient (HQ) – The ratio of a single substance exposure level for a specified time period to a reference dose of that substance derived from a similar exposure period. It is defined by the ratio of the average daily dose (ADD) of a chemical to the reference dose (RfD) for the chemical, or the ratio of the exposure concentration to the reference concentration (RfC).

Hazard Ranking System (HRS) – A scoring system used by the USEPA to assess the relative threat associated with actual or potential releases of hazardous substances at contaminated sites.

Hazardous substance – Any substance that can cause harm to human health or the environment whenever excessive exposure occurs.

Hazardous waste – Wastes that are ignitable, explosive, corrosive, reactive, toxic, radioactive, pathological, or have some other property that produces substantial risk to life. It is that byproduct which has the potential of causing detrimental effects on human health and/or the environment if not managed efficiently.

Heavy metals – Members of a group of metallic elements which are recognized as toxic and generally bioaccumulative. The term arises from the relatively high atomic weights of these elements.

'Hot-spot' – Term used to denote zones where contaminants are present at much higher concentrations than the immediate surrounding areas. It represents a relatively small area which is highly contaminated within a study area.

Human equivalent dose – A dose which, when administered to humans, produces effects comparable to those produced by a dose in experimental animals.

Human health risk – The likelihood (or probability) that a given exposure or series of exposures to a hazardous substance will cause adverse health impacts on individual receptors experiencing the exposures.

Hydraulic conductivity – A measure of the ability of earth materials to transmit fluid, which is dependent on the type of fluid passing through the material.

Hydrocarbon – Organic chemicals/compounds, such as benzene, that contain atoms of both hydrogen and carbon.

Hydrophilic – Having greater affinity for water, or 'water-loving'. Hydrophilic compounds tend to become dissolved in water.

Hydrophobic – Tending *not* to combine with water, or less affinity for water. Hydrophobic compounds tend to avoid being dissolved in water and are more attracted to nonpolar liquids (e.g., oils) or solids.

Incineration – A thermal treatment/degradation process by which contaminated materials are exposed to excessive heat in a variety of incinerator types. The incineration process typically involves the thermal destruction of contaminants by burning under controlled conditions. Depending on the intensity of the heat, the contaminants of concern are volatilized and/or destroyed during the incineration process.

Incompatible wastes – Wastes, which when mixed with other materials without controls, may create fire, explosion, or other severe hazards.

Indicator species – A species that is surveyed or sampled for analysis (usually in ecological risk assessments) because it is believed to represent the biotic community, some functional or taxonomic group, or some population that cannot be easily sampled or surveyed.

Individual excess lifetime cancer risk – An upper-bound estimate of the increased cancer risk, expressed as a probability, that an individual receptor could expect from exposure over a lifetime.

Individual risk – Probability that an individual person in a population will experience an adverse effect from exposures to some hazards.

Ingestion – An exposure type whereby chemical substances enter the body through the mouth, and into the gastrointestinal system.

Inhalation – The intake of a substance by receptors through the respiratory tract system.

Initiating event – The specific primary action that results in a risk being incurred.

Initiator – A chemical substance that causes subtle alteration of DNA or proteins within larger cells, rendering such cells capable of becoming cancerous.

Intake – The amount of material inhaled, ingested, or dermally absorbed during a specified time period. It is a measure of exposure, expressed in mg/kg-day.

Integrated Risk Information System (IRIS) – A USEPA database containing verified reference doses (*RfDs*) and slope factors (*SFs*), and up-to-date health risk and EPA regulatory information for numerous chemicals. It serves as a source of toxicity information for health and environmental risk assessment.

Interim action – A preliminary action that initiates remediation of a contaminated site, but may also constitute part of the final remedy.

Investigation-generated wastes (IGWs) (also, investigation-derived wastes, IDWs) – Wastes generated in the process of collecting samples during a remedial investigation or environmental characterization activity. Such wastes must be handled according to all relevant and applicable regulatory requirements. The wastes may include soil, groundwater, used personal protective equipment, decontamination fluids, and disposable sampling equipment.

K_d *(soil/water partition coefficient)* – Provides a soil- or sediment-specific measure of the extent of chemical partitioning between soil or sediment and water, unadjusted for the dependence on organic carbon.

K_{oc} *(organic carbon adsorption coefficient)* – Provides a measure of the extent of chemical partitioning between organic carbon and water at equilibrium. It is a measure that indicates the extent to which a compound will sorb to the solid organic content of geologic media in the subsurface. It is computed as the ratio of the amount of chemical sorbed per unit weight of organic carbon in the soil to the concentration of the chemical in solution at equilibrium.

K_{ow} *(octanol/water partition coefficient)* – Provides a measure of the extent of chemical partitioning between water and octanol at equilibrium. It is a measure that indicates the extent to which a compound is attracted to an organic phase (for which octanol is a proxy) and hence the compound's tendency to sorb to subsurface materials. It is computed by dividing the amount that will dissolve in octanol by the amount that will dissolve in water.

K_w *(water/air partition coefficient)* – Provides a measure of the distribution of a chemical between water and air at equilibrium.

Landfill – A controlled site for the disposal of wastes on land, generally operated in accordance with regulatory safety and environmental compliance requirements.

Latent period – The time between the initial induction of a health effect from first exposures to a chemical and the manifestation or detection of actual health effects.

LC_{50} *(mean lethal concentration)* – The lowest concentration of a chemical in air or water that will be fatal to 50% of test organisms living in that media, under specified conditions.

LD_{50} *(mean lethal dose)* – The median lethal dose, i.e., the single dose (ingested or dermally absorbed) required to kill 50% of a test animal group.

Leachate – Aqueous, often-contaminated, liquid generated when water percolates or trickles through waste materials or contaminated sites and collects components of those wastes. Leaching usually occurs at landfills as a result of infiltration of rainwater or snowmelt, and may result in hazardous chemicals entering soils, surface water, or groundwater.

Lifetime average daily dose (LADD) – The exposure, expressed as mass of a substance contacted and absorbed per unit body weight per unit time, averaged over a lifetime. It is usually used to calculate carcinogenic risks; it takes into account the fact that, whereas carcinogenic risks are determined with an assumption of lifetime exposure, actual exposures may be for a shorter period of time. *LADD* may be derived from the *ADD* to reflect the difference between the length of the exposure period and the exposed person's lifetime, viz.:

$$LADD = ADD \times \frac{\text{exposure period}}{\text{lifetime}}$$

Lifetime exposure – The total amount of exposure to a substance that a human would be subjected to in a lifetime.
Lifetime risk – Risk which results from lifetime exposure to a chemical substance.
Light nonaqueous phase liquids (LNAPLs) – Organic fluids that are lighter than water, capable of forming an immiscible layer that floats on the water table; also referred to as 'floaters' (e.g., gasoline and fuel oil).
Lipophilic – Refers to substance having an affinity for fat and high lipid solubility. Also, a physico-chemical property which describes a partitioning equilibrium of solute molecules between water and an immiscible organic solvent, which favors the latter.
LOAEC (lowest-observed-adverse-effect-concentration) – The lowest concentration in an exposure medium in a study that is associated with an adverse effect on the test organisms.
LOAEL (lowest-observed-adverse-effect level) – The chemical dose rate causing statistically or biologically significant increases in frequency or severity of adverse effects between the exposed and control groups. It is the lowest dose level, expressed in mg/kg body weight/day, at which adverse effects are noted in the exposed population.
$LOAEL_a$ – *LOAEL* values adjusted by dividing by one or more safety factors.
LOEL (lowest-observed-effect-level) – The lowest exposure or dose level of a substance at which effects are observed in the exposed population; the effects may or may not be serious.

Matrix (or medium) – The predominant material comprising the environmental sample being investigated (e.g., soils, water, air).
Maximum daily dose (MDD) – The maximum dose calculated for the duration of receptor exposure, and used to estimate risks for subchronic or acute noncarcinogenic effects of environmental contaminants.
Mitigation – The process of reducing or alleviating a problem situation.
Modeling – The use of mathematical equations to simulate and predict real events and processes.
Monitoring – The measurement of concentrations of chemicals in environmental media or in tissues of humans and other biological receptors/organisms.
Monte Carlo simulation – A technique in which outcomes of events or variables are determined by selecting random numbers subject to a defined probability law. It is a technique used to obtain information about the propagation of uncertainty in mathematical simulation models.
Multi-hit models – Dose–response models that assume more than one exposure to a toxic material is necessary before effects are manifested.
Multistage models – Dose–response models that assume there are a given number of biological stages through which a carcinogenic agent must pass, without being deactivated, for cancer to occur. A *linearized multistage model* is a derivation of the multistage model for which the data are assumed to be linear at low doses.

Neurotoxic – Toxic effect on any aspect of the central or peripheral nervous system.
Neurotoxicity – Hazard effects that are poisonous to the nerve cells.
NOAEC (no-observed-adverse-effect-concentration) – The highest concentration in an

exposure medium in a study that is *not* associated with an adverse effect on the test organisms.

NOAEL (no-observed-adverse-effect level) – The chemical intakes at which there are no statistically or biologically significant increases in frequency or severity of adverse effects between the exposed and control groups (meaning statistically significant effects are observed at this level, but they are not considered to be adverse). It is the highest level at which a chemical causes no observable adverse effect in the species being tested or the exposed population.

NOAEL$_a$ – NOAEL values adjusted by dividing by one or more safety factors.

NOEL (no-observed-effect level) – The dose rate of chemical at which there are no statistically or biologically significant increases in frequency or severity of any effects between the exposed and control groups. It is the highest level at which a chemical causes no observable changes in the species or exposed populations under investigation.

Nonaqueous phase liquid (NAPL) – Organic compounds in the liquid phase, that is not completely miscible in water.

Nonparametric statistics – Statistical techniques whose application is independent of the actual distribution of the underlying population from which the data were collected.

Nonthreshold chemical – Also called *zero threshold chemical*; refers to a substance which is known, suspected, or assumed to potentially cause some adverse response at any dose above zero.

Off-site – Areas outside the boundaries or limits of a presumed contaminated site.

One-hit model – Dose–response model of the form $P(d) = [1 - e^{-cd}]$, where $P(d)$ is the probability of cancer death from a continuous dose rate (d) and c is a constant. The one-hit model is based on the concept that a tumor can be induced after a single susceptible target or receptor has been exposed to a single effective unit dose of an agent.

On-site – Areas within the boundaries or limits of a presumed contaminated site.

Organic carbon content of soils or sediments – This reflects the amount of organic matter present, and generally correlates with the tendency of chemicals to accumulate in the soil or sediment. The accumulation of chemicals in soils or sediments is frequently the result of adsorption onto organic matter. Thus, in general, the higher the organic carbon content of the soil or sediment, the more a contaminant will be adsorbed to the soil particles, rather than be dissolved in the water or gases permeating the soil or sediment.

Partition coefficient – A term used to describe the relative amount of a substance partitioned between two different phases, such as a solid and a liquid. It is the ratio of the chemical's concentration in one phase to its concentration in the other phase.

Partitioning – A chemical equilibrium condition in which a chemical's concentration is apportioned between two different phases according to the partition coefficient.

Pathway – Any specific route which environmental contaminants take in order to travel away from the source and to reach potential receptors or individuals.

PEL (permissible exposure limit) – A maximum (legally enforceable) allowable level for a chemical in workplace air.

Permeability – A measure of a material's ability to transmit fluid. It is the capacity of a porous medium to transmit a fluid subjected to an energy gradient (the hydraulic gradient, in the case of water).

Persistence – Attribute of a chemical substance which describes the length of time that such substance remains in a particular environmental compartment before it is physically removed, chemically modified, or biologically transformed.

pH – A measure of the acidity or alkalinity of a material or medium.

Pharmacokinetics – Study of changes in a toxicant (e.g., via absorption, distribution, biotransformation, and excretion) in parts of the mammalian body over time.

Photodegradation – Any chemical breakdown reaction that is initiated by ultraviolet rays (or light) from sunlight.

Pica – The behavior in children and toddlers (usually under age 6 years) involving the intentional eating/mouthing of large quantities of dirt and other objects.

Plume – A zone containing predominantly dissolved (or vapor phase) contaminants and sorbed contaminants in equilibrium within the transport medium. A plume usually will originate from the contaminant source areas. It represents the pathway taken by pollutants released by point sources in air or water.

PM-10, PM_{10} – Particulate matter with physical/aerodynamic diameter < 10 μm. It represents the respirable particulate emissions.

Pollutant – A potentially harmful physical, chemical, or biological agent occurring in the environment, in consumer products, or at the workplace as a result of human activities.

Pollution – The release of a physical, chemical, or biological agent into the environment, that has the potential to impact human and/or ecological health.

Population-at-risk (PAR) – A population subgroup that is more susceptible to hazard or chemical exposures. It represents that group which is more sensitive to a hazard or chemical, than is the general population.

Population excess cancer burden – An upper-bound estimate of the increase in cancer cases in a population as a result of exposure to a carcinogen.

Porosity – The ratio of the volume of void space in earth materials to the total volume of the material. The wider the range of grain sizes, the lower the porosity.

Potency – A measure of the relative toxicity of a chemical.

Potentially responsible party (PRP) – Generally, refers to those identified by the USEPA as potentially liable under its hazardous waste site law (viz., CERCLA or Superfund) for cleanup costs at specified waste sites.

Potentiation – The effect of a chemical which enhances the toxicity of another chemical.

ppb (parts per billion) – An amount of substance in a billion parts of another material; also expressed by μ/kg or μg/L.

ppm (parts per million) – An amount of substance in a million parts of another material; also expressed by mg/kg or mg/L.

ppt (parts per trillion) – An amount of substance in a trillion parts of another material; also expressed by ng/kg or ng/L.

Practical quantitation limit (PQL) – Also called *sample quantitation limit (SQL)*. It is the lowest level that can be reliably achieved within specified limits of precision and accuracy during routine laboratory operating conditions. It represents a detection limit that has been corrected for sample characteristics, sample preparation, and analytical adjustments such as dilution. Typically, the *PQL* or *SQL* will be about 5 to 10 times the chemical-specific detection limit.

Preliminary assessment (PA) – A survey and evaluation whereby sites are characterized to determine their potential to release significant amounts of contaminants into the environment.

Preliminary site appraisal – Process used for quick assessment of a site's potential to adversely affect the environment and/or public health.

Probability – The likelihood of an event occurring, numerically represented by a value between 0 and 1; a probability of 1 means an event is certan to happen, whereas a probability of 0 means an event is certain *not* to happen.

Probit model – A probit, or probability unit, is obtained by modifying the standard variate of the standardized normal distribution. This transformation can then be used in the analysis of dose–response data used in a risk characterization.

Promoter – A chemical that, when administered after an initiator has been given, promotes the change of an initiated cell culminating in a cancer.

Proxy concentration – Assigned contaminant concentration value for situations where sample data may not be available, or when it is impossible to quantify accurately.

Qualitative – Description of a situation without numerical specifications.

Quality assurance (QA) – A system of activities designed to assure that the quality control

Appendix B: Glossary of selected terms and definitions 469

system is performing adequately. It consists of the management of investigations data to assure that they meet the data quality objectives.

Quality control (QC) – A system of specific efforts designed to test and control the quality of data obtained in an investigation. It consists of the management of activities involved in the collection and analysis of data to assure they meet the data quality objectives. It is the system of activities required to provide information as to whether the quality assurance system is performing adequately.

Quantitation limit (QL) – The lowest level at which a chemical can be accurately and reproducibly quantitated. It usually is equal to the instrument detection limit *(IDL)* multiplied by a factor of 3 to 5, but varies for different chemicals and different samples.

Quantitative – Description of a situation that is presented in exact numerical terms.

Reasonable maximum exposure (RME) – A concept that attempts to identify the highest exposure (and, therefore, the greatest risk) that could reasonably be expected to occur in a given population.

Receptor – Member of a potentially exposed population, such as persons or organisms that are potentially exposed to concentrations of a particular chemical compound.

Reference concentration (RfC) – A concentration of a chemical substance in an environmental medium to which exposure can occur over a prolonged period without expected adverse effects. The medium in this case is usually air, with the concentration expressed in mg of chemical per m^3 of air.

Reference dose (RfD) – The maximum amount of a chemical that the human body can absorb without experiencing chronic health effects, expressed in mg of chemical per kg body weight per day. It is the estimate of lifetime daily exposure of a noncarcinogenic substance for the general human population (including sensitive receptors) which appears to be without an appreciable risk of deleterious effects, consistent with the threshold concept.

Remedial action – Those actions consistent with a permanent remedy in the event of a release of a hazardous substance into the environment, meant to prevent or minimize the effects of such releases so that they do not migrate further to cause substantial danger to present or future public health or welfare or the environment.

Remedial action objective – Restoration objectives that specify the level of cleanup, area of cleanup (or area of attainment), and the time required to achieve cleanup (i.e., the restoration time-frame).

Remedial alternative – An action considered in a feasibility study, that is intended to reduce or eliminate significant risks to human health and/or the environment at a contaminated site.

Remedial investigation (RI) – The field investigations of hazardous waste sites to determine pathways, nature, and extent of contamination, as well as to identify preliminary alternative remedial actions. It focuses on data collection and site characterization to identify and assess threats or potential threats to human health and the environment posed by a site.

Remediation – The process of cleaning up a potentially contaminated locale, in order to prevent or minimize the potential release and migration of hazardous substances from the impacted media that could cause adverse impacts to present or future public health and welfare, or the environment.

Removal action – An action that is implemented to address a direct threat to human health or the environment.

Representative sample – A sample that is assumed *not* to be significantly different from the population of samples available.

Residual risk – The risk of adverse consequences that remains after corrective actions have been implemented.

Respirable fraction (of dust) (or, respirable particulate matter) – Fraction of dust particles that enter the respiratory system because of their size distribution. Generally, the size of these particles corresponds to an aerodynamic diameter of ≤ 10 μm.

Response (toxic) – The reaction of a body or organ to a chemical substance or other physical, chemical, or biological agent.
Restoration time-frame – Time required to achieve requisite cleanup levels or environmental restoration goals.
Retardation coefficient – A measure of how quickly a contaminant moves through the ground compared to the average groundwater velocity, resulting from sorption effects. It is computed as the ratio of the total contaminant mass in a unit aquifer volume to the contaminant mass in solution.
Risk – The probability or likelihood of an adverse consequence from a hazardous situation or hazard, or the potential for the realization of undesirable adverse consequences from impending events. It is a measure of the probability and severity of an adverse effect to health, property, or the environment. *Individual risk* is the frequency at which an individual may be expected to sustain a given level of harm from the realization of specified hazards. *Societal risk* is the relationship between frequency and number of people suffering from a specified level of harm in a given population as a result of the realization of specified hazards.
Risk acceptability – A willingness to take or deal with risk 'pretty much as it is'.
Risk acceptance – The willingness of an individual, group, or society to accept a specific level of risk in order to obtain some gain or benefit.
Risk appraisal – The assessment of whether existing or potential biologic receptors are presently, or may in the future, be at risk of adverse effects as a result of exposures to specific hazards.
Risk assessment – A methodology that combines exposure assessment with health and environmental effects data to estimate risks to human or environmental target organisms which are the result of exposure to hazards. Simply, it is an evaluation of the potential for exposure to contaminants and the associated hazard effects.
Risk-based concentration – A contaminant concentration determined from an evaluation of the compound's overall risk to human health upon exposure.
Risk characterization – The estimation of the incidence and severity of the adverse effects likely to occur in a human population or ecological group due to actual or predicted exposure to a substance or hazard.
Risk control – The process to manage risks associated with a hazard situation. It may involve the implementation, enforcement, and re-evaluation of the effectiveness of corrective measures from time to time.
Risk decision – The process used for making complex public policy decisions relating to the control of risks associated with hazardous situations.
Risk determination – The evaluation of the environmental and health impacts of contaminant releases or hazard propagation.
Risk estimate – A description of the probability that a potential receptor exposed to a specified dose of a chemical or hazard will develop an adverse response.
Risk estimation – The process of quantifying the probability and consequence values for a hazard situation. It is the process used to determine the extent and probability of adverse effects of the hazards identified, and to produce a measure of the level of health, property, or environmental risks being assessed.
Risk evaluation – The complex process (incorporating value judgments) of developing acceptable levels of risk to individuals or society.
Risk group – A real or hypothetical exposure group composed of the general or specific population groups.
Risk management – The steps and processes taken to reduce, abate, or eliminate the risk that has been revealed by a risk assessment. It is an activity concerned with decisions about whether an assessed risk is sufficiently high to present a public health concern, and about the appropriate means for controlling the risks judged to be significant.
Risk perception – The magnitude of the risk as it is perceived by an individual or population. It consists of the measured risk and the preconceptions of the observer.

Appendix B: Glossary of selected terms and definitions

Risk reduction – The action of lowering the probability of occurrence and/or the value of a risk consequence, thereby reducing the magnitude of the risk.
Risk-specific dose (RSD) – An estimate of the daily dose of a carcinogen which, over a lifetime, will result in an incidence of cancer equal to a given risk level. It is the dose associated with a specified risk level.
Risk tolerability – A willingness to 'live with' a risk, and to keep it under review. 'Tolerances' refer to the extent to which different groups or individuals are prepared to tolerate identified risks.

Sample blank – Blanks are samples considered to be the same as the environmental samples of interest except with regard to one factor whose influence on the samples is being evaluated. Sample blanks are used to ensure that contaminant concentrations actually reflect site conditions, and are not artifacts of the sampling and sample handling processes. The blank consists of laboratory distilled, deionized water that accompanies the empty sample bottles to the field as well as when the samples are returning to the laboratory, where it is not opened until both blank and the actual site samples are ready to be analyzed.
Sample duplicate – Two samples taken from the same source at the same time and analyzed under identical conditions for QA/QC purposes.
Sample quantitation limit (SQL) – Also called *practical quantitation limit (PQL)*. It is the lowest level that can be reliably achieved within specified limits of precision and accuracy during routine laboratory operating conditions. It represents a detection limit that has been corrected for sample characteristics, sample preparation, and analytical adjustments such as dilution. Typically, the *PQL* or *SQL* will be about 5 to 10 times the chemical-specific detection limit.
Sampling and analysis plan (SAP) – Documentation that consists of a quality assurance project plan (QAPP) and a field sampling plan (FSP), designed to facilitate environmental investigations.
Saturated zone – An underground geologic formation in which the pore spaces or interstitial spaces in the formation are filled with water under pressure, equal to or greater than atmospheric pressure.
Sediment – Soil that is normally covered with water. It generally is considered to provide a direct exposure pathway to aquatic life.
Sensitive receptor – Individual in a population who is particularly susceptible to health impacts due to exposure to a chemical pollutant.
Sensitivity analysis – A method used to examine the operation of a system by measuring the deviation of its normal behavior due to pertubations in the performance of its components from their normal values. In risk assessment, this may involve an analysis of the relationship of individual factors (such as chemical concentration, population parameter, exposure parameter, and environmental medium) to variability in the resulting estimates of exposure and risk.
Site assessment – Process used to identify toxic substances that may be present at a site and to present site-specific characteristics that influence the migration of contaminants.
Site categorization – A classification of sites to reflect the uniqueness of each site.
Site characterization – A process that attempts to identify the types and sources of contaminants present at a site and the site's hydrogeologic characteristics.
Site cleanup – The decontamination of a site initiated as a result of the discovery of contamination at a site or property.
Site mitigation – The process of cleaning up a contaminated site in order to return it to an environmentally acceptable state.
Skin adherence – The property of a material which causes it to be retained on the surface of the epidermis (i.e., adheres to the skin).
Skin (or dermal) permeability coefficient – A flux value, normalized for concentration, that represents the rate at which a chemical penetrates the skin (cm/h).
Soil gas – The vapor or gas found in the unsaturated soil zone.

Solubility – A measure of the ability of a substance to dissolve in a fluid.
Sorption – The processes that remove solutes from the fluid phase and concentrate them on the solid phase of a medium.
Source area – Sections at a contaminated locale containing contamination that remains in place. The source area may stretch beyond the original contaminant release location; included in the common definition of source area are regions along the contaminant flowpath where contaminants are present in precipitated or nonaqueous-phase liquid form.
Standard – A general term used to describe legally established values above which regulatory action will be required.
Standard deviation – The most widely used statistical measure to describe the dispersion of a data set, defined for a set of *n* values as follows:

$$s = \left[\frac{\sum_{i=1}^{n} (X_i - X_m)^2}{n-1} \right]^{0.5}$$

where X_m is the arithmetic mean for the data set of *n* values.
Stochasticity – Variability in parameters or in models that contain such parameters, resulting from the inherent variability of the system under consideration.
Stressor (also, agent) – Any physical, chemical, or biological entity that can induce an adverse response.
Stressor–response profile – The product of characterization of ecological effects in the analysis phase of an ERA. The stressor–response profile summarizes the data on the effects of a stressor and the relationship of the data to the assessment endpoint.
Subchronic – Relating to intermediate duration, usually used to describe studies or exposure levels spanning 5 to 90 days duration.
Subchronic daily intake (SDI) – The exposure, expressed in mg/kg-day, averaged over a portion of a lifetime.
Subchronic exposure – The short-term, high-level exposure to chemicals, i.e., the maximum exposure to or doses of a chemical over a portion of a lifetime.
Synergism – An interaction of two or more chemicals that results in an effect that is greater than the sum of their effects taken independently. It is the effects from a combination of two or more events, efforts, or substances that are greater than would be expected from adding the individual effects.
Systemic – Relating to the whole body, rather than individual parts of exposed individuals or receptors.

Terrestrial – That relating to land – as distinct from air or water.
Threshold – The lowest dose or exposure of a chemical at which a specified measurable effect is observed and below which such effect is not observed. *Threshold dose* is the minimum exposure dose of a chemical that will evoke a stipulated toxicological response. *Toxicological threshold* refers to the concentration at which a compound begins to exhibit toxic effects.
Threshold chemical (also, nonzero threshold chemical) – Refers to a substance which is known or assumed to have no adverse effects below a certain dose.
Threshold limit – A chemical concentration above which adverse health and/or environmental effects may occur.
Tolerance limit – The level or concentration of a chemical residue in media of concern above which adverse health effects are possible, and above which levels corrective action should therefore be undertaken.
Toxic – Harmful, or deleterious with respect to the effects produced by exposure to a chemical substance.
Toxicant – Any synthetic or natural chemical with an ability to produce adverse health effects. It is a poisonous contaminant that may injure an exposed organism.
Toxicity – The harmful effects produced by a chemical substance. It is the quality or degree

of being poisonous or harmful to human or ecological receptors. It represents the property of a substance to cause any adverse physiological effects (on living organisms).

Toxicity assessment – Evaluation of the toxicity of a chemical based on all available human and animal data. It is the characterization of the toxicological properties and effects of a chemical substance, with special emphasis on the establishment of dose–response characteristics.

Toxic substance – Any material or mixture that is capable of causing an unreasonable threat to human health or the environment.

Treatment – A change in the composition or concentration of a waste substance so as to make it less hazardous, or to make it acceptable at disposal and re-use facilities. It involves the application of technological process(es) to a contaminant or waste in order to render it nonhazardous or less hazardous or more suitable for resource recovery.

Trip blank – A trip blank is transported just like actual samples, but does not contain the chemicals to be analyzed. The purpose of this blank is to evaluate the possibility that a chemical could seep into samples (to adulterate them) during transportation to the laboratory.

Tumor – Growth of tissue forming an abnormal mass. *Neoplasm* relates to a genetically altered, relatively autonomous growth of tissue; it is composed of abnormal cells, the growth of which is more rapid than that of other tissues and is not coordinated with the growth of other tissues.

Uncertainty – The lack of confidence in the estimate of a variable's magnitude or probability of occurrence.

Uncertainty factor (UF) – Also called *safety factor*, refers to a factor that is used to provide a margin of error when extrapolating from experimental animals to estimate human health risks.

Underground storage tank (UST) – A tank fully or partially located below the ground surface, that is designed to hold gasoline or other petroleum products, or indeed other chemical products.

Unit cancer risk (UCR) – The excess lifetime risk of cancer due to a continuous lifetime exposure/dose of one unit of carcinogenic chemical concentration (caused by one unit of exposure in the low exposure region).

Unit risk (UR) – A measure of the carcinogenic potential of a substance when a dose is received through the inhalation pathway, that is based on several assumptions. It is an upper-bound estimate of the probability of contracting cancer as a result of constant exposure over the individual lifetime to an ambient concentration of 1 $\mu g/m^3$.

Upper-bound estimate – The estimate not likely to be lower than the true (risk) value.

Upper confidence limit, 95% (95% UCL) – The upper limit on a normal distribution curve below which the observed mean of a data set will occur 95% of the time. This is also equivalent to stating that there is at most a 5% chance of the true mean being greater than the observed value. It is a value that equals or exceeds the true mean 95% of the time.

Vadose zone – Also called the *unsaturated soil zone*, this is the zone between the ground surface and the top of the groundwater table.

Volatile organic compound (VOC) – Any organic compound that has a great tendency to vaporize, and is susceptible to atmospheric photochemical reactions. It volatilizes (evaporates) relatively easily when exposed to air.

Volatility – A measure of the tendency of a compound to vaporize from the liquid state.

Volatilization – The transfer of a chemical from the liquid into the gaseous phase.

Water table – The top of the saturated zone where confined groundwater is under atmospheric pressure.

Appendix C

Some Basic Definitions and Concepts in Probability Theory and Statistics

A basic knowledge of probability theory and statistics is essential to the understanding and appreciation of risk analysis methods, particularly probabilistic risk assessment. Therefore, a summary of the fundamental notations and theorems pertaining to some probability and statistical definitions and concepts commonly used in probabilistic risk analyses are given below in this appendix.

- *Boolean algebra commutative laws*, are defined by:

$$A + B = B + A \quad \text{and} \quad AB = BA$$

- *Boolean algebra associative laws*, are defined by:

$$(A + B) + C = A + (B + C) \quad \text{and} \quad (AB)C = A(BC) = ABC$$

- *Boolean algebra distributive laws*, are defined by:

$$A(B + C) = AB + AC \quad \text{and} \quad A + BC = (A + B)(A + C)$$

- *Probability distributions* are mathematical equations or graphical representations of the relationship between all possible values (or outcomes) a variable can have and the likelihood (expressed as a number between zero and unity) that the variable will have a particular value.

  Probability distributions can be discrete or continuous. *Discrete distributions* (also referred to by *probability mass function* (pmf)) can be represented as bar charts that describe the probabilities of a specific finite number of values a variable can have; *continuous distributions* (also referred to by *probability density function* [pdf]) are represented as smooth curves and describe probabilities for variables that can have a continuous range and infinite number of possible values. The area under the smooth curve between two points represents the probability that the true value of the variable lies between those points.

  Another type of probability distribution of significant interest is the *cumulative distribution function (cdf)* – which shows the probability of a variable being equal to or less than each value within the appropriate range for that variable.

- *Joint probability distributions*. Also known as *multivariate probability distributions*, are probability distributions used to describe the joint behavior of a number of random variables. The random variables may or may not be statistically independent.

- *Arithmetic mean (also, average)*. A measure of central tendency for data from a normal distribution, defined for a set of n values, by the sum of values divided by n, viz.:

$$X_m = \frac{\sum_{i=1}^{n} X_i}{n}$$

- *Geometric mean.* A measure of the central tendency for data from a positively skewed distribution (lognormal), given by:

$$X_{gm} = [(X_1)(X_2)(X_3) \ldots (X_n)]^{1/n}$$

or,

$$X_{gm} = \text{antilog}\left[\frac{\sum_{i=1}^{n} \log X_i}{n}\right]$$

- *Standard deviation.* The most widely used measure to describe the dispersion of a data set, defined for a set of n values as follows:

$$s = \sqrt{\left[\frac{\sum_{i=1}^{n}(X_i - X_m)^2}{n-1}\right]}$$

where X_m is the arithmetic mean for the data set of n values. The higher the value of this descriptor, the broader is the dispersion of the data set about the mean.

- *Coefficient of variation (CoV).* A nondimensional form of the standard deviation, defined as follows:

$$CoV = \frac{S}{X_m}$$

- *Coefficient of skewness (CoS).* A descriptor of the skewness or assymmetry of the probability density function (*pdf*). As a nondimensional measure of skewness, the *CoS* is zero if the pdf is symmetric about the mean value. However, a zero value for the *CoS* does not necessarily imply a symmetric pdf. Positive values of the *CoS* usually indicate that the pdf is skewed to the right (i.e., the pdf has a longer tail on the right side of the mean value), and a negative *CoS* usually indicates a longer tail on the left side of the mean value.

- *Confidence interval (CI).* A statistical parameter used to specify a range and the probability that an uncertain quantity falls within this range.

- *Upper confidence limit, 95% (95% UCL).* The upper limit on a normal distribution curve below which the observed mean of a data set will occur 95% of the time. This is also equivalent to stating that there is at most a 5% chance of the true mean being greater than the observed value.

- *Confidence limits, 95 percent (95% CL).* The limits of the range of values within which a single estimation will be included 95% of the time. For large samples (i.e., $n > 30$),

$$95\% \; CL = X_m \pm \frac{1.96\sigma}{\sqrt{n}}$$

where *CL* is the confidence level, and σ is the estimate of the standard deviation of the mean (X_m). For a limited number of samples ($n \leq 30$), a confidence limit or confidence interval may be estimated from

$$CL = X_m \pm \frac{ts}{\sqrt{n}}$$

where t is the value of the Student t-distribution (refer to standard statistical texts) for the desired confidence level and degrees of freedom, $(n-1)$.

- *Statistical concepts in additive models.* In the application of statistical concepts (of combining parameter uncertainty) in additive models, consider an additive expression of the form: $Y = X_1 + X_2 + \ldots + X_n$.

  If the variables are statistically independent, the arithmetic mean of Y (i.e., $\mu(Y)$) is given by the sum of the arithmetic means of the input parameters, i.e.,

  $$\mu(Y) = \mu(X_1) + \mu(X_2) + \ldots + \mu(X_n)$$

  and the arithmetic standard deviation (i.e., $\sigma(Y)$) of the output is estimated from the sum of the variances of the inputs as follows:

  $$\sigma(Y) = \sqrt{[\sigma^2(X_1) + \sigma^2(X_2) + \ldots + \sigma^2(X_n)]}$$

- *Statistical concepts in multiplicative models.* In the application of statistical concepts (of combining parameter uncertainty) in multiplicative models, consider a multiplicative expression of the form:

  $$Y = \frac{X_1 \times X_2 \times \ldots \times X_{n-1}}{X_n} \equiv X_1 \times X_2 \times \ldots \times X_{n-1} \times \left[\frac{1}{X_n}\right]$$

  If the variables are statistically independent and uncorrelated, the geometric mean of the product (i.e., $GM(Y)$) is given by the product of the geometric means of the input parameters, i.e.,

  $$GM(Y) = GM(X_1) \times GM(X_2) \times \ldots \times GM(X_{n-1}) \times GM(1/X_n)$$

  and the geometric standard deviation (i.e., $GSD(Y)$) of the product/output can be estimated from the geometric standard deviation of the inputs as follows:

  $$GSD(Y) = \exp\left\{\sqrt{([\ln(GSD(X_1))]^2 + [\ln(GSD(X_2))]^2 + \ldots + [\ln(GSD(X_{n-1}))]^2 + [\ln(GSD(1/X_n))]^2)}\right\}$$

  If the input parameters $X_1, X_2, \ldots, X_n$ are described by a lognormal distribution or can be approximated by a lognormal distribution, then the arithmetic mean value of Y, $\mu(Y)$, is given by:

  $$\mu(Y) = GM(Y) \times \exp\left\{\frac{[\ln(GSD(Y))]^2}{2}\right\}$$

  and the arithmetic standard deviation of Y, $\sigma(Y)$, is given by:

  $$\sigma(Y) = \mu(Y) \times \sqrt{\left[\frac{\mu(Y)^2}{GM(Y)^2} - 1\right]}$$

- *Unconditional probability*, $Pr\{A\}$, is the fraction of items resulting in event A, among the complete set of all items.
- *Conditional probability*, $Pr\{A/B\}$, is the probability of occurrence of event A, given that event B has already occurred. This is the proportion/fraction of items resulting in event A amongst the total set of items that give rise to event B. This is expressed by

  $$Pr\{A/B\} = \frac{Pr\{A \cap B\}}{Pr\{B\}}$$

  where $Pr\{--\}$ is the probability of the specified event, and $\{A \cap B\}$ denotes the intersection of events A and B.

- *Joint probability*, $Pr\{A \text{ and } B\}$ is the fraction of items giving rise to the simultaneous occurrence of events A and B, among the complete set of all items. Thus,

$$Pr\{A \cap B\} = Pr\{B\} \times Pr\{A/B\}$$

- *Independence*. Event A is said to be independent of event B if, and only if $Pr\{A/B\} = Pr\{A\}$. This means, the probability of event A is unaffected by the occurrence of event B and vice versa, so that,

$$Pr\{A \text{ and } B\} = Pr\{A \cap B\} = Pr\{A\} \times Pr\{B\}$$

- *Mutually exclusive events*, are events that cannot occur simultaneously. Thus,

$$Pr\{A \cap B\} = 0$$

where $Pr\{--\}$ is the probability of the specified event, and $\{A \cap B\}$ denotes the intersection of events A and B (i.e., the elements common to both events A and B).
- *Inclusive probability*, is defined as

$$Pr\{A \text{ or } B\} = Pr\{A \cup B\} = Pr\{A\} + Pr\{B\} - Pr\{A \cap B\}$$

where $Pr\{--\}$ is the probability of the specified event, and $\{A \cup B\}$ denotes the union of events A and B.
- *Addition theorem*, provides an alternative way of calculating the probability of the union of two events as the sum of their probabilities minus the probability of their intersection, expressed by

$$P(A + B) = P(A) + P(B) - P(AB)$$

This can be extended to three or more events, e.g.,

$$P(A + B + C) = P(A) + P(B) + P(C) - P(AB) - P(AC) - P(BC) + P(ABC)$$

$$P(A + B + C + D) = P(A) + P(B) + P(C) + P(D) - P(AB)$$
$$- P(AC) - P(AD) - P(BC) - P(BD) - P(CD)$$
$$+ P(ABC) + P(ABD) + P(BCD) + P(ACD) - P(ABCD)$$

- *Total probability theorem*, states that, given a set of mutually exclusive collectively exhaustive events $B_1, B_2, \ldots, B_n$, the probability $Pr\{A\}$ of another event A can be expressed by

$$Pr\{A\} = Pr\{A \cap B_1\} + Pr\{A \cap B_2\} + \ldots + Pr\{A \cap B_n\}$$

$$= \sum_{i=1}^{n} Pr\{A \cap B_i\} = \sum_{i=1}^{n} Pr\{A/B_i\} \times Pr\{B_i\}$$

- *Bayes's Theorem*, which follows from the total probability theorem, states that the conditional probability of B_j given the event A is given by

$$Pr\{B_j/A\} = Pr\{B_j \cap A\}/Pr\{A\} = Pr\{A \cap B_j\}/Pr\{A\} = \frac{Pr\{A/B_j\}Pr\{B_j\}}{\sum_i [Pr\{A/B_i\} \times Pr\{B_i\}]}$$

Thus, Bayes's Theorem allows updating of prior probabilities, $Pr\{B_j\}$, to yield posterior probabilities, $Pr\{B_j/A\}$, given new information A.

Appendix D

Toxicological Information for Selected Environmental Chemicals

Carcinogenic and noncarcinogenic toxicity indices relevant to the estimation of human health risks – represented by the cancer slope factor and reference dose, respectively – are presented in Table D.1 for selected chemical constituents that may be encountered in the environment. A more complete and up-to-date listing may be obtained from a variety of toxicological databases – such as the Integrated Risk Information System (IRIS), developed and maintained by the USEPA (see Appendix E.2).

Table D.1 Toxicological parameters for selected environmental chemicals

Environmental chemical	Toxicity index			
	Oral SF (1/mg/kg-day)	Inhalation SF (1/mg/kg-day)	Oral RfD (mg/kg-day)	Inhalation RfD (mg/kg-day)
Inorganic chemicals				
Aluminum (Al)			1.00E+00	
Antimony (Sb)			4.00E−04	
Arsenic (As)	1.75E+00	1.20E+01	3.00E−04	5.00E+01
Barium (Ba)			7.00E−02	1.40E−04
Beryllium (Be)	4.30E+00	8.40E+00	5.00E−03	
Cadmium (Cd)		1.50E+01	5.00E−04	
Chromium (Cr–total)			1.00E+00	
Chromium VI (Cr+6)		4.10E+01	5.00E−03	
Cobalt (Co)			2.90E−04	2.90E−04
Cyanide (CN) – free			2.00E−02	
Manganese (Mn)			5.00E−03	1.40E−05
Mercury (Hg)			3.00E−04	8.60E−05
Molybdenum (Mo)			5.00E−03	5.00E−03
Nickel (Ni)		9.10E−01	2.00E−02	
Selenium (Se)			5.00E−03	
Silver (Ag)			5.00E−03	
Thallium (Tl)			8.00E−05	
Vanadium (V)			7.00E−03	
Zinc (Zn)			3.00E−01	
Organic compounds				
Acetone			1.00E−01	1.00E−01
Alachlor			1.00E−02	
Aldicarb			2.00E−04	
Anthracene			3.00E−01	
Atrazine			5.00E−03	

continues overleaf

Table D.1 (continued)

Environmental chemical	Toxicity index			
	Oral SF (1/mg/kg-day)	Inhalation SF (1/mg/kg-day)	Oral RfD (mg/kg-day)	Inhalation RfD (mg/kg-day)
Benzene	2.90E-02	2.90E-02		
Benzo(a)anthracene	1.20E+00	3.90E-01		
Benzo(a)pyrene [BaP]	1.20E+01	3.90E+00		
Benzo(b)fluoranthene	1.20E+00	3.90E-01		
Benzo(k)fluoranthene	1.20E+00	3.90E-01		
Benzoic acid			4.00E+00	4.00E+00
Bis(2-ethylhexyl)phthalate	1.40E-02	1.40E-02	2.00E-02	2.20E-02
Bromodichloromethane	1.30E-01	1.30E-01	2.00E-02	2.00E-02
Bromoform	7.90E-03	3.90E-03	2.00E-02	2.00E-02
Carbon disulfide			1.00E-01	2.90E-03
Carbon tetrachloride	1.30E-01	1.30E-01	7.00E-04	
Chlordane	1.30E+00	1.30E+00	6.00E-05	
Chlorobenzene			2.00E-02	
Chloroform	3.10E-02	1.90E-02	1.00E-02	1.00E-02
2-Chlorophenol			5.00E-03	
Chrysene	1.20E-01	3.90E-02		
m-Cresol [3-Methylphenol]			5.00E-02	
o-Cresol [2-Methylphenol]			5.00E-02	
Cyclohexanone			5.00E+00	
1,4-Dibromobenzene			1.00E-02	
Dibromochloromethane	8.40E-02	8.40E-02	2.00E-02	2.00E-02
1,2-Dibromomethane [EDB]	8.50E+01	7.70E-01		
1,2-Dichlorobenzene			9.00E-02	
Dichlorodifluoromethane			2.00E-01	
p,p'-Dichlorodiphenyldichloroethane [DDD]	2.40E-01			
p,p'-Dichlorodiphenyldichloroethylene [DDE]	3.40E-01			
p,p'-Dichlorodiphenyltrichloroethane [DDT]	3.40E-01	3.40E-01	5.00E-04	
1,1-Dichloroethane	5.70E-03	5.70E-03		
1,2-Dichloroethane	7.00E-02	7.00E-02		
1,1-Dichloroethene	6.00E-01	1.80E-01	9.00E-03	9.00E-03
cis-1,2-Dichloroethene			1.00E-02	1.00E-02
trans-1,2-Dichloroethene			2.00E-02	2.00E-02
2,4-Dichlorophenol			3.00E-03	
Dieldrin	1.60E+01	1.60E+01	5.00E-05	
Di(2-ethylhexyl)phthalate [DEHP]	1.40E-02		2.00E-02	2.20E-02
Diethyl phthalate			8.00E-01	
2,4-Dimethylphenol			2.00E-02	
2,6-Dimethylphenol			6.00E-04	
3,4-Dimethylphenol			1.00E-03	
m-Dinitrobenzene			1.00E-04	
1,4-Dioxane	1.10E-02			
Endosulfan			5.00E-05	
Endrin			3.00E-04	
Ethylbenzene			1.00E-01	2.90E-01
Ethyl chloride				2.90E+00
Ethyl ether			2.00E-01	
Ethylene glycol			2.00E+00	
Fluoranthene			4.00E-02	4.00E-02
Fluorene			4.00E-02	

Appendix D: Toxicological information for selected environmental chemicals

Table D.1 (continued)

Environmental chemical	Toxicity index			
	Oral SF (1/mg/kg-day)	Inhalation SF (1/mg/kg-day)	Oral RfD (mg/kg-day)	Inhalation RfD (mg/kg-day)
Formaldehyde		4.50E–02	2.00E–01	
Furan			1.00E–03	
Heptachlor	4.50E+00	4.50E+00	5.00E–04	
Hexachlorobenzene	1.60E+00	1.60E+00	8.00E–04	
Hexachlorodibenzo-*p*-dioxin [HxCDD]	6.20E+03	6.20E+03		
Hexachloroethane	1.40E–02	1.40E–02	1.00E–03	
n-Hexane				5.72E–02
Indeno(1,2,3-cd)pyrene	1.20E+00	3.90E–01		
Isobutyl alcohol			3.00E–01	
Lindane [gamma-HCH]			3.40E–04	
Malathion			2.00E–02	
Methanol			5.00E–01	
Methyl mercury			3.00E–04	
Methyl parathion			2.50E–04	
Methyl ethyl ketone [MEK]			6.00E–01	2.90E–01
Methyl isobutyl ketone [MIBK]			8.00E–02	2.30E–02
Methylene chloride [Dichloromethane]	7.50E–03	1.65E–03	6.00E–02	
Mirex			2.00E–06	
Nitrobenzene			5.00E–04	
n-Nitroso-di-*n*-butylamine	5.40E+00	5.40E+00		
n-Nitroso-di-*n*-methylethylamine	2.20E+01			
n-Nitroso-di-*n*-propylamine	7.00E+00			
n-Nitrosodiethanolamine	2.80E+00			
n-Nitrosodiethylamine	1.50E+02	1.50E+02		
n-Nitrosodimethylamine	5.10E+01	5.10E+01		
n-Nitrosodiphenylamine	4.90E–03			
Pentachlorobenzene			8.00E–04	
Pentachlorophenol	1.80E–02	1.80E–02	3.00E–02	
Phenol			6.00E–01	
Polychlorinated biphenyls [PCBs]	7.70E+00			
Pyrene			3.00E–02	3.00E–02
Styrene			2.00E–01	2.90E–01
1,2,4,5-Tetrachlorobenzene			3.00E–04	
1,1,1,2-Tetrachloroethane	2.60E–02	2.60E–02	3.00E–02	
1,1,2,2-Tetrachloroethane	2.70E–01	2.70E–01		
Tetrachloroethene	5.10E–02	2.10E–02	1.00E–02	1.00E–02
2,3,4,6-Tetrachlorophenol			3.00E–02	
Toluene			2.00E–01	1.40E+00
Toxaphene	1.10E+00	1.10E+00		
1,2,4-Trichlorobenzene			1.00E–02	
1,1,1-Trichloroethane			9.00E–02	2.90E–01
1,1,2-Trichloroethane	5.70E–02	5.60E–02	4.00E–03	4.00E–03
Trichloroethene	1.50E–02	1.00E–02	6.00 E-03	6.00E–03
1,1,2-Trichloro-1,2,2-trifluoroethane [CFC-113]			3.00E+01	
Trichlorofluoromethane			3.00E–01	
2,4,5-Trichlorophenol			1.00E–01	
2,4,6-Trichlorophenol	1.10E–02	1.10E–02		
1,1,2-Trichloropropane			5.00E–03	

continues overleaf

Table D.1 (*continued*)

Environmental chemical	Toxicity index			
	Oral SF (1/mg/kg-day)	Inhalation SF (1/mg/kg-day)	Oral RfD (mg/kg-day)	Inhalation RfD (mg/kg-day)
1,2,3-Trichloropropane			6.00E–03	
Triethylamine				2.00E–03
1,3,5-Trinitrobenzene			5.00E–05	
2,4,6-Trinitrotoluene [TNT]	3.00E–02		5.00E–04	
o-Xylene			2.00E+00	2.00E–01
Xylenes (mixed)			2.00E+00	
Others				
Asbestos (different set of units applied)*		2.3E-1 per fibers/mL		
Hydrazine	3.00E+00	1.70E+01		
Hydrogen chloride				2.00E–03
Hydrogen cyanide			2.00E–02	
Hydrogen sulfide			3.00E–03	2.60E–04

* Notes: (1) 2.3E-1 per fibres/mL $\equiv$ 2.3E-7 per fibres/m^3
 (2) It is noteworthy that regulatory agencies (such as the California EPA) use a significantly more restrictive value of 1.9 per fibres/mL ($\equiv$ 1.9E-6 per fibres/m^3) as the inhalation SF for asbestos.

Appendix E

Selected Environmental Tools and Databases of Potential Relevance to Risk Assessment and Environmental Management Programs

A variety of decision analytical tools and logistics, as well as computer databases and information libraries, may find useful applications in risk assessment and environmental management programs designed to address environmental contamination problems. Selected examples of such logistical tools and database systems are presented below in this appendix.

E.1 SELECTED DECISION SUPPORT TOOLS AND LOGISTICAL COMPUTER SOFTWARE

Often a variety of scientific and analytical tools are employed to assist the decision-maker with various issues associated with the management of an environmental contamination problem. A select number of application tools (consisting of scientific models and database systems) appropriate for such purposes are highlighted below; a primary communication system to consider in order to obtain further information on the listed softwares (and indeed others) would be the on-line service of the *Internet* – the most widely used international network communication service; otherwise libraries and telephone directories may provide the necessary up-to-date contacts. This listing is by no means complete and exhaustive; other similar logistical tools can indeed be used to support environmental management programs, in order to arrive at informed decisions on environmental contamination problems. In fact, recent years have seen a proliferation of software systems for a variety of environmental management programs. Care must therefore be exercised in the choice of an appropriate tool for specific problems.

E.1.1 AERIS (Aid for Evaluating the Redevelopment of Industrial Sites)

AERIS is an expert system consisting of a multimedia risk assessment model used to generate site-specific cleanup guidelines. It consists of a computer program capable of deriving cleanup guidelines for industrial sites where re-development is being considered.

AERIS serves as a useful remediation model for identifying cleanup objectives. It can be used to identify the factors that are likely to be major contributors to potential exposures and concerns at sites, and those aspects of a re-development scenario with the greatest need for better site-specific information.

AERIS is designed to evaluate situations where the soil had been contaminated sufficiently long enough to establish equilibrium or near-equilibrium conditions. Thus, it is not suitable for evaluating recent spill sites or locations.

Further information on AERIS may be obtained from the following sources:

- Decommissioning Steering Committee, Canadian Council of Resource and Environment Ministers (CCREM), Canada.
- SENES Consultants Ltd, Richmond Hill, Ontario, Canada.

E.1.2 AIRTOX (Air TOXics risk management framework)

AIRTOX is a decision analysis model for air toxics risk management. The framework consists of a structural model that relates emissions of air toxics to potential health effects, and a decision-tree model that organizes scenarios evaluated by the structural model.

AIRTOX can be used to evaluate the magnitude of health risks to a population, a specific source's contribution to the total health risk, and the cost-effectiveness of current and future emission control measures.

Further information on AIRTOX may be obtained from the following:

- EPRI (Electric Power Research Institute), Palo Alto, California, USA.

E.1.3 API DSS (Decision Support System for exposure and risk assessment)

The American Petroleum Institute (API)'s exposure and risk assessment Decision Support System (DSS) is a software system designed to assist environmental professionals in the estimation of human exposures and risks from sites contaminated with petroleum products. The DSS is a user-friendly tool that can be used to estimate site-specific exposures and risks; identify the need for site remediation; develop and negotiate site-specific cleanup levels with regulatory agencies; and efficiently and effectively evaluate the effects of parameter uncertainty and variability on estimated risks, using Monte Carlo techniques. It estimates receptor point concentrations by executing fully incorporated unsaturated zone, saturated zone, air emission, air dispersion, and particulate emission models.

The computational modules of the DSS can be implemented in either a deterministic or Monte Carlo mode; the latter is used to quantify the uncertainty in the exposure and risk values that could result from uncertainties in the input parameters.

From physical, chemical, and toxicological property data provided in the DSS databases, risk assessments can be conducted for 16 hydrocarbons, 6 petroleum additives, and 3 metals. The databases can also be expanded to include up to 100 other constituents.

Further information on API DSS may be obtained from the following:

- American Petroleum Institute (API), Washington, DC, USA.
- Geraghty & Miller, Inc., Millersville, Maryland, USA.

E.1.4 API PRDF (Petroleum Release Decision Framework)

The American Petroleum Institute (API)'s Petroleum Release Decision Framework (PRDF) has been designed to provide a logical approach for site characterization. It is used to facilitate such activities as collecting and archiving field data (focusing on key decision parameters), and developing and evaluating potential corrective action plans.

The PRDF is structured to provide a consistent, comprehensive, and systematic way to identify possible contaminant sources, to characterize the site and contaminants potentially present, and to help assess actual or potential exposures.

Further information on API PRDF may be obtained from the following:

- American Petroleum Institute (API), Washington, DC, USA.
- Geraghty & Miller, Inc., Millersville, Maryland, USA.

Appendix E: Selected environmental tools and databases 487

E.1.5 CalTOX

CalTOX is a multimedia, multipathway risk assessment model that allows stochastic simulation to be performed. The model is based on the principles of conservation of mass and chemical equilibrium; it calculates the gains and losses in each environmental compartment over time, by accounting for both transport from one compartment into another, and also chemical biodegradation and transformation. A noteworthy feature is that the model makes the distinction between the environmental concentration and the exposure concentration.

CalTOX is a risk assessment model that mathematically relates the concentration of a chemical in the soil to the theoretical dose a person may receive. It presents an increased accuracy in the evaluation of human health risks from hazardous waste sites and permitted facilities, because it incorporates an appropriate fate and transport model and stochastic risk assessment process.

CalTOX represents a fugacity model for evaluating the time-dependent movement of contaminants in various environmental media. The model is designed to be used for stochastic analyses, but can be used deterministically as well.

CalTOX quantitatively addresses both uncertainty and variability by allowing the presentation of both the risks and the calculated cleanup goals as probability distributions – facilitating a clearer distinction between the risk assessment and risk management steps in site remediation decisions.

CalTOX find applications/uses in the calculation of risks, and in the calculation of cleanup goals for contaminated sites. Risks and cleanup levels are calculated for on-site receptors only, and for one chemical (organics only) at a time. It is generally applicable at sites where the primary source of contamination is the soil. CalTOX is limited to modeling long-term (several months to years) exposures – i.e., it cannot be used to assess short-term risks or for recent releases.

Further information on CalTOX may be obtained from the following:

- California Environmental Protection Agency (Cal EPA), Department of Toxic Substances Control (DTSC), Sacramento, California, USA.
- Lawrence Livermore National Laboratory, Berkeley, California, USA.

E.1.6 CAMEO (Computer-Aided Management of Emergency Operations)

CAMEO addresses acute exposures and risks, such as those associated with accidental releases. They are generally assumed to last for hours or, at most, days. Concentrations that are immediately dangerous to life or health (IDLH) are then the relevant index of hazard.

As the name implies, CAMEO is designed as a tool to help emergency responders (such as fire departments) and planners (such as local emergency planning committees) deal with the acute risks posed by the storage and transport of hazardous chemicals.

Some of the most interesting features of CAMEO include: the 'Response Information Data Sheets' database, containing information on chemical properties, acute hazards, and appropriate emergency handling for some 3000 chemicals; a mapping module that provides base maps, and allows the user to add information on local facilities, transportation routes, etc.; and the ability to perform screening-level analyses of the areas potentially affected by release of chemicals into air, and to display these areas on the maps.

Further information on CAMEO may be obtained from the following:

- Environmental Health Center, National Safety Council, Washington, DC, USA.

E.1.7 Crystal Ball

Crystall Ball is a user-friendly, graphically oriented forecasting and risk analysis program that helps minimize the uncertainty associated with decision-making. The system employs standard spreadsheet models in its application.

Through the use of Monte Carlo simulation techniques, Crystal Ball forecasts the entire range of results possible for a given situation. It also shows the confidence levels, in order that the analyst will know the likelihood of any specific event taking place. The tool further allows sensitivity evaluations to be carried out in a very effective manner, by the use of sensitivity charts; Crystal Ball calculates sensitivity by computing rank correlation coefficients between every assumption and every forecast cell during simulation.

With an intuitive graphical interface, Crystal Ball gives users powerful capabilities to perform uncertainty analyses based on Monte Carlo simulations.

Further information on Crystal Ball may be obtained from the following:

- Decisioneering, Inc., Boulder, Colorado, USA.

E.1.8 GEOTOX

GEOTOX is a multimedia environmental transport and transformation model. It consists of a computer program designed to calculate time-varying chemical concentrations in air, soil, groundwater, and surface water. The model can be applied to constant or time-varying chemical sources.

The GEOTOX model uses two sets of input data – one providing the properties of the environment or landscape receiving the contaminants, and the other describing the properties of the contaminants. Model output is in the form of environmental concentrations, and intake by various exposure pathways, as well as total intake.

GEOTOX can be used to predict/derive contaminant concentrations in various multimedia compartments, which are subsequently combined with appropriate human inhalation and ingestion rates, and absorption factors to calculate exposure. Relative health risk for a number of chemicals can be calculated.

Further information on GEOTOX may be obtained from the following:

- Lawrence Livermore National Laboratory, Berkeley, California, USA.

E.1.9 IEUBK (Integrated Exposure Uptake Biokinetic model for lead in children)

The Integrated Exposure Uptake Biokinetic (IEUBK) model for lead in children is a menu-driven, user-friendly model designed to determine exposure from lead in air, water, soil, dust, diet, and paint and other sources with pharmacokinetic modeling to predict blood levels in children 6 months to 7 years of age.

Further information on IEUBK may be obtained from the following:

- USEPA's Office of Emergency and Remedial Response, Washington, DC, USA.

E.1.10 LEADSPREAD

LEADSPREAD provides a methodology for evaluating exposure and the potential for adverse health effects resulting from multipathway exposure to inorganic lead in the environment. Each pathway is represented by an equation relating incremental blood lead increase to a concentration in a medium, using contact rates and empirically determined ratios. The contributions via all pathways are added to arrive at an estimate of median blood lead concentration resulting from the multipathway exposure. The method is adapted to a computer spreadsheet. It can be used to determine blood levels associated with multiple pathway exposures to lead associated with an environmental contamination problem.

LEADSPREAD basically consists of a mathematical model for estimating blood lead concentrations as a result of contacts with lead-contaminated environmental media. A distributional approach is used, allowing estimation of various percentiles of blood lead concentration associated with a given set of inputs.

Further information on LEADSPREAD may be obtained from the following:

- Office of Scientific Affairs, Department of Toxic Substances Control (DTSC), California EPA, Sacramento, California, USA.

E.1.11 MEPAS (Multimedia Environmental Pollutant Assessment System)

MEPAS (Multimedia Environmental Pollutant Assessment System) is an analytical model that has been developed to address problems at hazardous waste sites. It is a versatile tool that can handle a variety of different types of source terms. MEPAS couples contaminant release, migration and fate for environmental media (groundwater, surface water, air) with exposure routes (inhalation, ingestion, dermal contact, external dose) and risk/health consequences for radiological and nonradiological carcinogens and noncarcinogens.

MEPAS includes a sector-averaged Gaussian plume algorithm that simulates the atmospheric transport of contaminants; simulates groundwater transport using a three-dimensional algorithm; uses a simplistic approach to modeling the surface water pathway; and includes foodchains as an integral part of its exposure–dose component. It can model both on-site and off-site contaminant exposures.

MEPAS represents an integrated, site-specific, multimedia environmental assessment. It can simulate the transport and distribution of contaminants over time and space within air, water, soil, and foodchain pathways. It estimates long-term health effects at receptor locations, as well as normalized maximum hourly concentrations for determining acute effects.

MEPAS integrates and evaluates transport and exposure pathways for chemicals and radioactive releases according to their potential human health impacts. It takes the nontraditional approach of combining all major exposure pathways into a multimedia computational tool for public health impact.

Further information on MEPAS may be obtained from the following:

- Battelle – Pacific Northwest National Laboratory, Richland, Washington, USA.

E.1.12 MMSOIL

MMSOIL (The Multimedia Contaminant Fate, Transport, and Exposure Model) is a multimedia screening tool developed for the 'relative comparison' of hazardous waste sites. The model performs a mass balance for the air and groundwater pathways separately, relative to the initial source term.

MMSOIL includes a sector-averaged Gaussian plume algorithm that simulates the atmospheric transport of contaminants; has a complex groundwater modeling component; includes a simplistic approach to modeling the surface water pathway; and considers foodchain modeling 'externally'/separately. It can model both on-site and off-site chemical exposures.

MMSOIL was designed specifically to simulate the release of toxic chemicals from underground storage tanks, surface impoundments, waste piles, and landfills. It can model the fate and transport of chemicals, and calculates human exposure and health risk, as well as concentration in all important media.

Further information on MMSOIL may be obtained from the following:

- Office of Research and Development, USEPA, Washington, DC, USA.

E.1.13 MULTIMED (MULTIMEDia exposure assessment model)

MULTIMED is a computer model for simulating the transport and transformation of contaminants released from a hazardous waste disposal facility into the multimedia environment. The MULTIMED model simulates releases into air and soil – including the unsaturated (vadose) and saturated zones, and possible interception of the subsurface contaminant plume by a surface stream. It further simulates movement through the air, soil, groundwater, and surface water media to contact humans and other potentially affected receptors. Uncertainties in parameter values used in the model are quantified using Monte Carlo simulation techniques.

MULTIMED is typically used to simulate the movement of contaminants leaching from a waste disposal facility. It is intended for general exposure and risk assessments of waste facilities, and for the analyses of the impacts of engineering and management controls.

Further information on MULTIMED may be obtained from the following:

- Environmental Research Laboratory, Office of Research and Development, USEPA, Athens, Georgia, USA.

E.1.14 RAAS (Remedial Action Assessment System)

RAAS (Remedial Action Assessment System) is a sophisticated, Windows-based software package that serves as a tool for analyzing and evaluating the tradeoffs necessary to select a preferred approach for restoring a contaminated site. It helps the user to develop a detailed site description; estimate baseline and residual risks to public health posed by the site; identify and screen applicable environmental restoration technologies; formulate and evaluate technically feasible, complete remedial response alternatives; and assess and compare a remedial response alternative across some set criteria to provide a basis for final remedy selection.

RAAS can be used to quantitatively evaluate the effectiveness of several remedial alternatives in terms of concentration, risk reduction, and effect on media properties. In addition, the methodology enables the user to assess and cross-compare those remedial response alternatives across some set critera (e.g., cost, human health, risk reduction, time) to provide a basis for comparisons among them.

Further information on RAAS may be obtained from the following:

- Battelle – Pacific Northwest National Laboratory, Richland, Washington, USA.

E.1.15 RBCA (Risk-Based Corrective Action) spreadsheet system

The RBCA (Risk-Based Corrective Action) spreadsheet system/tool kit is a complete step-by-step package for the calculation of site-specific risk-based soil and groundwater cleanup goals, which will then facilitate the development of site remediation plans. The system includes fate and transport models for major and significant exposure pathways (i.e., air, groundwater, and soil), together with an integrated chemical/toxicological library of a number of chemical compounds (i.e., over 80, and also expandable by the user).

RBCA is a standardized approach to designing remediation strategies for contaminated sites. It was developed by the American Society for Testing and Materials (ASTM) to help prioritize sites according to the urgency and type of corrective action needed to protect human health and the environment.

The RBCA process allows for the calculation of baseline risks and cleanup standards, as well as for remedy selection and compliance monitoring at petroleum release sites. The user simply provides site-specific data to determine exposure concentrations, average daily intakes, baseline risk levels, and risk-based cleanup levels.

Further information on the RBCA tool kit/spreadsheet system may be obtained from the following:

- ASTM (American Society for Testing and Materials), Philadelphia, Pennsylvania, USA.
- Groundwater Services, Inc., Houston, Texas, USA.
- Environmental Systems and Technologies, Inc., Blacksburg, Virginia, USA.

Appendix E: Selected environmental tools and databases

E.1.16 RISC (Risk Identification of Soil Contamination)

RISC (Risk Identification of Soil Contamination) is a knowledge-based framework for risk identification and evaluation of sites with contaminated soils. It consists of computer modules that facilitate site investigations, risk analyses, and priority-ranking for former industrial facilities.

The RISC framework uses expert information on the fate and behavior of contaminants in soil systems to predict potential risks to human health and the environment that could result from contaminated site problems. Dutch, English, and German versions of this computer model system are available.

Further information on the RISC computer model system may be obtained from the following:

- Van Hall Institute, Groningen, The Netherlands.

E.1.17 RISK*ASSISTANT

RISK*ASSISTANT provides an array of analytical tools, databases, and information-handling capabilities for human health risk assessment. It has the ability to tailor exposure and risk assessments to local conditions.

The RISK*ASSISTANT software is designed to assist the user in rapidly evaluating exposures and human health risks from chemicals in the environment at a particular site. It is designed to evaluate human health risks associated with *chronic* exposures to chemicals. The user need only provide measurements or estimates of the concentrations of chemicals in the air, surface water, groundwater, soil, sediment, and/or biota.

Further information on RISK*ASSISTANT may be obtained from the following:

- Hampshire Research Institute, Alexandria, Virginia, USA.
- USEPA, Research Triangle Park, North Carolina, USA.
- California EPA, Sacramento, California, USA.
- New Jersey Department of Environmental Protection, Trenton, New Jersey, USA.
- Delaware Department of Natural Resources and Environmental Control, Dover, Delaware, USA.

E.1.18 RISKPRO

RISKPRO is a complete software system designed to predict the environmental risks and effects of a wide range of human health-threatening situations. It consists of a multimedia/multipathway environmental pollution modeling system, providing for modeling tools to predict exposure from pollutants in the air, soil, and water.

RISKPRO is used to evaluate receptor exposures and risks from environmental contaminants. It graphically represents its results through maps, bar charts, wind-rose diagrams, isopleth diagrams, pie charts, and distributional charts. Its mapping capabilities can also allow the user to create custom maps showing data and locations of environmental contaminant plumes.

Further information on RISKPRO may be obtained from the following:

- General Sciences Corporation (GSC), Laurel, Maryland, USA.

E.1.19 SITES (The contaminated Sites risk management system)

SITES is a flexible interactive PC computerized decision-support tool for organizing relevant information needed to conduct risk management analyses for contaminated sites. It has the dimensionality to model multiple chemicals, pathways, population groups, health effects, and remedial actions. The model uses information from diverse sources, such as site investigations, transport and fate modeling, behavioral and exposure estimates, and toxicology.

SITES is indeed a computer-based integrating framework used to help evaluate and compare site investigation and remedial action alternatives in terms of health and environmental effects and total economic costs/impacts. The user completely defines the scope of the analyses. Both deterministic and probabilistic analyses are possible.

The decision-tree structure in SITES allows for explicit examination of key uncertainties and the efficient evaluation of numerous scenarios. The model's design and computer implementation facilitates quick and extensive sensitivity analyses.

Further information on SITES may be obtained from the following:

- EPRI (Electric Power Research Institute), Palo Alto, California, USA.

E.1.20 TOXIC

TOXIC is a microcomputer program that calculates the incremental risk to the hypothetical maximum exposed individual from hazardous waste incineration. It calculates exposure to each pollutant individually, using a specified dispersion coefficient (which is the ratio of pollutant concentration in air – in $\mu g/m^3$ to pollutant emission rate in g/s).

TOXIC is used in hazardous waste facility risk analysis. It is a flexible and convenient tool for performing inhalation risk assessments for hazardous waste incinerators. It gives point estimates of inhalation risks.

Further information on TOXIC may be obtained from the following:

- Rowe Research and Engineering Associates, Alexandria, Virginia, USA.

E.2 SELECTED DATABASES AND INFORMATION LIBRARIES WITH IMPORTANT RISK INFORMATION FOR RISK ASSESSMENT AND ENVIRONMENTAL MANAGEMENT

Several databases relating to numerous chemical substances exist within the scientific community that may find extensive useful applications in the management of environmental contamination problems. Example databases of general interest to environmental contamination management problems – one for its international appeal, and the others for their wealth of risk assessment support information – are presented below. The presentation is meant to demonstrate the overall wealth of scientific information that already exists, and that should generally be consulted to provide the relevant support to environmental management programs.

E.2.1 The International Register of Potentially Toxic Chemicals (IRPTC) database

In 1972, the United Nations Conference on the Human Environment, held in Stockholm, recommended the setting up of an international registry of data on chemicals likely to enter and damage the environment. Subsequently, in 1974, the Governing Council of the United Nations Environment Programme (UNEP) decided to establish both a chemicals register and a global network for the exchange of information that the register would contain. The definition of the register's objectives was subsequently elaborated as follows:

- make data on chemicals readily available to those who need it;
- identify and draw attention to the major gaps in the available information and encourage research to fill those gaps;
- help identify the potential hazards of using chemicals and improve people's awareness of such hazards; and

Appendix E: Selected environmental tools and databases

- assemble information on existing policies for control and regulation of hazardous chemicals at national, regional, and global levels.

In 1976, a central unit for the register, named the International Register of Potentially Toxic Chemicals (IRPTC), was created in Geneva, Switzerland, with the main function of collecting, storing, and disseminating data on chemicals, and also to operate a global network for information exchange. IRPTC network partners, the designation assigned to participants outside the central unit, consist of National Correspondents appointed by governments, national and international institutions, national academies of science, industrial research centres, and specialized research institutions. Chemicals examined by the IRPTC have been chosen from national and international priority lists. The selection criteria used include the quantity of production and use, the toxicity to humans and ecosystems, persistence in the environment, and the rate of accumulation in living organisms.

IRPTC stores information that would aid in the assessment of the risks and hazards posed by a chemical substance to human health and environment. The major types of information collected include that relating to the behavior of chemicals, and information on chemical regulations. Information on the behavior of chemicals is obtained from various sources such as national and international institutions, industries, universities, private databanks, libraries, academic institutions, scientific journals and United Nations bodies such as the International Programme on Chemical Safety (IPCS). Regulatory information on chemicals is largely contributed by IRPTC National Correspondents. Specific criteria are used in the selection of information for entry into the databases. Whenever possible, IRPTC uses data sources cited in the secondary literature produced by national and international panels of experts to maximize reliability and quality. The data are then extracted from the primary literature. Validation is performed prior to data entry and storage on a computer at the United Nations International Computing Centre (ICC).

E.2.1.1 Types of information in the IRPTC databases

The complete IRPTC file structure consists of databases relating to the following subject matter and areas of interest: Legal; Mammalian and Special Toxicity Studies; Chemobiokinetics and Effects on Organisms in the Environment; Environmental Fate Tests, and Environmental Fate and Pathways into the Environment; and Identifiers, Production, Processes and Waste.

The IRPTC *Legal* database contains national and international recommendations and legal mechanisms related to chemical substances control in environmental media such as air, water, wastes, soils, sediments, biota, foods, drugs, consumer products, etc. This setup allows for rapid access to the regulatory mechanisms of several nations and to international recommendations for safe handling and use of chemicals.

The *Mammalian Toxicity* database provides information on the toxic behavior of chemical substances in humans; toxicity studies on laboratory animals are included as a means of predicting potential human effects. The *Special Toxicity* databases contain information on particular effects of chemicals on mammals, such as mutagenicity and carcinogenicity, as well as data on nonmammalian species when relevant for the description of a particular effect.

The *Chemobiokinetics and Effects on Organisms in the Environment* databases provide data that will permit the reliable assessment of the hazard of chemicals present in the environment to man. The absorption, distribution, metabolism, and excretion of drugs, chemicals, and endogenous substances are described in the Chemobiokinetics databases. The Effects on Organisms in the Environment databases contain toxicological information regarding chemicals in relation to ecosystems and to aquatic and terrestrial organisms at various nutritional levels.

The *Environmental Fate Tests, and Environmental Fate and Pathways into the Environment* databases assess the risk presented by chemicals to the environment.

The *Identifiers, Production, Processes and Waste* databases contain miscellaneous information about chemicals, including physical and chemical properties; hazard classification for chemical production and trade statistics of chemicals on a worldwide or regional basis;

information on production methods; information on uses and quantities of use for chemicals; data on persistence of chemicals in various environmental compartments or media; information on the intake of chemicals by humans in different geographical areas; sampling methods for various media and species, as well as analytical protocols for obtaining reliable data; recommendable methods for the treatment and disposal of chemicals; etc.

E.2.1.2 The role of IRPTC in risk assessment and environmental management

The IPRTC, with its carefully designed database structure, provides a sound model for national and regional data systems. More importantly, it brings consistency to information exchange procedures within the international community. The IPRTC is serving as an essential international tool for chemical hazards assessment, as well as a mechanism for information exchange on several chemicals. The wealth of scientific information contained in the IRPTC can serve as an invaluable database for environmental management programs.

E.2.1.3 Sources of information on IRPTC

Further information on, and access to IRPTC may be obtained from the following sources:

- National Correspondent to the IRPTC and scientific bodies/institutions such as a country's National Academy of Sciences. (Also, following the successful implementation of the IRPTC databases, a number of countries created National Registers of Potentially Toxic Chemicals (NRPTCs) that are completely compatible with the IRPTC system.)

E.2.2 The Integrated Risk Information System (IRIS) database

The Integrated Risk Information System (IRIS), prepared and maintained by the Office of Health and Environmental Assessment of the USEPA is an electronic database containing health risk and regulatory information on several specific chemicals. It is an on-line database of chemical-specific risk information; it is also a primary source of EPA health hazard assessment and related information on a number of chemicals of environmental concern.

IRIS was originally developed for EPA staff in response to a growing demand for consistent risk information on chemical substances for use in decision-making and regulatory activities. The information in IRIS is accessible to those without extensive training in toxicology, but with some rudimentary knowledge of health and related sciences.

E.2.2.1 Types of information in IRIS

The IRIS database consists of a collection of computer files covering several individual chemicals. To aid users in accessing and understanding the data in the IRIS chemical files, the following supportive documentation is provided:

- Alphabetical list of the chemical files in IRIS and list of chemicals by Chemical Abstracts Service (CAS) number.
- Background documents describing the rationales and methods used in arriving at the results shown in the chemical files.
- A user's guide that presents step-by-step procedures for using IRIS to retrieve chemical information.
- An example exercise in which the use of IRIS is demonstrated.
- Glossaries in which definitions are provided for the acronyms, abbreviations, and specialized risk assessment terms used in the chemical files and in the background documents.

The chemical files contain descriptive and numerical information on several subjects – including oral and inhalation reference doses (RfDs) for chronic noncarcinogenic health effects, and oral and inhalation cancer slope factors (SFs) and unit cancer risks (UCRs) for chronic exposures to carcinogens. It also contains supplementary data on acute health hazards and physical/chemical properties of the chemicals.

E.2.2.2 The role of IRIS in risk assessment and environmental management

IRIS is a tool which provides hazard identification and dose–response assessment information, but does not provide problem-specific information on individual instances of exposure. It is a computerized library of current information that is updated periodically. Combined with specific exposure information, the data in IRIS can be used to characterize the public health risks of a chemical of potential concern under specific scenarios, which can then facilitate the development of effective corrective action decisions designed to protect public health. The information in IRIS can indeed be used to develop corrective action decision for potentially contaminated sites, such as via the application of risk assessment and risk management procedures.

E.2.2.3 Sources of information on IRIS

Further information on, and access to IRIS may be obtained from the following:

- IRIS User Support, USEPA, Environmental Criteria and Assessment Office, Cincinnati, Ohio, USA.
- Chemical Information System [CIS] (Commercial vendor), Baltimore, Maryland, USA.
- Dialog Information Services, Inc. [DIALOG] (Commercial Vendor), Palo Alto, California, USA.
- National Library of Medicine [NLM], Bethesda, Maryland, USA.

E.2.3 RAMAS Library of Ecological Software

RAMAS is a software library for building ecological models. RAMAS programs incorporate species-specific data to predict the future changes in the population and assess the risk of population extinction or explosion and chances of recovery from a disturbance. All programs have user-friendly menu systems and context-sensitive, on-line help facilities.

The family of population simulators provided by the RAMAS Library consists of models that cover a wide range of circumstances, and provides a very good demonstration for the principles of population dynamics.

E.2.3.1 Types of information in the RAMAS Library

Some of the major ecological software programs contained in the RAMAS Library include:

- *RAMAS/age, for modeling fluctuations in age-structured populations.* It is an interactive simulator for age-structured population dynamics. The program predicts how many individuals there will be in an age class in future years and estimates the probability that the population will exceed or fall below a specified level of abundance.
- *RAMAS/ecoBound, for ecological boundary delineation.* It provides a selection of traditional and contemporary methods to draw ecological boundaries in a single software package.
- *RAMAS/ecotoxicology, for population-level risk assessment.* It translates individual-level impacts to population-level risk assessment. The program imports data from standard laboratory bioassay experiments, incorporates these data into the parameters of a population model, and performs a risk assessment by analyzing population-level differences between control and impacted samples.
- *RAMAS/GIS, linking landscape data with population viability analysis* (i.e., linking GIS with metapopulation dynamics). It is a comprehensive extinction risk assessment system, in which the user can run the metapopulation model to predict the risk of species extinction, expected occupancy rates, and metapopulation abundance. The program combines geographic and demographic data for risk assessment.

- *RAMAS/metapop, for viability analysis for stage-structured metapopulations.* It is an interactive program that allows the user to build models for species that live in multiple patches, and incorporates the spatial aspects of metapopulation dynamics. The program can be used to predict extinction risks and explore management options such as reserve design and translocations, and to assess human impact on fragmented populations.
- *RAMAS/space, consisting of spatially-structured models for conservation biology.* It is an interactive program of metapopulation modeling that incorporates the spatial aspects of metapopulation dynamics. It is used in conservation biology to predict extinction risks and explore management options such as reserve design, translocations, and reintroductions. The program predicts how many individuals there will be in the metapopulation and in each of its populations in future years, estimates the probability that the species will go extinct or will fall below or exceed a specified total abundance, and calculates the distribution of extinction times.
- *RAMAS/stage, for analyzing stage-structured populations.* It allows a user to build, run, and analyze discrete-time models for species with virtually any life history. It is especially useful for modeling species with complex life histories. The program estimates the chance that a population will go extinct or suffer a decline; it also estimates the chance that the population will grow to some level, and/or the risks that an abundance will fall below, or above, any threshold.
- *RAMAS/time, for ecological time series analysis.* It is designed to analyze time series and to model immediate and delayed density-dependent feedbacks that affect organism numbers. It also determines the qualitative type of dynamical behavior that characterizes the studied population.
- *Risk Calc, for calculating bounds on point estimates.* It allows the user to determine the uncertainties in the risk estimates, so that reliability of any predictions can be determined. The program uses intervals and fuzzy numbers to represent uncertainty, and provides an environment to compute with these numbers in which all uncertainties are carried forward automatically.

E.2.3.2 Sources of information on RAMAS Library of Software

Further information on RAMAS Library of Software may be obtained from the following sources:

- Applied Biomathematics (Ecological Research and Software), Setauket, New York, USA.

E.2.4 Other Miscellaneous Information Sources

A variety of other information sources are available to facilitate various risk assessment and/or environmental management tasks, including the following:

- *The Chemical Substances Information Network (CSIN).* The Chemical Substances Information Network (CSIN) is not a database, but rather an interactive network system that links together a number of databases relating to several chemical substances. The CSIN accesses data on chemical nomenclature, composition, structure, properties, toxicological information, health and environmental effects, production and uses, regulations, etc. The CSIN and the databases it accesses are in the public domain. However, users have to make independent arrangements with vendors of those databases in the network that need to be used for specific assignments. Further information on CSIN may be obtained from CIS, Baltimore, Maryland, USA.
- *MMEDE (Multimedia-Modeling Environmental Database and Editor).* The Multimedia-Modeling Environmental Database and Editor (MMEDE) is a user-friendly database interface for physical, chemical, and toxicological parameters typically associated with environmental assessments. The parameters for evaluation of impacts of hazardous and radioactive materials can be viewed, estimated, modified, printed, deleted, and exported.

Appendix E: Selected environmental tools and databases 497

Information in the database include physical parameters, dose factors, toxicity factors, environmental transfer factors, environmental decay half times, and other parameters of relevance. A source-of-information citation is used for every parameter value. Further information on MMEDE may be obtained from Battelle – Pacific Northwest National Laboratory, Richland, Washington, USA.

Further listings may generally be available on the Internet, which serves as a very important and contemporary international network communication system.

Appendix F

Selected Units of Measurements and Noteworthy Expressions

MASS/WEIGHT UNITS

g	gram(s)
ton (metric)	tonne = 1×10^6 g
kg	kilogram(s) = 10^3 g
mg	milligram(s) = 10^{-3} g
μg	microgram(s) = 10^{-6} g
ng	nanogram(s) = 10^{-9} g
pg	picogram(s) = 10^{-12} g

VOLUMETRIC UNITS

cm³ or cc	cubic centimeter(s) = 10^{-3} L
mL	milliliter(s) = 10^{-3} L
L	liter(s) = 10^3 cm³
m³	cubic meter(s) = 10^3 L

ENVIRONMENTAL CONCENTRATION UNITS

ppm	parts per million
ppb	parts per billion
ppt	parts per trillion

These are used for expressing/specifying the relative masses of contaminant and medium. Note that, because water is assigned a mass of 1 kilogram per liter, mass-to-mass and mass-to-volume measurements are interchangeable for this medium.

CONCENTRATION EQUIVALENTS

1 ppm	≡	1 mg/kg or mg/L
1 ppb	≡	1 μg/kg or μg/L
1 ppt	≡	1 ng/kg or ng/L

Concentrations in soils or other solid media:

mg/kg	mg chemical per kg weight of sampled medium

Concentrations in water or other liquid media:

mg/L	mg chemical per liter of total liquid volume

Concentrations in air media:

mg/m³   mg chemical per m³ of total fluid volume

CONVERSION FACTORS

- To convert from ppm to mg/m³, use the following conversion relationship:

$$[mg/m^3] = [ppm] \times \frac{[\text{molecular weight of substance, in g/mol.}]}{24.45}$$

- To convert from ppm to µg/m³, use the following conversion relationship:

$$[\mu g/m^3] = [ppm] \times [\text{molecular weight of substance, in g/mol.}] \times 40.9$$

Note: the above conversion relationships assume standard temperature and pressure (STP), i.e., temperature of 25°C and barometric pressure of 760 mmHg or 1 atm.

UNITS OF CHEMICAL INTAKE AND DOSE

mg/kg-day = milligrams of chemical exposure per unit body weight of exposed receptor per day

TYPICAL EXPRESSIONS COMMONLY USED IN ENVIRONMENTAL MANAGEMENT PROGRAMS

- *'Order of Magnitude'*. Reference to each 'order of magnitude' means a ten-fold difference, i.e., the base parameter may vary by a factor of 10. For example, three orders of magnitude may be used to describe the difference between 3 and 3000 ($= 3 \times 10^3$). It is often used in reference to the calculation of environmental quantities or risk probabilities.
- *Exponentials denoted by 10^κ*. Superscripts refer to the number of times '10' is multiplied by itself. For example, $10^2 = 10 \times 10 = 100$; $10^3 = 10 \times 10 \times 10 = 1000$; $10^6 = 10 \times 10 \times 10 \times 10 \times 10 \times 10 = 1\,000\,000$.
- *Exponentials denoted by $10^{-\kappa}$*. The negative superscript is equivalent to the reciprocal of the positive term, i.e., $10^{-\kappa}$ equals $1/10^\kappa$. For example, $10^{-2} = 1/10^2 = 1/(10 \times 10) = 0.01$; $10^{-3} = 1/10^3 = 1/(10 \times 10 \times 10) = 0.001$; $10^{-6} = 1/10^6 = 1/(10 \times 10 \times 10 \times 10 \times 10 \times 10) = 0.000001$.
- *Exponentials denoted by $X.YZ\ E+\kappa$*. The number after the 'E' indicates the power to which 10 is raised, and then multiplied by the preceding term (i.e., the number of times '10^κ' is multiplied by preceeding term, or $X.YZ \times 10^\kappa$). For example, $1.00E-01 = 1.00 \times 10^{-1} = 0.1$; $1.23E+04 = 1.23 \times 10^{+4} = 12\,300$; $4.44E+05 = 4.44 \times 10^5 = 444\,000$.
- *'Conservative assumption'*. In exposure and risk assessment, refers to the selection of assumptions (when real-time data are absent) that are unlikely to lead to underestimation of exposure or risk.
- *'Risk of 1×10^{-6} (or simply, 10^{-6})'*. Also written as 0.000001, or one in a million, means that one additional case of cancer is projected in a population of one million people exposed to a certain level of chemical X over their lifetimes. Similarly, a risk of 5×10^{-3} corresponds to 5 in 1000 or 1 in 200 persons; and a risk of 2×10^{-6} means two chances in a million of the exposure causing cancer.

Appendix G

Suggested Reference Journals

The Air Pollution Consultant, Elsevier, Amsterdam, The Netherlands.
American Journal of Industrial Medicine, J. Wiley & Sons, Chichester, UK.
Applied and Environmental Microbiology, American Society for Microbiology, Washington, DC.
Aquatic Conservation: Marine and Freshwater Ecosystems, J. Wiley & Sons, Chichester, UK.
Aquatic Toxicology, Elsevier Science B.V., Amsterdam, The Netherlands.
Archives of Environmental Contamination and Toxicology, Springer-Verlag, New York.
Archives of Environmental Health, Allen Press, Inc., Lawrence, KS.
Archives of Toxicology, Springer-Verlag, New York.
Biodegradation, Kluwer Academic Publishers, Dordrecht, The Netherlands.
Boston College Environmental Affairs Law Review, Boston College Law School, Newton Centre, MA.
Bulletin of Environmental Contamination and Toxicology, Springer-Verlag, New York.
Civil Engineering, ASCE, Reston, VA.
Chemosphere, Elsevier, Amsterdam, The Netherlands.
Clean Air (and Environmental Protection), National Society for Clean Air and Environmental Protection, Brighton, UK.
Clean Air, The Journal of the Clean Air Society of Australia and New Zealand, Eastwood, NSW, Australia.
Colorado Journal of International Environmental Law and Policy, University Press of Colorado, Niwot, CO.
Columbia Journal of Environmental Law, School of Law, Columbia University, New York, NY.
Conservation, Ecology, Environmental Planning, Plenum Press, New York.
Critical Reviews in Environmental Science and Technology, CRC Press, Inc., Boca Raton, FL.
Critical Reviews in Toxicology, CRC Press, Inc., Boca Raton, FL.
Current Issues in Public Health, Chapman & Hall, London, UK.
Ecological Abstracts, Elsevier, Amsterdam, The Netherlands.
Ecological Applications: A Publication of the Ecological Society of America, Washington, DC.
Ecological Modelling: International Journal on Ecological Modelling and Systems Ecology, Elsevier Science B.V., Amsterdam, The Netherlands.
Ecological Research, The Ecological Society of Japan, Blackwell Science (Australia) Pty. Ltd., Victoria, Australia.
Ecology Law Quarterly, Boalt Hall School of Law, University of California, Berkeley, CA.
Eco-Management and Auditing, J. Wiley & Sons, Chichester, UK.
Ecotoxicology, Chapman & Hall, London, UK.
Ecotoxicology and Environmental Safety, Academic Press, Inc., San Diego, CA.
Environment and Development Economics, Cambridge University Press, Cambridge.
Environment International, Elsevier, Amsterdam, The Netherlands.
Environmental Conservation, Cambridge University Press, Cambridge.
Environmental and Ecological Statistics, Chapman & Hall, London, UK.
Environmental and Engineering Geoscience, AEG, Texas A&M University, College Station, TX.
Environmental Geochemistry and Health, Chapman & Hall, London, UK.

Environmental Health Perspectives Supplements, National Institute of Environmental Health Sciences, Research Triangle Park, NC.
Environmental Health Perspectives: Journal of the National Institute of Environmental Health Sciences, Research Triangle Park, NC.
Environmental Health and Pollution Control, Elsevier, Amsterdam, The Netherlands.
Environmental Impact Assessment Review, Elsevier, Amsterdam and New York.
Environmental Law, Northwestern School of Law of Lewis & Clark College, Portland, OR.
Environmental Law and Management, J. Wiley & Sons, Chichester, UK.
Environmental Management, Springer-Verlag, New York.
Environmental Manager, J. Wiley & Sons, Chichester, UK.
Environmental Modelling and Software, Elsevier, Amsterdam, The Netherlands.
Environmental and Molecular Mutagenesis, J. Wiley & Sons, Chichester, UK.
Environmental Monitoring and Assessment: An International Journal, Kluwer Academic Publishers, Dordrecht, The Netherlands.
Environmental Pollution, Elsevier, Amsterdam, The Netherlands.
Environmental Policy and Law, International Council of Environmental Law (ICEL), Bonn, Germany.
The Environmental Professional, NAEP, Blackwell Science, Inc., Cambridge, MA.
Environmental Quality Management, J. Wiley & Sons, Chichester, UK.
Environmental Regulation and Permitting, J. Wiley & Sons, Chichester, UK.
Environmental Science and Technology, American Chemical Society, Washington, DC.
Environmental Research: A Journal of Environmental Medicine and the Environmental Sciences, Academic Press, Inc., San Diego, CA.
Environmental Toxicology and Chemistry: An International Journal, Pergamon, Elsevier Science Ltd, Oxford, UK.
Environmental Toxicology and Pharmacology, Elsevier, Amsterdam, The Netherlands.
Environmental Toxicology and Water Quality, J. Wiley & Sons, Chichester, UK.
The Environmentalist, Chapman & Hall, Hampshire, UK.
EnvironMetrics, John Wiley & Sons, Northampton, UK.
EPA Journal, US Environmental Protection Agency, Office of Communications, Education, and Public Affairs, Washington, DC.
European Environment, J. Wiley & Sons, Chichester, UK.
European Water Pollution Control, Elsevier, Amsterdam, The Netherlands.
The George Washington Journal of International Law and Economics, George Washington University, Washington, DC.
Global Environmental Change, Elsevier, Amsterdam, The Netherlands.
The Harvard Environmental Law Review, Harvard University, Cambridge, MA.
Harvard Journal of Law and Public Policy, Harvard Law School, Cambridge, MA.
Hastings International and Comparative Law Review, Universtiy of California, Hastings College of the Law, San Francisco, CA.
The Hazardous Waste Consultant, Elsevier, Amsterdam, The Netherlands.
Hazardous Waste and Hazardous Materials, Mary Ann Liebert, Inc., Publishers, Larchmont, NY.
Human and Ecological Risk Assessment, CRC Press, Boca Raton, FL.
Human and Experimental Toxicology: An International Journal, British Toxicology Society, Macmillan Press Ltd, Hampshire, UK.
Hydrological Processes: An International Journal, J. Wiley & Sons, Chichester, UK.
Indoor Environment, KARGER, Rothenfluh, Switzerland.
International Archives of Occupational and Environmental Health, Springer-Verlag, New York.
International Journal of Climatology, J. Wiley & Sons, Chichester, UK.
International Journal of Environment and Pollution, Inderscience Enterprises Ltd/UNESCO, Geneva, Switzerland.
International Journal of Environmental Health Research, CARFAX Publishing Company, Oxfordshire, UK.

International Journal of Geographical Information Science, Taylor & Francis, London, UK.
International Journal of Toxicology, Taylor & Francis, London, UK.
Issues in Science and Technology, J. Wiley & Sons, Chichester, UK.
Journal of the Air and Waste Management Association, AWMA, Pittsburgh, PA.
Journal of Applied Toxicology, J. Wiley & Sons, Chichester, UK.
Journal of Arid Environments, Academic Press, London, UK.
Journal of Clean Technology and Environmental Sciences, Princeton Scientific Publishing Company, Inc., Princeton, NJ.
Journal of Environmental Economics and Management, Academic Press, San Diego, CA.
Journal of Environmental Engineering, American Society of Civil Engineers, ASCE, New York.
Journal of Environmental Health, National Environmental Health Association, Denver, CO.
Journal of Environmental Law, Oxford University Press, Oxford, UK.
Journal of Environmental Management, Academic Press, San Diego, CA.
Journal of Environmental Pathology, Toxicology, and Oncology, Begell House, Inc., New York.
Journal of Environmental Medicine, J. Wiley & Sons, Chichester, UK.
Journal of Environmental Planning and Management, Carfax Publishing Ltd, Oxfordshire, UK.
Journal of Environmental Quality, ASA/CSSA/SSSA, Madison, WI.
Journal of Environmental Science and Health, Marcel Dekker, Inc., Monticello, NY.
Journal of Environmental Science and Health, Part A: Environmental Science and Engineering and Toxic and Hazardous Substance Control, Marcel Dekker, Inc., Monticello, NY.
Journal of Environmental Science and Health, Part B: Pesticides, Food Contaminants, and Agricultural Wastes, Marcel Dekker, Inc., Monticello, NY.
Journal of Environmental Science and Health, Part C: Environmental Carcinogenesis and Ecotoxicology Reviews, Marcel Dekker, Inc., Monticello, NY.
Journal of Environmental Systems, Baywood Publishing Company, Inc., Amityville, NY.
Journal of Environmental Technology, Publications Division, Selper Ltd, London, UK.
Journal of Hazardous Materials, Elsevier, Amsterdam, The Netherlands.
Journal of the Institution of Water and Environmental Management, IWEM, London, UK.
Journal of Risk Research, E&FN Spon/Routledge, London, UK.
Journal of Risk and Uncertainty, Kluwer Academic Publishers, Norwell, MA.
Journal of Soil Contamination, Lewis Publishers/CRC Press, Boca Raton, FL.
Journal of Toxicology and Environmental Health, Taylor & Francis, London, UK.
Land Degradation and Rehabilitation, J. Wiley & Sons, Chichester, UK.
Law and Policy, Blackwell Publishers Ltd, Oxford, UK.
Natural Toxins, J. Wiley & Sons, Chichester, UK.
P2: Pollution Prevention Review, J. Wiley & Sons, Chichester, UK.
Pesticide Science, J. Wiley & Sons, Chichester, UK.
Pollution, Environmental Hazards, Environmental Toxicology, Plenum Press, New York.
Process Safety and Environmental Protection, Institution of Chemical Engineers, London, UK.
Regulatory Toxicology and Pharmacology, Academic Press, London, UK.
Remediation – The Journal of Environmental Cleanup Costs, Technologies and Techniques, J. Wiley & Sons, Chichester, UK.
Restoration Ecology: The Journal of the Society for Ecological Restoration, Blackwell Science, Inc., Cambridge, MA.
Risk Analysis: An International Journal, Plenum Press, New York.
Risk, Decision and Policy, Routledge, London, UK.
The Science of the Total Environment (An International Journal for Scientific Research into the Environment and its Relationship with Man), Elsevier Science B.V., Amsterdam, The Netherlands.
Sustainable Development, J. Wiley & Sons, Chichester, UK.
Toxic Exposure Advisory, Taylor & Francis, London, UK.

Toxic Substance Mechanisms, Taylor & Francis, London, UK.
Toxicology and Ecotoxicology News, Taylor & Francis, London, UK.
Toxicology and Industrial Health: An International Journal, Princeton Scientific Publishing Company, Inc., Princeton, NJ.
Toxicology, Elsevier Science B.V., Amsterdam, The Netherlands.
Toxicology Letters, Elsevier Science Ireland Ltd, Shannon, Ireland.
UCLA Journal of Environmental Law and Policy, UCLA School of Law, Los Angeles, CA.
Waste Management, Pergamon/Elsevier Science Ltd, Oxford, UK.
Waste Management and Research, ISWA, Copenhagen, Denmark.
Water, Air and Soil Pollution: An International Journal of Environmental Pollution, Kluwer Academic Publishers, Dordrecht, The Netherlands/London, UK.
Yale Law and Public Policy. Yale Law School, New Haven, CT.

Index

Note: Page references in *italics* refer to Figures; those in **bold** refer to Tables and Boxes

acceptable daily intake (ADI) 267
  for carcinogens 358–9
  for noncarcinogens 359
  see also reference dose (RfD)
acceptable soil concentration (ASC) 369–70, 371, 373
acceptable water concentration (AWC) 373–4, 375, 376
action level (AL) 345
addition theorem 478
additive models, statistical concepts in 477
AERIS 485–6
Agenda 21 5
air pathway exposure analysis (APEA) 140, 437–8
air sampling schemes
  instantaneous (grab) sampling 40
  integrated air sampling 40
air toxics risk assessment 437–8
AIRTOX 486
allowable daily intakes (ADIs) 348
ambient water quality criteria (AWQC) 305
analysis-of-variance (ANOVA) 168–9
API DSS 486
API PRDF 486
area sources of air emissions 138
arithmetic mean
  definition 475
  use of 169
asbestos, airborne exposure to, risk evaluation 383–6
  management decision 386
  risk estimation 384–5
  types of asbestos 383
AT123D **126**
attenuation–dilution factors 361

ballpark estimate 117
Bayes's Theorem 479
beef products ingestion exposure 250
beneficial re-use of contaminated materials 443–4

benefit–cost analysis 418
benefits assessment 418
best available techniques not entailing excessive cost (BATNEEC) 56
bioconcentration factor (BCF) 112, 148
biodegradation 114
body burden (BB) 218
Boolean algebra
  associative laws 475
  commutative laws 475
  distributive laws 475
BOXMOD **118**
BTEX compounds, evaluation for 402

CalTOX 487
CAMEO 487
cancer risk (CR) 223, 351
carcinogen classification systems 209–13
  strength-of-evidence classification 209, 212–13, **213**
  weight-of-evidence classification 209–12, **210, 211**
carcinogenesis
  categorization 208–9
  determination of 208–9
  threshold versus nonthreshold concepts 208
carcinogenic chronic daily intake (CDI) 258–60, 261–3, 267–8
cause-consequence analysis (CCA) 335
CemoS/Air **120**
CemoS/Plume **121**
CemoS/Soil **136**
CemoS/Water **125**
censored data 172–3
chain-of-custody documentation 36
chemical degradation 115
Chemical Substances Control Act (1973) (Japan) 60
Chemical Substances Information Network (CSIN) 496
chemicals of potential ecological concern (COPEC) 305

chronic daily intake (CDI) 193
　carcinogenic 258–60, 261–3, 267–8
chronic hazard index 287
Clean Air Act (CAA) (1963; 1970) (USA)
　52–3
　Amendments (1990) 53
Clean Water Act (CWA) (1977) (USA) 53
CMLS (Chemical Movement in Layered
　Soils) model **129**
co-carcinogen, definition 208
Comprehensive Environmental Response,
　Compensation and Liability Act
　(CERCLA) (1980) (USA) 53–4
conceptual models, design of 181–4
conceptual site model (CSM)
　design **184**
　diagrammatic representation *183*
　elements of 184
conditional probability 477
confidence interval 476
confidence limits (95 percent) 476
contaminant migration, modeling of
　algorithms 137–48
　estimation of contaminant concentrations
　　in air 137–45
　　　airborne dust concentrations 141–2
　　　airborne vapor concentrations
　　　　142–4
　　　classification of air emission sources
　　　　138–9
　　　determination of air contaminant
　　　　emission potential 139–40
　　　dispersion modeling and 140–1
　　　household air contamination due to
　　　　volatilization from water 144–5
　　　volatilization to shower air 144
　　in animal products 147–8
　　　contaminant bioconcentration in
　　　　meat and dairy products 148
　　in fish products 148
　　in soils 145
　　in vegetation 146–7
　　in water 145–6
　factors affecting contaminant fate and
　　transport in the environment 149–50
　model selection 117
　selected model 117–37, **118–36**
　utility and application of environmental
　　models 116–17
contaminant properties 107–15
　degradation 114–15
　diffusion 108
　dispersion 108
　partitioning and the partition coefficients
　　109–12

physical state 107
sorption and retardation factor 112–14
volatilization 108–9
　boiling point 109
　Henry's Law constant 108, 109
　vapor pressure 109
　water solubility 107–8
contaminant release, causes and mechanisms
　of 155, **157–8**
contaminant transfer between environmental
　compartments 103–6, *105, 106*
contaminant transport and fate analysis
　189
contaminant vapors, migration into building
　structures 438–9
contaminants of concern, selecting (COCs)
　174–5, **175**, *176*
contaminated site
　cleanup goals 440–1
　problem 18, *20*
continuous distributions 475
Control of Pollution (Special Waste)
　Regulations (1980) (UK) 58
Control of Pollution Act (CoPA) (1974)
　(UK) 57
corrective action response programs
　design of 439–40
　evaluation of ecological risk issues in
　　442–3
cost-benefit analysis 418
cost-effectiveness analysis 419
cradle-to-grave concept 52, 53
crop ingestion exposure 250
Crystall Ball 487–8
cumulative distribution function 475

dairy products ingestion exposure 250
data quality objectives (DQOs) 36
DDT 10
decision analysis, application to
　environmental management programs
　418–25
　multi-attribute decision analysis and utility
　　theory 419, 421–5, *423*
　risk–cost–benefit optimization and
　　tradeoffs analysis 418, 420–1, *422*
degradation 114–15, 149
Deposit of Poisonous Waste Act (DPWA)
　(1972) (UK) 57
dermal exposures 186–7, 190, 264–7
　average daily dose (ADD) for
　　noncarcinogenic effects 268–70
　lifetime average daily dose (LADD) for
　　carcinogenic effects 267–8
dilution–attenuation factor (DAF) 146

Index   507

dioxins, use of toxicity equivalence factors
    279
discrete distributions 475
DNAPLs 106
dosage, hazardous level and 11
dose–response relationships 81, 204–6, *206*,
    216–18, 232
DREAM 122

ecological effects of environmental
    contamination 27–30, **28–9**
ecological quotient 223
ecological risk assessments (ERA)
  basic tasks **301**
  ecological hazard evaluation 302–7, 316
    ecosystem types 302–3
    evaluation of habitats and community
        structure 303–7
      assessment endpoints 304
      chemicals of potential ecological
          concern (COPEC) 305
      focused examination of select species
          303–4
      indicator species 304–5
      lower trophic levels 303
      nature and effects of environmental
          contaminants 306–7, **306**
      selection of target species 305, **305**
  ecological risk characterization 311–13,
      316
    uncertainty analysis 312–13, **312**
  exposure assessment 307–10, 316
    calculation of chemical intakes
        308–10
    conceptual site models (CSM) 307–8,
        308, 309
    wildlife or game exposure to
        environmental chemicals 310,
        310
  ecotoxicity assessment 310–11, 316
  fundamental elements **301**
  general framework 313–15, *313*
    problem formulation phase 313, 314
    problem analysis phase 313, 314–15
    risk characterization 313, 315
  cf human health endangerment
      assessments 320–2, **321**
  methodology 299–313, *300*
  in practice 315–20, *317*
    general considerations 319–20
    general purpose 318–19
    selection criteria **318**
ecological risk quotient (ErQ) 311–2
Endangered Species Act (1973; 1988) (USA)
    54

ENPART **132**
environmental assessment 439
environmental benchmarks, use of 363–4
environmental characterization process
    35–46, **36**, 439
  designing a program 44–5
  health and safety plan (HSP) 41–3
  implementation of a program 45–6
  quality assurance (QA) 43
  quality assurance/quality control (QA/QC)
      plan 43–4
    blind replicates 44
    equipment blanks 44
    field blank 44
    spiked samples 44
    trip blank 43–4
  quality control (QC) 43
  sampling and analysis plan (SAP)
      35–43
    checklist for developing protocols **37**
    elements of **37**
    laboratory and analytical program
        requirements 40–1, **41**
    sampling plan checklist **38**
    sampling protocols 38–9, **39**
    sampling strategies 39–40
environmental contamination problems,
    minimizing 6–10
environmental hazards, sources of 155,
    156
environmental impact assessments (EIAs) 5,
    435–7
environmental investigation 156–74
  data quality objectives (DQOs) 158–9
  data collection and analysis 161–6
    background sampling considerations
        163–5
      evolution of data 165
      factors affecting design 164–5
    quality control (QC) samples 165–6
    strategies 162–3
  evaluation of environmental sampling
      data 166–74
    statistical analysis 167–72
      hypothesis testing 168–9
      parametric vs nonparametric
          statistics 168
      selection of averaging techniques
          169–72
    treatment of censored data 172–4
      derivation and use of 'proxy'
          concentrations 173–4
      statistical evaluation of nondetect
          values 174
  program design 159–61, **160**

environmental management and audit scheme (EMAS) 56–7
environmental management programs
 decision issues **436**
environmental management strategies, development of 444–5
 risk assessment as cost-effective tool in 445–7
environmental management units (EMUs) 45
environmental pollutants, sources of 15–18
 examples of exposure to hazardous materials **22–6**
 industry 17–18, **17**, **19**
Environmental Protection Act (1990) (UK) 58
Environmental Protection Agency (USA) 52
environmental quality criteria (EQC) 345–7, *346*, 363–4, 443
environmental regulations
 in different parts of the world 59–60
 Federal Republic of Germany 59
 France 58–9
 Netherlands 58
 in North America 52–5
 paradigm of 60–2
 UK 57–8
 within the European Community 55–9
environmental restoration programs, design of 439–40
European Community (EC)
 Economic Treaty 55
 environmental laws 55–9
 directives 56
European Environment Agency 57
event-tree modeling and analysis (ETA) 326–30, 429–30
 conceptual elements 327–30, *328*
 pathway probability concept 329
 risk costs and impacts assessment 329–40
EXAMS-II **123**
exposed population analysis **189**
exposure analysis 181
exposure assessment process 187–98
 chronic versus subchronic exposures 193–4
 clinical intake versus dose 193
 development *188*
 example scenario *192*
 exposure parameters **191**
 general procedural elements **189**

generic model 194–8, **194**
 averaging exposure estimates 197–8
  over population age groups 197
  over population age subgroups 198
  within a population group 197–8
 incorporating contaminant degradation into exposure calculations 195–7
 receptor age adjustments to human exposure factors 195, **196**
 multimedia and multipathway exposure modeling 190–2
 tabular illustration **189**
 utility of exposure characterization 198–9
exposure point concentration (*EPC*) 167–8
exposure scenarios, development of 185–7
 evaluation flowchart *185*
 nature and spectrum 185–7

failure modes and effects analysis (FMEA) 335
failure modes, effects, and criticality analysis (FMECA) 335
failure probability 325–6
fault-tree analysis (FTA) 326, 330–5
 illustrative elements 332–5, *333*, *334*
 selected symbols *331*
FDM **119**
feasibility study (FS) 440
Federal Food, Drug, and Cosmetic Act (USA) 54
Federal Insecticide, Fungicide, and Rodenticide Act (FIFRA) (1978) (USA) 54
Federal Waste Disposal Act (FRG) 59
Federal Water Pollution Control Act (FWPCA) 53
Fish and Wildlife Conservation Act (1980) (USA) 55
Frank effect level (FEL) 269

Geographic Information Systems (GISs), utilization in environmental management programs 426–8
 applications in environmental management 427–8
 integration of environmental and risk models 426–7
geometric mean
 definition 476
 use of 169
GEOTOX 488

global environmental protection 3–6
  harmonizing policies 5–6
  socio-economic context 4–5
ground-level concentration (GLC) 138

half-lives, chemical 114
hazard
  definition 67
  identification and accounting 68
hazard analysis (HAZAN) 68–70, *69*, 335
hazard and operability study (HAZOP) 335
hazard categorization system 30–2
hazard disaggregation factor 355
hazard effects assessment, utility of 218–19
hazard index (HI) 286, 352
hazard quotient (HQ) 286
hazard quotient/hazard index estimates 223
Hazardous and Solid Waste Amendments (HSWA) (1984) (USA) 53
Hazardous Material Transport Act (1975) (USA) 55
Hazardous Substance Response Trust Fund 54
health-based criteria for carcinogens 347–50
  action levels for carcinogenic chemicals 348–9
    example calculations 348–9
  action levels for noncarcinogenic chemicals/systemic toxicants 349–50
    example calculations 349–50, **350**
health implications 18–30
  effects of environmental contamination 27–30, **28–9**
  factors affecting organ system targeted 27, **29**
health-protective chemical concentrations in consumer products 355–9
  assessing the chemical safety of consumer products 358–9
  determination of tolerable chemical concentrations 358–9
  formulation of potential consumer exposures 356–8
  dietary exposures to chemicals 357–8
  inhalation exposures to particulates 357
  inhalation exposures to volatiles 356–7
  oral exposure 356
  skin exposures 356

health-protective risk-based chemical concentrations 350–3
  RBCs for carcinogenic constituents 351–2
  RBCs for noncarcinogenic effects of chemical constituents 352–3
health risk assessment
  definition 243
  dermal exposures 254–6
    contaminated soil 255, **256**
    contaminated water 255, **256**
    dermal contact with waters and seeps 255
    soils contact/dermal absorption 255
  cf ecological risk assessments 320–2, **321**
  ingestion exposures 250–4
    food 251, **252**
      animal products 252, **254**
      mother's milk 252–3, **254**
      pica and ingestion of soil/sediment 253–4, **255**
      plant products 251, **253**
      seafood 251–2, **253**
    water
      drinking water 251, **251**
      ingestion during swimming 251, **252**
  inhalation exposures 247–50
    to particulates from contaminated fugitive dust 247, **248**
    to vapor-phase contaminants 247–50, **248**, **249**
  methodology 243–5, *244*
  potential receptor exposures to environmental contaminants 245–67, *246*, 295
    case-specific parameters **257**, **259**
    computation of intake factors 256–67
    dermal exposure 264–7
      average daily dose (ADD) for noncarcinogenic effects 268–70
      lifetime average daily dose (LADD) for carcinogenic effects 267–8
    ingestion exposures 261–4
      ADD for noncarcinogenic effects 263–4
      LADD for carcinogenic effects 261–3
    inhalation exposures 258–61
      ADD for noncarcinogenic effects 260–1
      LADD for carcinogenic effects 258–60
  in practice 295–6

health risk assessment *cont.*
  risk characterization 279–95, 296
    absorption adjustments 280–1
    aggregate effects of chemical mixtures 282–3
    carcinogenic risk calculations 288–91
      contaminants in soils 289–91
      contaminants in water 288–9
      estimation of carcinogenic risks to human health 283–5
      individual cancer risk 285
      linear low-dose model 283–4, **284**
      one-hit model 283–4, **284**
      population excess cancer burden 285
      risk for inhalation pathways 283
      risk for noninhalation pathways 283
    estimation of noncarcinogenic risks to human health 286–7, **286**
      chronic versus subchronic noncarcinogenic effects 287
    noncarcinogenic hazard calculations 291–5
      contaminants in soils 294–5
      contaminants in water 291–4
    risk computations 288–95, **290**, **293**
    toxicological parameters use in 267–79, 295
      for carcinogenic effects 273–6
        derivation of SFs and URFs 273–4
        inhalation potency factor 275
        inter-conversion of carcinogenic toxicity parameters 274–6
        oral potency factor 275–6
      for noncarcinogenic effects 267–73
        derivation of RfCs 270–1
        derivation of RfDs 268–70, **271**
        inter-conversions of noncarcinogenic toxicity parameters 271–3
        use of surrogate toxicity parameters 276–9
          carcinogenic effects 279
          noncarcinogenic effects 278–9
          route-to-route extrapolation 277–9
          toxicity equivalence factors 279
heat, recovery from incinerators 7
heavy metals, recycling 7
Henry's Law constant 139, 141, 142, 149
horizontal bar charts of risk summary data 224, *226*
human health risk assessment *see* health risk assessment

IEUBK 488
inclusive probability 478
independence 478
indicator species 303, 304–5
ingestion exposure 186, 190, 250–4, 261–4
  average daily dose (ADD) for noncarcinogenic effects 263–4
  lifetime average daily dose (LADD) for carcinogenic effects 261–3
inhalation exposures 186, 190, 247–50, 258–61
  average daily dose (ADD) for noncarcinogenic effects 260–1
  lifetime average daily dose (LADD) for carcinogenic effects 258–60
initiator, definition 208
INPOSSM **134**
integrated exposure analysis **189**
Integrated Risk Information System (IRIS) database 282, 494–5
International Register of Potentially Toxic Chemicals (IRPT) database 492–4
IRIS (Integrated Risk Information System) 282, 494–5
ISCLT **118**
ISCST **118**

joint probability 478
joint probability distributions 475

LADD 267
landfill in France 59
lead 10
LEADSPREAD 488–9
linearized multistage model (MLE) 216, 217, 273–4
LNAPLs 104–6
Love Canal 51
lowest-observed-adverse-effect level (LOAEL) 205, 267, 268, 269
  determination of RfC using 270
  determination of RfD using 270
lowest-observed-effect level (LOEL) 267
LULU 428

Maastricht Treaty 55
margin of exposure (MOE) 272, 273–4
Marine Mammal Protection Act (1972) (USA) 55
mass balance analyses 361–2
materials recycling and re-use 7, 9
maximum individual risk 285
MCPOSSM **134**
mechanistic models 216
MEPAS 489

mercury 10
metals, mobility of, in soils 104
Migratory Bird Treaty Act (1972) (USA) 55
MLTT **120**
MMEDE 496–7
MMSOIL **489**
MOC **128**
modifying factors (MFs) 268, 269
  selecting, in derivation of RfDs **271**
Monte Carlo simulation techniques 226, 234
  application to probabilistic analysis 235–7, *236*, **237**, *238*
multi-attribute decision analysis 419, 421–5
multi-hit model 216
MULTIMED 489–90
multimedia contaminant release analysis **189**
multimedia mathematical models 190–2
multipathway exposure models 190–2
multiplicative, statistical concepts in 477
multivariate probability distributions 475
municipal solid waste (MSW) incinerators 437
mutually exclusive events 478
MYGRT **127**

National Ambient Air Quality Standard 141
National Environmental Protection Act (1969) (USA) 52
National Environmental Quality Control Council (NEQCC) (USA) 52
NIABY 428
NIMBY (not-in-my-backyard) 7, 428
NIMTO 428
NOAELs 232
'no-further-action' decisions (NFAD) 415–16
no-observed-adverse-effect level (NOAEL) 205, 267, 268, 269
  determination of RfC using 270
  determination of RfD using 269–70
no-observed-effect level (NOEL) 203, 204, 267, 268, 269
nonaqueous phase liquids (NAPLs) 104
  light (LNAPLs) 104–6
  dense (DNAPLs) 106
noncancer hazard index 355
noncancer (systemic) toxicity 207
noncarcinogenic chronic daily intake 260–1, 263–4, 268–70
nondetects (*ND*s) 172–3
  statistical evaluation of 174

nonpoint (diffuse) sources 16
nonthreshold processes 207, 208

Occupational Safety and Health Administration (OSHA) 42
  Hazardous Waste Operations and Emergency response Activities (HAZWOPER) training 42
octanol/water partition coefficient 108, 110
oils, recovery of 7
OML **119**
one-hit model 216
optimum hazard 425, *425*
order-of-magnitude estimate 117
organic carbon adsorption coefficient 110–11
organo-lead, evaluation for 402–3

partitioning coefficients 149
  use of 360–1
pathway probability (PWP) 223, 329
PDM (Probabilistic Dilution Model) **125**
petroleum-contaminated sites, corrective action response plan development 400–7
  contaminant release analysis 404–5
  corrective action and risk management decision process 405–7
  contamination assessment 406
  site categorization 405–6
  site restoration program 406–7
  evaluation of petroleum product constituents 402–3
  fate and behavior of petroleum constituent releases 403–4
photodegradation 115
photolysis 115
physiologically based pharmacokinetic (PB-PK) model 217, 278
pica behavior 253–4, **255**, 369
pie charts of risk summary data 224, *225*
plausible upper-bound estimate 95, 96
point sources 15–16
  of air emissions 138
polluter pays principle 59
pollution abatement strategies and policies 9–10
Pollution Prevention Act (1990) (USA) 54–5
polychlorinated aromatic hydrocarbons (PCBs) 10
population-at-risk (PAR) 190
population cancer burden 285
POSSM **132–3**
preliminary remediation goals (PRGs) 441

probabilistic analysis 235–7
probabilistic risk assessments
  methodology 325–36
    cause-consequence analysis (CCA) 335
    conceptual representation *327*
    event-tree modeling and analysis (ETA) 326–30
      conceptual elements 327–30, *328*
      pathway probability concept 329
      risk costs and impacts assessment 329–30
    failure modes and effects analysis (FMEA) 335
    failure modes, effects, and criticality analysis (FMECA) 335
    fault-tree analysis (FTA) 326, 330–5
    hazard analysis (HAZAN) 335
    hazard and operability study (HAZOP) 335
    reliability block diagrams (RBDs) 335
  in practice 336–40
    analysis of hazardous materials transportation risks 337
    demonstration problem 338–40
    hazardous materials TSDF design and failure investigations 336–7
    quasi-PRA applications 337–8
  utility of 340–1
probability density function 475
probability distributions 475
probability mass function 475
probit model 217
promoter, definition 208
'proxy' concentrations 173–4
Public Health Act (1936) (UK) 57

quality adjusted life expectancy (QALE) 419

RAAS 490
RAMAS Library of Ecological Software 495–6
RBCA 490
REACHSCA **123**
reasonably maximum exposure (RME) 193
reference concentration (RfC) 268
  derivation 270–1
reference dose (RfD) 267–8
  derivation 268–70
regulatory dose (RgD) 272
relation plots of risk summary data 224, *228*-9
reliability block diagrams (RBDs) 335

remedial action
  alternatives for contaminated sites 441–2
  evaluation of human health risks associated with 442
remedial action plan (RAP) development 440–1
remedial investigation (RI) 440
remedial investigation/feasibility study (RI/FS) 440
Resource Conservation and Recovery Act (1965; 1976) (USA) 53
Resource Recovery Act (1970) (USA) 53
retardation factor 112–14, 149
RISC 491
risk
  acceptable 92–4
    de manifestis 92
    de minimis (acceptable) 92–4
    risk acceptability criteria 92–4
  analysis 68–70, *69*
  basis for measuring 70–3
  definition 67–8
  individual versus group 91–2
  risk perception 94
  risk tolerability 94, **95**
risk aggregation factor 354
risk assessment process
  attributes 75–6
  conservatism in 95–6
  cost-effectiveness of environmental management 11–13
  definition 72–3
  as a diagnostic tool 82–4
    baseline risk assessments 83–4
  elements of 79–82
    exposure assessment and analysis 81–2
    exposure-response evaluation 81
    hazard identification and accounting 80–1
    risk characterization and consequence determination 82
  as holistic tool for environmental management 87–8
  in practice 84–7
    fundamental procedural components *85*
    outline for a report **86–7**
  predictive 75
  purpose 74–5, **74**
  retrospective 75
  uncertainty in 96–7
  versus risk management 97
RISK* ASSISTANT 491

Index 513

risk-based chemical concentrations (RBCs) 350–3
  for carcinogenic constituents 351–2
  degradation rates and 362–3
  modified, for carcinogenic chemicals 354–5
  modified, for noncarcinogenic constituents 355
  for noncarcinogenic effects of chemical constituents 352–3
risk-based remediation goals
  cleanup decision in site restoration programs 377–8
  factors affecting development of 367–9, **368**
  soil cleanup levels 369–73
    for carcinogenic contaminants 370–1, **370**
    for noncarcinogenic effects of site contaminants 371–3, **372**
  water cleanup levels 373–7
    for carcinogenic contaminants 374–5, **374**
    for noncarcinogenic effects of site contaminants 375–7, **376**
risk–benefit analysis 418–19
risk–benefit–cost factor 410
risk communication 428–31, **430**
  designing an effective program 429–31
risk cost (RC) 329
risk–cost–benefit optimization 418, 420–1
risk–decision matrix 413, *414*
risk management program
  framework for 416, *417*
  general nature 409–16
    conditions for 'no-further-action' decisions 415–16
    design elements 411, **411**
    examples 410
    interim corrective action program 414–15
    risk reduction versus costs *410*
    system for establishing risk management needs 412–13
      environmental assessment 412
      mitigation study 413
      preliminary appraisal 412
      risk appraisal 412–13
      risk determination 413
risk management units (RMU) 45
risk management versus risk assessment 97
risk mitigation, definition 431

risk presentation 223–9
  graphical 224, *238*
  uncertain risks 224–9
  utility of 237
risk prevention, definition 431
risk reduction, definition 431
risk summarization 223–9
  graphical presentation 224, *238*
  utility of 237
RISKPRO 491
risk-specific concentrations
  in air 360
  in water 360
RITZ **130–1**
Rome, Treaty of 55

Safe Drinking Water Act (SDWA) (1974) (USA) 53
  Amendments (1986) 53
safety factors 268
sample collection methods 36
sample holding times 36
sample preservation techniques 36
sample quantitation limit (SQL) 173
sample shipment methods 36
Sanitary Acts (UK) 57
SARAH **124**
sediment quality criteria (SQC) 305
sediment-biota partitioning coefficients 360
sediment-water partitioning coefficients 360
sensitivity analysis 231, 235
SESOIL **131**
site cleanup limits 361
site restoration, hypothetical illustration
  conceptual model diagram *395*
  identification of site contaminants 388–9, **389**, **390**
  introduction and background 386–7
  objective and scope 387
  risk characterization 390–400, **392–4**
    downgradient residential population exposure to groundwater 397–400, **399**
    on-site worker 395–7, **396**
    site construction worker 397, **398**
  risk management decision 400
  screening for chemicals of potential concern 389–90, **391**
  summary **401**
  technical elements of diagnostic risk assessment process 387–8, *388*
SITES 491–2
skewness, coefficient of 476

slope factor (SF) 232, 273–4
SMCM **135**
socio-economic development problems
    21
soil ingestion exposure 250
soil sampling schemes
    depth sampling 40
    surface sampling 40
soil–water partition coefficient (soil/water
    distribution coefficient) 111–12
soils, movement of contaminant through
    104
Solid Waste Disposal Act (1965) (USA)
    53
solubility in water 149
SOLUTE **126**
solvents, recovery of 7
sorption 112–13, 114
species diversity index 303
standard deviation 476
stochasticity *see* variability
stressor–response profile 205, *206*, 310
structure–activity analysis 219
subchronic daily intake (SDI) 193
subchronic hazard index 287
SWAG **131**
systemic toxicants 207
systemic toxicity 207

threshold chemicals 205
threshold processes 207, 208
time-to-occurrence models 216
tolerable concentration
    for carcinogens 358–9
    for noncarcinogens 359
tolerance distribution models 216
total petroleum hydrocarbons (TPH),
    evaluation for 402, 403
total probability theorem 478
TOXIC 492
toxic substances
    categorization of human toxic effects
        207–9
        basis for threshold versus nonthreshold
            concepts 208
        determination of chemical
            carcinogenicity 208–9
    identification 203–6
        dose–response relationships 204–6
        manifestations of toxicity 203–4
            incidence 204
            reversibility 204
            seriousness 204
Toxic Substances Control Act (TSCA)
    (1976) (USA) 54

toxicity, chemical, evaluation 213–18
    dose–response assessment 216–18
        dose–response quantification 217–18
        extrapolation models 216–17
    hazard effects assessment 213–15
        animal bioassays 214
        case clusters 214
        clinical case studies 215
        ecotoxicological studies 215
        epidemiologic studies 215
        epidemiology 214
        laboratory animal studies 214–15
        reasons for conducting **215**
        structural toxicology 214
toxicity equivalence factors (TEFs) 279
toxicological endpoints **207**
toxicology, definition 203
tradeoffs analysis 420–1
treatment, storage, and disposal facilities
    (TSDFs) 7, 16, 155, 325

uncertainty analysis 97, **189**, 233–5
    need for 229–30
    qualitative 234
    quantitative 234–5
uncertainty factors (UFs) 268, 269
    selecting, in derivation of RfDs **271**
uncertainty in risk assessment 96–7,
    229–30
    in human health risk assessments **280**
    types and nature of
        completeness/scenario uncertainties
            230
        modeling uncertainties 230
        parameter uncertainties 230
        sources of, in endangerment
            assessments 231–3
            associated with toxicity of chemical
                mixtures 232
            background exposures 232
            general extrapolations 231
            limitations in model form 232
            quantitative extrapolations of
                adjustments in dose–response
                evaluation 232
            representativeness of sampling data
                232–3
unconditional probability 477
underground storage tanks (USTs) 400, 404
unit risk factor (URF) 273–4
Untied Nations Conference on Environment
    and Development (UNCED) (1992) 4
upper confidence limit (*UCL*)
    definition 476
    use of 170

Index 515

utility-attribute analysis 421
utility functions, preferences and evaluation of 421–4
utility optimization 424–5, *424*
utility theory 419, 421–5, *423*

vapor pressure 139, 149
variability analysis in risk assessment 229–30
  need for 229–30
variation, coefficient of 476
vertical bar charts of risk summary data 224, *227*
VHS (Vertical and Horizontal Spread model) **127**
VIP (Vadose zone Interactive Process) **129**
virtually safe doses (VSDs) *see* health-based criteria for carcinogens
VLEACH **130**

volatile organic chemicals (VOCs) 138
  data collection and analysis 162
waste exchange 77
Waste Management Law (1970) (Japan) 59
waste management program, components 7–9, *8*
waste minimization 7
waste recycling and re-use 7, 9
water ingestion exposure 250, 251
Water Quality Act (1987) (USA) 53
water sampling schemes
  composite samples 40
  continuous flowing samples 40
  grab samples 40
water/air partition coefficient 110
Wild and Scenic Rivers Act (1972) (USA) 55
worst-case scenario 95, 96
WTRISK **124**

Index compiled by Annette Musker